# Methods in Molecular Biology™

AF768418

*Series Editor*
**John M. Walker**
**School of Life Sciences**
**University of Hertfordshire**
**Hatfield, Hertfordshire, AL10 9AB, UK**

For further volumes:
http://www.springer.com/series/7651

# Nanoimaging

## Methods and Protocols

Edited by

**Alioscka A. Sousa**

*National Institute of Biomedical Imaging and Bioengineering, National Institutes of Health, Bethesda, MD, USA*

**Michael J. Kruhlak**

*National Cancer Institute, National Institutes of Health, Bethesda, MD, USA*

Humana Press

*Editors*
Alioscka A. Sousa
National Institute of Biomedical Imaging
and Bioengineering
National Institutes of Health
Bethesda, MD, USA

Michael J. Kruhlak
National Cancer Institute
National Institutes of Health
Bethesda, MD, USA

ISSN 1064-3745 ISSN 1940-6029 (electronic)
ISBN 978-1-4939-5961-7 ISBN 978-1-62703-137-0 (eBook)
DOI 10.1007/978-1-62703-137-0
Springer New York Heidelberg Dordrecht London

© Springer Science+Business Media, LLC 2013
Softcover re-print of the Hardcover 1st edition 2013
This work is subject to copyright. All rights are reserved by the Publisher, whether the whole or part of the material is concerned, specifically the rights of translation, reprinting, reuse of illustrations, recitation, broadcasting, reproduction on microfilms or in any other physical way, and transmission or information storage and retrieval, electronic adaptation, computer software, or by similar or dissimilar methodology now known or hereafter developed. Exempted from this legal reservation are brief excerpts in connection with reviews or scholarly analysis or material supplied specifically for the purpose of being entered and executed on a computer system, for exclusive use by the purchaser of the work. Duplication of this publication or parts thereof is permitted only under the provisions of the Copyright Law of the Publisher's location, in its current version, and permission for use must always be obtained from Springer. Permissions for use may be obtained through RightsLink at the Copyright Clearance Center. Violations are liable to prosecution under the respective Copyright Law.
The use of general descriptive names, registered names, trademarks, service marks, etc. in this publication does not imply, even in the absence of a specific statement, that such names are exempt from the relevant protective laws and regulations and therefore free for general use.
While the advice and information in this book are believed to be true and accurate at the date of publication, neither the authors nor the editors nor the publisher can accept any legal responsibility for any errors or omissions that may be made. The publisher makes no warranty, express or implied, with respect to the material contained herein.

Printed on acid-free paper

Humana Press is a brand of Springer
Springer is part of Springer Science+Business Media (www.springer.com)

# Preface

For more than a century, microscopy has been a centerpiece of extraordinary discoveries in biology. Along the way, remarkable imaging tools have been developed so that we can now observe biological processes at the subcellular level in live cells, visualize detailed intracellular architecture in three dimensions, quantitate the elemental composition of subcellular components, and study the structure and dynamics of isolated molecular complexes.

A key attribute of many of the different microscopy techniques is their ability to generate images at spatial resolutions down to the nanoscale, which is the length scale pertinent to individual macromolecular assemblies and subcellular ultrastructure. In *Nanoimaging: Methods and Protocols*, we therefore present a diverse collection of techniques and methodologies in microscopy capable of providing images of isolated molecular complexes, cells, and tissues at nanoscale spatial resolution. The book is divided into four parts, each containing several chapters pertaining to a given field. *Part I* covers light microscopy techniques with a special emphasis on imaging at spatial resolutions that go beyond the historically and physically important Abbe limit (diffraction limit). *Part II* contains several established and powerful techniques in electron microscopy for high-resolution imaging of molecules, cells, and tissues, in both two and three dimensions. *Part III* highlights the unique methods that reveal the advantages of scanning probe microscopy for imaging and probing macromolecular complexes and membrane surface topography. *Part IV* closes with chapters on correlative microscopy techniques, soft X-ray tomography, and secondary ion mass spectrometry imaging. Keeping up with the tradition of the Methods in Molecular Biology series, chapters are largely composed of detailed protocols describing a specific experimental procedure.

We regret that this volume simply could not possibly contain all of the latest advances on nanoscale imaging that have abounded in the literature in recent years (e.g., all the newest modalities of super-resolution optical microscopy and the many different categories of scanning probe microscopy). We also had to leave behind established techniques in optical microscopy as we consider them not strictly "nano" in the context of this volume.

The new surge of interest in microscopy that we now witness has once again elevated the importance of the microscope to biomedical research, a level that surely provides a new foundation for exciting new discoveries. One can only begin to imagine the extraordinary ways with which molecules, cells, and tissues will be "dissected" at the nanoscale with ever-increasingly powerful imaging tools. We thus believe that *Nanoimaging* is a timely volume, and we hope it will achieve its mission of highlighting many of the most exciting possibilities in microscopy for the investigation of biological structures at the nano length molecular scales.

***Bethesda, MD, USA***

***Alioscka A. Sousa***
***Michael J. Kruhlak***

# Contents

## Part II Electron Microscopy



## Part III Scanning Probe Microscopy



## Part IV Complementary Techniques

## Contributors


CHRISTOPHER J. ACKERSON • *Department of Chemistry, Colorado State University, Fort Collins, CO, USA*

KEVIN P. ARMENDARIZ • *Ralph N. Adams Institute for Bioanalytical Chemistry, University of Kansas, Lawrence, KS, USA*

MARIA A. ARONOVA • *Laboratory of Cellular Imaging and Macromolecular Biophysics, National Institute of Biomedical Imaging and Bioengineering, National Institutes of Health, Bethesda, MD, USA*

DANIEL BILLINGSLEY • *Department of Oral Biology, University of Leeds, Leeds, UK; School of Physics and Astronomy, University of Leeds, Leeds, UK*

PATRIZIA BOCCACCI • *MicroscoBio Research Center, Department of Computer Science (DISI), University of Genoa, Genoa, Italy*

ROSANNE M. BOUDREAU • *Anatomy Department, University of California San Francisco, San Francisco, CA, USA; National Center for X-ray Tomography, Berkeley, CA, USA*

JOSÉ MARÍA CARAZO • *Biocomputing Unit, National Center of Biotechnology (CSIC), Universidad Autónoma de Madrid, Madrid, Spain*

DANIEL CHOQUET • *Interdisciplinary Institute for Neuroscience, Université de Bordeaux, Bordeaux, France; CNRS, IINS UMR 5297, Bordeaux, France*

LILIA A. CHTCHEGLOVA • *Center for Advanced Bioanalysis GmbH (CBL), Linz, Austria*

LAURENT COGNET • *Université de Bordeaux, Talence, France; CNRS & Institut d'Optique, Talence, France*

KATIA CORTESE • *MicroscoBio Research Center, Department of Experimental Medicine (DIMES), University of Genoa, Genoa, Italy*

MARION CREMER • *Biocenter, Department of Biology II, Ludwig Maximilians University Munich(LMU), Germany*

J. CUENCA • *Biocomputing Unit, National Center of Biotechnology (CSIC), Universidad Autónoma de Madrid, Madrid, Spain*

J.M. DE LA ROSA TREVÍN • *Biocomputing Unit, National Center of Biotechnology (CSIC), Universidad Autónoma de Madrid, Madrid, Spain*

ALBERTO DIASPRO • *MicroscoBio Research Center, Department of Physics (DIFI), University of Genoa, Genoa, Italy; IIT—Italian Institute of Technology, Genoa, Italy*

DAMJANA DROBNE • *Department of Biology, Biotechnical Faculty, University of Ljubljana, Ljubljana, Slovenia*

ROBERT C. DUNN • *Ralph N. Adams Institute for Bioanalytical Chemistry, University of Kansas, Lawrence, KS, USA*

JOHANN ENGELHARDT • *Department of Optical Nanoscopy, German Cancer Research Center, Heidelberg, Germany*

LINDSAY R. EPPERLY • *Anatomy Department, University of California San Francisco, San Francisco, CA, USA; National Center for X-ray Tomography, Berkeley, CA, USA*

KIERAN FINAN • *Department of Biotechnology and Biophysics, Julius-Maximilians University Würzburg, Würzburg, Germany*

BENJAMIN FLOTTMANN • *Heidelberg Collaboratory for Image Processing, University of Heidelberg, Heidelberg, Germany*
TAKASHI FUNATSU • *Laboratory of Bio-Analytical Chemistry, Graduate School of Pharmaceutical Sciences, The University of Tokyo, Tokyo, Japan*
MARIA CRISTINA GAGLIANI • *MicroscoBio Research Center, Department of Experimental Medicine (DIMES), University of Genoa, Genoa, Italy*
KATHARINA GAUS • *Centre for Vascular Research, Lowy Cancer Research Centre, University of New South Wales, Sydney, Australia*
GRÉGORY GIANNONE • *Interdisciplinary Institute for Neuroscience, Université de Bordeaux, Bordeaux, France; CNRS, IINS UMR 5297, Bordeaux, France*
J. GÓMEZ-BLANCO • *Biocomputing Unit, National Center of Biotechnology (CSIC), Universidad Autónoma de Madrid, Madrid, Spain*
KRISTEN M. HARRIS • *Center for Learning and Memory, The University of Texas at Austin, Austin, TX, USA; Section of Neurobiology, The University of Texas at Austin, Austin, TX, USA*
MIKE HEILEMANN • *Department of Biotechnology and Biophysics, Julius-Maximilians University Würzburg, Würzburg, Germany; Heidelberg Collaboratory for Image Processing, University of Heidelberg, Heidelberg, Germany*
CHRISTINE L. HEINECKE • *Department of Chemistry, Colorado State University, Fort Collins, CO, USA*
PETER HINTERDORFER • *Institute for Biophysics, University of Linz, Linz, Austria*
ERIC HOSY • *Interdisciplinary Institute for Neuroscience, Université de Bordeaux, Bordeaux, France; CNRS, IINS UMR 5297, Bordeaux, France*
HEATH A. HUCKABAY • *Ralph N. Adams Institute for Bioanalytical Chemistry, University of Kansas, Lawrence, KS, USA*
MICHELLE S. ITANO • *Department of Cell and Developmental Biology, University of North Carolina at Chapel Hill, Chapel Hill, NC, USA*
HALEY A. KLITZING • *Department of Chemistry, University of Illinois, Urbana, IL, USA*
ROMAN I. KONING • *Department of Molecular Cell Biology, Leiden University Medical Center, Leiden, The Netherlands*
ABRAHAM J. KOSTER • *Department of Molecular Cell Biology, Leiden University Medical Center, Leiden, The Netherlands*
MARY L. KRAFT • *Departments of Chemistry and Chemical and Biomolecular Engineering, University of Illinois, Urbana, IL, USA*
MICHAEL J. KRUHLAK • *Experimental Immunology Branch, National Cancer Institute, National Institutes of Health, Bethesda, MD, USA*
MASAAKI KUWAJIMA • *Center for Learning and Memory, The University of Texas at Austin, Austin, TX, USA*
MARION LANG • *Department of Optical Nanoscopy, German Cancer Research Center, Heidelberg, Germany*
CAROLYN A. LARABELL • *Anatomy Department, University of California San Francisco, San Francisco, CA, USA; National Center for X-ray Tomography, Berkeley, CA, USA; Physical Biosciences Division, Lawrence Berkeley National Laboratory, Berkeley, CA, USA*
MARK A. LE GROS • *National Center for X-ray Tomography, Berkeley, CA, USA;*

*Physical Biosciences Division, Lawrence Berkeley National Laboratory, Berkeley, CA, USA*
RICHARD D. LEAPMAN • *Laboratory of Cellular Imaging and Macromolecular Biophysics, National Institute of Biomedical Imaging and Bioengineering, National Institutes of Health, Bethesda, MD, USA*
DANIEL LEVY • *Institut Curie, Paris, France*
MEREWYN K. LODER • *Section of Cell Biology, Division of Diabetes Endocrinology and Metabolism, Department of Medicine, Faculty of Medicine, Imperial College London, London, UK*
ASTRID MAGENAU • *Lowy Cancer Research Centre, Centre for Vascular Research, University of New South Wales, Sydney, Australia*
ROBERTO MARABINI • *Escuela Politecnica Superior, Universidad Autónoma de Madrid, Madrid, Spain*
YOLANDA MARKAKI • *Biocenter, Department of Biology II, Ludwig Maximilians University Munich(LMU), Germany*
GERRY MCDERMOTT • *Anatomy Department, University of California San Francisco, San Francisco, CA, USA; National Center for X-ray Tomography, Berkeley, CA, USA*
JOHN M. MENDENHALL • *Center for Learning and Memory, The University of Texas at Austin, Austin, TX, USA*
PIERRE-EMMANUEL MILHIET • *Institut National de la Santé et de la Recherche Médicale, Montpellier, France; Centre National de la Recherche Scientifique, Centre de Biochimie Structurale, Université de Montpellier,, Montpellier, France*
U. VALENTIN NÄGERL • *Interdisciplinary Institute for Neuroscience (IINS), Université de Bordeaux, Bordeaux, France; Centre National de la Recherche Scientifique (CNRS), UMR 5297, Bordeaux, France*
WILLIAM H. NEWHART • *Ralph N. Adams Institute for Bioanalytical Chemistry, University of Kansas, Lawrence, KS, USA*
J. OTÓN • *Biocomputing Unit, National Center of Biotechnology (CSIC), Universidad Autónoma de Madrid, Madrid, Spain*
DYLAN M. OWEN • *Lowy Cancer Research Centre, Centre for Vascular Research, University of New South Wales, Sydney, Australia*
DILWORTH Y. PARKINSON • *Advanced Light Source, Lawrence Berkeley National Laboratory, Berkeley, CA, USA*
A. QUINTANA • *Biocomputing Unit, National Center of Biotechnology (CSIC), Universidad Autónoma de Madrid, Madrid, Spain*
GUY A. RUTTER • *Section of Cell Biology, Division of Diabetes Endocrinology and Metabolism, Department of Medicine, Faculty of Medicine, Imperial College London, London, UK*
SERGIO SANTOS • *Department of Oral Biology, University of Leeds, Leeds, UK; School of Physics and Astronomy, University of Leeds, Leeds, UK; Laboratory for Energy and Nanosciences, Masdar Institute of Science and Technology, Abu Dhabi, UAE*
LOTHAR SCHERMELLEH • *Department of Biochemistry, University of Oxford, Oxford, UK*
ROMAN SCHMIDT • *Department of NanoBiophotonics, Max Planck Institute for Biophysical Chemistry, Göttingen, Germany*
YOSHITAKA SHIRASAKI • *Laboratory for Immunogenomics, RIKEN Research*

*Center for Allergy and Immunology, Kanagawa, Japan*
JEAN-BAPTISTE SIBARITA • *Interdisciplinary Institute for Neuroscience, Université de Bordeaux, Bordeaux, France; CNRS, IINS UMR 5297, Bordeaux, France*
DANIEL SMEETS • *Biocenter, Department of Biology II, Ludwig Maximilians University Munich(LMU), Germany*
CARLOS OSCAR SORZANO • *Biocomputing Unit, National Center of Biotechnology (CSIC), Universidad Autónoma de Madrid, Madrid, Spain*
ALIOSCKA A. SOUSA • *National Institute of Biomedical Imaging and Bioengineering, National Institutes of Health, Bethesda, MD, USA*
CHRISTIAN STEINHAUER • *Institut für Physikalische und Theoretische Chemie—NanoBioScience, Braunschweig University of Technology, Braunschweig, Germany*
CARLO TACCHETTI • *MicroscoBio Research Center, Department of Experimental Medicine (DIMES), University of Genoa, Genoa, Italy; Imaging Research Center, Foundation San Raffaele Scientific Institute, Milan, Italy*
NEIL THOMSON • *Department of Oral Biology, University of Leeds, Leeds, UK; School of Physics and Astronomy, University of Leeds, Leeds, UK*
PHILIP TINNEFELD • *Institut für Physikalische und Theoretische Chemie—NanoBioScience, Braunschweig University of Technology, Braunschweig, Germany*
JAN TØNNESEN • *Interdisciplinary Institute for Neuroscience (IINS), Université de Bordeaux, Bordeaux, France; UMR 5297, Centre National de la Recherche Scientifique (CNRS), Bordeaux, France*
TAKASHI TSUBOI • *Department of Life Sciences, Graduate School of Arts and Sciences, The University of Tokyo, Tokyo, Japan*
J. VARGAS • *Biocomputing Unit, National Center of Biotechnology (CSIC), Universidad Autónoma de Madrid, Madrid, Spain*
J.J. VEGA • *Biocomputing Unit, National Center of Biotechnology (CSIC), Universidad Autónoma de Madrid, Madrid, Spain*
GIUSEPPE VICIDOMINI • *IIT—Italian Institute of Technology, Genoa, Italy*
PETER K. WEBER • *Lawrence Livermore National Laboratory, Glenn T. Seaborg Institute, Livermore, CA, USA*
SARAH M. WILDGEN • *Ralph N. Adams Institute for Bioanalytical Chemistry, University of Kansas, Lawrence, KS, USA*
DAVID J. WILLIAMSON • *Lowy Cancer Research Centre, Centre for Vascular Research, University of New South Wales, Sydney, Australia*
MAI YAMAGISHI • *Laboratory for Immunogenomics, RIKEN Research Center for Allergy and Immunology, Kanagawa, Japan*
CHRISTOPHER M. YIP • *Department of Chemical Engineering and Applied Chemistry, Terrence Donnelly Centre for Cellular and Biomolecular Research, University of Toronto, Toronto, ON, Canada*
A. ZALDÍVAR-PERAZA • *Biocomputing Unit, National Center of Biotechnology (CSIC), Universidad Autónoma de Madrid, Madrid, Spain*

# Chapter 1

# Introduction: Nanoimaging Techniques in Biology

**Alioscka A. Sousa and Michael J. Kruhlak**


## Abstract

To dissect the astonishing complexity of the biomolecular machinery functioning within a cell, imaging has been an integral tool in biology, allowing researches to "view" the detailed molecular biology responsible for coordinating cellular life. To visualize the molecular components of cellular structures requires, in particular, imaging techniques capable of reaching nanoscale spatial resolutions. Such nanoimaging techniques are the focus of this volume. Chapters in the current volume are divided into four parts and include specialized techniques in the areas of light, electron, and scanning probe microscopy, as well as methodologies employing combinatorial and complementary imaging approaches.


## 1. Introduction

Despite the diverse physiological properties of different cell types, they are composed, generally, of the same biochemical and molecular components. The coordinated assembly of these components provides the functional characteristics of the cell and, ultimately, the viability of the organism. A common goal in biology is, therefore, to dissect the astonishing complexity of the cellular molecular machinery, seeking to understand how individual components of this machinery assemble to form higher-order structures, navigate the cell with synchronized dynamics, and interact to produce a given biological outcome. To realize this common goal, imaging has been an integral tool in biology, allowing researches to "view" the detailed molecular biology responsible for coordinating cellular life.

To directly observe the molecular components of a cell, either in isolated form or within a cellular context, biologists have at their disposal an array of established techniques in the major areas of light, electron, and scanning probe microscopy. While pivotal changes in biology have repeatedly occurred following key developments in microscopy, the quest to explore the inner workings of

Alioscka A. Sousa and Michael J. Kruhlak (eds.), *Nanoimaging: Methods and Protocols*, Methods in Molecular Biology, vol. 950, DOI 10.1007/978-1-62703-137-0_1, © Springer Science+Business Media, LLC 2013

cells at ever-increasing levels of detail has been, at the same time, a tremendous stimulus to the development of novel and ingenious approaches to imaging. Thus, to some extent, biology has led to a continuous reinvention of the light, electron, and scanning probe microscopes, in addition to nurturing the development of completely new tools in microscopy. Today, advanced imaging modalities enable the visualization of biological specimens at a range of length scales and resolutions, in multiple dimensions (both time and space), with chemical or molecular specificity, and from the surface or bulk of the specimen, revealing the secrets of the molecular bionetwork within the cell.

Starting with the optical microscope, one of the most important historical developments in this field can be assigned to Ernst Abbe, whose rigorous description of wave-based optical theory supporting image formation in the microscope provided the rationale for the limitation of resolution and, thereby, the expected barrier to observing fine details of the molecular machinery within a specimen. The resolution limit as imposed by diffraction still holds true, but advances in instrumentation and a greater understanding of the physical photochemistry of labeling fluorophores have allowed microscopists to achieve resolutions beyond Abbe's diffraction limit. This includes the recent development of numerous ingenious super-resolution optical imaging methods that reveal specimen details on the nanometer scale.

The legendary work by Ernst Ruska and colleagues used Abbe's wave-based optical theory to great advantage in developing the electron microscope. The dramatically shorter wavelength of electrons employed in the electron microscope allowed biologists for the first time to examine the internal architecture of cells with nanoscale spatial resolution. The conventional view of electron microscopy, however, is that thin and exhaustively processed specimens can be observed only with rather nonspecific contrasting agents, thereby limiting the specificity of the molecular information obtained. Modern electron microscopes are based on Ruska's prototype but have developed specialized imaging arrangements that provide greater contextual information about the sample while overcoming some of the traditional limitations of specimen preparation. The diverse modes of operation of the modern electron microscope are able to provide images of native cellular structures, in three dimensions, from large sample volumes, and containing relevant quantitative information.

Contrary to light and electron microscopy in which a magnified image of a specimen is generated by a lens-based optical system, techniques in scanning probe microscopy rely on the general principle of raster-scanning a sharp probe across a surface to produce a magnified view of the specimen topography. The hallmark scanning probe instrument is the atomic force microscope (AFM) that differs significantly from conventional lens-based microscopes in the underlying principle of image formation, and not surprisingly offers its own set of attributes and opportunities, such as the possibility to probe specific chemical and biophysical interactions between a functionalized scanning tip and the underlying sample surface. Near-field scanning optical microscopy (NSOM) is another type of scanning probe technique of current use in biology. It involves scanning with a small probe positioned less than a wavelength of light from the sample whereby light is transmitted through a small aperture in the probe tip, thus providing a means to circumvent the diffraction-limited far-field resolution of optical microscopes.

Several other specialized nanoscale imaging techniques exist that do not fall within the more general categories of light, electron, and scanning probe microscopy as highlighted above. One such technique, soft X-ray microscopy, has a history that dates back several decades, although only more recently has it become more adept at imaging biological specimens. Today, soft X-ray tomography is emerging as a powerful method to visualize entire eukaryotic cells in three dimensions with a spatial resolution that bridges that of the light and electron microscope. Another specialized technology, imaging mass spectrometry, combines attributes of both secondary ion mass spectrometry and imaging. This relatively new technique offers the exciting possibility of obtaining very specific compositional information about a sample's surface with a lateral resolution in the tens of nanometers.

A feature common to all imaging techniques is the need to produce contrast, and in fact the success of some of the most modern techniques relies on ingenious ways in which structures can be induced to reveal themselves. Of course, an important characteristic of any imaging modality is the achievable spatial resolution, and here a diverse range of methodologies exist that cover resolutions from the nano to the macro length scales. To visualize cellular molecular components and substructures requires in particular imaging techniques capable of reaching nanoscale spatial resolutions. Such nanoimaging techniques are therefore the focus of this volume.

Chapters in the current volume are divided into four parts and include specialized techniques in the areas of light, electron, and scanning probe microscopy (Parts 1, 2, and 3, respectively), as well as methodologies employing combinatorial and complementary imaging approaches (Part 4). Nanoimaging of biological samples involves precise logistical considerations that are more acute than for conventional imaging, such as illumination timing and intensity, signal information sampling, specimen preparation and labeling, to list a few. Given the added complexity of imaging biological specimens on the nanometer scale, we hope that the chapters comprising this volume provide valuable information from successful application of a broad range of important imaging methods for the nanoscale visualization of biological structures.

## 2. Part 1: Light Microscopy

Light microscopy is invaluable to biologists, but its application to resolving molecular details is limited by the physical properties of light, principally the diffraction of light upon passing through a fixed aperture. Diffraction of light results in the blurring of a point object in the image and is dependent on the wavelength of light and the numerical aperture of the collecting lens. The expanded observed intensity profile of the point is known as the point spread function, with the actual object located at the center of the intensity profile. The dimensions of the point spread function are larger than the physical size of the individual molecular components within the specimen. To truly image specimen molecular details, the diffraction barrier needs to be surpassed. A collection of super-resolution optical imaging methods have been able to attain resolutions beyond the diffraction limit by employing sophisticated, and elegant, illumination parameters, optical configurations, or fluorophores with unique biophysical properties. A number of methods exploit nonlinearity in the reflected light from the specimen and often employ a timing mechanism to achieve the improved resolution.

**Part 1** of this volume contains several chapters describing optical imaging methods that achieve spatial resolution beyond the practical resolution limit of conventional light microscopy. The volume begins with a description of total internal reflection fluorescence (TIRF) microscopy, which uniquely provides improved axial spatial resolution by illuminating the sample using an evanescent wave to excite fluorophores within a limited depth. Establishing the evanescent wave illumination relies on adjusting the angle of the excitation light and refractive index differences between the support substrate, usually glass coverslip, and the growth medium of the sample. The details of imaging biological samples with a TIRF microscope are highlighted in Chapter 2. Another impressive configuration that improves the axial resolution of the light microscope by aligning two opposing objective lenses in the optical path and, thereby, effectively doubling the numerical aperture is known as 4Pi microscopy. The specialized optical configuration of 4Pi microscopy enables a substantial reduction of the dimensions of the point spread function in the optical axis, and Chapter 3 describes the use of this unique optical arrangement to image subcellular architecture.

A number of relatively new developments in light microscopy are presented in Chapters 4 through 8, all of which achieve resultant spatial resolution beyond the diffraction limit as defined by the Abbe criterion. The successful implementation of the illumination-based super-resolution imaging methods of structured illumination microscopy (SIM) and stimulated emission depletion

(STED) microscopy are described in Chapters 4 and 5, respectively. Among sub-diffraction limited techniques, structured illumination holds the distinction of being one of the only methods that can work with simple reflectance without the need for special dyes or fluorescence. Multiple spatially modulated illumination patterns are used to double the effective numerical aperture and, thereby, improve the lateral resolution. STED imaging uses nonlinear multistep optical illumination strategy to reduce the size of the point spread function and achieve super-resolution in the nanometer scale. The following three chapters discuss probe-based super-resolution imaging techniques based on the principle of single-molecule localization: photoactivatable localization microscopy (PALM) (Chapter 6), light microscopy with points accumulation for imaging in nanoscale topography (uPAINT) (Chapter 7), and blink microscopy (Chapter 8). Collectively, the probe-based techniques rely on the mathematically known location of the actual object within the observed intensity profile of the point spread function. Using time to acquire individual PSFs from multiple areas within the sample, the location of the objects within each PSF can be localized with high precision, and with multiple PSFs collected over time, an image map can be constructed with the location of each object determined with high spatial resolution in the final image. Several other exciting variations of the single-molecule localization method exist, but we regret that not all could possibly be included in this volume.

The successful implementation of most of the existing super-resolution imaging methods depends in great part on the biophysical properties of the fluorophores used to highlight specific molecular complexes in the cell. Therefore, the selection of appropriate fluorophores suitable, both for the sample to be imaged and the method used for imaging, is an important topic concerning super-resolution light microscopy and is presented in Chapter 9.

Being able to accurately pinpoint the location of discrete fluorescent objects is not only used to construct super-resolution images of static specimens, but can also be employed to track the trajectory of particular molecular complexes in living cells. By tracking single particles with high precision, the location of the particle at each time point can be used to build a map of the molecule's history, and provide insight into its function. Chapter 10 details how this principle can be applied to track the dynamics of ribonucleoprotein complexes in live cells.

Collectively, the optical imaging methods, and the successful application of the methods to image biologically relevant samples, described in **Part 1** of the volume highlight the many exciting possibilities available to obtain nanometer scale information from molecular assemblies within the cell.

## 3. Part 2: Electron Microscopy

The electron microscope has been indisputably a primary nanoscale imaging tool in biology. Traditionally, however, electron microscopy (EM) has been somewhat restricted to specialized laboratories, due to factors such as cost of instrumentation and challenges with sample preparation. Nevertheless, today some of the most advanced aspects of the technique have become readily accessible to the biological electron microscopist, thanks to technical developments in instrumentation, sample preparation and data processing. Part 2 of this volume presents some of the most important methodologies in the field of electron microscopy, with an emphasis on techniques that go beyond just basic implementations of EM.

**Part 2** begins with a discussion on single-particle cryo-electron microscopy (cryo-EM), an advanced technique for generating high-resolution three-dimensional images of macromolecular assemblies. Single-particle cryo-EM does not require molecular crystals, being instead applicable to ensembles of isolated molecular complexes. This unique feature of the technique allows the three-dimensional imaging of molecular machines assuming specific functional states or having particular interactions with other molecules. Processing single-particle cryo EM data to generate a three-dimensional volume of the specimen requires careful consideration of many parameters. Chapter 11 therefore lays down a protocol to analyze single-particle cryo-EM data, guiding the reader towards each step of the data processing and analysis workflow.

Next in **Part 2**, Chapter 12 underlines the distinctive capability of the electron microscope in making accurate measurements of molecular mass from isolated macromolecular assemblies. Compared to other techniques such as mass spectrometry, the method of molecular mass measurement in the electron microscope is unique in that it enables determination of the mass per length of filamentous specimens or the mass per area of planar structures. Chapter 12 highlights in particular how quantitative mass per length measurements can be employed to gain important structural information on Alzheimer's amyloid fibrils.

Chapter 13 provides another example of the capability of the electron microscope in generating quantitative information from biological specimens. Specifically, in the techniques of energy-filtering transmission electron microscopy (EFTEM) and electron energy-loss spectroscopy (EELS), electrons that have traversed a thin sample can be collected and their energies analyzed by an electron spectrometer to yield quantitative information on the nanoscale distribution of specific elements in the specimen. Chapter 13 explains some of the experimental and computational steps in

EFTEM and EELS leading to the formation of quantitative elemental maps. Because phosphorus is present in abundance in DNA, these approaches enable imaging the intrinsic distribution of nucleic acids without the use of nonspecific heavy-metal stains.

Significantly, the last several years have witnessed a remarkable explosion in the application of electron tomography, a technique that allows the generation of three-dimensional images of cellular ultrastructure at nanoscale spatial resolution. While recording conventional electron tomograms from heavy-metal stained specimens is somewhat straight-forward using modern electron microscopes, imaging samples in their native state (i.e., frozen-hydrated) can be considerably more challenging. Cryo-electron tomography of native, frozen-hydrated eukaryotic cells is therefore the focus of Chapter 14.

While electron tomography can produce unique nanoscale three-dimensional views of subcellular architecture, the technique has a fundamental limitation in that samples must be electron transparent (i.e., less than about 400-nm thick). Thus, only a very small fraction of an entire eukaryotic cell or piece of tissue is amenable to analysis by electron tomography. Recently, however, there has been growing interest in developing approaches for the nanoscale reconstruction and analysis of large volumes of cellular and tissue ultrastructure. As an application for this methodology, three-dimensional EM data obtained from across several microns of brain tissue can be used to explore how patterns of circuit connectivity relate to neuronal function. Chapter 15 describes in detail how serial-section electron microscopy can be optimally implemented to generate large-volume three-dimensional reconstructions of brain tissue.

Another approach to visualizing large volumes of biological specimens in three dimensions is based on the hybrid technique of focused ion beam/scanning electron microscopy (FIB/SEM). The FIB/SEM method is the topic of Chapter 16, which illustrates primarily how a FIB can be used as a nano-scalpel to reveal subsurface microstructures that can then be examined by SEM imaging.

Despite the innumerous advantages as a nanoimaging tool, EM has an important limitation in that it does not produce images with intrinsic molecular specificity. Chapter 17 therefore underlines the use of ultrasmall gold nanoparticle conjugates as labels to localize specific intracellular proteins within sectioned cells or identify specific subunits within multi-subunit molecular assemblies. In particular, Chapter 17 explains how to synthesize a type of ultrasmall and uniform gold nanoparticle, and, in addition, how to attach antibodies to the nanoparticle allowing it to be used as a label for EM imaging of biological specimens.

## 4. Part 3: Scanning Probe Microscopy

Scanning probe microscopy offers unparalleled possibilities for imaging and probing model and native membranes as well as macromolecular complexes. **Part 3** of this volume contains four chapters covering two distinct types of scanning probe instruments, namely the atomic force microscope and the near-field scanning optical microscope.

**Part 3** starts with Chapter 18, which highlights the AFM as a tool for high resolution imaging of macromolecular complexes. Examples of imaging single DNA molecules and DNA-protein interactions are presented, together with an ample discussion of the important specimen variables and operational parameters that affect image quality and lead to reproducible imaging conditions.

The high lateral and vertical spatial resolution afforded by the AFM combined with its ability to operate in liquids has turned this instrument into a distinctive tool for the high-resolution imaging of two-dimensional crystals of membrane proteins. However, membrane proteins that are difficult to crystallize and/or express and purify are not easily amenable to high resolution AFM imaging. Chapter 19 therefore presents an attractive approach for preparing specimens of membrane protein for high resolution AFM imaging, whereby transmembrane proteins are incorporated within supported lipid bilayers at only picomolar concentrations.

Chapter 20 describes novel developments in atomic force microscopy for the high spatial resolution mapping of specific receptors on native membranes. The basic principle of this approach involves raster-scanning a functionalized tip onto the specimen and recording force spectroscopy measurements at every pixel of the topographic image. Chapter 20 explains the implementation of this exciting approach, termed simultaneous topography and recognition imaging technique (TREC), showing as an example how it can be used to map the nanoscale distribution of Fcγ receptors on macrophages.

**Part 3** ends with a chapter on near-field scanning optical microscopy (NSOM). In this technique, an optical probe of subwavelength diameter is raster-scanned across a sample keeping very close proximity to its surface. Lateral resolution becomes independent of far-field diffraction, being instead a function of the diameter of the probe aperture and its distance to the sample surface. Chapter 21 explains how scanning probe microscopes can be modified in-house to implement NSOM, and provides examples of NSOM fluorescence imaging of model and natural biological membranes.

## 5. Part 4: Complementary Techniques

The imaging techniques discussed so far possess distinct advantages in obtaining specific information from biological samples. Combining multiple imaging methods provides a means to acquire specific information within a contextual structure, thus enhancing the collective information from the sample. For example, the specificity of the labeling used in fluorescence microscopy can be used to great advantage to complement the architectural information available from electron microscopy. Correlative fluorescence and electron microscopy methods not only provide specificity along with ultrastructural detail but also help reduce the burden of screening the sample for imaging, thereby increasing the probability of obtaining information from samples in the desired growth state, or biochemical process.

A multitude of correlative microscopy combinations have been developed with a few interesting examples described in **Part 4** of the volume. A description of two correlative fluorescence and electron microscopy techniques are presented in the first two chapters of **Part 4**, with the combination of fluorescence microscopy and energy-filtering transmission electron microscopy used to reveal the ultrastructure and biochemical composition of interesting subnuclear structures in Chapter 22, and the three-dimensional high data output correlative light and electron microscopy imaging of cryosections in Chapter 23.

Uniquely, the combination of fluorescence microscopy with atomic force microscopy provides the specificity of fluorophore labeling with the fine topographical details available using AFM. The description of an integrated AFM and fluorescence instrument is presented in Chapter 24 that has been successfully employed in imaging specific interactions occurring at membrane interfaces. Interestingly, the multimodal instrument is amenable to combining super-resolution optical methods with scanning probe microscopy, with great promise for data-rich nanoimaging in biology.

An exciting method for imaging small prokaryotes or eukaryotic cells in three dimensions has emerged by applying the principle of soft X-ray microscopy in a tomographic imaging configuration. Soft X-ray microscopy uses illuminating photons in the water window of the electromagnetic spectrum, meaning that these photons are much more efficiently absorbed by organic mass than water. The technique enables imaging intact frozen-hydrated cells with a spatial resolution in the tens of nanometers, bridging an important gap between light and electron microscopy. Chapter 25 describes experimental and computational procedures in soft X-ray tomography leading to the nanoscale three-dimensional reconstruction of frozen-hydrated cells.

The ability to obtain chemically specific information at nanoscale spatial resolution within cellular or model membranes is an exceptional feature of secondary ion mass spectrometry imaging. In this technique, a focused ion beam raster scans across a specimen producing secondary ions that are then analyzed by a mass spectrometer. A map of the nanoscale distribution of specific membrane lipids can be generated by detecting the secondary ion signal from those lipid species that have been isotopically labeled. In Chapter 26, basic protocols and methodologies are described for the secondary ion mass spectrometry imaging of sphingolipids in the plasma membrane.

## 6. Conclusion

Collectively, the nanoimaging methods presented in this volume of Methods in Molecular Biology we expect will provide valuable information on how to apply and adapt the methods for imaging other samples of interest, and guide the use of nanoimaging in the exploration of the molecular machinery responsible for regulating cellular life.

# Part I

## Light Microscopy

# Chapter 2

# Live-Cell Imaging of Vesicle Trafficking and Divalent Metal Ions by Total Internal Reflection Fluorescence (TIRF) Microscopy

**Merewyn K. Loder, Takashi Tsuboi, and Guy A. Rutter**

## Abstract

Total internal reflection fluorescence (TIRF) microscopy is an especially powerful tool for visualizing live cellular events. Fluorescent molecules alone provide broad information about the expression and localization of proteins and other molecules; however, the temporal and spatial resolution is confounded by signal from outside the area of interest and the intensity of the illumination required. TIRF overcomes this limitation by using the reflective properties of a laser beam to illuminate a narrow (<100 nm) strip at the surface of a cell with a relatively low powered evanescent wave, thus making it possible to measure events occurring specifically at the plasma membrane such as exocytosis, single molecule interactions, and ionic changes during signal transduction. Here we describe some of the methods for using TIRF microscopy to study the processes involved in exocytosis from excitable cells (i.e., neurons, endocrine, neuroendocrine, and exocrine cells) and the release of physiologically active substances (i.e., neurotransmitters, hormones, and mucus).

The failure of regulated exocytosis is associated with various diseases such as allergy, brain dysfunction, and endocrine illness. Diabetes mellitus, which is due to an absolute (type I) or relative (type II) deficiency of insulin secretion from pancreatic β-cells, is a major area of therapeutic interest. Insulin is stored in dense core vesicles with $Zn^{2+}$ ions in pancreatic β-cells. Insulin secretion is regulated by plasma glucose concentration which acts through intracellular metabolism to influence intracellular $[Ca^{2+}]$. However, the precise molecular mechanisms controlling insulin granule movement towards, and fusion at, the plasma membrane remain only partially understood. To tackle this problem, we have used live cell imaging techniques to image regulated exocytosis in single living β-cells alongside intracellular $Ca^{2+}$ and $Zn^{2+}$ concentrations.

**Key words:** Exocytosis, Insulin, Calcium, Zinc, Pancreatic β-cells, Total internal reflection fluorescence microscopy

## 1. Introduction

Optical live cell imaging techniques represent a powerful tool for detecting molecular events such as cell adhesion, secretion of hormones, and intracellular signal transduction in live cells.

Alioscka A. Sousa and Michael J. Kruhlak (eds.), *Nanoimaging: Methods and Protocols*, Methods in Molecular Biology, vol. 950, DOI 10.1007/978-1-62703-137-0_2, © Springer Science+Business Media, LLC 2013

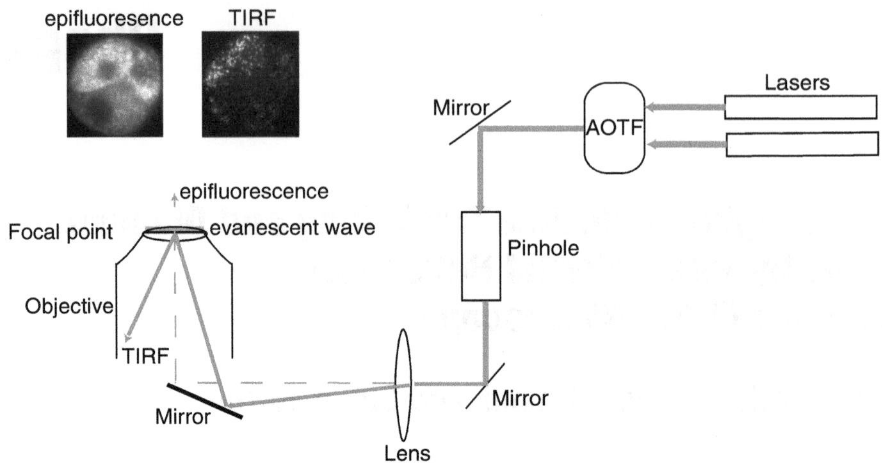


Fig. 1. Schematic of a typical TIRF microscope (system 2) and representative images of a cluster of three β-cells under epifluorescent (*left*) and TIRF (*right*) illumination.

Epifluorescence imaging of live cells can provide information on the dynamics of protein localization and molecular interactions, as well as the intracellular concentrations of signal transduction molecules and ions. However, at high objective magnification under epifluorescence, there is poor vertical spatial resolution and fluorescence from outside the focal plane can overwhelm the images limiting the use of this technique for studying active cellular processes (Fig. 1). To overcome this limitation, total internal reflection fluorescence (TIRF) microscopy was developed by Daniel Axelrod in the early 1980s (1, 2).

TIRF microscopy employs an angled excitation laser light and excites fluorophores just beneath a glass-water interface with extremely high vertical spatial resolution. A laser beam, hitting a substance of different refractive index to the one in which it is travelling, bends. In TIRF microscopy, the laser irradiates the specimen at an angle of incidence greater than or equal to the critical angle of refraction whereby the excitation laser light is totally reflected. The reflection generates a very thin electromagnetic field (~200 nm) in the specimen. This field, called the evanescent wave or field, undergoes exponential intensity decay with increasing distance from the surface. Thus, only those fluorophores within 200 nm of the specimen interface are excited, providing high vertical spatial resolution fluorescence images suitable for studying rapid active biological events. Release of hormones via regulated exocytosis in response to a physiological stimulus is a fast, dynamic and complex process requiring the trafficking of cargo to the plasma membrane, reorganization of the cytoskeleton and actin networks (3–5), and the fusion of vesicles with the plasma membrane through protein recruitment to the soluble NSF attachment protein receptors (SNARE) complex (6). Many diseases, from depression to diabetes, involve perturbations of these processes. Thus, understanding of the behavior of secretory granules at the plasma membrane and of the regulatory molecules involved is an essential goal of modern cell biology.

With high speed data acquisition and ultra sensitive charge-coupled device (CCD) cameras, TIRF microscopy has been used to observe the exocytotic release of glutamate in astrocytes (7), to measure changes in $Ca^{2+}$ concentration beneath the plasma membrane (8), and to investigate dense-core vesicle behavior (4, 9–13). We have employed TIRF to visualize insulin exocytosis and intracellular $Ca^{2+}$ and $Zn^{2+}$ signaling in pancreatic β-cells and here we describe the details of this technique.

## 2. Materials

### *2.1. Equipment*

There are many types of TIRF configurations including the use of either a prism or illumination "through the lens"; the latter is the most common in molecular biology and requires a ultra high numerical aperture (NA) objective. Here, we describe two configurations used in our laboratories: System 1 a Nikon TIRF microscope and System 2 based on a Zeiss Axiovert microscope. System 1 was largely used to generate the granule movement data described whereas system 2 was used for the plasma membrane divalent cations studies. However, the methods described can be used interchangeably.

#### *2.1.1. System 1: Nikon TIRF Microscope*

1. An ultra high numerical aperture (NA) objective lens (CFI Apo TIRF 100× H, NA = 1.49; Nikon) is mounted on a nosepiece of a Ti-E inverted microscope . Immersion oil (Cargille Laboratories) of a high refractive index ($n$ = 1.515) is used to couple the lens to the glass dish. Excitation laser light from a diode pumped solid-state (DPSS) laser (wave length = 488 nm, 30 mW; Spectra Physics) is introduced to an optical fiber. The light from the optical fiber is focused with two illumination lenses on the back focal plane near the extreme edge of the objective lens. By theoretical calculation, the space constant for exponential decay of the evanescent field was estimated to be 100 nm (see Note 1).
2. To excite the green fluorescent protein (GFP), or the yellow fluorescent protein (YFP) variant Venus, we use a 488 nm DPSS laser. For imaging of green fluorescence, we use a dichroic mirror (T510LPXRXT, Chroma) which divides the blue and green components of the images. We then use a 515–550 nm band-pass filter (ET535/50m, Chroma) to obtain the green fluorescence

component. A computer-controlled electromagnetically driven shutter (VMM-D3J, Uniblitz) is placed between the optical fiber and illumination lenses to prevent photobleaching of the cells between image acquisitions. The shutter is opened synchronously with an electron multiply charge-coupled device (EM-CCD) camera (DU897E-CS0-#BV, Andor Technology) and exposure controlled with MetaMorph software (version 7.7, Molecular Devices).

#### *2.1.2. System 2: Zeiss Axiovert TIRF Microscope*

1. An ultra high numerical aperture objective lens (Apo 100× H, NA = 1.45) is mounted on the nosepiece of an Axiovert 200M inverted microscope. Immersion oil (Olympus) of a high refractive index ($n$ = 1.517) is used to couple the lens to a glass coverslip. Cells are excited using a DPSS laser (473 nm, 80 mW; Crystalaser). Emitted light is separated using a double dichroic (590/22bs), magnified with an 1.6× optovar. For two color imaging, the emitted light is split into two channels using a Dualview (Optical Insights) image splitter with a dichroic cutoff at 585 nm (T585LP) and band-pass filters 595–670 nm (ET630-75) and 485–550 nm (ET517/65M) (all filters from Chroma). Separate images of the two channels are taken on an iXonEM + EM-CCD camera (Andor Scientific) controlled by Andor IQ software. Images shown are acquired at 20 Hz and 50 ms exposure at 100 nm/pixel. The light from the lasers is focused through a pinhole via two illumination lenses on to the back focal plane near the extreme edge of the objective lens. By theoretical calculation, the space constant for exponential decay of the evanescent field is estimated to be 100 nm (see Note 1).
2. For high resolution granule movement studies using system 2, we use a higher numerical aperture lens (1.65 NA 100× ApoTIRF, Olympus), special immersion oil with a very high refractive index ($n$ = 1.78 SPI supplies), and TIRF coverslips (035055-9-U990, Olympus).

### 2.2. Cell Preparation (See Note 2)

#### *2.2.1. MIN6 Cells*

1. Dulbecco's modified Eagle's medium (DMEM, high glucose) supplemented with 10% fetal bovine serum (FBS), 100 U/mL penicillin, 100 μg/mL streptomycin, and 30 μM 2-mercaptoethanol.
2. Phosphate-buffered saline (PBS): pH 7.4, without $CaCl_2$ and $MgCl_2$.
3. Trypsin 0.25% with 1 mM ethylenediaminetetraacetic acid (EDTA) solution.
4. Poly-L-lysine (PLL): 1 mg/mL.
5. Glass-bottom 35-mm dishes treated with PLL in PBS for 30 min.

6. MIN6 cells (passage # 19–30; a gift from Prof. Junichi Miyazaki, Osaka University, Osaka, Japan) (14).
7. Lipofectamine 2000 (Invitrogen)

*2.2.2. Primary Pancreatic β-Cells*

1. Female CD1 mice age 12–16 weeks.
2. RPMI 1640 medium: 10 mM glucose and 2 mM glutamine supplemented with 10% FBS, 100 U/mL penicillin, and 100 μg/mL streptomycin.
3. $Ca^{2+}$-free buffer: 138 mM NaCl, 5.6 mM KCl, 1.2 mM $MgCl_2$, 5 mM HEPES, 1 mM EDTA, 100 U/mL penicillin, and 100 μg/mL streptomycin, pH 7.35. Alternatively: 0.05% Trypsin/0.02% EDTA.
4. Type IV collagenase.
5. Poly-L-lysine (PLL): 1 mg/mL.
6. Glass-bottom 35-mm dishes or 22 mm coverslips treated with PLL in PBS for 30 min.

### 2.3. Probes Used for the Labeling of Dense Core Vesicles

*2.3.1. Neuropeptide Y Plasmid Used for Labeling MIN6 Cells*

Human prepronuropeptide Y (NPY) cDNA lacking a TGA stop codon was amplified by PCR using a forward primer, 5′-GAATTCATGCTAGGTAACAAGCGACT-3′ (primer #1; *Eco*RI site underlined), and reverse primer, 5′-GGATCCCCA CATTGCAGGGTCTTCA-3′ (primer #2; *Bam*HI site underlined), where TGA is replaced by TGG. The PCR product was subcloned into a pCR2.1 vector (Invitrogen). The NPY cDNA fragment lacking a stop codon was cleaved with *Eco*RI/*Bam*HI and subcloned into the *Eco*RI/*Bam*HI site of the multiple cloning site of a yellow fluorescent protein variant expression vector (pVenus-N1) (15).

*2.3.2. Adenoviral Probe Used for Labeling Primary Cultured β-Cells*

Recombinant adenovirus encoding NPY–Venus (NPY–Venus) was prepared using the AdEasy system (16, 17). A DNA fragment encoding fused NPY and Venus cDNAs was generated using PCR with the NPY–Venus plasmid as template. The forward primer, 5′-GTCGAC*GACGAG*ATGCTAGGTAACAAGCGA included a *Sal*I restriction site (underlined) and Kozak sequence (italicized), and the reverse primer, 5′-TGCGGCCGCTTACTTGTACAGCTC, contained the Venus stop codon and a *Not*I restriction site (underlined). The *Sal*I–*Not*I fragment, comprising full-length NPY–Venus, was purified and ligated into plasmid pShuttle-CMV under the control of the human cytomegalovirus promoter. Co-transformation of competent $RecA^+$ bacteria (XL1-B) with the above construct and the adenoviral pAdEasy-1 vector generated recombinant adenoviral plasmid, which was purified and transfected into HEK293 cells. The transfected cells were readily monitored through Venus fluorescence, and adenoviral particles obtained by the extraction from the cells 10 days after transfection. The virus was further amplified and purified by centrifugation on a CsCl gradient. Stocks of $10^9$–$10^{10}$ infectious particles per milliliter were retained and used in experiments. Titration of the virus was performed by infecting HEK293 cells with serially diluted viral stocks, and counting positive cells through Venus fluorescence (18).

### *2.4. Reagents for TIRF Imaging*

1. Krebs-Ringer bicarbonate (KRB) solution: 125 mM NaCl, 3.5 mM KCl, 1.5 mM $CaCl_2$, 0.5 mM $MgSO_4$, 0.5 mM $KH_2PO_4$, 25 mM $NaHCO_3$, 3 mM glucose, and 10 mM HEPES; pH 7.4 equilibrated with 95:5 $O_2/CO_2$.
2. High glucose-containing KRB: KRB with 25 mM glucose.
3. High-KCl-containing KRB: Stimulation with KCl is achieved by perfusion of 50 mM KCl-containing KRB (NaCl is reduced to maintain the osmolarity).
4. Trypsin 0.25% with 1 mM ethylenediaminetetraacetic acid (EDTA) solution
5. Poly-L-lysine (PLL): 1 mg/mL
6. Glass coverslips and 6-well plates
7. RPMI medium, supplemented with 10% FBS and 3 mM glucose
8. 2 μM Fura Red-AM: 2 mM stock solution Fura Red-AM in DMSO, diluted in KRB solution, and sonicated for 10 s to ensure resuspension.
9. 3 μM Fluozin-3 AM

## 3. Methods

### *3.1. Cleaning and Coating of Glass-Bottom 35-mm Dishes*

1. Clean glass-bottom 35-mm dishes by first placing them into a 500 mL beaker containing 300 mL of Milli-Q water, and then place the beaker in an ultrasound bath for 1 h.
2. Dip the glass-bottom 35-mm dishes into another 500 mL beaker containing 300 mL of 70% ethanol for 1 h (see Note 3).
3. Take out glass-bottom 35-mm dishes from the beaker inside a cell culture clean bench, then sterilize them using a UV lamp for 30 min in the clean bench.
4. Apply 100 μL of PLL solution onto each glass-bottom 35-mm dish.
5. After 30 min, remove the PLL solution and wash three times with 1 mL of sterile PBS.

### *3.2. Transfection of Dense Core Vesicle Marker to MIN6 Cells*

1. Culture cells in a 10-cm Petri dish and incubate in 5% $CO_2$ at 37°C.
2. For TIRF imaging, plate the cells in PLL-coated glass-bottom 35-mm dishes.

3. Transfect the MIN6 cells with 3 μg of NPY–Venus vector using Lipofectamine 2000 according to the manufacturer's instructions. All TIRF experiments are performed 2 days after transfection.

#### 3.3. Adenoviral Infection to Primary Pancreatic β-Cells

1. *Preparation of mouse β-cells*: For preparation of mouse islets of Langerhans, a detailed isolation procedure by collagenase digestion is described by Ravier and Rutter (19). Briefly, islets are isolated by collagenase digestion of the pancreas of female CD1 mice and selected by hand-picking or histopaque gradient centrifugation. Islets are then cultured in RPMI 1640 for 1 day. Islets are dissociated after incubation in $Ca^{2+}$-free buffer or Trypsin EDTA for 5 min. After brief centrifugation, the solution is replaced with RPMI 1640, and the islets disrupted by pipetting through a glass pipette. After this procedure, the disrupted cells are then cultured for 1 day on PLL-coated glass-bottom 35-mm dishes or 22-mm coverslips. β-cells are likely to represent 60–80% of the cells in preparations used and can be further selected by identifying by eye the largest cells within the field.
2. *Infection*: 1 day before imaging, cells are infected with adenoviruses at a multiplicity of infection of 30–100 infectious particles per cells for 4 h, and then the culture medium is changed.

#### 3.4. Imaging of Exocytosis

1. Change the culture medium to KRB for 30 min before imaging at 37°C.
2. Transfer the glass-bottom dishes to the thermostat-controlled heating stage (37°C) of the TIRF microscope (system 1).
3. Find fluorescent NPY–Venus-expressing MIN6 cells or infected dispersed primary β-cells.
4. Adjust the laser incident angle to obtain total internal reflection (2).
5. Precisely focus on the cell surface.
6. Lower the laser power or use a neutral density filter to reduce phototoxicity.
7. Increase EM-CCD sensitivity.
8. Acquire images at 30- to 300-ms intervals.
9. Stimulate the cells by changing buffers from KRB to either high glucose- or high-KCl-containing KRB.
10. Continue imaging over 20 min. after changing buffers.

#### 3.5. Labeling Cells for Intracellular $Zn^{2+}$ and $Ca^{2+}$

1. After overnight culture, disperse islets using Trypsin–EDTA (20–50 islets per well) and plate them on to glass coverslips coated with PLL in 6-well plates.
2. After 24 h in RPMI + 10% FBS, keep cells for 1 h in RPMI containing 3 mM glucose.

3. Load cells with 2 μM Fura Red-AM.
4. For double cation imaging, change media after 25 min to KRB-containing 3 μM fluozin-3 AM and incubate cells for 5 min at 37°C before transferring to a chamber for imaging. At physiological levels of $Zn^{2+}$ there is no significant effect on Fura Red fluorescence (20).

***3.6. Imaging of Intracellular $Zn^{2+}$ Concentration***

1. Change the culture medium to KRB for 30 min before imaging at 37°C.
2. Transfer the coverslip to a chamber on the stage of the TIRF microscope (system 2) with continuous perfusion with KRB at 37°C.
3. Find a cluster of dispersed primary β-cells loaded with FluoZin-3 (see Note 4).
4. Adjust the laser incident angle to obtain total internal reflection.
5. Precisely focus on the cell surface "footprint."
6. Lower the laser power or use a neutral density filter to reduce phototoxicity.
7. Acquire images at 30- to 100-ms intervals.
8. Stimulate the cells by changing buffers from KRB to either high glucose- or high-KCl-containing KRB.
9. Continue imaging for 20 min after changing buffers.
10. Obtain the maximum and minimum signal for calibration by changing from KRB to KRB + 5 μM pyrithione or 50 μM TPEN solutions, respectively.

***3.7. Imaging of Intracellular $Ca^{2+}$ Concentration (See Note 5)***

1. Change the culture medium to KRB for 30 min before the imaging at 37°C.
2. Transfer the coverslip to a chamber on the stage of the TIRF microscope (system 2) with continuous perfusion with KRB at 37°C.
3. Find a cluster of dispersed primary β-cells loaded with Fura Red.
4. Adjust the laser incident angle to obtain a total internal reflection.
5. Precisely focus on the cell surface "footprint."
6. Lower the laser power or use neutral density filter to reduce phototoxicity.
7. Acquire images at 30- to 100-ms intervals.
8. Stimulate the cells by changing buffers from KRB to either high glucose- or high-KCl-containing KRB. Continue imaging for 20 min.

### 3.8. Offline Image Analysis

To analyze fusion, single exocytotic events are selected manually, and the average fluorescence intensity of individual vesicle in a $0.7 \times 0.7$ μm square placed over the vesicle center calculated by using MetaMorph software. To distinguish between fusion events and vesicle movement (i.e., vesicles pause at the plasma membrane and then move back inside the cell without fusing), we monitor the fluorescence changes just before the disappearance of fluorescent signals. When there is a fusion event, a rapid transient increase in fluorescence intensity (to a peak intensity 1.5 times greater than the original fluorescence intensity within 1 s) is observed, whereas when vesicles move, the fluorescence intensity gradually decreases to the background level. The number of fusion events during stimulation period is counted manually based on the above criteria (21). Sequences can be exported as single TIFF files or converted into QuickTime or AVI movies.

### 3.9. Example of Exocytosis Imaging by TIRF

TIRF microscopy conditions can be used which allow us to observe fluorescence deriving from the fusion of a single vesicle, detect molecules in small numbers and report their function at high temporal and spatial resolution in living cells (Fig. 2a). Induction of a $Ca^{2+}$ influx into the cells by high glucose-containing KRB causes NPY–Venus-containing spots to brighten and spread suddenly, which is consistent with the exocytosis of the fluorescently labeled NPY (Fig. 2b). As shown in Fig. 2b right, the fluorescence increases first in a central region, then spreads into the surrounding area and finally declines as dye diffuses away from the exocytotic site (22). Very rarely, we also observe transient increases in fluorescence that are not associated with the complete disappearance of the fluorescence (Fig. 2c). These events typically exhibit a slower time course (~3 s to reach peak fluorescence) than those described above (Fig. 2b, right). Interestingly, much larger vesicle cargoes, such as tissue-type plasminogen activator (tPA), which are found in chromaffin cells (23), show quite different fluorescence intensity changes (Fig. 2d): after an initial increase, tPA-GFP fluorescence describes a much slower time course than the slow disappearance of NPY–Venus (i.e., Fig. 2c) in both PC12 (12) and clonal MIN6 β-cells (24), presumably reflecting either slow release of tPA or vesicle resealing and reacidification (Fig. 2d). In contrast, the slow disappearance of NPY–Venus may reflect the approach of vesicles to the plasma membrane without exocytosis (i.e., vesicle retrieval or vesicle recycling process). In each case TIRF microscopy allows one to track the fate of individual dense-core vesicles before, during, and after exocytosis in living cells.

### 3.10. Example of $Zn^{2+}$ and $Ca^{2+}$ Imaging by TIRF Imaging

Divalent cations are important signaling molecules in most cell types. Increases in intracellular free ($Ca^{2+}$) are required for exocytosis (8, 9) and $Zn^{2+}$ release is an important modulator of cell excitability in both neurons (25) and pancreatic α-cells (26).

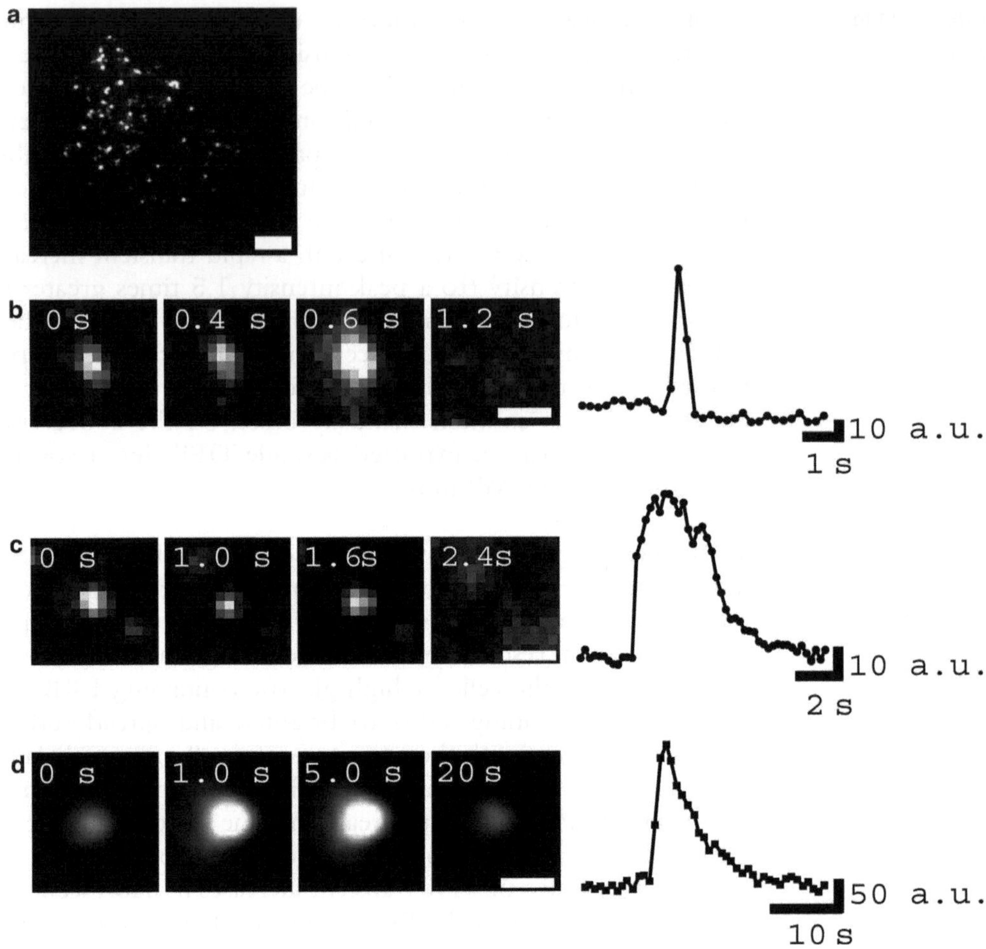


Fig. 2. Effect of high [K+] stimulation on exocytosis as reported with NPY-Venus-expressing MIN6 cells. (**a**) Typical TIRF image of the cellular distribution of NPY-Venus vesicles in a MIN6 cell. Scale bar: 5 μm. (**b**, *left*) Sequential images of a single NPY-Venus expressing vesicle observed after application of high [K+] stimulation in MIN6 cells. Third image (0.6 s) shows a diffuse cloud of the NPY-Venus fluorescence, and final image (1.2 s) shows an abrupt disappearance of the fluorescent spot. (**b**, *right*, **c**, *right*; **d**, *right*) Time course of the fluorescence changes measured in the circles placed in the center of fluorescent spots. The ordinate shows the arbitrary units of brightness. Note that the time scale in (**d**) differs from those in (**b**) and (**c**). (**c**, *left*) Sequential images of a single vesicle observed after high [K+] stimulation without showing any diffuse cloud of the NPY-Venus fluorescence. The third image (1.6 s) does not show any cloud of the dye, while the final image (2.4 s) shows a slow disappearance. (**d**, *left*) Sequential images of a single vesicle observed after high [K+] stimulation of the tPA-EGFP fluorescence. Note that tPA-EGFP fluorescence increased initially, and then dimmed slower than those in (**b**) and (**c**). Scale bars: 1 μm (modified from our previous data (21, 22)).

Furthermore, polymorphisms in the *SLC30A8* gene encoding the β-cell specific zinc transporter ZnT8 are associated with type II diabetes and defective insulin granule formation (27). Intracellular free $Ca^{2+}$ has been shown to display both global cytosolic increases and plasma membrane "hot spots" required for normal insulin secretion (28). TIRF microscopy can be used to examine the concentration of $Ca^{2+}$ and $Zn^{2+}$ at the plasma membrane simultaneously

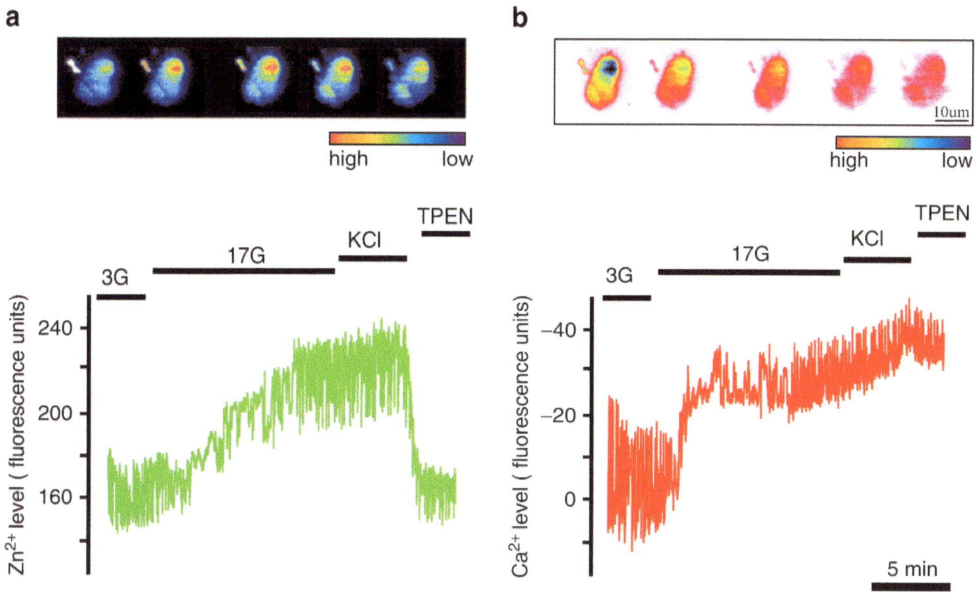


Fig. 3. Effect of glucose and KCl on sub-plasma membrane $Zn^{2+}$ and $Ca^{2+}$ concentrations in primary beta cells. (**a**, *top*) Typical representative pseudo-colored images of a primary beta cell at the time points on the trace below. (**a**, *bottom*) representative trace of the plasmalemmal $Zn^{2+}$ concentration in response to changes in either glucose (17 mM), KCl, or TPEN. (**b**, *top*) Typical pseudo-colored images of a primary beta cell $Ca^{2+}$ at the time points on the trace below. (**b**, *bottom*) representative baseline corrected trace of the sub-plasmalemmal $Ca^{2+}$ concentration in response to glucose (17 mM), KCl, or TPEN. Note that the scale is inverted as Fura Red fluorescence at 480 nm excitation decreases in response to increased $Ca^{2+}$ concentration. Scale bar: 10 μm.

in primary beta cells during glucose stimulation (Fig. 3). The cell footprint is defined in the IQ-capture software and the fluorescence intensity measured throughout the time course of the experiment. Figure 3a shows that in response to elevated glucose sub-plasma membrane $Zn^{2+}$ increases, this increase is steady reaching a maximum after about 10 min. The approximate concentration of $Zn^{2+}$ can be calculated from the maximum fluorescence in the presence of 50 μM ZnCl and 5 μM pyrithione and the minimum in the presence of 50 μM TPEN with the $K_D$ of the probe (in the case of FluoZin-3 $K_D = 15$ nM) using the following formula:

$$[Zn^{2+}] = K_D\left(\frac{F_{min} - F}{F - F_{max}}\right), \tag{1}$$

where $F$ is the observed fluorescence at a given time. In β-cells the concentration of $Zn^{2+}$ beneath the membrane after glucose stimulation is 1.9 ± 0.18 nM (mean ± S.E.M.). This is at the limit of the sensitivity of the method and equivalent to a two- to threefold increase over the resting cytosolic range of 400–600 pM (29)—well within the expected range for activating a physiological signaling mechanism. Glycinergic receptors, for example, have a zinc affinity of 27–540 nM (30) and zinc finger proteins, such as those found in many transcription factors, have affinities in the nM–pM range (31). Figure 3b shows the increase in $Ca^{2+}$ in the same cell. The increase in [$Ca^{2+}$] occurs at the same time as the increase in [$Zn^{2+}$], but appears to reach a maximum value more quickly, although both are sensitive to the L-type calcium channel blocker nifedipeine.

## 4. Conclusion

Here we have described methods for measuring insulin granule movement and sub-plasma membrane divalent cation concentrations using TIRF microscopy. These techniques can be used alongside fluorescently labeled proteins to study the role of each in the control of insulin secretion. By providing near single molecule resolution and dynamic real-time information in single cells these approaches can be combined with the use of healthy, diseased, or genetically modified tissues as a powerful approach to dissecting the regulatory pathways driving exocytosis in both health and disease.

## 5. Notes

1. During the excitation of fluorescent-labeled vesicles in a single cell (refractive index, $n_2 = 1.38$) adherent to a glass-bottom 35-mm dish (refractive index, $n_1 = 1.53$), wave fronts from a laser light pass through the glass and are reflected from the glass to cell interface at a critical angle $\alpha$, generating an evanescent wave. The critical angle ($\alpha$) at which total internal reflection of a laser light hitting the interface between two different refractive indexes substances (i.e., glass and culture medium for TIRF imaging) is determined by:

$$\alpha \geq \sin^{-1}\left(\frac{n_2}{n_1}\right), \quad n_1 > n_2,$$

   where $n_1$ is the refractive index of the glass, $n_2$ is the refractive index of cells (1.38).

   When we use an ultra high NA objective lens (NA = 1.49) to monitor exocytosis, the critical angle for total internal reflection is:

$$\alpha = \sin^{-1}\left(\frac{1.38}{1.53}\right) = 64.41^{\circ}$$

The maximal achievable angle for total internal reflection by using this objective lens (i.e., $n_2$ equals NA of objective lens) with glass is:

$$\alpha = \sin^{-1}\left(\frac{1.49}{1.53}\right) = 76.86^\circ$$

Thus, the laser light has to be introduced into the middle of these angles. The decay depth of evanescent field ($d$) can be calculated theoretically using the formula:

$$d = \frac{\lambda}{4\pi\sqrt{n_1^2 \sin^2\theta - n_2^2}},$$

where $\theta$ is the angle of incidence, $\lambda$ is the wavelength of the laser. For $\theta = 70$ ° and $\lambda = 488$ nm, decay constant ($d$) becomes 96.3 nm, indicating that a layer of 100–200 nm area from the glass and cell interface is the illumination area of the evanescent wave (32).

2. Other cell types can also be used for imaging; for example, bovine chromaffin cells, PC12 cells, and enteroendocrine cell line GLUTag cells have been tested. The only requirement is that the cells can tightly attach to the glass-bottom 35-mm dishes or coverslips.
3. Imperfect washing of the glass-bottom 35-mm dishes causes the increase of fluorescence background and obstructs the high resolution granule movement and intracellular divalent cations studies.
4. Very high levels of the dye can be toxic. Avoid cells with punctual staining as this can indicate unhealthy cells.
5. Simultaneous measurements of $Zn^{2+}$ and $Ca^{2+}$ are obtained by loading both dyes and using the beam splitter.

## References


1. Axelrod D (1981) Cell-substrate contacts illuminated by total internal reflection fluorescence. J Cell Biol 89:141–145
2. Axelrod D, Burghardt TP, Thompson NL (1984) Total internal reflection fluorescence. Annu Rev Biophys Bioeng 13:247–268
3. Ivarsson R, Jing X, Waselle L, Regazzi R, Renstrom E (2005) Myosin 5a controls insulin granule recruitment during late-phase secretion. Traffic 6:1027–1035
4. Kasai K, Ohara-Imaizumi M, Takahashi N et al (2005) Rab27a mediates the tight docking of insulin granules onto the plasma membrane during glucose stimulation. J Clin Invest 115:388–396
5. Tomas A, Meda P, Regazzi R, Pessin JE, Halban PA (2008) Munc 18-1 and granuphilin collaborate during insulin granule exocytosis. Traffic 9:813–832
6. Wang Z, Thurmond DC (2009) Mechanisms of biphasic insulin-granule exocytosis—roles of the cytoskeleton, small GTPases and SNARE proteins. J Cell Sci 122:893–903
7. Cali C, Marchaland J, Regazzi R, Bezzi P (2008) SDF 1-alpha (CXCL12) triggers glutamate exocytosis from astrocytes on a millisec-

ond time scale: imaging analysis at the single-vesicle level with TIRF microscopy. J Neuroimmunol 198:82–91

8. Becherer U, Moser T, Stuhmer W, Oheim M (2003) Calcium regulates exocytosis at the level of single vesicles. Nat Neurosc 6:846–853
9. Lang T, Wacker I, Steyer J et al (1997) $Ca^{2+}$-triggered peptide secretion in single cells imaged with green fluorescent protein and evanescent-wave microscopy. Neuron 18:857–863
10. Ohara-Imaizumi M, Nishiwaki C, Kikuta T, Nagai S, Nakamichi Y, Nagamatsu S (2004) TIRF imaging of docking and fusion of single insulin granule motion in primary rat pancreatic beta-cells: different behaviour of granule motion between normal and Goto-Kakizaki diabetic rat beta-cells. Biochem J 381:13–18
11. Steyer JA, Almers W (1999) Tracking single secretory granules in live chromaffin cells by evanescent-field fluorescence microscopy. Biophys J 76:2262–2271
12. Taraska JW, Perrais D, Ohara-Imaizumi M, Nagamatsu S, Almers W (2003) Secretory granules are recaptured largely intact after stimulated exocytosis in cultured endocrine cells. Proc Natl Acad Sci USA 100:2070–2075
13. Tsuboi T, Zhao C, Terakawa S, Rutter GA (2000) Simultaneous evanescent wave imaging of insulin vesicle membrane and cargo during a single exocytotic event. Curr Biol 10:1307–1310
14. Miyazaki J, Araki K, Yamato E et al (1990) Establishment of a pancreatic beta cell line that retains glucose-inducible insulin secretion: special reference to expression of glucose transporter isoforms. Endocrinology 127:126–132
15. Nagai T, Ibata K, Park ES, Kubota M, Mikoshiba K, Miyawaki A (2002) A variant of yellow fluorescent protein with fast and efficient maturation for cell-biological applications. Nat Biotechnol 20:87–90
16. He TC, Zhou S, da Costa LT, Yu J, Kinzler KW, Vogelstein B (1998) A simplified system for generating recombinant adenoviruses. Proc Natl Acad Sci USA 95:2509–2514
17. Diraison F, Motakis E, Parton LE, Nason GP, Leclerc I, Rutter GA (2004) Impact of adenoviral transduction with SREBP1c or AMPK on pancreatic islet gene expression profile: analysis with oligonucleotide microarrays. Diabetes 53(Suppl 3):S84–S91
18. Ainscow EK, Zhao C, Rutter GA (2000) Acute overexpression of lactate dehydrogenase-A perturbs beta-cell mitochondrial metabolism and insulin secretion. Diabetes 49:1149–1155
19. Ravier MA, Rutter GA (2010) Isolation and culture of mouse pancreatic islets for ex vivo imaging studies with trappable or recombinant fluorescent probes. Methods Mol Biol 633: 171–184
20. Haugland RP (2009) Hand book of fluorescent probes and research products. Molecular Probes, Eugene
21. Tsuboi T, Ravier MA, Parton LE, Rutter GA (2006) Sustained exposure to high glucose concentrations modifies glucose signaling and the mechanics of secretory vesicle fusion in primary rat pancreatic beta-cells. Diabetes 55: 1057–1065
22. Tsuboi T, Rutter GA (2003) Multiple forms of "kiss-and-run" exocytosis revealed by evanescent wave microscopy. Curr Biol 13:563–567
23. Parmer RJ, Mahata M, Mahata S, Sebald MT, O'Connor DT, Miles LA (1997) Tissue plasminogen activator (t-PA) is targeted to the regulated secretory pathway. Catecholamine storage vesicles as a reservoir for the rapid release of t-PA. J Biol Chem 272:1976–1982
24. Tsuboi T, McMahon HT, Rutter GA (2004) Mechanisms of dense core vesicle recapture following "kiss and run" ("cavicapture") exocytosis in insulin-secreting cells. J Biol Chem 279:47115–47124
25. Bancila V, Nikonenko I, Dunant Y, Bloc A (2004) Zinc inhibits glutamate release via activation of pre-synaptic K channels and reduces ischaemic damage in rat hippocampus. J Neurochem 90:1243–1250
26. Bloc A, Cens T, Cruz H, Dunant Y (2000) Zinc-induced changes in ionic currents of clonal rat pancreatic-cells: activation of ATP-sensitive $K^+$ channels. J Physiol 529(Pt 3):723–734
27. Nicolson TJ, Bellomo EA, Wijesekara N et al (2009) Insulin storage and glucose homeostasis in mice null for the granule zinc transporter ZnT8 and studies of the type 2 diabetes-associated variants. Diabetes 58:2070–2083
28. Collins SC, Hoppa MB, Walker JN et al (2010) Progression of diet-induced diabetes in C57BL6J mice involves functional dissociation of Ca2(+) channels from secretory vesicles. Diabetes 59:1192–1201
29. Vinkenborg JL, Nicolson TJ, Bellomo EA, Koay MS, Rutter GA, Merkx M (2009) Genetically encoded FRET sensors to monitor intracellular $Zn^{2+}$ homeostasis. Nat Methods 6:737–740
30. Miller PS, Beato M, Harvey RJ, Smart TG (2005) Molecular determinants of glycine receptor alphabeta subunit sensitivities to $Zn^{2+}$-mediated inhibition. J Physiol 566:657–670
31. Hanas JS, Larabee JL, Hocker JR (2005) Zinc finger interactions with metal and other small molecules. In: Iuchi S, Kuldell N (eds) Zinc Finger proteins: from atomic contact to cellular function. Landes Bioscience, Austin
32. Ravier MA, Tsuboi T, Rutter GA (2008) Imaging a target of $Ca^{2+}$ signalling: dense core granule exocytosis viewed by total internal reflection fluorescence microscopy. Methods 46:233–238

# Chapter 3

# 4Pi Microscopy

## Roman Schmidt, Johann Engelhardt, and Marion Lang

## Abstract

Optical microscopy has become a key technology in the life sciences today. Its noninvasive nature provides access to the interior of intact and even living cells, where specific molecules can be precisely localized by fluorescent tagging. However, the attainable 3D resolution of an optical microscope has long been hampered by a comparatively poor resolution along the optic axis. By coherent focusing through two objective lenses, 4Pi microscopy improves the axial resolution by three- to fivefold. This primer is intended as a starting point for the design and operation of a 4Pi microscope of type A.

**Key words:** 4Pi microscopy, Optical imaging, Superresolution

## 1. Introduction

In a conventional optical microscope, light is focused onto the sample by a single objective lens. The size and shape of the resulting focus depend on the solid angle $\Omega$ that is covered by the incoming light cone. For a single lens, $\Omega$ is restricted to a limit of $2\pi$, which results in a focus that is elongated along the optic axis.

4Pi microscopy (1, 2) overcomes this limitation by employing two opposing objective lenses to coherently illuminate the sample from both sides. As a consequence, $\Omega$ is enlarged and approaches the limit of $4\pi$ — an illumination from all spatial directions. Interference of the counter propagating wave fronts leads to a sharpening of the focus and thus to a significantly improved axial resolution (see Fig. 1).

4Pi microscopes have already been successfully utilized for a range of studies, including the Golgi complex (3), histone chromatin structures (4), the nuclear pore complex (5), and promyelocytic leukemia nuclear bodies (6). Since the 4Pi concept only relies on extended focusing, it synergistically augments superresolution

Alioscka A. Sousa and Michael J. Kruhlak (eds.), *Nanoimaging: Methods and Protocols*, Methods in Molecular Biology, vol. 950, DOI 10.1007/978-1-62703-137-0_3, © Springer Science+Business Media, LLC 2013

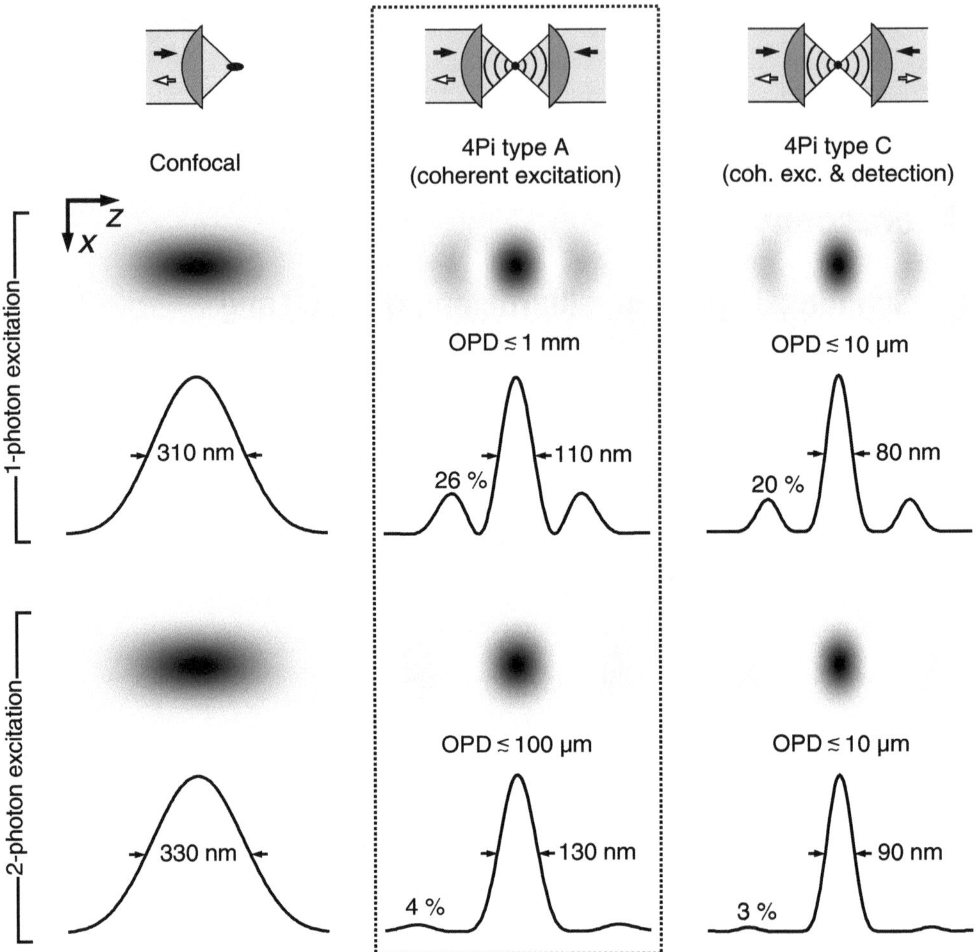


Fig. 1. Overview of the modes of operation of a 4Pi microscope. The effective point spread functions are characterized by their full width at half maximum (FWHM) and the relative side lobe amplitude of their center axial profiles. Coherent excitation by two opposing objective lenses (4Pi type A) yields a substantial resolution increase over a conventional confocal microscope (Confocal). By using linear (single photon, 1PE) excitation with continuous-wave lasers that feature a long coherence length, the limit of tolerable optical path length difference (OPD) between the 4Pi interferometer arms is governed by geometrical requirements of the 4Pi unit (e.g., due to beam scanning). A further resolution increase is obtained by additional coherent detection (4Pi type C), which lowers the OPD limit to the short coherence length of the fluorescence light and requires a dispersion correction that is not covered here. Optional two-photon excitation (2PE) suppresses the side lobes in both modes but constrains the allowed OPD to the spatial pulse width of the 2PE light. Calculated PSFs for excitation wavelengths $\lambda_{1PE} = 488$ nm and $\lambda_{2PE} = 780$ nm, detection centered at $\lambda_d = 525$ nm and a semi-aperture angle of $\alpha = 74°$.

techniques that operate beyond the limits of diffraction (7) to provide true 3D resolution at the nanometer scale (8–10).

The resolving power of a linear imaging system is characterized by its point-spread-function (PSF), which quantifies the blurring of a point object by the imaging process. The PSF of a 4Pi microscope features a sharp central peak that is flanked by secondary interference maxima along the optic axis. These so-called "side lobes" manifest themselves as shifted ghost images of the acquired object. As a rule-of-thumb, ghost images can subsequently be removed by deconvolution if their amplitude stays below 50% of the image that is generated by the central maximum (11). This ratio is primarily governed by the semi-aperture angle $\alpha$ of the objective lenses used and becomes sufficiently small above $\alpha \approx 60°$; full coverage of $\Omega$ which corresponds to $\alpha = 90°$ is therefore not required.

Excitation by two-photon absorption (12) offers additional side lobe suppression (13, 14) in cases where sufficiently low side lobes cannot be obtained by focusing alone due to sample conditions. The 2PE probability scales with the square of the illumination intensity which lowers the relative side lobe amplitudes accordingly. The extended stokes shift pushes the side lobes away from the confocal detection volume, which further reduces their contribution.

Figure 1 provides an overview of the imaging modalities of a 4Pi microscope. Here we discuss the implementation of the comparatively simple case of coherent excitation only, referred to as 4Pi type A (2), in combination with oil immersion lenses and TDE embedded (15) samples. This scheme (16) is compatible with commercial beam scanning 4Pi systems and becomes particularly robust by utilizing 2PE. Furthermore, imaging with single-photon excitation is generally feasible which renders sophisticated laser sources dispensable.

In the following sections, we cover the 4Pi specific basic aspects of the microscope, sample preparation, alignment procedures, and image processing, assuming that the reader is familiar with the design and operation of a conventional confocal microscope. This especially includes safety guidelines regarding the work with lasers that must be followed at all times.

## 2. Materials

The materials listed below have been successfully employed for 4Pi imaging (6, 16, 17). Commercial equivalents may work as well.

1. Leica TCS SP2 (Leica Microsystems).
2. 4Pi unit, custom built (see Fig. 2).
3. 2 paired objective lenses, HCX PL APO CS 100/1.46 Oil, $\alpha = 74°$ (Leica Microsystems).
4. Type F immersion liquid, $n_e^{23} = 1.5180$ (Leica Microsystems).
5. Sample holder, custom built (see Fig. 3).
6. Coverslips No. 1.5, manually selected $d = 170\ \mu m \pm 10\ \mu m$, small Ø (e.g., Ø = 12 mm) (Menzel Gläser).

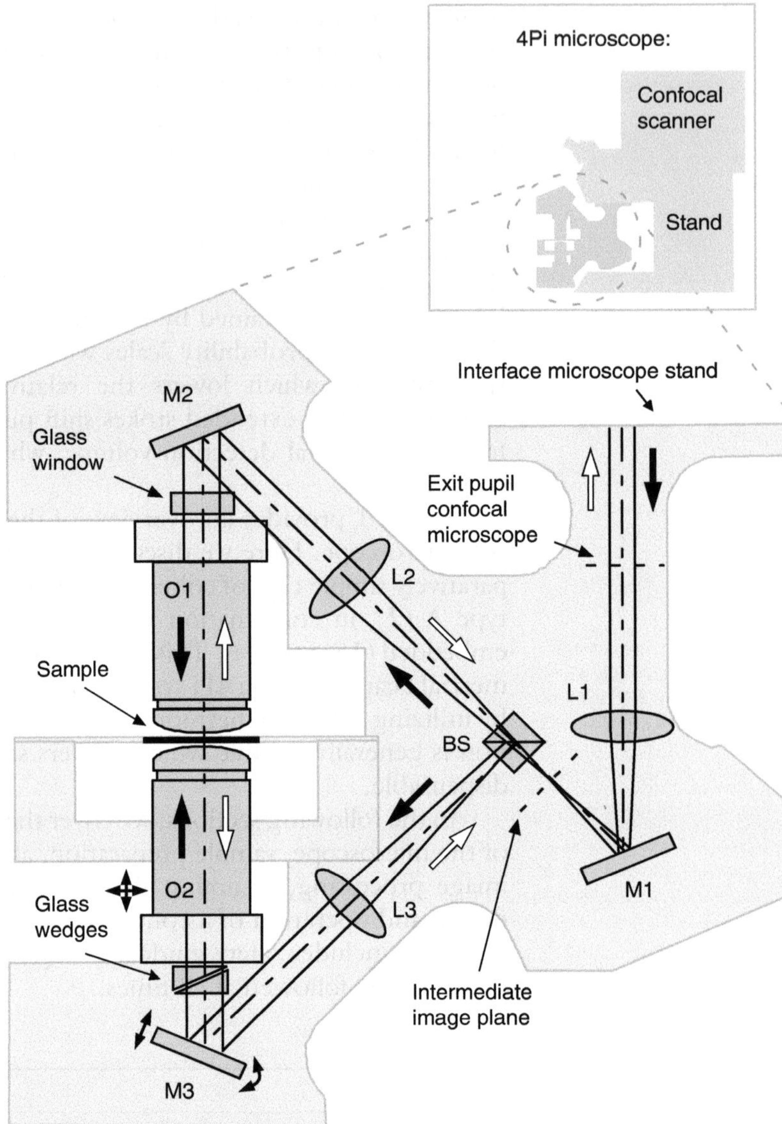


Fig. 2. Sketch of a 4Pi microscope. Excitation light (solid arrows) originating from the microscope stand is divided by the beam splitter BS and focused onto the same spot by the opposing objective lenses O1 and O2. The lenses L1, L2, and L3 and the mirrors M1, M2, and M3 form the intermediate optical system, ensuring that the two scanning pivotal points coincide with the entrance pupils of the two objective lenses. Fluorescence (hollow arrows) is collected by both lenses, recombined at BS, and directed toward the microscope stand. Scanning is performed by a standard confocal microscope unit. Movable optical wedges and a glass window may be installed in the respective interferometer arms (BS-Sample) for 4Pi modalities that use interfering fluorescence (type B and C) and thus require dispersion compensation over a large wavelength range. Adapted from ref. 18 with permission.

Fig. 3. 4Pi sample holder. In order to be accessible for both objective lenses, the sample is sealed in medium between two coverslips (a, Ø 18 mm and b, Ø 30 mm) that rest on a steel ring (c). A one-quarter mirror coating on the inward face of b facilitates aberration correction by reflecting the individual excitation beams back towards the detector.

7. Coverslips No. 1.5, manually selected $d = 170\ \mu m \pm 10\ \mu m$, large Ø (e.g., Ø = 30 mm) (Menzel Gläser), with custom Al mirror coating (see Fig. 3).
8. 2,2-Thiodiethanol (TDE), highest purity (Sigma-Aldrich).
9. Phosphate-buffered saline (PBS), pH 7.4.
10. Refractometer.
11. FluoSpheres fluorescent microspheres, Red (580/605, for 1PE) or Yellow-green (505/515, for 2PE), Ø = 100 nm (Invitrogen).
12. Absolute ethanol.
13. Nail polish.

## 3. Methods

The coherent use of opposing lenses involves a special sample preparation protocol and introduces additional parameters that have to be controlled during the imaging process. This section discusses the design of a beam scanning 4Pi microscope and describes alignment procedures in chronological order of the workflow: One-time procedures that are used to set up the microscope, and routine adjustments that typically have to be executed repeatedly between measurements. Finally, a simple deconvolution approach is introduced as a means to free the acquired images from ghosting.

### 3.1. The 4Pi Microscope Set-Up

As an interferometric device, a 4Pi microscope is susceptible to temperature changes. Ensure that the installation site of the microscope is stabilized at the specified operation temperature of the objective lenses ($T=23°C$) to about ±1°C by proper air-conditioning. Due to the increased axial resolution, vibration isolation by an optical table with pneumatic damping is highly recommended.

To enable 4Pi imaging on a conventional (beam scanning) confocal microscope (18), the objective turret and the sample stage are replaced by a module that accommodates the objective lens pair, a suitable sample stage and the necessary optics. The standard layout of this "4Pi unit" is outlined in Fig. 2. It exhibits six additional degrees of freedom that require active fine control, typically in form of piezo actuators.

Translation of one objective lens (O2) in all three dimensions allows keeping both lenses pointing at the same spot in order to create a diffraction-limited common focus.

Variation of the optical path length difference (OPD) between the two interferometer arms by a moveable mirror (M3) changes the phase ($\varphi_{4Pi}$) between the two incident wave fronts at the focus and is used to set up constructive interference at the focal center ($\phi_{4Pi}=0$).

Finally, two more degrees of freedom are introduced by beam scanning, which addresses lateral image coordinates by the scan angle of the beam at the entrance pupil of the objective lens. The mirror symmetry of the 4Pi unit ensures that during image acquisition, the same scan angle is applied to both objective lenses and thereby both foci remain coincident and interfere. However, if the pivot points of the beams do not rest at corresponding positions within each pupil, the planes of constant phase (PCP) between the incident wave fronts are no longer parallel to the focal plane, and $\varphi_{4Pi}$ thereby becomes a function of the scan angle. Rotational freedom of M3 allows aligning both pivot points, which eliminates this undesired dependency.

#### 3.1.1. Objective Lenses and Sample Optics

The crucial optical elements of a 4Pi microscope are the sample itself and the pair of objective lenses. The removal of ghost images requires sufficiently low amplitude of the side maxima of the PSF, which is governed primarily by the semi-aperture angle $\alpha$ of the objective lenses used. Large $\alpha$ that enable 4Pi type A imaging with single-photon excitation are provided by oil immersion lenses with a high numerical aperture (NA) that have become commercially available for the use in total internal reflection microscopy.

Since aberrations can quickly spoil the benefits of these lenses (see Note 1), care has to be taken to employ them under their design conditions which cover temperature ($T$), slide thickness, ($d$) and the refractive index (RI) of the immersion oil ($n$), which is temperature dependent.

The RI of the immersion oil has to be matched by the RI of the mounting medium, since due to the 4Pi geometry at least one lens will be focusing deep into the sample at any time. Mounting in TDE (15) was originally developed for this reason and is described in Subheading 3.2.1.

Spherical aberrations that are already introduced by small deviations from these specifications can be assessed by evaluating the reflection from a mirror that is scanned through the focus, and compensated by the correction collars of the objective lenses, if present (see Subheading 3.4.1).

For the use as a 4Pi pair, two individual lenses have to be selected or factory-adjusted such that their axial chromatic shifts $\Delta_{1,2}$ between the excitation line $\lambda_e$ and the mean of the detection band $\lambda_d$ are small (~<100 nm) to maintain confocality for both wavelengths (see Note 2).

Furthermore, a residual magnification mismatch of the lens pair will limit the field of view (FOV) that can be attained by beam scanning without additional corrections. The tolerable lateral separation between both foci amounts to about half of their lateral full width at half maximum (FWHM, ~100 nm). A magnification mismatch of 5 ‰ would introduce this separation over a FOV of 10 μm diameter.

### 3.2. Sample Preparation with TDE Mounting Medium

TDE (2,2-thiodiethanol) is an inexpensive, nontoxic glycol derivative that is used as a mounting medium (15, 19) for technical and biological samples (see Note 3). Pure TDE (see Note 4) exhibits a refractive index of $n_D^{23} = 1.521$ (at 589 nm and 23°C), and can be precisely tuned to the RI of immersion oil ($n = 1.518$) by the addition of water ($n = 1.333$) (see Note 5) or buffer to maintain the optimal pH for the dye in use.

#### 3.2.1. Preparing TDE Mounting Medium

To prepare TDE embedding medium that matches the refractive index of immersion oil:

1. Mix 97 ml 2,2-thiodiethanol (TDE) with 3 ml PBS (see Note 6).
2. Stir for at least 15 min.
3. Adjust pH to 7.5 with HCl/NaOH (see Note 7).
4. Check refractive index with refractometer and adjust refractive index to 1.518 by addition of 100% TDE or PBS, if necessary (see Note 5).

Direct application of TDE medium to an aqueous cell sample involves the risk of distorting the cellular structure by osmotic shock. Therefore, a dilution series is required to ensure a slow exchange:

#### 3.2.2. Embedding the Sample in TDE Medium

1. Prepare TDE dilution series with PBS: 10% (v/v) TDE, 25% (v/v) TDE, 50% (v/v) TDE, and 97% (v/v) TDE.
2. Incubate labeled sample in 10% (v/v) TDE for 10 min.

3. Incubate labeled sample in 25% (v/v) and 50% (v/v) TDE for 5 min each.
4. Incubate labeled sample in TDE medium twice for 5 min each.

*3.2.3. Mounting the Sample*

In order to be accessible for both objective lenses, the sample is mounted in a sandwich fashion, between two coverslips of different sizes. A smaller coverslip (see Note 8) carries the prepared sample (e.g., the cells), whereas a larger coverslip attaches to the sample holder and provides clearance for the lateral movement of the sample stage (see Fig. 3). Furthermore, the larger coverslip is covered with a sparse layer of fluorescent beads and a mirror coating to facilitate the alignment of the microscope (see Note 9).

1. Prepare sample on small coverslip.
2. Embed sample in TDE medium (see Subheading 3.2.2).
3. Place mirrored coverslip on flat and stable ground, with mirror facing upwards.
4. Clean mirrored coverslip with ethanol (see Note 10).
5. Dilute beads in ethanol by 1:1,000.
6. Apply 10 μl of diluted bead suspension to mirrored coverslip. Allow to dry on air.
7. Mount sample on small coverslip with TDE medium upside down on large coverslip.
8. Apply gentle pressure, remove excess medium.
9. Seal with nail polish (see Note 11).

## 3.3. One-Time Alignment Procedures

*3.3.1 Basic Confocal Alignment*

The flatness and apochromatic corrections that are necessary for a large field of view over a broad wavelength spectrum require an elaborate lens design. In case of the Leica objective lenses used, the resulting complexity is reduced by shifting a small amount of color-dependent magnification corrections to the tube lens of the microscope. A tilt between these lens groups thus introduces a lateral chromatic shift, which renders precise beam alignment mandatory (see Note 12).

1. Apply standard alignment procedures for a confocal microscope to each interferometer arm of the 4Pi unit individually. Block the arm that is not used.
2. Check the beam collimation at both back focal planes with a shear plate.

*3.3.2. Orientation of the Focal Planes*

1. Start a live xz reflection scan of the mirror and adjust the distance of both objective lenses until two instances of the mirror appear in the image, one for each lens.
2. Switch to a live xy scan of the mirror using the fixed lens.

3. Parallelize its focal plane and the mirror by adjusting the orientation of the sample stage such that linear intensity gradients in the image are minimized. A potentially remaining radial gradient expresses the flatness error of the field, which limits the usable FOV to the uniform area at its center.
4. Repeat this procedure for the adjustable lens, but tilt the lens instead of the sample stage.

#### 3.3.3. Superposition of the Fields of View

Henceforth, the FOV of both objective lenses have to be matched in order to scan the sample with a common focus:

1. Superimpose their focal planes by axial fine adjustment of the adjustable lens O2 (Fig. 2) during a live xz scan of the mirror.
2. Move the sample to remove the mirror from the FOV, stop scanning and lock the beam to a low scan angle.
3. Move the adjustable lens laterally until the outgoing beam that is transmitted through the sample becomes parallel to the incoming beam. The FOV of both lenses should now overlap and generate a double image of the same sample area, shifted by a few tens of micrometers.
4. Identify corresponding imperfections of the mirror (like scratches, holes, etc.) and fine-tune the position of the adjustable lens in all dimensions until their images coincide.

#### 3.3.4. Optical Path Difference

In order to provide coherent illumination of the sample, the optical path difference (OPD) between the two interferometer arms may not exceed the coherence length of the excitation laser or, for pulsed systems (as used for 2PE), its spatial pulse width.

1. Translate M3 (Fig. 2) while keeping both FOV superimposed until interference fringes appear in the reflection image (see Note 13), and maximize modulation depth. OPDs up to a few millimeters can be compensated this way.

#### 3.3.5. Planes of Constant Phase (Coarse)

Image post-processing is facilitated by use of a constant PSF. This implies a constant $\phi_{4\mathrm{Pi}}$ over the imaged region of interest that can be set up by orienting the PCP parallel to the focal plane:

1. A dependency of $\phi_{4\mathrm{Pi}}$ on the scan angle manifests as a fringe pattern on an xy image of the sample mirror. Identify this pattern and minimize the number of fringes by tilting BS (Fig. 2) until a centered circular pattern is observed.
2. Check for a lateral chromatic shift that has potentially been introduced and counteract by tilting M3 (Fig. 2).
3. Repeat this procedure until further adjustment of BS does not yield any improvement.

### 3.4. Routine Adjustments

#### 3.4.1. Spherical Aberrations

Use the correction collars to compensate sample induced spherical aberrations by optimizing the axial profile of a reflection image of the mirror:

1. Start a live xz 4Pi scan of the mirror, close the confocal pinhole below 0.5 airy units and increase the laser power as necessary to obtain a good signal-to-noise ratio.
2. Raise the adjustable lens until the reflexes seen by each lens are separated by a few microns to avoid interference artifacts.
3. Adjust each correction collar to maximize the amplitude (and minimize the FWHM) of the main peak of the corresponding profile (see Note 14). The relative amplitude of side maxima should not exceed about 15%.

#### 3.4.2. Planes of Constant Phase (Fine)

1. Visualize the orientation of PCP as described in Subheading 3.3.5.
2. Center the ring pattern by fine-tuning the tilt of M3 (Fig. 2).

#### 3.4.3. Point Spread Function

Visualize the PSF to fine adjust the alignment of the objective lenses and set up constructive interference ($\phi_{4\text{Pi}} = 0$):

1. Switch to fluorescence detection and move the sample region of interest into view.
2. Locate a nearby reference such as a bead or a point like structure of the sample, and superimpose the FOV of both objective lenses (see Subheading 3.3.3.) while running live xz and xy scans.
3. Watch for an increase in local image brightness as both foci approach each other and maximize brightness.
4. Verify alignment by alternately blocking each interferometer arm and comparison of the individual images.
5. The xz image of the reference structure in focus should now feature a deep modulation (see Notes 15 and 16) and resembles the PSF of the microscope, however, at a random $\varphi_{4\text{Pi}}$. Set up a symmetrical PSF with a single central maximum ($\phi_{4\text{Pi}} = 0$, see Fig. 4, inset) by translation of M3. The alignment of the microscope is now complete.

It is good practice to evaluate the PSF of the microscope before actual measurements are begun, in order to verify that resolution and side lobe requirements are met. Side lobe height typically exceeds the theoretical predictions by about 10% points (see Fig. 4) due to residual aberrations. In this regard, it is important to note that the averaged axial profile (z-response) of the 4Pi PSF differs from its center profile along the optic axis which results in stronger ghost images for laterally extended objects (see Fig. 4).

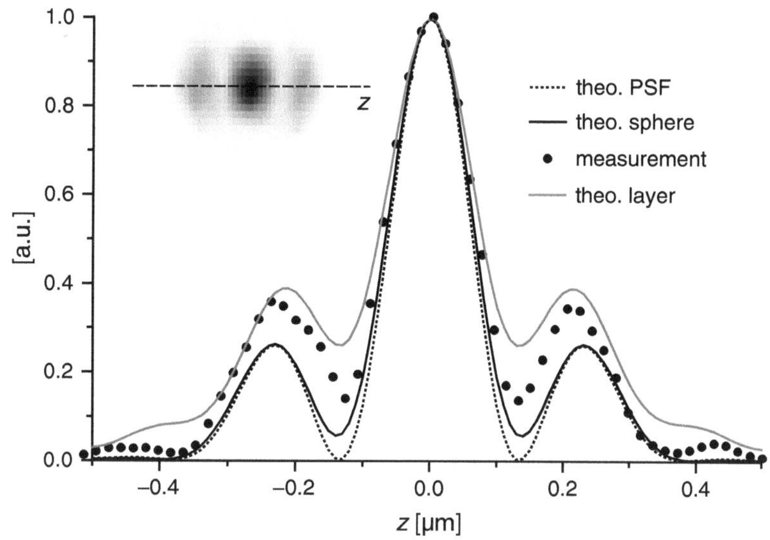


Fig. 4. PSF assessment by means of a fluorescent bead. Theoretical predictions of the center axial profiles for the PSF (*dotted line*), the image of a Ø = 100 nm sphere (*solid line*) and a 100 nm thick lateral layer (*gray line*). The image of the sphere closely resembles the PSF, while the image of the layer exhibits comparatively stronger ghosting and reduced modulation depth due to lateral averaging over the PSF. The measured profile of a 100 nm fluorescent bead (data points and inset) displays a FWHM of 140 nm and side lobes of 35%, in good agreement to theory (*solid line*). $\lambda_e = 568$ nm, $\lambda_d = 605$ nm, $\alpha = 74°$.

### 3.5. Image Post-Processing

The imaging process can be described as the convolution of the object with the PSF of the microscope. The side lobes of the 4Pi PSF thereby give rise to ghost images of the object that can be removed by image deconvolution to aid in the interpretation of the acquired data.

Image deconvolution is an inverse problem that is ill-posed due to the limited passband of the PSF and the presence of noise (see Note 17); typically multiple solutions exist that are consistent with the raw data. While regularized, nonlinear, iterative methods like the classical Richardson-Lucy (20, 21) algorithm can be quite powerful, they are also computationally expensive, require a-priori assumptions about the object for proper regularization and may lead to artifacts, for example, due to the amplification of noise.

Fortunately, the removal of the ghost images in 4Pi data does not require such a comprehensive approach but can be accomplished by the much simpler process of what is commonly called three-point deconvolution. It takes advantage of the fact that the 4Pi PSF can be approximated by a convolution of a peak function and a one-dimensional comb function along the optic axis that defines the amplitudes and relative positions of the main maximum and side lobes (see Fig. 5a, b). Ghost images can thus be removed by simply convolving the image with the inverse of the comb function (11) (see Fig. 5c, d). However, if the relative side lobe

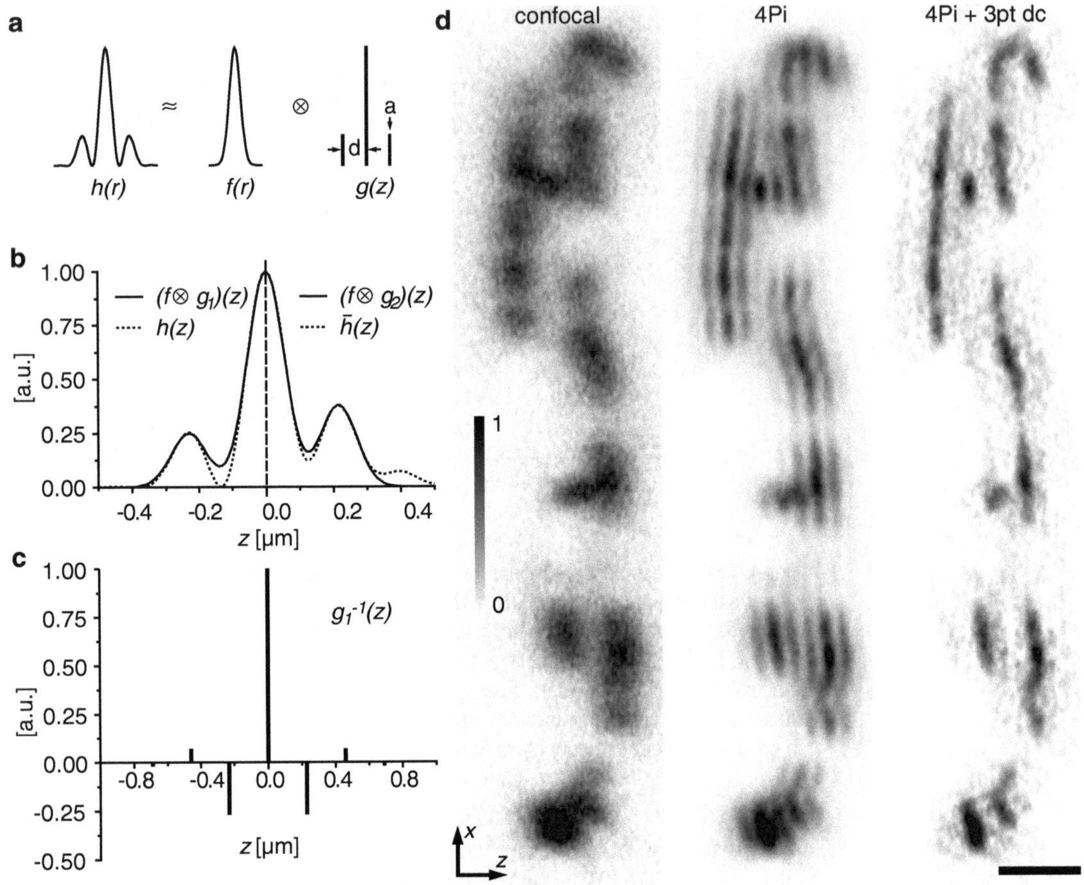


Fig. 5. Three-point deconvolution of a 4Pi image of Golgi stacks. (**a**) Approximation of the PSF h by the convolution of a shape function f with a comb function g. (**b**) Fit of the central axial profile $h(z)$ and the $z$-response $\bar{h}(z)$ of the 4Pi PSF with parameter sets $g_1$ ($a=25\%$, $d=230$ nm) and $g_2$ ($a=38\%$, $d=215$ nm). In both cases, the shape function f is chosen as a Gaussian that corresponds to the resolution of the microscope (127 nm FWHM). (**c**) Inverse filter as obtained by the inversion of $g_1$. (**d**) Three-point deconvolution applied to 4Pi raw data of the distribution of Cy3-immunolabelled cis-Golgi protein GM 130 in a Vero cell renders a virtually ghosting-free image. $\lambda_e = 568$ nm, $\lambda_d = 605$ nm, $\alpha = 74°$, scale bar 1 µm.

amplitude approaches 50%, information content that is crucial for this procedure is lost from the image, reflected by an inverse filter that eventually extends over the whole image and residual ghost images in the deconvolved data.

For best results, ensure that neither saturation nor thresholding occurs during data acquisition. Since ghost image amplitude and position in practice depend on the (local) extensions of the object (see Fig. 4), tailor the comb parameters to a best fit of the actual raw data. Consider 2PE (see Fig. 6) if experimental constraints render 1PE imaging with sufficiently low side lobes not feasible. Three-point-deconvolution increases the noise level of the image. This may lead to negative pixel values (see Note 18) that may be neglected (set to zero) if the absolute image brightness is of little concern.

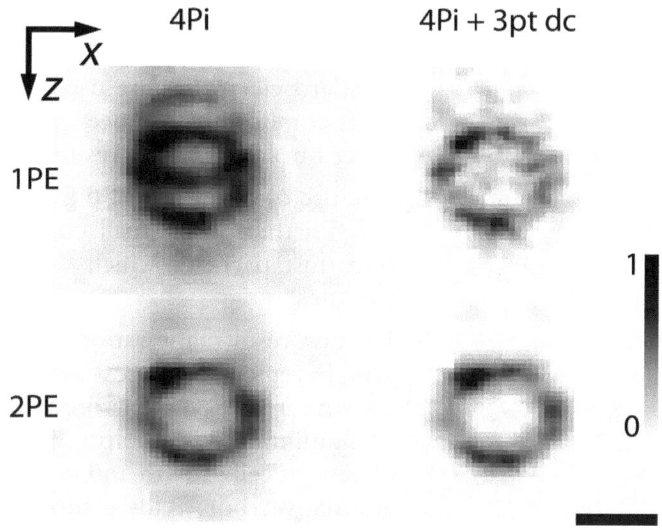


Fig. 6. Promyelocytic leukemia (PML) nuclear bodies imaged by 4Pi microscopy using linear (1PE) and two-photon (2PE) excitation. The globular structure of PML protein in raw (4PI) and three-point-deconvolved data (4Pi+3 pt dc) appears cleaner in the 2PE case since ghost images are suppressed more efficiently by the confocal pinhole. Scale bar 500 nm.

## 4. Notes

1. Spherical aberration scales with the fourth power of $\alpha$ (22).
2. Prefocusing of the infra red excitation beam is often used to reduce the axial chromatic shifts $\Delta_{1,2}$ for 2PE. However, this affects both lenses equally and will not compensate a potential difference that introduces an offset of $|\Delta_1 - \Delta_2|/2$ between the focal planes for $\lambda_e$ and $\lambda_d$.
3. The properties of some fluorophores are changed in TDE: Especially the absorption and emission spectra, but also the quantum efficiency and bleaching properties may be altered when the samples are mounted in TDE.
4. Commercially available TDE may contain impurities. It can be further purified by distillation (boiling temperature 164–166°C at 27 hPa).
5. TDE is hygroscopic. The refractive index of undiluted TDE ($n = 1.522$) may vary due to its water content. Since water and TDE are fully miscible, the attainable RIs of a solution cover the whole range from $n = 1.33$ to $n = 1.522$.
6. Typically TDE is used in combination with a phosphate buffer, but other aqueous buffers also work. Buffer concentrations too high may cause crystallization

7. Due to the viscosity of TDE, the measurement of its pH is difficult. Either a special pH electrode has to be used or, if standard electrodes are used, one has to wait for a long time until equilibrium has been reached. Indicator paper provides faster but less accurate results.
8. The use of coverslips of different sizes facilitates the final sealing. Furthermore, small coverslips are less likely to bend or trap dust particles which would result in an uneven sample thickness.
9. For best results, aberrations should be measured as close to the sample region of interest as possible. To this end, cells can be grown directly on a mirrored coverslip. However, the presence of aluminum in cell culture has been reported to have adverse effects on cell growth and proliferation. Gold can be used as an alternative, but yields a more delicate coating that tends to flake off.
10. While care should be taken not to destroy the mirror coating, a few deliberate scratches can be helpful for the alignment of the microscope.
11. Nail polish is often used to seal microscopy samples. Colored or glittering nail polish should be avoided. Certain solvents in the nail polish may quench the fluorescence.
12. If a lateral chromatic shift is neglected, only the fluorescence collected by a single objective will pass the confocal pinhole which reduces the available signal-to-noise ratio by $1/\sqrt{2}$.
13. The effective OPD is doubled for reflection imaging.
14. If the axial profile stays blurred, verify that the reflected light hits the confocal pinhole.
15. The achievable modulation depth also depends on the axial extension of the imaged structure. While this property can be exploited as a sensitive ruler (3), PSF measurements require structures that are smaller than the expected (axial) resolution.
16. Air bubbles in the beam path, either within the immersion medium or the sample itself, are a common source of error. They can easily be identified by looking at the light that is transmitted through either objective lens.
17. While readout noise can be reduced by means of better detectors, this does not apply to the shot noise which is due to the fact that only a limited number of photons is available for detection.
18. Negative values beyond what could be noise hint to the PSF being not well described by the chosen comb.

## References


1. Hell SW (1990) Double-scanning confocal microscope, European Patent 0491289.
2. Hell SW, Stelzer EHK (1992) Properties of a 4pi confocal fluorescence microscope. J Opt Soc Am A Opt Image Sci Vis 9:2159–2166
3. Egner A, Jakobs S, Hell SW (2002) Fast 100-nm resolution three-dimensional microscope reveals structural plasticity of mitochondria in live yeast. Proc Natl Acad Sci 99: 3370–3375
4. Bewersdorf J, Bennett BT, Knight KL (2006) H2AX chromatin structures and their response to DNA damage revealed by 4Pi microscopy. Proc Natl Acad Sci 103:18137–18142
5. Hüve J, Wesselmann R, Kahms M, Peters R (2008) 4Pi microscopy of the nuclear pore complex. Biophys J 95:877–885
6. Lang M, Jegou T, Chung I et al (2010) Three-dimensional organization of promyelocytic leukemia nuclear bodies. J Cell Sci 123:392–400
7. Hell SW (2007) Far-field optical nanoscopy. Science 316:1153–1158
8. Schmidt R, Wurm CA, Jakobs S, Engelhardt J, Egner A, Hell SW (2008) Spherical nanosized focal spot unravels the interior of cells. Nat Meth 5:539–544
9. Shtengel G, Galbraith JA, Galbraith CG et al (2009) Interferometric fluorescent super-resolution microscopy resolves 3D cellular ultrastructure. Proc Natl Acad Sci 106: 3125–3130
10. Hell SW, Schmidt R, Egner A (2009) Diffraction-unlimited three-dimensional optical nanoscopy with opposing lenses. Nat Phot 3:381–387
11. Nagorni M, Hell SW (2001) Coherent use of opposing lenses for axial resolution increase in fluorescence microscopy. i. comparative study of concepts. J Opt Soc Am A Opt Image Sci Vis 18:36–48
12. Denk W, Strickler JH, Webb WW (1990) 2-photon laser scanning fluorescence microscopy. Science 248:73–76
13. Hell SW, Stelzer EHK (1992) Fundamental improvement of resolution with a 4Pi-confocal fluorescence microscope using two-photon excitation. Opt Commun 93:277–282
14. Lang MC, Staudt T, Engelhardt J, Hell SW (2008) 4Pi microscopy with negligible sidelobes. New J Phys 10:1–13
15. Staudt T, Lang MC, Medda R, Engelhardt J, Hell SW (2007) 2,2′-thiodiethanol: a new water soluble mounting medium for high resolution optical microscopy. Microsc Res Tech 70:1–9
16. Lang M, Muller T, Engelhardt J, Hell SW (2007) 4Pi microscopy of type A with 1-photon excitation in biological fluorescence imaging. Opt Express 15:2459–2467
17. Schmidt R, Wurm CA, Punge A, Egner A, Jakobs S, Hell SW (2009) Mitochondrial cristae revealed with focused light. Nano Lett 9:2508–2510
18. Gugel H, Bewersdorf J, Jakobs S, Engelhardt J, Storz R, Hell SW (2004) Cooperative 4pi excitation and detection yields sevenfold sharper optical sections in live-cell microscopy. Biophys J 87:4146–4152
19. Wurm CA, Neumann D, Schmidt R, Egner A, Jakobs S (2010) Sample preparation for STED microscopy. In: Papkovsky DB (ed) Methods in molecular biology. Humana Press, New York, USA, pp 185–199
20. Richardson WH (1972) Bayesian-based iterative method of image restoration. J Opt Soc Am 62:55–59
21. Lucy LB (1974) Iterative technique for rectification of observed distributions. Astron J 79:745–754
22. Born M, Wolf E (2002) Principles of optics, 7th edn. Cambridge University Press, Cambridge

# Chapter 4

# Fluorescence In Situ Hybridization Applications for Super-Resolution 3D Structured Illumination Microscopy

**Yolanda Markaki, Daniel Smeets, Marion Cremer, and Lothar Schermelleh**

## Abstract

Fluorescence in situ hybridization on three-dimensionally preserved cells (3D-FISH) is an efficient tool to analyze the subcellular localization and spatial arrangement of targeted DNA sequences and RNA transcripts at the single cell level. 3D reconstructions from serial optical sections obtained by confocal laser scanning microscopy (CLSM) have long been considered the gold standard for 3D-FISH analyses. Recent super-resolution techniques circumvent the diffraction-limit of optical resolution and have defined a new state-of-the-art in bioimaging. Three-dimensional structured illumination microscopy (3D-SIM) represents one of these technologies. Notably, 3D-SIM renders an eightfold improved volumetric resolution over conventional imaging, and allows the simultaneous visualization of differently labeled target structures. These features make this approach highly attractive for the analysis of spatial relations and substructures of nuclear targets that escape detection by conventional light microscopy. Here, we focus on the application of 3D-SIM for the visualization of subnuclear 3D-FISH preparations. In comparison with conventional fluorescence microscopy, the quality of 3D-SIM data is dependent to a much greater extent on the optimal sample preparation, labeling and acquisition conditions. We describe typical problems encountered with super-resolution imaging of in situ hybridizations in mammalian tissue culture cells and provide optimized DNA-/(RNA)-FISH protocols including combinations with immunofluorescence staining (Immuno-FISH) and DNA replication labeling using click chemistry.

**Key words:** Structured illumination microscopy, Super-resolution imaging, Fluorescence in situ hybridization, 3D-FISH, RNA-FISH, Immuno-FISH, DNA replication labeling, Click chemistry

## 1. Introduction

Fluorescence in situ hybridization on 3D preserved nuclei (3D-FISH) and consecutive imaging with state-of-the-art wide-field, or confocal laser scanning microscopy is a powerful and widely used approach to detect specific RNA transcripts, or DNA target sites, such as entire chromosome territories, subchromosomal domains, or specific gene loci within the nuclear space (1–4).

Alioscka A. Sousa and Michael J. Kruhlak (eds.), *Nanoimaging: Methods and Protocols*, Methods in Molecular Biology, vol. 950,
DOI 10.1007/978-1-62703-137-0_4, © Springer Science+Business Media, LLC 2013

3D-FISH utilizes the ability of short, labeled single-stranded DNA sequences to hybridize to complementary nucleotide target sequences and to provide information about their spatial arrangement within fixed cells. Hybridizing probes have a typical size of 30–1,000 bp and may be derived from isolated chromosomes, chromosomal fragments, bacterial artificial chromosome (BAC) libraries or represent specific PCR-amplified sequences. Labeling of larger DNA probes with modified nucleotide analogues that carry either a fluorescent dye or a haptene (e.g., Biotin, Digoxigenin) is typically performed by random primed label-PCR or nick translation, while small oligonucleotides are typically 5′ end labeled. Haptene labeling requires immunodetection with fluorescent conjugates (5–9).

On the detection side, however, the resolving power of optical microscopy, despite all technical progress, has long been restricted by physical constraints of the resolution being diffraction limited to ±250 nm in lateral (x, y) and ±600 nm in axial (z) direction (10). Hence, all objects within a specimen with a distance less than about half of the wavelength of the light employed merge to one larger object, resulting in an overall blurred image with restrained level of detail (Fig. 1, left panel).

Three-dimensional structured illumination microscopy (3D-SIM) is one of a number of recently emerging cutting-edge optical imaging techniques that circumvent the long-standing resolution barrier (11–14). In this approach, the sample is illuminated with multiple interfering light beams instead of a plane wide-field illumination. This entails the emitted light to contain higher frequency image information encoded as observable modulations of the image, in a manner similar to the formation of Moiré patterns. From a series of 15 raw images per plane recored with shifted and rotated illumination pattern, this extra information is decoded computationally to reconstruct a 3D image stack with a lateral resolution of ±120 nm and an axial resolution of ±300 nm, about equivalent to an eightfold improvement in volumetric resolution compared to conventional light microscopy (15–17). An important benefit of 3D-SIM is the ability to acquire optical serial sections of specimens up to 20 μm in depth and to simultaneously image multiple colors using a broad range of standard fluorescent dyes. To date, a growing number of publications have demonstrated the general applicability and versatility of 3D-SIM to address a wide range of cell biological questions (17–22). To this end, however, a careful readjustment of labeling schemes and protocols is required, in order to fully explore the potential of super-resolution imaging approaches (23).

The combination of FISH and 3D-SIM in three-dimensionally preserved specimen promises great potential (24) allowing the precise localization of selected RNA or DNA probes, in particular of very small sizes, in relation to structurally distinct chromatin

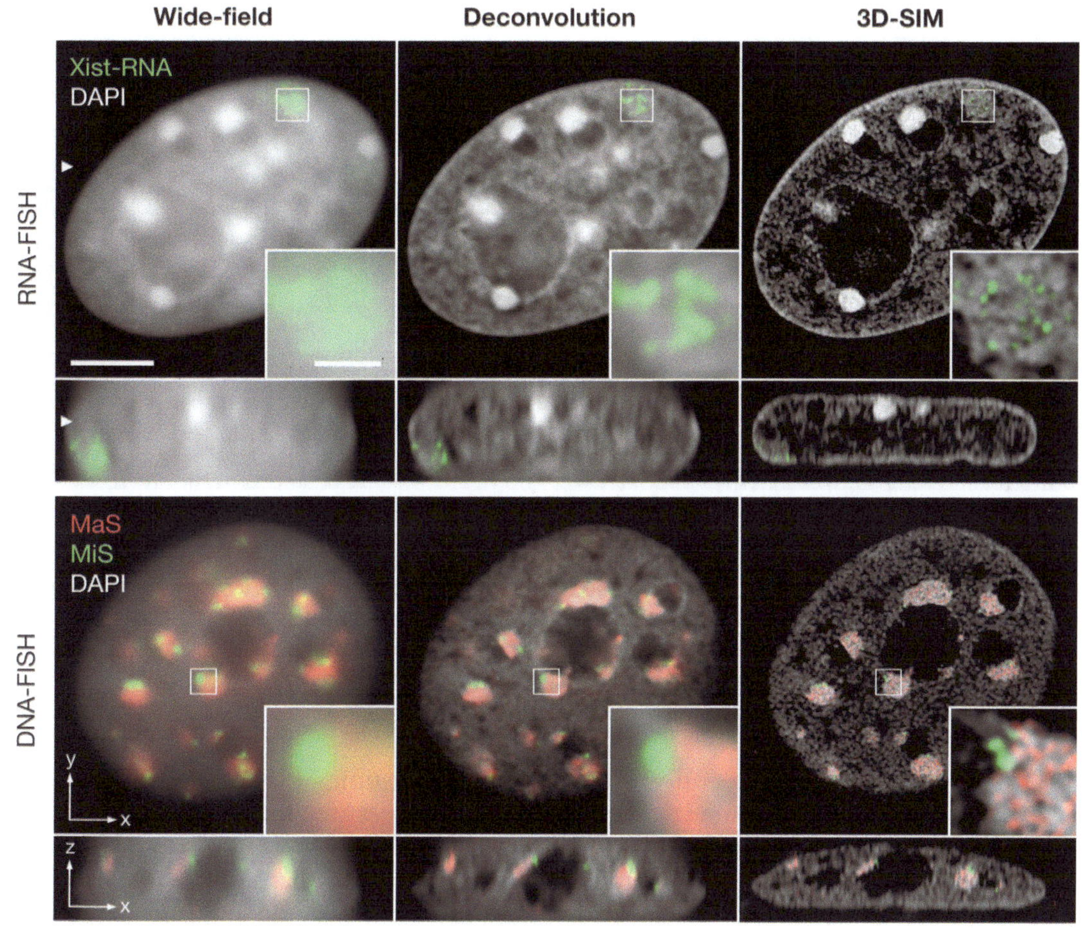


Fig. 1. Comparative analysis of RNA/DNA FISH visualized by wide-field fluorescence microscopy and 3D-SIM. *Upper panel*: RNA-FISH on a mouse C2C12 cell nucleus with biotin-labeled cDNA-probes against the noncoding Xist-RNA, detected with Alexa 488-conjugated Streptavidin (*green*). DNA has been counterstained with DAPI (*gray*). Depicted are the same optical midsection as wide-field, deconvolved wide-field, and 3D-SIM image. The selected magnified region shows one of the two inactive X-chromosomes (Xi) typically seen in these polyploid cells. Orthogonal midsections highlight the increased axial resolution. Further, 3D-SIM allows to discern the greater compaction of the Xi chromatin, as reflected by the DAPI staining and to find Xist-RNA as distinct structures preferentially located in the less intense DAPI regions of the Xi. *Lower panel*: DNA-FISH on a C127 cell nucleus hybridized with probes against the pericentromeric major satellite repeats (MaS) detected with Alexa 594 (*red*), the centromeric minor satellite repeats (MiS) detected with Alexa 488 (*green*) and counterstained with DAPI (*gray*). Note the spatial separation of MiS domains from individual sister chromatids of replicated chromosomes (inset), resolvable both in lateral and axial direction only by 3D-SIM. Scale bar represents 5 μm (inset 1 μm).

domains, while being able to discern dedicated features of nuclear morphology (Fig. 1, right panel). It is worth considering, however, that harsh denaturation of the double stranded DNA template is necessary to achieve hybridization, raising questions on the preservation of subnuclear structures on the level of resolution obtained (Fig. 2) (25).

We focus here on the preparation of high-quality samples. It is important to keep in mind that the sample itself is an integral part

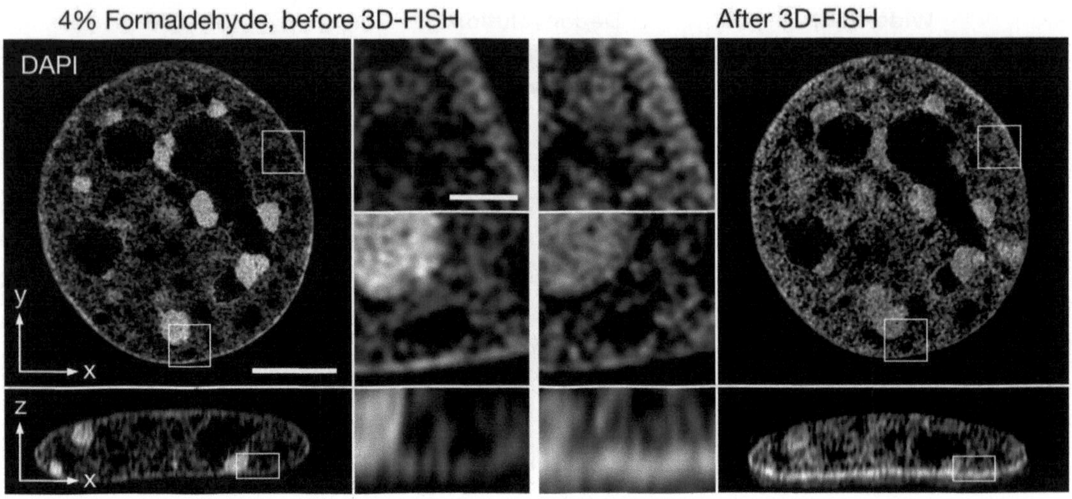


Fig. 2. Effect of FISH treatment on the nuclear morphology. *Left panel*: Mouse C127 cell nucleus fixed with 4% formaldehyde, permeabilized with 0.5% Triton/PBS and counterstained with DAPI and imaged with 3D-SIM. *Right panel*: The same cell nucleus relocated after demounting, FISH and re-stained with DAPI for comparison. The selected magnifications indicate the degree of preservation of eu- and heterochromatic chromatin ultrastructure on the level of resolution obtained with 3D-SIM. While subtle changes are noticeable (e.g., slight swelling of euchromatin, less prominently stained chromocenter) the overall morphology appears well preserved. Scale bar represents 5 μm (inset 1 μm).

of the optical system. At the same time, it is subject to the largest variation in quality, but also the most amenable to the investigator. Accordingly, the optical quality of the sample needs to be matched as well as possible to avoid any evitable aberration, e.g., by refractive index mismatch or impurities (10, 26). Further emphasis should also be put on an optimal signal-to-background level, which is pivotal for 3D-SIM as any additional sources of light scattering and/or blur reduce the modulation contrast of the structured illumination and compromise the reconstruction quality. Finally, bleaching of the sample during recording of a large number of raw images per dataset needs to be minimized to no more than ~30% to avoid reconstruction artifacts. To fulfill these criteria, an overall thorough sample preparation procedure is required, where utmost care has to be taken in all steps to preserve sample integrity and keep the preparation as clean as possible, prerequisites that are essential in order to acquire high-quality data reconstructions. While 3D-FISH protocols have been well established for conventional imaging approaches (2), we aim to provide optimized protocols that meet the special requirements for optimal super-resolution 3D-SIM imaging.

In the following we focus on a protocol for 3D-FISH of genomic DNA targets, which can also be used for the delineation of robust nuclear RNA transcripts such as Xist-RNA. For elaborated methods to detect single RNA transcripts we refer to refs. 27–30.

## 2. Materials

### 2.1. Chemicals

1. Formaldehyde, 37% in $H_2O$ stabilized with ~10% methanol or freshly depolymerized para-formaldehyde.
2. 1 M Tris–HCl pH 7.
3. 1 M Na-ascorbate.
4. 100 mM $CuSO_4$.
5. Hydrochloric acid (HCl).
6. Glycerol.
7. Dextran sulfate.
8. Tween 20.
9. Triton X-100.
10. Glycine.
11. Bovine serum albumin (BSA).
12. Formamide.
13. Fish skin gelatine, 45% (FSG).
14. Sodium azide.
15. 200 mM ribonucleoside vanadyl complexes (RVC; Sigma).

### 2.2. Cells and Cell Labeling Reagents

1. Cultured cells grown at ~60% confluency on borosilicate coverslips of appropriate size (e.g., 18 × 18 mm) and thickness of 0.170 ± 0.005 mm (#1.5) in 6-well culture plates.
2. Primary antibodies: anti-DIG (digoxigenin); anti-DNP (dinitrophenyl); anti-Digoxigenin-Alexa 488; for the detection of biotin (Strept)Avidin-Alexa 488.
3. Secondary antibodies conjugated to fluorescent dyes. We recommend CF405M (Biotium) for excitation with 405 nm and Alexa 488, Alexa 568, and Alexa 594 (Life technologies) for excitation with 488, 561, and 592 nm laser lines, respectively.
4. 5-Ethenyl-2′-deoxyuridine (EdU, Life technologies) (optional).
5. Azide dye, e.g., Alexa 488 azide, Alexa 594 azide (Life technologies) (optional).
6. 4′,6-Diamidino-2-phenylindole (DAPI).
7. Non-hardening antifade mounting medium (e.g., Vectashield, Vector Laboratories).

### 2.3. Buffers and Solutions

Use preferably $H_2O$ from a distillation or deionization unit with an electric conductance value of <18 mS for all buffers ($ddH_2O$). Additionally, filtration of solutions is recommended to remove precipitates (see Note 1). In case of RNA-FISH, add 2 mM RVC to the blocking and antibody solutions.

1. 20× phosphate-buffered saline (20× PBS): 2.74 M NaCl, 53.7 mM KCl, 130 mM $Na_2HPO_4$, 35 mM $KH_2PO_4$; for working dilutions into 1× PBS add 0.05% Na-azide.
2. Fixation solution: 2 or 4% formaldehyde in 1× PBS prepared and used freshly.
3. Permeabilization buffer: 0.5% Triton X-100 in 1× PBS.
4. PBST: 1× PBS, 0.02% Tween 20.
5. PBST blocking solution: 2% BSA, 0.5% FSG, PBST, stored at 4°C for up to 1 week.
6. 20% glycerol solution in PBS.
7. 0.1 N HCl.
8. 20× saline-sodium citrate (20× SSC) pH 7: 3 M NaCl, 0.3 M sodium citrate, pH 7.
9. 4× SSCT: 20× SSC diluted 1:5 with $ddH_2O$, and supplemented with 0.02% Tween 20 and 0.05% sodium azide.
10. 2× SSC: 20× SSC diluted 1:10 with $ddH_2O$, and supplemented with 0.05% sodium azide.
11. 0.1× SSC: 2× SSC diluted 1:200 with $ddH_2O$, and supplement with 0.05% sodium azide.
12. Hybridization buffer: 20% dextran sulfate, 4× SSC.
13. 4× SSCT blocking solution: 2% bovine serum albumin (BSA), 0.5% FSG, 4× SSCT, stored at 4°C for up to 1 week.
14. 50% formamide in 2× SSC: 50 mL 20× SSC + 250 mL formamide + 200 mL $ddH_2O$, adjust pH to 7, store at −20°C.
15. EdU stock solution: 10 mM in DMSO (stored at −20°C).
16. DAPI stock solution: 5 mg/mL in $ddH_2O$ (stored at −20°C).
17. DAPI working solution: 2.5 μg/mL in PBST, freshly prepared (protect from light).

### 2.4. General Laboratory Equipment

1. Clean glass microscope slides.
2. Fine-tip forceps.
3. Diamond cutter (optional).
4. Dewar container for liquid $N_2$ transport.
5. Heating block flat surface for slides.
6. Tube heating block or PCR machine.
7. Water baths.
8. Heated incubation chamber.
9. Humidified chamber.
10. Floating tin box.
11. Shaking platform.
12. Soft tissue paper (e.g., Kimwipes green).

13. Rubber cement (e.g., Fixogum).
14. Transparent nail polish (base coat).

#### 2.5. 3D-SI Microscopy

DeltaVision OMX V3 imaging system (Applied Precision) equipped with 405, 488, and 592 nm lasers, a UPlan APO 100× 1.40 oil objective (Olympus), back-thinned EMCCD cameras (Photometrics Cascade II:512), and computer workstation for image processing. The system optionally comes with a second wide-field microscope (Personal DV) used for pre-screening the sample by eye and storing *xy*-sample positions to be relocated on the OMX stage (for a detailed technical description see refs. 31, 32).

## 3. Methods

#### 3.1. Preparation of Coverslips

For fluorescence imaging with 3D-SIM of fixed cells, adherent tissue culture cells are typically grown on borosilicate coverslips (#1.5) suited for use with high NA oil objectives. Coverslips typically have a surface size of 18 × 18 mm, but can be cut to smaller sizes using a diamond cutter (see Note 2).

Coverslips should be thoroughly cleaned from dust by rinsing with deionized $H_2O$ and stored in pure ethanol. Do not flame or autoclave the coverslips for sterilization, as they tend to bend and deform. Preferably, take coverslips directly out of their storage container and place them into 6-well culture dishes under the laminar flow, initially vertically standing at the edge of the well until all the ethanol has evaporated. Then, coverslips can be flipped horizontally into the wells and cells seeded on top.

#### 3.2. Seeding Cells

Cells should be preferably ~60% confluent prior to fixation. Avoid working with dense cell layers. It is important to produce a single cell suspension and seed cells evenly onto coverslips, and not in patches. An evenly distributed, healthy cell population offers a great deal to the optimal visualization and minimizes nonspecific antibody binding problems caused by debris or other cellular impurities in the culture.

Non-adherent cells have also been successfully visualized when cross-linked, for instance, to poly-lysine coated cover glass ref. 2. However, cross-linkers tend to absorb antibodies or dyes nonspecifically, such as DAPI, and may therefore increase background fluorescence.

#### 3.3. Cell Fixation and Sample Preparation for Hybridization

Depending on the desired cellular structures to be labeled for imaging, different labeling strategies are needed, such as, immunofluorescence labeling of proteins before or after FISH, or detection/imaging of replication labeled DNA before FISH. All possible combinations of the protocols described below are summarized in Table 1. At this point, if an EdU pulse is to be performed,

**Table 1**
**Possible combinations of protocols**

**Protocol ⇒**

| | | | | | | | | | |
|---|---|---|---|---|---|---|---|---|---|
| 3D-FISH | 3.3 | 3.4 | 3.5 | 3.6 | | | | | |
| 3D-FISH combined with EdU pulses | EdU pulse 10–30 min | 3.3 | 3.4 | 3.5 | 3.8 | 3.6 | | | |
| 3D-FISH combined with IF | 3.3 | 3.4 | 3.5 | 3.7 | 3.6 | | | | |
| IF labeling before 3D-FISH treatment | 3.3.1–3.3.8 | 3.7 | 3.6.1–3.6.2 | 3.3.9–3.3.16 | 3.4 | 3.5 | 3.6 | | |
| Relocating a cell before and after 3D-FISH | 3.3.1–3.3.8 | 3.6.3–3.6.8, or 3.7 and 3.6 | Image | Demount | 3.3.9–3.3.16 | 3.4 | 3.5 | 3.6 | |
| Relocating an EdU-pulsed cell before and after 3D-FISH treatment | EdU pulse 10–30 min | 3.3.1–3.3.8 | 3.8 | Image | Demount | 3.3.9–3.3.16 | 3.4 | 3.5 | 3.6 |

incubate cells for the desired time period in 10 μM EdU containing growth medium (see Subheading 3.8).

1. Wash cells twice with PBS (tissue culture grade, no sodium azide added before fixation). At this point move to a fresh culture dish where PBS has been added. This is done to avoid scraping off cells attached to the culture dish, which may stick to coverslips and interfere with the cleanliness of the preparation.
2. Fix cells in 4% formaldehyde/PBS for 10 min at room temperature (see Note 3).
3. Exchange the fixative stepwise with PBST. This prevents the sample from drying, thus avoiding unwanted shrinkage or deformed cellular morphology and other artifacts. The best way to achieve the fixative removal is by adding an equal amount of washing buffer (PBST) directly to the fixative and gradually exchanging the fixative with washing buffer. This is done by removing most of the solution from the well while leaving small amounts behind, enough to keep the coverslip covered with buffer, topping up with fresh buffer and repeating this step for 2–3 times, till the fixative has been completely exchanged with washing buffer.
4. Wash with PBST (see Note 4).
5. Quench with freshly prepared 20 mM glycine dissolved in PBS for 10 min to "block" free aldehydes.
6. Wash with PBST.
7. Permeabilize with 0.5% Triton X-100/PBS for 10 min. In case of RNA-FISH, add 2mM Ribonucleoside Vanadyl Complex (RVC) (Sigma) to the permeabilization buffer and perform this step on ice to inhibit RNases and avoid RNA digestion.
8. Wash with PBST.
9. Incubate in 20% glycerol/PBS for at least 1 h at room temperature.
10. Dip the coverslip for approximately 6 s into a liquid $N_2$-filled container using fine forceps, remove, place shortly on a Kimwipe tissue and wait a few seconds to defreeze. Then, return to the well containing 20% glycerol/PBS. Repeat this freezing/thawing step twice in order to offer a better penetration and accessibility to the probe at target DNA sequences. Take care not to leave coverslips drying out while thawing.
11. Wash twice with PBS for 2 min.
12. In case of DNA-FISH incubate samples in 0.1 N HCl for 5 min. This step is to remove cytoplasmic proteins, that would otherwise contribute to nonspecific background signals and improve probe penetration. Times can be shortened or increased, but 5 min is a good compromise to maintain nuclear integrity and achieve cytoplasmic deproteination.

13. Wash twice with PBS for 2 min.
14. At this point change to SSC buffer and wash twice with 2× SSC for 2 min.
15. Incubate samples in 50% Formamide/2× SSC pH 7 at 4°C for at least 6 h, or overnight (see Note 5). Some cell types may require pepsinization before hybridization for better probe access. Pepsin mainly removes cytoplasmic proteins; however, it also promotes detachment of cells from the glass. Therefore, this process has to be monitored under the microscope and stopped before cells start to detach. A post-fixation step after pepsinization is recommended (for an exact protocol we refer to ref. 2).

### 3.4. Setup of Hybridization Solution, Denaturation of Probe and Target DNA and Hybridization

#### 3.4.1. Setup of Hybridization Solution

We recommend using 1–5 ng DNA/μL hybridization solution for repetitive probes, such as major mouse satellite DNA delineating the pericentromeric regions of mouse chromosomes, and 20–40 ng DNA/μL for single or low copy probes, such as BAC DNA for the delineation of a specific gene locus (see Note 6). The latter probes require unlabeled competitor DNA (e.g., Cot-1 DNA) added to the probe for suppression of nonspecific hybridization. The amount of the competitor DNA should be around 10- up to 50-fold of the concentration of the probe DNA. As a rule of thumb, a hybridization area covered by 18×18 mm coverslip requires 5–8 μL of hybridization mixture. In case of smaller or larger hybridization areas the required amount is adjusted accordingly.
Mix in 1.5 ml reaction tube:

1. The labeled probe: ~30 ng/μL hybridization solution of BAC DNA, or ~5 ng/μL hybridization solution of a major mouse satellite. In case of single/low copy probes (e.g., BAC probes) add unlabeled competitor DNA, such as Cot-1 DNA in a 10 to 50-fold excess compared to probe DNA.
2. 2 μg unlabeled salmon sperm DNA for efficient precipitation (especially important for precipitation of small amounts of DNA).
3. Mix probe DNA with ice-cold 100% EtOH (3× volume) for at least 30 min, but preferably overnight at −20 or −80°C.
4. Spin down at 15,000×*g* for 20 min.
5. Discard supernatant and dry the pellet (preferably using vacuum centrifuge).
6. Resuspend the pellet in 50% formamide/2× SSC/10% dextran sulfate as follows: resolve the pellet in the appropriate amount (one half of the final volume) of 100% formamide (shake at 37°C, can take up to a few hours) and then add an equal volume of 4× SSC/20% dextran sulfate (=hybridization solution). Briefly mix and incubate at 37°C for 10 min.
7. Hybridization probes can be stored at −20°C for up to several years or immediately used for hybridization set-up.

*3.4.2. Simultaneous Denaturation of Probe and Target DNA and the Hybridization Setup*

We recommend simultaneous denaturation of nuclear DNA and probe DNA as it is quick, simple, and optimal for retention of 3D morphology. Specific hybridization signals are assured by different hybridization kinetics of single (low) copy DNA and repetitive DNA sequences: repetitive sequences are hybridized with unlabeled Cot1-DNA prior to (locus specific) hybridization of labeled single (low) copy sequences contained in the respective probe. However, as a precaution to ensure complete denaturation of probe DNA and at the same time keep the time for denaturation of target DNA at a minimum, we suggest to denature the probe solution separately before adding it to the cells. Thus, steps 1–3 are optional.

1. Denature probe for 2 min at 86°C in a water bath or heat block (see Note 7).
2. Immediately immerse tubes on ice for 2 min.
3. Spin down briefly in order to prevent loss of evaporated probe on the lid.
4. Add denatured probe solution to a glass slide (e.g., 3 μL of probe/12 × 12 mm coverslip surface or 6 μL probe/18 × 18 mm coverslip surface).
5. Take the coverslip out of the formamid solution, remove excess liquid by touching the edge onto a Kimwipe tissue and mount onto a drop containing the probe. Try to remove all excess solution, since even 1 μL of leftover solution is effectively a 1:3 dilution of your probe. Work quickly to avoid drying of cells.
6. Seal with rubber cement.
7. Air-dry specimen until rubber cement has completely set.
8. Denature specimens on a heat block for 2 min at 76°C (This step is omitted for RNA-FISH detection).
9. Incubate in a floating tin box in a 37°C water bath for at least 15 h for repetitive sequences and RNA-FISH, and at least 2–3 days for single/low copy probes (see Note 8).

***3.5. Detection Procedure***

1. Remove rubber cement carefully with the help of a drop of 2× SSC buffer. Make sure that you do not smudge your specimen, but rather softly drag the coverslip onto the drop of buffer over the edge of the slide. Then lift it up and transfer it to a new culture dish.
2. Wash 3× for 5 min with 2× SSC in a heated chamber at 37°C on a shaking platform. Use mild shaking at a speed that the buffer swirling is barely seen and without the coverslips moving in the wells. Shaking is optional and should be done at this very low frequency in order to preserve sample integrity.

3. Wash 3× for 5 min with 0.1× SSC preheated to 60°C and incubate in a 37°C heated chamber under mild shaking. This stringent washing step is omitted in the case of RNA detection (see Note 9).
4. Wash twice with 4× SSCT at 37°C under mild shaking.
5. Block with 2% BSA/0.5% FSG in 4× SSCT for 30–60 min.
6. Dilute primary antibodies to an appropriate concentration in blocking solution containing 2% BSA/0.5% FSG in 4× SSCT.
7. Incubate cells in a dark humidified chamber for 1 h on a stripe of Parafilm well attached onto a smooth surface (i.e., plastic or glass), placed within the chamber. To incubate the specimens, place a drop of antibody solution (~50–100 μL) onto the Parafilm surface, blot-dry the coverslip thoroughly and place it cell-side down onto the solution drop. Incubations are usually carried out at room temperature. Overnight incubation at 4°C should be avoided as this tends to increase nonspecific binding of many antibodies. After antibody incubation, remove excess antibody solution by blot-drying coverslips prior to washing.
8. Wash 4 times with 4× SSCT. Washes should be carried out as in step 2.
9. Incubate with diluted secondary antibody solution accordingly for 1 h, as described above.
10. Wash 4 times with 4× SSCT at 37°C as in step 7.
11. Move to PBS buffer for any further step and wash 2× with PBST.

    For combined immunofluorescence or EdU detection move to Subheading 3.7 or 3.8, respectively. For troubleshooting of a poor outcome of hybridization we refer to ref. 2.

### *3.6. Post-fixation, DAPI Counterstaining and Mounting*

1. Postfix in 4% formaldehyde/PBS for 10 min at room temperature.
2. Exchange the fixative stepwise with PBST.
3. Counterstain DNA with DAPI (2 μg/mL in PBST) for 10 min at room temperature (vortex well, keep in dark).
4. Wash 4 times with PBST.
5. Rinse briefly, and only immediately prior to mounting, with $ddH_2O$. This step ensures the removal of salt precipitates left over from PBS.
6. Remove excess liquid by touching the coverslip edge onto a Kimwipe tissue.
7. Mount the coverslip on a drop of Vectashield placed in the center of a cleaned 26×76 mm microscope glass slide. Remove

excess Vectashield by softly pressing against the coverslip with a Kimwipe tissue and blot-dry across the edges. Optionally coverslips can be preincubated for 1 min on a drop (~10–20 μL) of Vectashield applied on Parafilm, then lifted and mounted as described above. This procedure ensures an embedding in undiluted medium with homogenously high refractive index (see Note 10).

8. Seal with nail polish and store at 4°C.

### 3.7. Immunofluorescence Procedure after FISH Hybridization

An optimal signal-to-background ratio (rather than utter brightness) and an absence of any floating fluorescent particles (e.g., by unbound antibody-complexes) are crucial factors for successful imaging with 3D-SIM, again stressing the importance of substantially greater care needed during preparation of the specimens.

1. Block nonspecific binding sites by incubating specimens for at least 1 h in 2% BSA/0.5% FSG/PBST (PBST-based blocking buffer) at room temperature (see Note 11).
2. Incubate primary antibodies in a dark humidified chamber as described in Subheading 3.5, step 6 (see Note 12).
3. Wash four times with PBST (see Note 4).
4. Secondary antibody incubation is carried out as in Subheading 3.5, step 6, but for 30–60 min.
5. Wash at least four times with PBST.
6. Proceed to 3.5 for post-fixation and DAPI staining.

Immunofluorescence for unstable epitopes can be performed prior to 3D-FISH, combined with post-fixation, which is also the case for relocation (see Table 1). However, we recommend the use of this protocol and trials on other combinations in case of failure to detect the immunolabeled proteins.

### 3.8. Detection of EdU Pulse Labeling

Pulse labeling with 5-ethenyl-2′-deoxyuridine (EdU) has been successfully applied to directly label replicating DNA in combination with FISH and immunofluorescence detection for 3D-SIM imaging, e.g., to determine cell cycle stage or to analyze the ultrastructure of replication foci (24). Unlike the classic labeling with 5-bromo-2′-deoxyuridin (BrdU), this approach does not rely on antibody detection upon denaturation of the DNA, but utilizes a Cu(I) catalyzed cycloaddition reaction ("click chemistry") to covalently attach a fluorescent dye containing a reactive azide group to the ethenyl-group of the nucleotide (33). Notably, the background is significantly reduced as no antibody reaction is involved. Moreover, the incorporated EdU as well as the azide-dye remain stable for subsequent detection after FISH procedure (see Note 13). Cells have been incubated with 10 μM EdU (added to the culture medium) for a pulse length of typically 10–30 min prior to fixation.

1. Blocking: as in Subheading 2.1.
2. Wash 2 times with PBS.
3. Freshly prepare appropriate amount of EdU reaction cocktail of 0.1 M Tris/HCl (pH 7) containing 4 mM $CuSO_4$, 50 mM Na-ascorbate (from a freshly prepared 1 M solution), and 20 μM azide dye. Incubate cells with 50 μL of reaction cocktail for 30 min at room temperature in a dark humidified chamber as in Subheading 3.5, step 6.
4. Remove excess liquid by blot-drying before placing into a 6-well plate containing PBST.
5. Wash thoroughly 3 times for 3 min with PBST.
6. Proceed to Subheading 3.5 for post-fixation and DAPI staining.

Alternatively, the Click-iT Labeling Kit (Life technologies) can be used. If cells need to be EdU pulsed, imaged, demounted and FISH treated and then relocated and reimaged, refer to Subheading 3.9. EdU replication labeling can be followed by immunofluorescence detection if required. In this case, see steps outlined in Table 1.

### *3.9. Relocating Cells Before and After FISH Treatment*

In some cases it may be useful to image individual cells before and after FISH treatment to monitor or circumvent potential destructive effects of the hybridization on a cellular structure of interest. With the DeltaVision OMX imaging system it is possible to image a cell of interest on a predefined surface (slide) and use the stored stage-position to relocate and image the same pinpointed cells in sequential rounds of detections. For instance, a replication labeled cell may be imaged with 3D-SIM after EdU detection (Subheading 3.8, steps 1–6), then demounted and processed for FISH. After remounting, the same cells can be scored by using the selected saved positions from the previous imaging round. A scheme of the relocation principle is presented in Fig. 5 and possible combinations of protocols in Table 1.

1. Seed cells on cleaned microscope slides instead of coverslips and grow them to ~60% confluency.
2. Mark the area that will be imaged on either side with the help of a diamond cutter. Wipe off remaining cells outside this region using Kimwipes tissues. All incubations and washing steps are performed in 4-well rectangular dishes.
3. Perform labeling (e.g., with antibodies, azide-dye, and/or DAPI) by placing a drop of labeling solution onto the predefined marked region, and cover it with an 18 × 18 mm piece of Parafilm to avoid dripping of solution off the slide. Perform all other steps as previously described.

4. Perform 3D-SIM imaging as described in Subheadings 3.10 and 3.11.
5. To demount and process for FISH, clean the top surface of the coverslip carefully from lens immersion oil with ethanol. With the use of fine-tip forceps lift gently the nail varnish at one corner and start peeling it off pulling upwards, while simultaneously adding a few drops of PBS to the already exposed sides on the coverslip underneath the varnish coat. Remove nail varnish completely, while trying not to move the coverslip. Add a generous amount of PBS to the coverslip and try lifting it gently while infusing PBS underneath. When the cells' surface is pretty well soaked, slide the coverslip smoothly to the edge of the slide and remove. Alternatively, slides can be immersed for 30 min in a Coplin jar filled with PBS and gently shaken on occasion until slide drops off.
6. Place the hybridization probe onto the marked region, cover it with an 18 × 18 mm coverslip and perform FISH as previously described.
7. For mounting these slides extra care needs to be taken removing excess solution. To avoid diluting the mounting media and compromised imaging, slides must be well blotted, in particular by applying the edges of precision wipes on to the surface of the slide up to the boarders of the marked region.

### 3.10. Mounting the Samples on the Microscope

Having a thoroughly cleaned coverslip is essential for optimal image acquisition. It is recommended to wipe coverslips with Kimwipes tissues soaked in 80% ethanol in repeating movements towards one direction to avoid smudging of nail varnish or residual oil onto the glass surface. After evaporation, coverslips should be swiped briefly with chloroform. Finally apply a drop of immersion oil with appropriate refractive index (see Note 14) and mount the sample on the microscope stage.

### 3.11. Acquisition of 3D-SIM Images and Data Processing

For acquiring a 3D-SI raw data set, three *z*-stacks with plane distances of 0.125 μm are collected with a different rotation angle of the striped illumination pattern. Each plane of a stack is recorded five times with the pattern being laterally shifted, thus summing up to 15 acquisitions per plane and color channel. To avoid bleed through in multicolor applications, the fluorescence channels are typically collected sequentially for every phase grating position and at every *z*-position. To reconstruct an optical section with super-resolution in *x*, *y* and *z*-direction, the algorithm uses the adjacent *z*-sections above and below, confining a minimal stack size of at least 1 μm. Depending on the nature of the specimen, the imaging depth can exceed 10 μm. However, as optical aberrations become more and more effective, the SI reconstruction quality deteriorates with deeper imaging. Importantly, the exposure times have to be set to meet the best compromise between signal strength, bleaching during acquisition and overall acquisition time, and are typically in the range of 20–100 ms with the laser not being attenuated ($T$= 100%). To provide sufficient modulation contrast of the induced interference, the aim is to archive best possible signal-to-background levels with at the same time acceptable bleaching rate over the time course of acquisition. In addition, to avoid the risk of mechanical drifts that would lead to reconstruction artifacts, an overall short data acquisition is desirable. For optimal reconstruction quality of the 3D-SIM data the use of regularly measured, channel-specific optical transfer functions (OTFs) of optimal quality are pivotal. For this purpose we recommend to measure on a regular basis the respective point spread functions (PSFs) using 175 nm diameter PS-Speck fluorescent microspheres (Life technologies).

## 4. Notes

1. It is recommended to filter solutions with a 0.45 μm syringe filter to remove precipitates which are very commonly encountered in particular in SSC-based buffers. These filters may be used several times, as they are expensive.
2. We advice the use of high precision cover glasses #1.5, equivalent to a thickness of 0.170 ± 0.005 mm (e.g., Marienfeld Superior) as the values of standard grade cover glasses can vary substantially from 0.160 to 0.190 mm, which in turn can negatively affect the super-resolution image quality. Hybridization is typically performed using small quantities (2–3 μL) of expensive DNA probes. In this case, coverslips may be cut with a diamond cutter with the help of a slide layered against the surface of the coverslip and pressed tight while cutting. Hand-cut coverslips should be rinsed from dust and residual glass chips with deionized $H_2O$ and stored in pure ethanol.
3. 2% Formaldehyde fixation is preferred for maintenance of the nuclear morphology visualized by 3D-SIM, while 4% formaldehyde concentration appears to result in a more well-preserved nuclear morphology in FISH-treated specimens (see Fig. 3). We have not noticed any difference between fixing 3D-FISH treated samples with freshly depolymerized PFA or diluted formalin (37% formaldehyde stabilized with 10% methanol). Most critical in the fixation and post-fixation step is to avoid even short drying of cells. This is ensured by a stepwise exchange of the fixative with PBST (see Subheading 3.3, step 3). Formalin should be kept in a well-airtight container and care should be taken as to avoid precipitation.
4. Washing steps, in particular after antibody incubations, are performed using abundant amounts of washing buffer, i.e.,

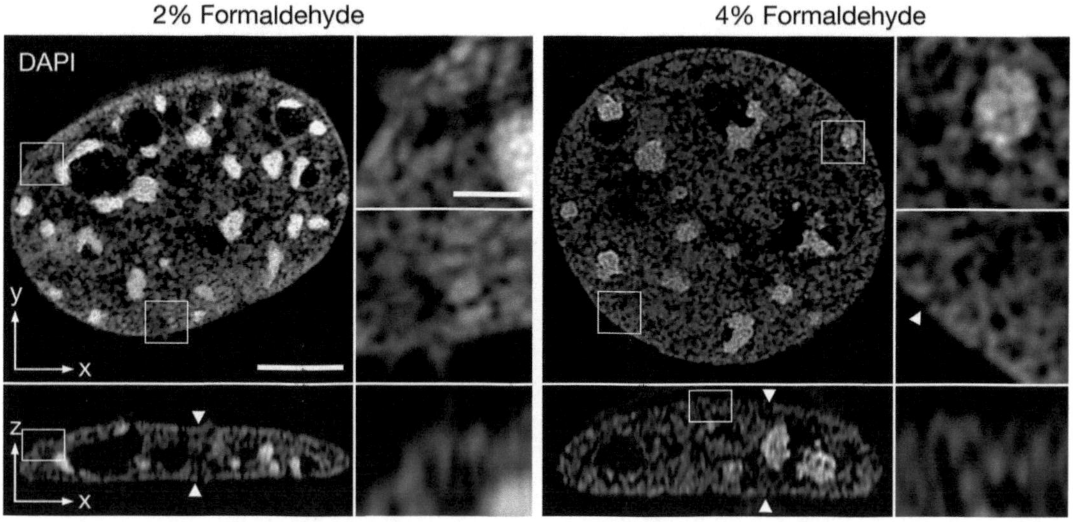


Fig. 3. Influence of the formaldehyde concentration on the nuclear morphology of FISH treated cells. C127 cell nucleus, fixed with 2% formaldehyde (*left panels*), or 4% formaldehyde (*right panels*), counterstained with DAPI and subsequently processed for FISH. Fixation with 2% formaldehyde renders nuclei more sensitive to disruption caused by the repeated liquid $N_2$ freezing and thawing (compare insets, *left panel*) and more prone to shrinkage due to the FISH procedure (*arrowheads* in the *xz*-sections). Chromatin voids underneath the nuclear pores can still be visualized after fixation with 4% formaldehyde (*arrowhead, right panel* inset). Overall chromatin conformation, especially in euchromatic areas, seems to be better preserved. Scale bar represents 5 μm (inset 1 μm).

few mL in a fresh 6-well plate. To our experience, it is more effective to perform more washing steps (4–5 changes of buffer), rather than incubating in the washing buffer for extended periods of time. To further increase washing efficiency coverslips may be dipped a few times into a beaker filled with a larger volume (e.g., 20 mL) of PBST between two transfer steps. Less adhering cells (e.g., mitotic cells), however, may be partially washed off by this procedure. When transferring coverslips with a fine forceps (e.g., from the incubation drop to the 6-well) excess solution should be removed by shortly touching the coverslip's edge onto a Kimwipe tissue, before placing in the well and continuing to the next step. The presence of the mild detergence Tween 20 in antibody incubation steps increases stringency for specific interactions, while in the subsequent washing steps reduces surface tension and ensures wetting of cells during transfer steps.

5. Formamide is toxic. Please work under a laminar flow hood and make sure of appropriate disposal of wastes. When making 50% formamide/2× SSC pH 7, the pH has to be readjusted, so do not simply dilute. Store this solution at −20°C and take out of the freezer, to reach ~4°C prior to use.
6. We have found that high concentrations of probes, which in the past were thought to result in a more effective labeling of

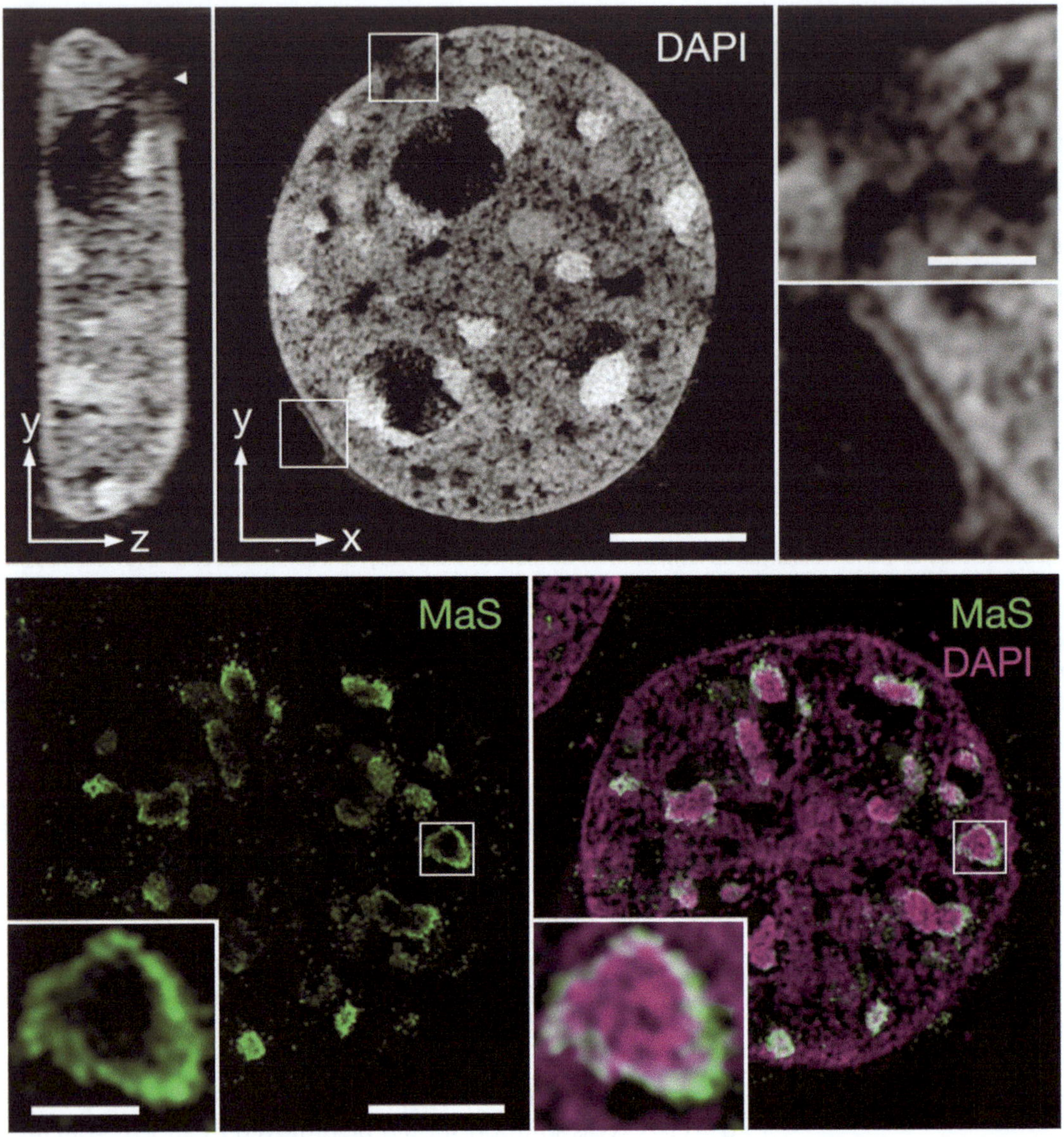


Fig. 4. Frequent 3D-FISH artifacts visualized by 3D-SIM. *Upper panel*: C127 cell nucleus stained with DAPI (*gray*) with evident lesions at the nuclear periphery (upper inset), and ectopic chromatin fragments expanding outside the nuclear periphery (lower inset). *Lower panel*: C127 cell nucleus hybridized with probes against murine major satellite repeats stained with Alexa 488 (MaS, *green*) and counterstained with DAPI (*magenta*). Too high concentrations of the FISH probe result in a limited penetration and a saturation of the hybridized sequences at the periphery of chromocenters that may easily be misinterpreted (compare Fig. 1, *lower panel*, for a hybridization with lower concentration). Scale bars represent 5 μm (insets 1 μm).

hybridized sequences can very often have the opposite effect. For instance, we observed rather limited access to highly abundant major satellite repeat target DNA in chromocenters of mouse cell nuclei, that is most likely caused by a probe oversaturation at the target sequences' surface (Fig. 4). This effect can be effectively avoided when using a tenfold dilution of the probe, leading to a much more even hybridization throughout the body of chromocenters. Similar results have been observed with the hybridized Cot-1 DNA, which when hybridized at high concentration concentrates at the nuclear periphery rather than binding throughout the entire nuclear volume. Hence, all probes need to be well titrated before deducting conclusion from initial observations. The same effect of limited accessibility of the hybridized probe has been observed on occasion when omitting the liquid nitrogen exchange step (24).

7. Avoid denaturing a probe more than once. It is preferable to transfer the amount required into a fresh reaction tube and store the remaining probe back at –20°C for future use.
8. Samples that will be processed for DNA FISH (even EdU pulsed) can be stored in 50% formamide/2× SSC for up to a couple of weeks. For RNA-FISH, DEPC treated $ddH_2O$ should be used for all buffers to prevent RNA degradation. From our experience, it is highly recommended to proceed with RNA-FISH hybridization immediately after fixation and cell pretreatment.
9. We typically perform detection of hapten (digoxigenin or dinitrophenyl) labeled probes by anti-DIG/DNP antibodies followed by appropriate fluorochrome-conjugated secondary antibodies; in case of biotin-labeled probes detection is performed with fluorochrome-conjugated (strept)avidine. In principle, also directly fluorochrome-labeled probes can be used. In this case one can directly proceed to DAPI staining (Subheading 3.6, step 3). However, Alexa Dye-conjugated dUTPs are very expensive and label efficiency has to be carefully pretested.
10. Glycerol-based mounting media, e.g., Vectashield (Vector Laboratories), preserve 3D morphology and perform well with all dyes typically used for 3D-SIM (34). In contrast, hardening mounting media, such as Prolong Gold (Life technology) despite having good antifading properties, flatten the cells substantially and compromise nuclear morphology.
11. Alternatively, a more targeted, although not in all cases more efficient, blocking can be achieved by supplementing PBST with 5–10% serum of the species where the secondary antibodies have been raised in, for example, normal horse serum can be used, if the secondary antibodies are raised in donkey. Finally, good results have also been accomplished with a HEPES-based buffer, or in other cases, by maintaining 0.2% Triton X-100 in the blocking buffer throughout immunodetection. Again, in cases of high nonspecific antibody binding and therefore high background signals, blocking times and different blocking agents have to be tested.

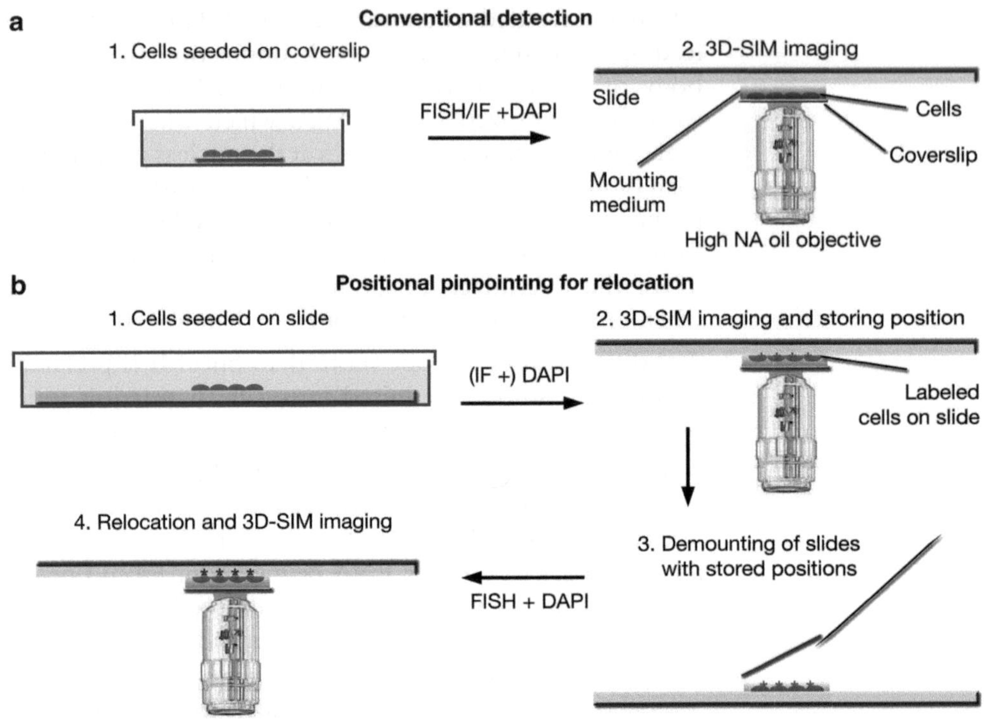


Fig. 5. Scheme for the pinpointing and relocation of individual cells. (**a**) Standard detection and imaging scheme. (**b**) Sample handling for imaging individual cells before and after FISH treatment as described in detail in Subheading 3.9.

12. Secondary antibodies should always be tested for their specificity and nonspecific binding before being routinely used. A general good rule is to start with a working dilution of 1–2 μg/ml, which seems to work fairly well in most cases. Too high concentrations increase the nonspecific background and should be avoided. Be careful not to expose fluorescently labeled antibodies to bright light and do not repeat several freeze and thaw rounds, but rather store small aliquots at −20°C, defreeze once and keep at 4°C, while always transferring on ice. Sometimes antibodies show high cross-reactivity and/or unspecific background caused by precipitation, which may be reduced by pre-clearing of antibody stock solutions by centrifugation at 15,000 × *g* for 15 min at 4°C and transferring the supernatant to a fresh tube, or by filtering the final diluted antibody solution through a 0.45 μm syringe filter. In this case a 1 ml syringe is more appropriate to avoid loss of solution.
13. We have found that the reactivity of incorporated EdU is well maintained after FISH treatment of specimens, while the fluorescent signal of the azide dye is very well preserved if the click chemistry is performed before FISH treatment, e.g., when applying the relocation scheme described in Subheading 3.9 (Fig. 5). Despite the increased distance of cells

growing on slides to the lens, the reconstructions acquired are satisfactory and comparable to standard conditions (compare cells in Figs. 1, 3, and 4 to relocated FISH treated cell in Fig. 2). The described relocation method is less tedious than e.g., using gridded coverslips of less well-defined thickness, and proved suitable when there is a need to image, de-mount the specimen, process and reimage the same cell. Consideration, however, should be taken here by the fact that a precise pixel by pixel comparison of the same optical plane before and after relocation is not possible due to inevitable small-scale distortions of the specimen and small $z$-differences in the acquisition start points (24).

14. The SI reconstruction algorithm relies on a symmetric point spread function (PSF) of the system. Variations in the refractive index of the mounting medium, cover glass thickness, temperature or imaging depth can affect the point spread behavior of the sample and deteriorate the quality of the reconstructed image. In this case, fluorescent dots within the sample would not appear as hourglass shaped in an orthogonal view of a wide-field image stack, but rather asymmetrically scattered towards either top or bottom. Notably, immersion oil with different refractive index can to some extent compensate for this effect and should be chosen after empirical testing. As a rule of thumb, if Airy diffraction rings are more pronounced at the coverslip side, the refractive index of the immersion oil must be lowered.

## Acknowledgments

The authors are indebted to H. Leonhardt and the BioImaging Network Munich for generous support. We thank J. Neumann for technical help and A. Nemeth for critical comments on the manuscript. This work was supported by DFG grants SCHE1596/2-1 and CR59/28-1 to M.C. and L.S and SFB 684 to Y.M.

### References

1. Cremer T, Cremer M (2010) Chromosome territories. Cold Spring Harb Perspect Biol 2: a003889
2. Cremer M, Grasser F, Lanctôt C et al (2008) Multicolor 3D fluorescence in situ hybridization for imaging interphase chromosomes. Methods Mol Biol 463:205–239
3. Foster HA, Bridger JM (2005) The genome and the nucleus: a marriage made by evolution. Genome organisation and nuclear architecture. Chromosoma 114:212–229
4. Misteli T (2010) Higher-order genome organization in human disease. Cold Spring Harb Perspect Biol 2:a000794
5. Volpi EV, Bridger JM (2008) FISH glossary: an overview of the fluorescence in situ hybridization technique. Biotechniques 45:385–386, 385–6, 388, 390 passim
6. Speicher MR, Gwyn BS, Ward DC (1996) Karyotyping human chromosomes by combinatorial multi-fluor FISH. Nat Genet 12: 368–375

7. Bolzer A, Kreth G, Solovei I et al (2005) Three-dimensional maps of all chromosomes in human male fibroblast nuclei and prometaphase rosettes. PLoS Biol 3:e157
8. Silverman AP, Kool ET (2007) Oligonucleotide probes for RNA-targeted fluorescence in situ hybridization. Adv Clin Chem 43:79–115
9. Blanco AM, Artero R (2010) A practical approach to FRET-based PNA fluorescence in situ hybridization. Methods 52:343–351
10. Pawley JB (2006) Handbook of biological confocal microscopy. Springer, New York, p 985
11. Schermelleh L, Heintzmann R, Leonhardt H (2010) A guide to super-resolution fluorescence microscopy. J Cell Biol 190:165–175
12. Huang B, Babcock H, Zhuang X (2010) Breaking the diffraction barrier: super-resolution imaging of cells. Cell 143:1047–1058
13. Toomre D, Bewersdorf J (2010) A new wave of cellular imaging. Annu Rev Cell Dev Biol 26:285–314
14. Hell SW (2009) Microscopy and its focal switch. Nat Methods 6:24–32
15. Gustafsson MGL, Shao L, Carlton PM et al (2008) Three-dimensional resolution doubling in wide-field fluorescence microscopy by structured illumination. Biophys J 94:4957–4970
16. Gustafsson MG (2000) Surpassing the lateral resolution limit by a factor of two using structured illumination microscopy. J Microsc 198:82–87
17. Schermelleh L, Carlton PM, Haase S et al (2008) Subdiffraction multicolor imaging of the nuclear periphery with 3D structured illumination microscopy. Science 320:1332–1336
18. Baddeley D, Chagin VO, Schermelleh L et al (2010) Measurement of replication structures at the nanometer scale using super-resolution light microscopy. Nucleic Acids Res 38:e8
19. Brown ACN, Oddos S, Dobbie IM et al (2011) Remodelling of cortical actin where lytic granules dock at natural killer cell immune synapses revealed by super-resolution microscopy. PLoS Biol 9:e1001152
20. Guizetti J, Schermelleh L, Mantler J et al (2011) Cortical constriction during abscission involves helices of ESCRT-III-dependent filaments. Science 331:1616–1620
21. Markaki Y, Gunkel M, Schermelleh L et al (2010) Functional nuclear organization of transcription and DNA replication: a topographical marriage between chromatin domains and the interchromatin compartment. Cold Spring Harb Symp Quant Biol 75:475–492
22. Elia N, Sougrat R, Spurlin TA, Hurley JH, Lippincott-Schwartz J (2011) Dynamics of endosomal sorting complex required for transport (ESCRT) machinery during cytokinesis and its role in abscission. Proc Natl Acad Sci USA 108:4846–4851
23. Bennett BT, Bewersdorf J, Knight KL (2009) Immunofluorescence imaging of DNA damage response proteins: optimizing protocols for super-resolution microscopy. Methods 48: 63–71
24. Markaki Y, Smeets D, Fiedler S, Schmid VJ, Cremer T, Cremer M (2012) The potential of 3D-FISH and super-resolution structured illumination microscopy for studies of 3D nuclear architecture. Bioessays 34(5):412–426
25. Solovei I, Cavallo A, Schermelleh L et al (2002) Spatial preservation of nuclear chromatin architecture during three-dimensional fluorescence in situ hybridization (3D-FISH). Exp Cell Res 276:10–23
26. Wallace W, Schaefer LH, Swedlow JR (2001) A workingperson's guide to deconvolution in light microscopy. Biotechniques 31(1076–8):1080, 1082 passim
27. Levsky JM, Braut SA, Singer RH (2003) Single cell gene expression profiling: multiplexed expression fluorescence in situ hybridization (FISH) application to the analysis of cultured cells. http://www.singerlab.org/protocols
28. Tam R, Shopland LS, Johnson CV, McNeil JA, Lawrence JB (2002) Applications of RNA FISH for visualizing gene expression and nuclear architecture. In: Beatty B, Mai S, Squire J (eds) FISH: a practical approach. Oxford University Press, New York, p 93
29. Brown JM, Buckle VJ (2010) Detection of nascent RNA transcripts by fluorescence in situ hybridization. Methods Mol Biol 659:33–50
30. Raj A, van den Bogaard P, Rifkin SA, van Oudenaarden A, Tyagi S (2008) Imaging individual mRNA molecules using multiple singly labeled probes. Nat Meth 5:877–879
31. Carlton PM, Boulanger J, Kervrann C et al (2010) Fast live simultaneous multiwavelength four-dimensional optical microscopy. Proc Natl Acad Sci USA 107:16016–16022
32. Dobbie IM, King E, Parton RM, Carlton PM, Sedat JW, Swedlow JR, Davis I (2011) OMX: a new platform for multimodal, multichannel wide-field imaging. Cold Spring Harb Protoc 2011:899–909
33. Salic A, Mitchison TJ (2008) A chemical method for fast and sensitive detection of DNA synthesis in vivo. Proc Natl Acad Sci USA 105:2415–2420
34. Cordes T, Maiser A, Steinhauer C, Schermelleh L, Tinnefeld P (2011) Mechanisms and advancement of antifading agents for fluorescence microscopy and single-molecule spectroscopy. Phys Chem Chem Phys 13:6699–6709

# Chapter 5

## Two-Color STED Imaging of Synapses in Living Brain Slices

**Jan Tønnesen and U. Valentin Nägerl**


### Abstract

STED microscopy is a novel fluorescence microscopy technique that breaks the classic diffraction barrier of optical microscopy. It offers the chance to investigate dynamic processes inside living cells with a spatial resolution well below 100 nm, possibly even down to a few nanometers, essentially without forgoing the benefits of conventional light microscopy, such as labeling specificity, sensitivity, and contrast. STED microscopy has already been exploited for several important neurobiological experiments. Given the tremendous potential as a transforming technology, it is important to understand how it works, and what its scope and limitations are. Here, we present a primer on STED microscopy, its basic principles and practical implementation, presenting a *how-to* guide on building and operating a STED microscope.


**Key words:** STED, Superresolution microscopy, Live-cell imaging, Diffraction barrier, Synapses, Neuronal morphology

## 1. Introduction

Stimulated emission depletion (STED) microscopy is a new fluorescence microscopy technique for diffraction-unlimited imaging (1, 2). In contrast, the spatial resolution of conventional optical microscopy is limited by the diffraction of light to around half the wavelength of the light used in the microscope (3), i.e., >250 nm for confocal and >350 nm for two-photon microscopy.

This limitation has been painfully true for synapses, which typically are too small to be properly resolved by conventional light microscopy, making it difficult to investigate their dynamic organization and regulation with fluorescence microscopy.

While the STED principle was first conceived in 1994 (4), its first successful implementation was published in 1999 (5), followed a year later by a demonstration inside living cells (6). In the context of neurobiology, STED microscopy was initially used only on fixed tissue preparations looking at immuno-stained distributions of

Alioscka A. Sousa and Michael J. Kruhlak (eds.), *Nanoimaging: Methods and Protocols*, Methods in Molecular Biology, vol. 950,
DOI 10.1007/978-1-62703-137-0_5, © Springer Science+Business Media, LLC 2013

synaptic proteins (7). However, in recent years STED microscopy has been increasingly exploited for live-cell imaging applications, imaging the motility of synaptic vesicles in nerve terminals (8), the endoplasmatic reticulum in living mammalian cells (9), morphological dynamics of dendritic spines in organotypic brain slices (10), as well as the dynamic distribution of actin inside spines (11).

STED microscopy retains many of the benefits of confocal microscopy regarding labeling, live-cell compatibility, speed, color contrast and optical sectioning and depth penetration. Like confocal microscopy, it can be readily combined on the same experimental setup with other powerful opto-physiological techniques, such as photo-uncaging, calcium imaging, optogenetics, and electrophysiology. Because STED microscopy offers the chance to close the gap between electron and light microscopy, it is attracting attention as a potentially game-changing imaging technology.

Here, we present a primer on STED microscopy, its basic principle and practical implementation. In addition, we give a *how-to* guide on building and operating a STED microscope for imaging of fluorescent proteins and organic dyes in two colors in living organotypic brain slices with a lateral resolution of around 50 nm. The microscope is entirely compatible with a setup designed for opto-physiological experiments, including two-photon uncaging and patch-clamp electrophysiology.

In general, the spatial resolution of a fluorescence microscope is defined by the minimal distance between two objects at which they can still be recognized as separate objects. For a laser-scanning microscope this distance is determined by the size of the focal spot because fluorescent objects that fall within the focal spot volume get excited at the same time, and thus cannot be distinguished from each other by the detector. Therefore, in order to increase the spatial resolution of a laser-scanning microscope it is necessary to decrease the size of the spatial extent over which dye molecules get co-excited.

Because of the diffraction of light, the microscope objective produces a focal spot of finite size, whose spatial extent is called the point-spread function (PSF). It is typically >250 nm for confocal microscopy and even larger for two-photon microscopy (>350 nm) because of the use of longer wavelength light.

The basic idea of STED microscopy is to reduce the size of the PSF, and hence improve spatial resolution, by actively inhibiting the fluorescence on the edge of the focal spot. This is achieved by the action of a second laser beam (called the STED laser), which has a doughnut-shaped or annular intensity distribution in the focal plane. The intense light of the doughnut inhibits the fluorescence on the outer edge of the PSF by the process of stimulated emission at a wavelength that is longer than the fluorescence, allowing the fluorescence signal to be separately detected (Fig. 1a–c). The current record in reducing the spot size is at 5.8 nm using

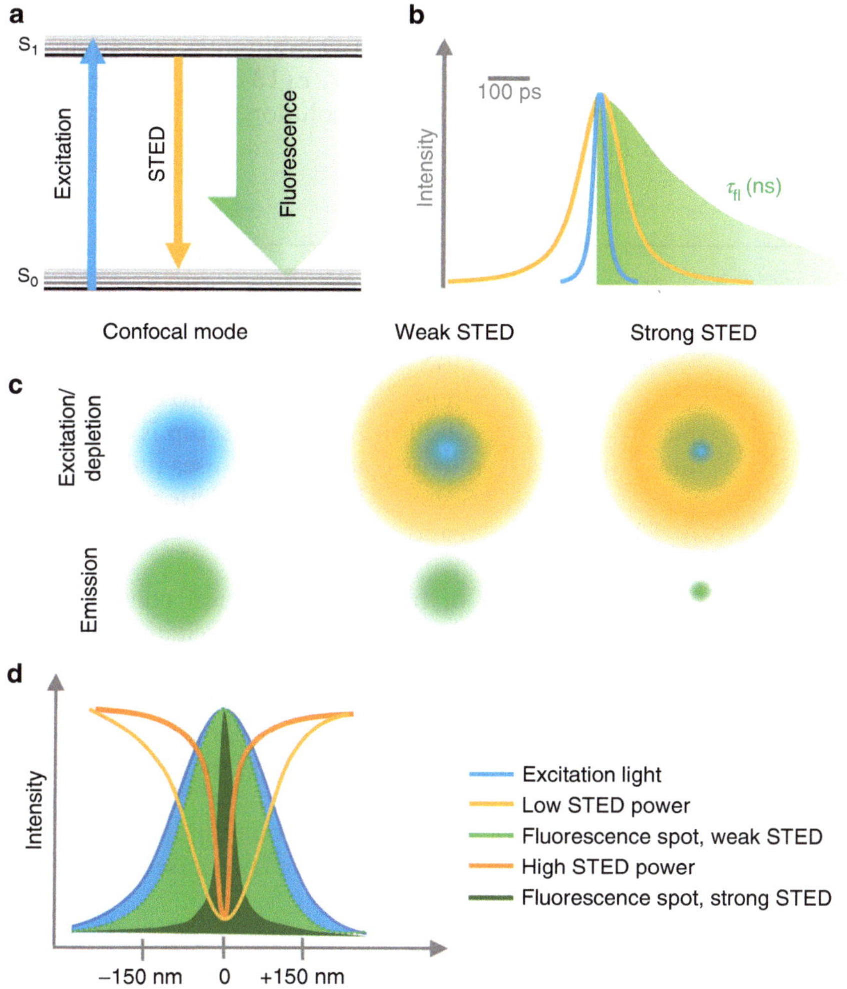


Fig. 1. Basic principle of STED microscopy. STED microscopy is based on fluorescence quenching by stimulated emission. (**a**) Jablonski diagram of the molecular events during fluorescence. Absorption of a blue photon excites the molecule from the ground state ($S_0$) to the first excited singlet state ($S_1$). Normally, the excited molecule returns back to the ground state by spontaneously emitting a photon after an average period defined by the fluorescence lifetime ($t$) of the fluorophore. In STED microscopy this process is actively prevented by returning the excited molecule back to the ground state via stimulated emission using a red-shifted laser (the STED laser). (**b**) An excitation laser pulse (*blue line*) causes fluorescence (in *green*) that decays over a period defined by the lifetime ($t$) of the fluorophore, typically a few nanoseconds. For the STED laser to efficiently intercept this process, the depletion pulses (*orange*) must temporally coincide with the excitation pulses. (**c**) Schematic of the STED doughnut effect on the size of the focal spot (PSF), and hence spatial resolution of the microscope. By increasing the STED laser power and depleting the excited state on a doughnut-shaped ring around the excitation spot, the area that is allowed to give off photons by spontaneous emission, i.e., fluorescence, can be made arbitrarily small. Therefore, the STED principle overcomes the diffraction barrier in a fundamental way. (**d**) Line profiles of the excitation and STED laser intensities and of the resultant size of the fluorescence spot.

diamond crystals, which is more than two orders of magnitude smaller than the wavelength of light that was used in the experiments (12).

The resolution of the STED microscope is expressed as $\Delta r \approx \lambda / \left(2\text{NA}\sqrt{1+I/I_s}\right)$, with NA denoting the numerical aperture of the lens, $\lambda$ the excitation beam wavelength and $I$ the intensity of the STED beam. $I_s$ is the STED beam intensity required to reduce the fluorescence by a factor of $1/e$ (13). Thus, there is in theory no longer a hard limit on spatial resolution, because $\Delta r$ approaches zero when $I$ is increased to infinity.

## 2. Materials

A schematic of a STED microscope including the most important components is shown in Fig. 2a and a photo of a functional STED microscope in Fig. 2b. The homebuilt setup is designed around a standard confocal microscope base and uses pulsed single-photon excitation at $\lambda = 485$ nm and pulsed depletion around $\lambda = 595$ nm for addressing green and yellow fluorophores.

### *2.1. Optical Components for Beam Steering*

A broad selection of high-quality optical components must be available for routing and shaping the excitation and STED laser beams. The list includes mirrors and adjustable mirror mounts, lens telescopes for expanding, decreasing and collimating laser beams, $\lambda/2$ and $\lambda/4$ plates for controlling the polarization of the STED beam, polarizing beam splitters for selecting a given polarization, and pinholes for beam alignment. Various optical diagnostic instruments (such as a beam profiler, powermeter, shear-plate, spectrometer, auto-correlator) are also essential for building a STED system.

Fig. 2. (continued) emitting at 795 nm. A phase mask introduces the intensity-zero in the center of the STED beam, and a $\lambda/4$-plate is used to optimize the doughnut-shaped intensity distribution of the STED laser in the focus. The STED beam is then aligned in space with the 485 nm excitation beam. The two pulsed lasers are synchronized and overlaid in time using a pulse delay. Using a fast nano-positioner for scanning along the *z*-axis, a beam scanner for *x,y* scanning and an inverted microscope objective configuration, there is easy experimental access to the slice from above, e.g., for electrophysiological patch-clamp recordings or dye loading. Optionally, other beams can be brought into the system, exemplified by an infrared wavelength range laser for two-photon uncaging and imaging. The fluorescence emission signal is descanned and routed into two photo-detectors separated by a 514 nm long-pass dichroic mirror, for two-color imaging. (**b**) Photo of a STED setup resembling the above scheme. The main components are: MaiTai Ti:sapphire lasers for STED and two-photon uncaging/imaging (1), OPO pumped by one of the MaiTai lasers (2), 485 nm excitation pulsed diode laser (3), beam scanner (4), microscope stage with electrophysiology gear (5), two APD detectors enclosed in a box (6), PMT detector (7). A flip-mounted pellicle mirror is used to send the light either to the APDs or to the PMT. (**c**) Close-up photo of the optical components required for merging and separating the beam paths. The edge of the microscope is visible on the right side. The linearly polarized STED beam (*orange*) is shaped by the vortex phase-plate (1), and aligned onto the excitation beam (*blue*) using a piezo-motorized mirror mount (2) by reflecting the STED beam on a 590 nm short-pass dichroic mirror that the excitation beam passes (3). The STED beam profile is optimized using a $\lambda/2$-plate (4) addressing the $\lambda/4$-plate (5) in front of the beam scanner (6). The fluorescence signal from the sample (*green*) is descanned through the same beam path and ultimately isolated using a 499 nm long-pass filter (7). Before the emission signal reaches the APD detectors it is cleaned by a $525 \pm 50$ nm band-pass filter (not in photo). A 680 nm short pass dichroic mirror is used to bring in a higher wavelength beam (*red*) for two-photon experiments (8).

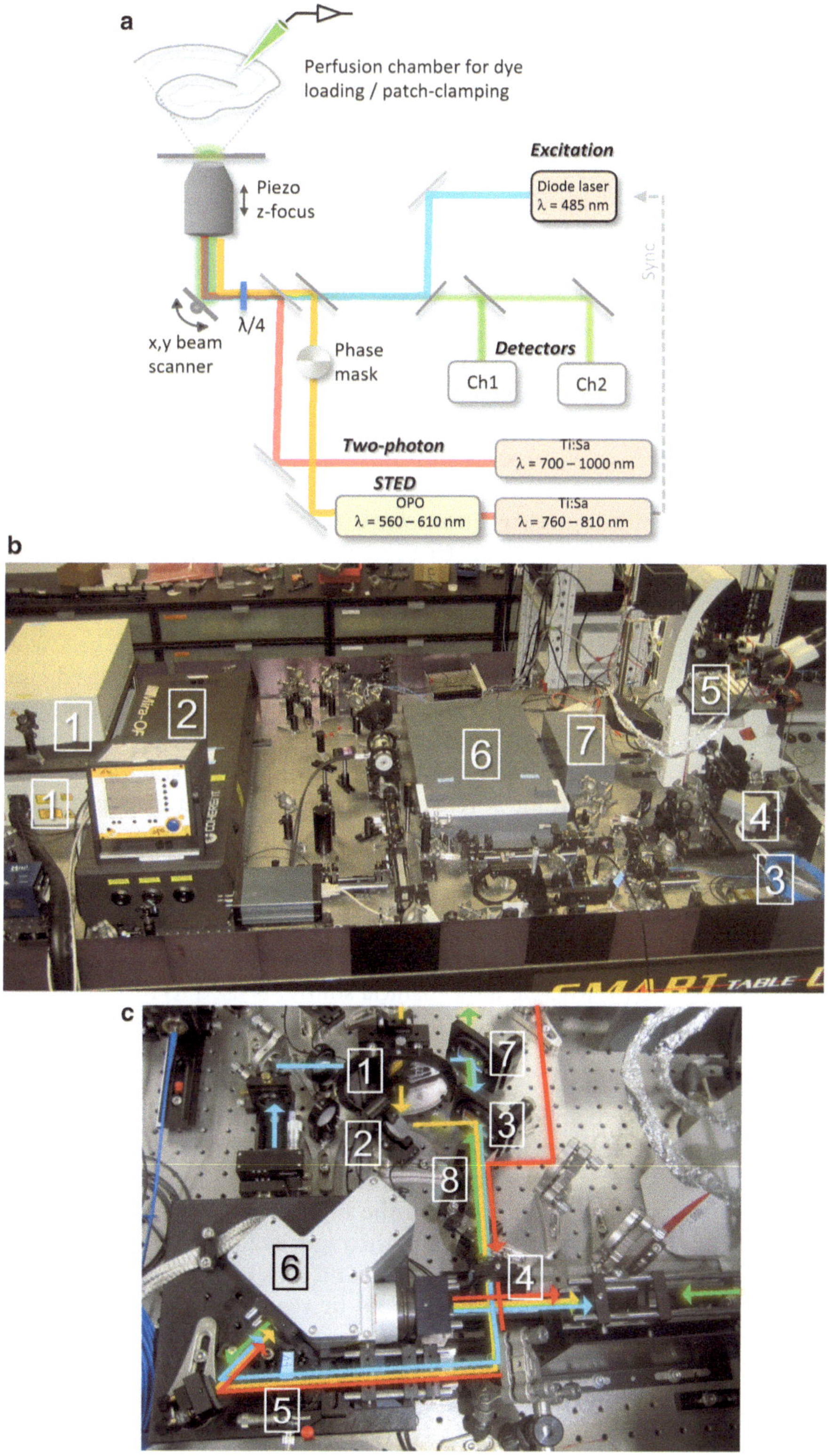


Fig. 2. Design of a homebuilt STED microscope (**a**) Schematic of a STED microscope setup used for imaging green fluorophores in living brain slices. The 595 nm STED beam is derived from a OPO, which is pumped by a Ti:sapphire laser

### 2.2. Microscope Base and Objective

A proven choice for building an inverted STED microscope for live-cell imaging is the base of the DMI 6000 CS Trino microscope (Leica, Mannheim, Germany) equipped with a high-numerical-aperture objective lens, e.g., PL APO, 100×, oil, NA 1.4 (9) or PL APO, 63×, glycerine w/correction ring (14) (both from Leica) (see Note 1). For building an upright system, the BX61 base from Olympus can be used. It is recommended that the scan and tube lenses, objective and microscope are all from the same manufacturer, to ensure that aberrations in the optics are properly corrected.

### 2.3. Excitation Laser Source

For excitation of green and yellow fluorophores a pulsed-laser diode (LDH-D-C-485; Picoquant, Berlin, Germany) delivering excitation pulses of 485 nm wavelength and of 90 ps duration, can be used (15). Laser pulsing can be externally triggered via the driver electronics and thus be synchronized with the STED laser.

### 2.4. STED Laser and Optical Parametric Oscillator

Although continuous wave (cw) lasers can be used instead of pulsed lasers, they need to be run at much higher average power than pulsed lasers to achieve a comparable gain in spatial resolution (16), which may pose problems for live-cell applications.

The $\lambda = 595$ nm STED beam for quenching yellow and green fluorescence can be derived from a Ti:sapphire pulsed laser (repetition rate ~80 MHz, pulse duration: 200 fs, wavelength $\lambda = 795$ nm; MaiTai HP, Spectra-Physics, Darmstadt, Germany) and converted to the desired wavelength by an optical parametric oscillator (OPO BASIC Ring fs RTP, APE, Berlin, Germany) (see Note 2) (9). In addition, a band-pass filter (e.g., $596 \pm 22$ nm) should be used to clean up the STED beam. For experiments in brain slices the power of the STED laser is typically between 5 and 20 mW at the back aperture of the objective. The intensity of the STED beam can be rapidly and continuously changed using a Pockels cell (Conoptics, Danbury, CT), whereas a mechanical shutter is used to turn the beam off.

### 2.5. STED Pulse Dispersion in Glass Rod and Polarization-Preserving Fiber

The 200 fs STED pulses coming from the OPO must be stretched in time to overlap with the 90 ps excitation pulses (Fig. 1b). Stretching is done by sending the beam through a polarization-preserving fiber of 100 m length (Schäfter & Kirchhoff, Hamburg, Germany), after which the pulses are around 300 ps long (9). To protect the fiber from the high peak powers inherent in fs pulses, the beam is first sent through a 25 cm glass rod (SF6), which broadens the pulses to around 2 ps (10). More than 50% transmission through the fiber can be readily achieved with standard coupling optics.

### 2.6. Pulse Delay Generator

For synchronizing the STED and excitation pulses and optimally coinciding them in time, the excitation laser is driven as a slave of the Ti:sapphire laser via an electronic pulse synchronizer with adjustable delay, for which custom-built or commercial solutions exist.

#### 2.7. Helical Vortex Phase Mask

To produce the doughnut intensity distribution, the STED beam is expanded to around 10 mm diameter, collimated, linearly polarized and passed through a polymeric phase mask (RPC Photonics, Rochester, NY), imprinting a helical-phase delay of exp ($i\varphi$), with $0 < \varphi < 2\pi$ onto the wave front (Fig. 3a). The phase mask produces a destructive interference pattern, such that the light intensity is zero all along the central axis of the beam (17). In addition to *x*–*y*, the spatial resolution can also be improved by the STED principle along the *z* axis by using another phase mask that imprints a $\pi$ delay. Both patterns can be imposed on a single beam by splitting it in two and passing each branch through the respective phase mask (18). Before entering the objective, the STED beam passes through a $\lambda/4$-plate, which circularly polarizes the light and improves the shape of the doughnut (19) (Fig. 3d).

#### 2.8. Dichroic Filters for Merging the Two Beams and Filtering Out the Emission Signal

The STED beam can be superimposed on the excitation beam by using a suitable short-pass filter (in our case with a cut-off wavelength of 590 nm), which passes the excitation light and reflects the STED beam (Fig. 3c). Immediately before the merge, the STED beam is reflected onto a piezo-controlled mirror, which is used for fine alignment of the STED and excitation beams (see Note 3) (Fig. 3b).

Using de-scanned, epi-fluorescence detection, the fluorescence signal follows the optical path in reverse direction of the STED and excitation beams back through the scanner. The emission signal can be isolated by another dichroic filter, e.g., a 499 nm long-pass in the case of GFP (Fig. 3g), and cleaned up with a band-pass filter in the range $525 \pm 60$ nm before reaching the detector(s) (10) (Fig. 3a).

#### 2.9. Beam Scanner and Fast Nano-Positioner of Microscope Objective

Whereas stage scanning is easier to implement, beam scanning can be much faster and is more compatible with electrophysiological recordings, as the sample and the stage do not move during scanning (20). The telecentric Yanus IV scanner (TILL Photonics, Gräfelfing, Germany) has proven useful for STED imaging (21) (Fig. 3f), in combination with standard data acquisition boards from National Instruments and the scanning software ImSpector (22). Rapid scanning in the *z*-axis can be done using a piezo-focuser (e.g., P-725, Physik Instrumente, Germany).

#### 2.10. Photodetectors

Photomultiplier tubes (PMTs) or avalanche photodiodes (APDs, e.g., SPCM-AQR-13; PerkinElmer, Fremont, CA) can be used for STED imaging, the latter having a much higher quantum efficiency (50–60% at $\lambda = 535$ nm), but considerably smaller dynamic range. In addition, hybrid detectors combining properties of PMTs and APDs may prove useful for STED imaging (23). A separate PMT can be used for aligning the optical system and optimizing the doughnut, by looking at the back reflection of gold beads (Fig. 3a). For two-channel spectral detection, a suitable dichroic mirror and

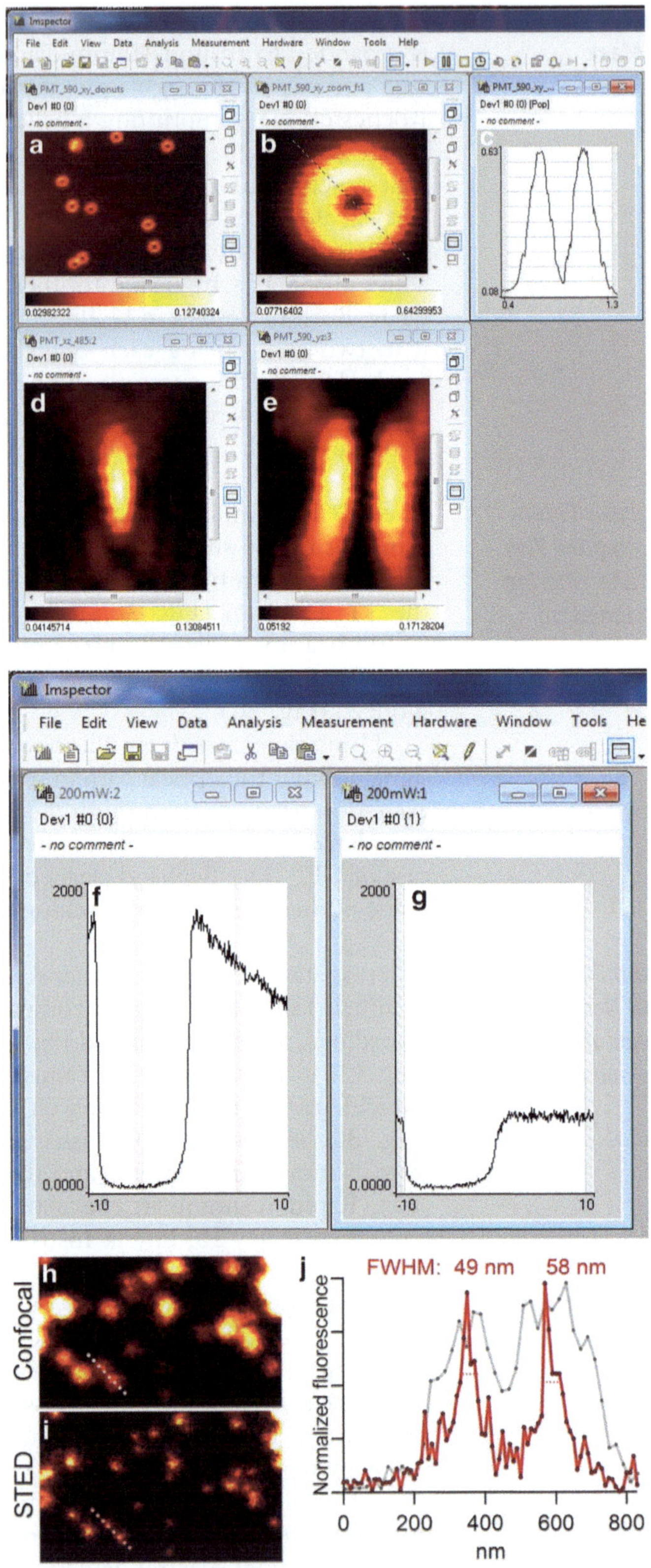
Inspector
File Edit View Data Analysis Measurement Hardware Window Tools Help
a
b
c
d
e
0.02982322
0.12740324
0.07716402
0.64299953
0.04145714
0.13084511
0.05192
0.17128204
Dev1 #0 {0}
- no comment -
200mW:2
200mW:1
Dev1 #0 {1}
2000
0.0000
-10
10
f
g
h
i
Confocal
STED
j
FWHM: 49 nm 58 nm
Normalized fluorescence
0
200
400
600
800
nm

a second APD are required (15) (Fig. 1a). The fluorescence signal is focused into a multimode optical fiber coupled to the APD. If the fiber has a diameter corresponding to the size of an Airy disk it serves as confocal pinhole.

### 2.11. Perfusion Recording Chamber

The brain slice is placed into an imaging/electrophysiological recording chamber equipped with temperature control and a perfusion system. A transparent chamber bottom is used for imaging on an inverted microscope (see Note 4). The glass bottom should be clean and without scratches, and the thickness should match the specifications for the given objective.

### 2.12. Organotypic Hippocampal Slice Cultures of Thy1-YFP Mice

Organotypic brain slice cultures from transgenic mice expressing yellow fluorescent protein (YFP) provide a convenient and powerful experimental preparation for studying synapses (10). The slices are prepared (according to the Gahwiler method (24)) on glass coverslips, which can be easily transferred to the perfused imaging chamber (10) (see Note 5).

### 2.13. Two-Photon Uncaging

A two-photon laser can be routed into the setup for two-photon photoactivation, bleaching or uncaging experiments (25, 26) (Fig. 2a), expanding the experimental possibilities to probe and manipulate synapses. The two-photon beam is brought in via a 680 nm short pass dichroic mirror (Fig. 3h) and aligned with the other two beams by using a dedicated piezo-adjustable mirror and by looking at the reflection from gold beads. In contrast to the MaiTai, the predecessor two-photon laser from Spectraphysics (Millenia/Tsunami) can be synchronized with the OPO, making it possible to use two-photon excitation for STED imaging.

Fig. 3. Evaluating the performance of a STED microscope. The figure represents the work flow for evaluating the STED microscope performance. (**a–g**) Screen-shots from experiments performed using the acquisition software ImSpector. (**a**, **b**) The STED beam doughnut profile is evaluated by reflecting the PSF on gold beads. (**b**, **c**) For effective depletion a symmetric doughnut with a good central null is crucial. (**c**) shows the line profile of the line through the doughnut in (**b**); the zero is around 5% of the max intensity. By scanning a single bead and switching between the excitation and the STED laser the two beams are brought into coincidence. (**d**) and (**e**) show *z*-profiles of the effective PSF of the excitation (**d**) and STED (**e**) beam. When the beams are brought into coincidence spatially the fluorescence quenching is measured by *z*-scans in a sea of fluorescence. (**f**) Excitation observed from scanning the excitation beam in the sea of fluorescence. (**g**) Repeating the scan with the STED beam superimposed on the excitation beam, fluorescence is efficiently reduced, here by around 75%. The sea of fluorescence is also used to determine the optimal delay of the STED pulses relative to the excitation pulses. When effective quenching is obtained in the sea of fluorescence, resolution is assessed by scanning fluorescent beads. (**h**) Fluorescent beads (diameter 40 nm) scanned in confocal mode. The frame is 3.5 μm wide. (**i**) The same frame imaged in STED mode with a clear resolution enhancement. (**j**) The line profiles of the (*dotted*) lines through two beads in the lower left of the frame reveal a resolution in STED mode of around 50 nm. Note that the FWHM values are calculated based on Gaussian fits through the line profiles, and the line profiles are based on 4-pixel wide lines.

The intensity and duration of the uncaging beam can be controlled by a Pockels cell and the acquisition software to allow for precise timing relative to the imaging and electrophysiology experiments.

## 3. Methods

### 3.1. General Alignment of Beams

It is essential that the laser beams are well aligned and well centered on all lenses and filters they go through. An optimal way of identifying the right beam paths is by backward alignment of the excitation, depletion and detection paths. To do this, the microscope objective is removed and a test diode laser beam is sent backward through the system, while all mirrors, filters and lenses are adjusted accordingly using pinholes to center the beam at all points along the optical path. Two or more pinholes are used as a reference for the optimal beam path, and a forward alignment is performed to route the respective beams through these pinholes. The fine alignment of detectors is done by scanning a fluorescent sample and adjusting to maximal signal (see Note 6). A shear-plate can be used to check for beam collimation, which is important for evaluating the beam after passing lenses and telescopes. The shear-plate only works for continuous wave (cw) lasers, and cannot be used with pulsed laser beams. However, some pulsed lasers (including the PicoQuant diode described earlier) have the option of switching between pulsed and cw operation. In any case, the shear-plate should be used during backward alignment to check the position of the lenses.

### 3.2. Making Doughnuts

The PSFs of the excitation and STED beams are visualized by scanning 150 nm gold beads in *x*–*y* and along the *z*-axis (Fig. 3a–e). Whereas the excitation beam usually does not require much tweaking, it can be challenging to create a good symmetric doughnut-shaped STED beam with a clear null (see Note 7) (Fig. 3b, c). A well-collimated and round beam passing through the vortex phase mask is essential, and a polarizing beam splitter before this can be useful for selecting linearly polarized light. The circular $\lambda/4$-plate should be mounted on a turnable stand (around the vertical axis of the plate), as well as in a rotation mount (rotation around the beam axis) between the phase mask and the objective (Fig. 3e). The turnable stand is important for shaping the doughnut and for avoiding back-reflections into the detectors and into the laser source. Experimenters should visually inspect that the interference null is centered in the STED beam immediately after the phase mask. If the doughnut is not symmetric, experimenters can try rotating the $\lambda/4$-plate to obtain maximum symmetry, and tilt it while maintaining the optimal rotation. These two steps can be iteratively applied (see Note 8). The symmetry of the doughnut in the planar and axial planes is optimized by adjusting the phase mask position and the $\lambda/4$-plate position on the fly while scanning a gold bead (Fig. 3b, e).

When symmetric doughnuts appear, the two PSFs are superimposed by moving the STED beam onto the excitation beam on the piezo-adjustable mirror. For alignment in the $z$-axis, the lens that collimates the STED beam as it emerges from the STED fiber is adjusted (Fig. 3d, e). The excitation beam path is usually more stable and it is a better choice as a reference beam than the STED beam, which is therefore chosen as the movable beam.

### 3.3. Two-Color STED Imaging

Several schemes have been described for two-color imaging of live cells, involving either additional lasers to address each fluorophore separately (27), or a single pair of lasers for excitation and fluorescence quenching. The latter approach, which is less complex and expensive, is based on discriminating the fluorophores either on the excitation side by using photo-switchable fluorophores (28), or on the emission side by spectral detection and linear unmixing (15). To image GFP and YFP by spectral detection a 514 nm long-pass filter can be used to split up the signals into two channels (Fig. 4b). The cross talk between the detection channels can be effectively diminished (<10%) by treating the images with linear unmixing algorithms, which can be easily implemented using ImageJ plugins.

### 3.4. Evaluating Microscope Performance

The microscope performance can be evaluated once a good doughnut-shaped STED beam PSF is obtained and aligned to the excitation beam in $x$, $y$, $z$. Two closely related parameters are of interest, namely *fluorescence quenching* and *resolution*. A saturating level of fluorescence quenching is necessary, but not sufficient, for achieving superresolution. The level of excited state depletion is readily evaluated by performing a 10–20 μm $z$-scan at the interface of the glass bottom of the chamber holding a drop of fluorophore solution (a sea of fluorescence).

Fluorescence quenching is expressed as the ratio between fluorescence obtained with and without the STED beam on (Fig. 3f, g). The sea of fluorescence is also used to find the optimal STED pulse delay relative to the excitation pulses and for aligning detectors. Depletion is most objectively reported without the phase mask in the beam, though for practical work the phase mask is usually left in place, and *doughnut depletion* is used as readout of performance. For the setup described here, doughnut depletion is approximately 10% lower than depletion without the null in the PSF. Useful reference values for STED power versus fluorescence quenching are presented in (15, 20).

The resolving power of the microscope is assessed by scanning fluorescent beads that are smaller than the achievable resolution.

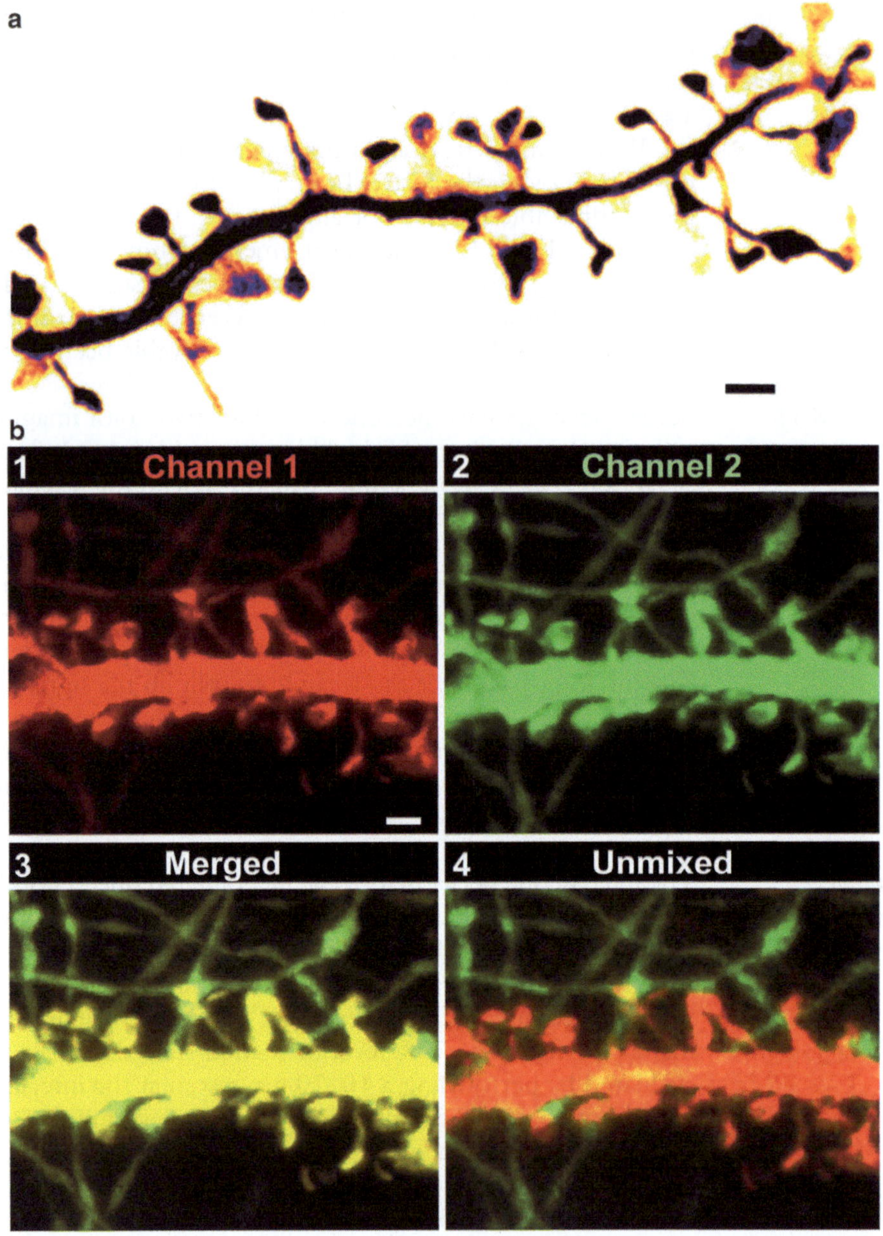


Fig. 4. Live-cell imaging of neuronal morphology in brain slices. (**a**) STED image of a dendritic segment acquired deep inside a living organotypic cultured brain slice, using Lifeact-YFP as a volume label, as described in ref. 14. (**b**) Two-color STED image of YFP labeled axons (*green*) forming synapses with spines on a dendrite labeled with Alexa Fluor 488 (*red*) in a living organotypic brain slice culture. Panels 1 and 2 show the raw images recorded in detection channels 1 and 2, respectively. Panel 3 shows the merged image from both channels, and panel 4 shows the result of unmixing on color separation, as described in ref. 15. Scale bars are 1 μm.

The full-width at half maximum (FWHM) of line profiles through the beads is a measure of the spatial resolution of the system (Fig. 3h–j). Generally 2- to 4-pixel thick lines should be used to avoid contributions of single outlier pixels (see Note 9).

### 3.5. Imaging Living Brain Slices

High resolution on beads usually translates into high resolution in live slices. However, several important issues must be considered when imaging live slices (see Note 10). Two general concerns in fluorescence microscopy are bleaching and phototoxicity, calling for experimenters to reduce excitation and STED beam powers as much as possible. Generally, when using freely diffusible volume labels in live neurons, bleaching is not an issue at moderate rates of STED time-lapse imaging (10). At higher acquisition rates, or when using higher excitation and depletion powers, photodynamic damage may become an issue (as evidenced for instance by membrane blebbing). In our experience, we observe less damage when using fluorescent proteins as compared with organic dyes. Bleaching and/or deleterious effects of imaging can be minimized in several ways, including reduced laser powers for excitation and depletion, imaging smaller frames, increasing pixel-size, decreasing pixel dwell-time or adding antioxidants, such as vitamin E (29), to the perfusion medium. Adjusting acquisition settings should be optimized on the fly depending on the size and observed fluorescence of the structure of interest.

For two-color STED imaging with a single beam pair and spectral detection, it is usually not necessary to change acquisition parameters, but obviously settings such as laser power and dwell time should be optimized for the given conditions.

When positioning the glass coverslip with the slice culture on the glass bottom of the perfusion chamber, care should be taken to have a clean interface, as any extra volume of liquid between these caused by trapped particles may be a source of aberrations, especially if the interface is not even. A light spring or lever can be used to clamp the glass coverslip down from the top, keeping it in position and minimizing the liquid interface between the coverslip and the chamber bottom. If imaging acute slices, they can be placed directly on the perfusion chamber bottom, which provides a single glass interface and reduces aberrations (20).

Integrating patch-clamp electrophysiology with STED imaging can be readily accomplished as the optics of the STED setup do not interfere with the movable microscope stage, and vice versa. Keeping in mind that small movements in the chamber caused by the heating system potentially distort images adjustments of the perfusion speed, perfusion flow path, bath volume or feedback settings of the heating system may be necessary. Some mechanical movement and drift between frame acquisitions can be corrected offline, e.g., using ImageJ plugins for image registration (30).

### 3.6. Image Analysis

STED microscopy provides direct optical superresolution without any need for offline resolution enhancing steps, and the general data processing does not differ from existing scanning microscopy techniques. To confirm that the structure of interest is indeed in focus, it is recommended to acquire a *z*-stack of images. Several free or commercially available software packages, including ImageJ, ImSpector, Matlab and others, are available for image analysis. However, user-defined scripts or programming may be necessary for automated analysis. Line profile plots through structures of interest are used to derive the FWHM, as described above for imaging beads.

To adhere to the Nyquist sampling theorem, the pixel size should be less than half the spatial resolution, which means around 20 nm per pixel.

## 4. Notes

1. The glycerine objective, which has a correction ring, allows for deeper imaging than the oil objective, but can be more tricky to work with. The back aperture of the oil objective is smaller than that of the glycerine objective (around 7 mm versus 10 mm, respectively), which makes it easier to backfill it. However, if enough STED power is available, one can use the wider beam required for using the glycerine objective for both, and simply switch between the two objectives with no further adjustments than minor routine beam alignment.
2. Here we present a single depletion wavelength value, however, an advantage of the OPO system is the tunable output wavelength, which potentially allows working with spectrally distinct fluorophores, provided that the optical filters on the setup are adjusted accordingly.
3. The excitation beam can serve as a very useful stable reference for the STED beam along their shared path. Make sure to have an excitation beam path that is centered on the optical components and perfectly hitting several permanently installed pinholes. These pinholes will provide a valuable tool for rough alignment of the STED beam and for the optional two-photon uncaging beam, as well as a backup reference for the excitation beam path.
4. Imaging through two glass interfaces with aqueous solution between them can be tricky due to differences in refraction indices. Outcomes may vary from slice to slice, or even region to region within a slice. If high resolution is obtained on fluorescent beads but not in tissue, the interfaces could well be the problem.
5. On an inverted setup, due to the limited working distance of the high NA objectives used, it may not be possible to get the surface of the slice into focus, making it difficult to see and steer

a patch-clamp pipette for recording or dye loading coming from above. However, it can quickly become routine to bring the patch pipette into a focal plane within the tissue, by moving the pipette from side to side above the slice and looking at its shadow through the slice, while carefully getting it into focus further into the tissue.

6. For a STED setup built for detecting green and yellow fluorophores it is a good idea to use a green laser diode for backward alignment, as this will pass filters and enter into the detection path. It may then be necessary to remove the STED beam band-pass cleanup filter to align all the way back to the STED fiber exit. When aligning backward using a 3–5 mW green laser, it is not possible to align backward on the other side of the STED fiber, as the light intensity is very low after passing the fiber. However, the green light will be visible by looking directly into the fiber, and this light is useful for aligning the beam onto the collimating lens that meets the STED beam right after the fiber in the forward direction.
7. Looking at STED beam reflections on gold beads, it is important to achieve a high-quality doughnut with a clean center. A contrast ratio of less than 5% between the center and the rim of the doughnut is necessary for acquiring crisp superresolved images in live tissue. Individual gold beads may behave differently, and it is advisable to try multiple ones before drawing any conclusions (this should also be kept in mind when optimizing the doughnut shape). It is acceptable to trust the best observed result, as false-positive good doughnuts are unlikely.
8. A $\lambda/2$-plate can be placed in front of the $\lambda/4$-plate and used to rotate the beam axially, which may help to obtain a good doughnut (Fig. 3d). Caution should be exercised when tilting the $\lambda/4$-plate to avoid possible back reflections of the beams, which may appear as sudden large increases in detection signal intensity.
9. A Gaussian fit through the line profile provides a more objective measure for the FWHM than relying on the plotted observed values (15). The imageJ plugin Sparkmaster is useful for assessing the FWHM of a large number of beads in a single frame, and thus for objectively evaluating the microscope´s performance (31).
10. In case the objective comes with a correction ring, switching from beads to live slices will almost certainly call for adjustments of the ring. To mimic more closely the settings of imaging in tissue, one can make a 3D preparation of gold and fluorescent beads in a 4% agar suspension.

## References


1. Toomre D, Bewersdorf J (2010) A new wave of cellular imaging. Annu Rev Cell Dev Biol 26:285–314
2. Hell SW (2007) Far-field optical nanoscopy. Science 316:1153–1158
3. Abbe E (1873) Beiträge zur Theorie des Mikroskops und der mikroskopischen Wahrnehmung. Archiv für Mikroskopische Anatomie 9:413–418
4. Hell SW, Wichmann J (1994) Breaking the diffraction resolution limit by stimulated emission: stimulated-emission-depletion fluorescence microscopy. Opt Lett 19:780–782
5. Klar TA, Hell SW (1999) Subdiffraction resolution in far-field fluorescence microscopy. Opt Lett 24:954–956
6. Klar TA, Jakobs S, Dyba M, Egner A, Hell SW (2000) Fluorescence microscopy with diffraction resolution barrier broken by stimulated emission. Proc Natl Acad Sci USA 97:8206–8210
7. Sieber JJ, Willig KI, Heintzmann R, Hell SW, Lang T (2006) The SNARE motif is essential for the formation of syntaxin clusters in the plasma membrane. Biophys J 90:2843–2851
8. Westphal V, Rizzoli SO, Lauterbach MA, Kamin D, Jahn R, Hell SW (2008) Video-rate far-field optical nanoscopy dissects synaptic vesicle movement. Science 320:246–249
9. Hein B, Willig KI, Hell SW (2008) Stimulated emission depletion (STED) nanoscopy of a fluorescent protein-labeled organelle inside a living cell. Proc Natl Acad Sci USA 105:14271–14276
10. Nägerl UV, Willig KI, Hein B, Hell SW, Bonhoeffer T (2008) Live-cell imaging of dendritic spines by STED microscopy. Proc Natl Acad Sci USA 105:18982–18987
11. Urban NT, Willig KI, Hell SW, Nägerl UV (2011) STED nanoscopy of actin dynamics in synapses deep inside living brain slices. Biophys J 101:1277–1284
12. Rittweger E, Han KY, Irvine SE, Eggeling C, Hell SW (2009) STED microscopy reveals crystal colour xcentres with nanometric resolution. Nat Photonics 3:144–147
13. Harke B, Keller J, Ullal CK, Westphal V, Schonle A, Hell SW (2008) Resolution scaling in STED microscopy. Opt Express 16:4154–4162
14. Urban NT, Willig KI, Hell SW, Nagerl UV (2011) STED nanoscopy of actin dynamics in synapses deep inside living brain slices. Biophys J 101:1277–1284
15. Tønnesen J, Nadrigny F, Willig KI, Wedlich-Soldner R, Nagerl UV (2011) Two-color STED microscopy of living synapses using a single laser-beam pair. Biophys J 101:2545–2552
16. Willig KI, Harke B, Medda R, Hell SW (2007) STED microscopy with continuous wave beams. Nat Methods 4:915–918
17. Klar TA, Engel E, Hell SW (2001) Breaking Abbe's diffraction resolution limit in fluorescence microscopy with stimulated emission depletion beams of various shapes. Phys Rev E Stat Nonlin Soft Matter Phys 64:066613
18. Wildanger D, Medda R, Kastrup L, Hell SW (2009) A compact STED microscope providing 3D nanoscale resolution. J Microsc 236:35–43
19. Engel E, Huse N, Klar TA, Hell SW (2003) Creating lambda/3 focal holes with a Mach-Zehnder interferometer. Appl Phys B 77:11–17
20. Ding JB, Takasaki KT, Sabatini BL (2009) Supraresolution imaging in brain slices using stimulated-emission depletion two-photon laser scanning microscopy. Neuron 63:429–437
21. Rankin BR, Moneron G, Wurm CA, Nelson JC, Walter A, Schwarzer D, Schroeder J, Colon-Ramos DA, Hell SW (2011) Nanoscopy in a living multicellular organism expressing GFP. Biophys J 100:L63–L65
22. Imspector.www.max-planck-innovation.de/de/industrie/technologieangebote/software/
23. Zipfel WR, Williams RM, Webb WW (2003) Nonlinear magic: multiphoton microscopy in the biosciences. Nat Biotechnol 21:1369–1377
24. Gahwiler BH (1984) Slice cultures of cerebellar, hippocampal and hypothalamic tissue. Experientia 40:235–243
25. Denk W (1994) Two-photon scanning photochemical microscopy: mapping ligand-gated ion channel distributions. Proc Natl Acad Sci USA 91:6629–6633
26. Matsuzaki M, Kasai H (2011) Two-photon uncaging microscopy. Cold Spring Harbor Protocol, Cold Spring Harbor
27. Pellett PA, Sun X, Gould TJ, Rothman JE, Xu MQ, Correa IR, Bewersdorf J (2011) Two-color STED microscopy in living cells. Biomed Opt Express 2:2364–2371
28. Willig KI, Stiel AC, Brakemann T, Jakobs S, Hell SW (2011) Dual-label STED nanoscopy of living cells using photochromism. Nano Lett 11:3970–3973
29. Matsuzaki M, Honkura N, Ellis-Davies GC, Kasai H (2004) Structural basis of long-term potentiation in single dendritic spines. Nature 429:761–766
30. ImageJ. http://rsbweb.nih.gov/ij/.
31. Picht E, Zima AV, Blatter LA, Bers DM (2007) SparkMaster: automated calcium spark analysis with Image J. Am J Physiol Cell Physiol 293: C1073–C1081

# Chapter 6

# Super-Resolution Imaging by Localization Microscopy


**Dylan M. Owen, Astrid Magenau, David J. Williamson, and Katharina Gaus**



## Abstract

Photoactivated localization microscopy (PALM) and the related technique of Stochastic optical reconstruction microscopy (STORM) are super-resolution imaging methods based on the precise localization of single molecules. Instruments based on these techniques are now commercially available and are capable of generating images with lateral resolutions in the tens of nanometers range. Here, we give an overview of the current state of this technology including live-cell and 3D PALM and provide an in-depth protocol for performing PALM experiments in a fixed cell monolayer. This includes both the instrumentation/acquisition aspects and the data analysis required for generating quantitative, super-resolution data of molecular distributions. In this example, the system under investigation will be fixed HeLa cells transfected with the photo-switchable fluorescent protein PS-CFP2 targeted to the plasma membrane by fusion to the N-terminus of the protein kinase Lck.



**Key words:** PALM, STORM, Cluster analysis, Lck


## 1. Introduction

Fluorescence microscopy has become an indispensible tool in the life sciences due to its specificity, minimally invasive nature, and ability to image dynamic processes on spatial scales spanning from subcellular structures to whole organisms. Advances in hardware including lens and detector design and image processing have pushed the resolution of conventional microscopes to the diffraction limit of light—typically around 300 nm in the lateral dimension. A major goal in microscopy research has, therefore, been to circumvent this limit and observe smaller structures (1). One of the major successes of this search has been the development of localization microscopy such as PALM, STORM, and other related methods (2–4).


Alioscka A. Sousa and Michael J. Kruhlak (eds.), *Nanoimaging: Methods and Protocols*, Methods in Molecular Biology, vol. 950, DOI 10.1007/978-1-62703-137-0_6, © Springer Science+Business Media, LLC 2013

The resolution limit stems from the inability to distinguish closely spaced fluorophores whose point spread functions (PSFs) overlap when imaged. Localization microscopy overcomes this by temporally separating the fluorescence signals—essentially by only allowing one molecule per diffraction-limited area to fluoresce at a time. This can be achieved by using photo-activatable or photo-convertible fluorescent proteins in the case of PALM (2) or small molecule dyes in the case of STORM (3, 4). As fluorescent molecules are sparsely distributed, individual PSFs are imaged by a sensitive camera. Using the high signal to noise ratio afforded by total internal reflection fluorescence (TIRF) imaging (5), it is possible to calculate the center of these PSF with a precision in the tens of nanometers range so that the coordinates of well separated fluorescent molecules is recorded. Molecules are typically imaged until they bleach, allowing a new set of sparse fluorophores to be activated and the process repeated. In this way, a complete fluorescence image is built up one molecule at a time. Typically this requires several thousand raw frames to be recorded.

There are now a wide variety of probes suitable for PALM/STORM; for an in-depth discussion, the reader is referred to Chapter 9 of this book. The typical fluorescent proteins are the photo-convertible proteins PS-CFP2 and mEos2. PS-CFP2 in its native state is excited at 405 nm and fluoresces around 470 nm. However, the 405 nm illumination also causes photo-conversion to a switched state after which excitation is at 488 nm and emission around 510 nm. PS-CFP2 therefore displays blue-to-green switching. In the case of mEos2 the switching (again caused by 405 nm radiation) is green-to-red. In the native state, excitation is performed with 488 nm illumination with subsequent fluorescence in the region of 520 nm. The red converted state has excitation at 561 nm and emission around 590 nm.

For STORM, measurements can now be performed using conventional small organic fluorophores in the related method of direct (d)STORM (4, 6). Unlike the fluorescent proteins previously discussed, these do not photo-convert, but instead can be made to blink on and off by switching to a long-lived, dark triplet state. This process typically requires the use of special reducing and oxygen scavenging buffers. Examples of dyes that can be used in this way include (among others) Cy5 and Alexa647.

Once the raw image frames have been recorded, they must be analyzed to identify the single-molecule PSFs over background noise and calculate their center. A common way to do this is to assume the PSFs are 2D Gaussian distributions and to use the Gaussian mask approach developed by Thompson et al. (7). This essentially calculates the center of mass of the PSF after smoothing with an idealized Gaussian function. The process returns the center coordinate (position of the molecule) and the localization precision. In the case of $N$ recorded photons, background variance $b$,

PSF width *s*, and camera pixel size relative to the PSF width *a*, the localization precision is given by:

$$\text{Precision} = \sqrt{\left(\frac{s^2}{N}\right) + \left(\frac{\frac{a^2}{12}}{N}\right) + \left(\frac{8\pi s^4 b^2}{a^2 N^2}\right)}$$

The centers of the PSFs can also be determined by fitting them to 2D Gaussian distributions using a least-squares approach. This is computationally more intensive, and a number of solutions now exist to speed up the process.

Once the coordinates of the fluorescent molecules have been determined they must be displayed in the form of an image and analyzed. For display, each molecular coordinate is typically convolved with an artificial Gaussian PSF with a width equal to the localization precision. Recalling the Nyquist sampling criterion, it should be noted that the resolution of such an image not only depends on these precisions, but also on the density of localized molecules in the region of interest.

Quantitative analysis of the molecular coordinates often involves the search for and characterization of molecular clustering. Several statistical methods exist for such analysis of point-pattern data and one of the most commonly employed is Ripley's *K*-function (8). This method has long been used in the analysis of spatial point-patterns derived from electron microscopy data and has recently been applied to coordinates derived from PALM and STORM experiments (9–11). A variation on this technique is Getis and Francklin's *L*-function in which each molecule is assigned a value representing how clustered it is, rather than the analyzed region as a whole (12). This allows the generation of "cluster maps" from which various cluster parameters can be derived (10, 11).

To generate images with the highest possible signal to noise ratios, PALM is typically performed in fixed cells imaged under TIRF illumination which limits the excitation volume to within 100 nm of the coverslip. This affords the highest possible resolution as the number of collected photons can be maximized by increasing the integration time, the background from out of focus light is minimized and the recorded molecular density kept high by imaging very large numbers of raw frames. This does, however, remove dynamic information from the data set and only allows molecular coordinates to be determined in 2D.

Recently, PALM has begun to be applied in live cell systems and in three dimensions. For live cell imaging, the integration time is typically reduced to increase the frame rate. Despite the resulting degradation in spatial resolution, this has allowed dynamic processed to be observed with time resolutions from several seconds to a minute (13).

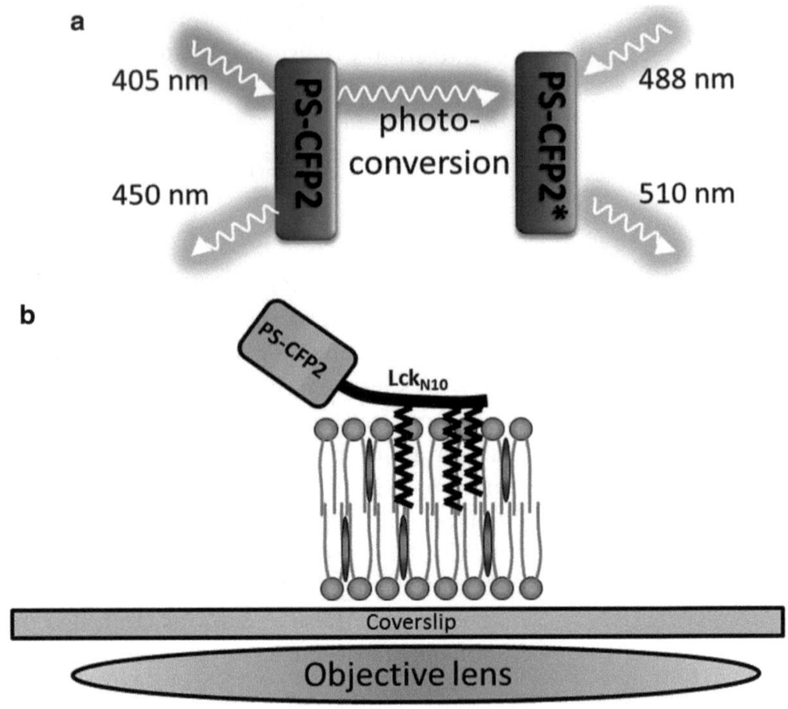


Fig. 1. (**a**) Scheme of the photo-switching behavior of PS-CFP2. The protein is initially excited at 405 nm and fluoresces in the blue region of the spectrum. 405 nm irradiation causes switching to a green fluorescent form. (**b**) Scheme of the sample setup. PS-CFP2 is fused to the 10 amino acid N-terminal of Lck, which contains sites for three posttranslational lipid modifications that target the construct to the cell plasma membrane.

In terms of 3D PALM, a number of approaches have been developed with various degrees of resolution and imaging depth and range. One approach is Biplane PALM (14). Here, two different focal planes are imaged simultaneously and the appearance of the molecule in each can be used to determine its *z*-position. This has been used to generate a *z*-resolution of 75 nm over a range of 1 μm. Another approach involves the use of a cylindrical lens to introduce optical astigmatism which causes the imaged PSFs to be elliptical (15). The aspect ratio of the ellipse contains information on the *z*-position of the particle and has been demonstrated with 50 nm *z*-resolution over 600 nm. Other methods to extract depth information are iPALM and PSF engineering, which are more difficult to implement (16).

In this chapter we will give a protocol for the 2D super-resolution imaging and subsequent quantitative cluster analysis of molecular distributions in the plasma membrane of fixed HeLa cells. The biological system under investigation will be the photo-switchable fluorescent protein PS-CFP2 fused to the 10 amino acid N-terminus of the T cell protein kinase Lck and subsequently transfected into HeLa cells (Fig. 1 a, b). The 10 amino acid sequence is posttranslationally modified with three fatty acids. These serve to target the fusion protein to the cell plasma membrane and more specifically to highly ordered lipid domains (lipid rafts) (10, 17, 18). These lipid rafts are thought to regulate the clustering and dynamics of membrane proteins and exist on tens of nanometers length scales, being therefore, not resolvable by conventional fluorescence microscopes.

## 2. Materials

### 2.1. Reagents

1. HeLa cells (American Type Culture Collection (ATCC)).
2. Full DMEM media: DMEM, 10% FCS, 2% L-glutamine.
3. Trypsin + EDTA.
4. 0.22 μm filtered phosphate buffered saline (PBS).
5. Lipofectamine LTX and OptiMEM (Invitrogen).
6. DNA plasmid ($Lck_{N10}$-PS-CFP2).
7. 100 nm diameter gold beads as fiducial markers.
8. 18 mm diameter Chamlide microscopy chamber (LCI Live Cell Instrument) and compatible #1.5 (0.17 mm) TIRF-suitable glass coverslips.
9. 16% paraformaldehyde solution. Dilute down to 4% using 1× DPBS.

### 2.2. Equipment

This chapter assumes the user is familiar with basic instrumentation for widefield and TIRF microscopy.

1. An inverted TIRF microscope capable of simultaneous excitation at 405 and 488 nm. It is recommended these wavelengths be provided by diode-pumped, solid-state lasers (DPSSLs). The imaging laser (488 nm) should have an output power at least in the 50 mW range and be delivered to the microscope via a single-mode optical fiber. The microscope should have a sensitive, cooled EMCCD camera with at least 256 × 256 pixels and be capable of imaging frame rates in the 2–5 ms range. The total magnification of the system should be such that the camera pixel size at the sample plane is around 100 nm. PALM of PS-CFP2 requires dichroics and filters for detection of fluorescence in the range of approximately 500–600 nm while exciting at 405 nm and 488 nm simultaneously. A similar method can be employed for mEos2 if a 561 nm laser is available for imaging the converted form. The microscope should be fully enclosed inside a light-proof, temperature-controlled incubation chamber.
2. Objective capable of TIRF illumination. TIRF requires a numerical aperture (NA) of at least 1.45 in order to generate

the high angle rays that will be totally internally reflected from the sample. The lens will therefore be oil-immersion and have a magnification in the range of 60–100 times and is typically specially designed for TIRF excitation.

3. Computer that can handle the large data sets and rapid data transfer rates—typical PALM datasets are around 5 GB in size. PALM processing and cluster analysis is computationally intensive so a powerful computer with Matlab (Mathworks Inc.) installed is recommended. A typical configuration is at least a 1 Tb hard—drive, 3 GHz quad—core processor and at least 16 GB of fast RAM.

## 3. Methods

### 3.1. Sample Preparation

1. Add 2 μL Lipofectamine LTX reagent to 100 μL OptiMEM in an Eppendorf tube. Add 1 μg of plasmid DNA (typically this will be ~1 μL in volume) to the mixture. Shake the tube and incubate at 37°C for 30 min (see Note 1).
2. Meanwhile, discard media from cell culture flask and wash the HeLa cells with 10 mL of 1× DPBS. Add 1 mL of 1× Trypsin–EDTA solution and rock the flask gently. Incubate at 37°C for 5 min until the cells detach. Resuspend the cells in 10 mL fresh, full media (DMEM + 10% FCS + 2% L-Glutamine).
3. Place TIRF suitable #1.5 (0.17 mm thickness) 18 mm diameter glass coverslips into a 12 well plate and wash three times with 0.22 μm filtered DPBS.
4. Add 1 mL of cell suspension (~ $0.5 \times 10^6$ cells) to each well. Rock the plate gently to ensure an even distribution.
5. Dropwise, add the Lipofectamine-OptiMEM-DNA solution to the well. Rock the plate gently to ensure an even distribution. Incubate for 24 h at 37°C in a 5% $CO_2$ atmosphere.
6. Remove the media from the well and discard. Wash the cells with 1× warm DPBS. Add 500 μL of 4% paraformaldehyde (PFA) in 1× DPBS. Incubate at 37°C for 15 min. Remove the PFA and wash three times with DPBS.
7. Mount the coverslip in an 18 mm diameter Chamlide microscope chamber. Add 20 μL of 100 nm gold colloidal bead solution and leave for 2 min to settle. Gently add 300 μL of 0.22 μm filtered DPBS.

### 3.2. Microscope Setup and Data Acquisition

1. Turn on all imaging equipment. Activate lasers at 405 and 488 nm. Set the camera (EMCCD) temperature to −60°C and the EM-gain to maximum (see Note 2).

2. Place a drop of immersion oil on the objective and place the sample on the microscope stage. Use the brightfield lamp and the eye-pieces to locate and focus on the cells.
3. Switch the detection channel to the camera. Using an integration time of 30 ms and 405 nm laser power of ~10 μW (at the sample) locate a transfected cell and focus on the basal membrane. Ensure that several of the gold bead fiducials are visible in the same field of view (see Note 3).
4. Switch the microscope to TIRF illumination mode, switch off the 405 nm laser, turn the excitation wavelength to 488 nm (typically ~15 mW at the sample) and the detection band to the green range (using filters which still allow simultaneous excitation at 405 nm). Fluorescence should still be visible due to previous photo-conversion by the 405 nm laser but this should quickly bleach (see Note 4).
5. Set the microscope to record a time series of roughly 15,000 frames and begin recording. Very few PSFs should be observed, which blink with a lifetime of just a few frames (see Note 5).
6. Turn on the 405 nm laser at a very low power (<10 μW at the sample). The density of PSFs detected by the camera should increase. It is important to keep this density low such that there are very few overlapping PSFs.
7. As the number of viable molecules is depleted over the course of the acquisition, it may be necessary to increase the power of the 405 nm converting later to increase the density of recorded PSFs. It may also be necessary to occasionally re-focus the microscope on the blinking molecules to compensate for drift (see Notes 6 and 7).
8. Save the data in a format compatible with the available data processing software.

### 3.3. Data Analysis

1. To detect single-molecule fluorescence events over background, calculate the center of these PSFs and, if possible, the precision of localization. There are now a number of different software packages available for obtaining single-molecule coordinates from time-series images. The following discussion gives some examples but is not exhaustive. A commercial PALM microscope is available from Zeiss (Elyra PS.1), which uses Zeiss Zen software to detect and centroid the PSFs. This has options to use the theory developed by Thompson et al. (7) and also full 2-D Gaussian fitting. This software also calculates the localization precision of the generated coordinates and can compensate for sample drift by tracking the immobile gold fiducials. Zen outputs a text file containing a list of the $x$ and $y$ coordinates and the precision of all detected molecules in the time-series. A freely available plug-in for ImageJ called QuickPALM is also

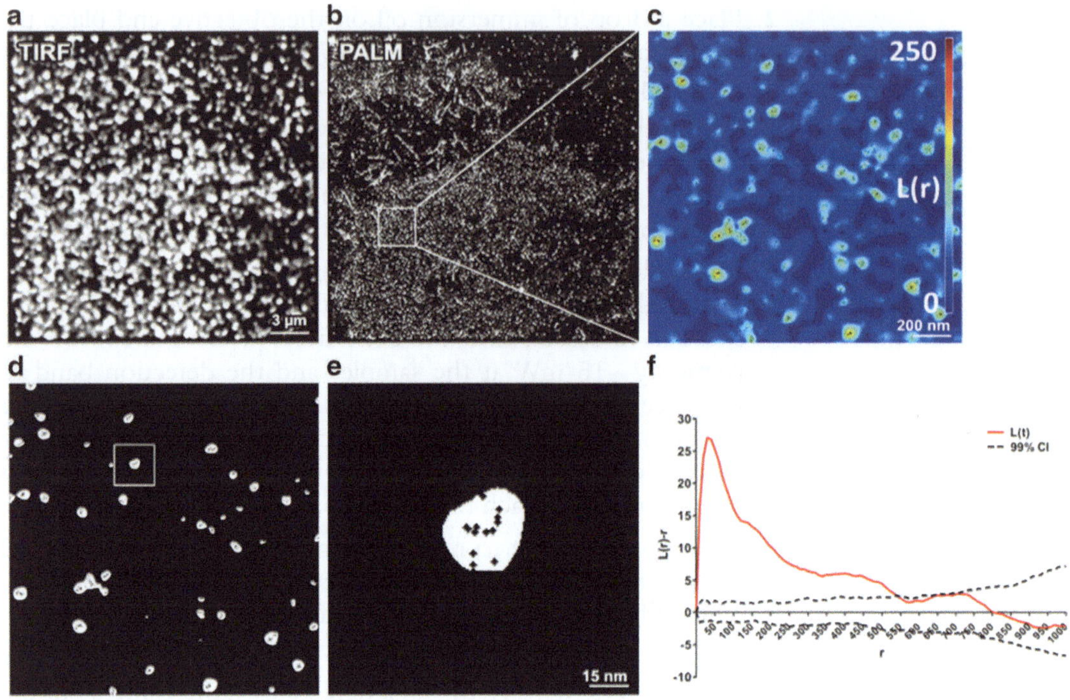


Fig. 2. (**a**, **b**) Examples of a TIRF image (**a**) and the higher spatial resolution offered by PALM (**b**). (**c**) A quantitative, color-coded cluster map generated from G&F's *L*-functions using *r* = 50 nm. Here, red values indicate local clustering above that expected of a random distribution. Since L(50) = 50 in the random case, values of L(50) = 250 have $(250/50)^2 = 25$ times the local molecular density than would be expected had the points in the ROI been randomly distributed (see the equation in step 5). (**d**, **e**) Binary cluster maps generated by thresholding the color-coded map at a user-defined value of L(50) which can then be used to extract cluster parameters. (**f**) Example of a Ripley's *K*-function curve with 99% confidence intervals indicating peak clustering in the 50–100 nm range.

available (19). This operates on a saved TIFF image sequence and outputs a list of the *x* and *y* coordinates in MS Excel format. New solutions are also being developed such as DAOSTORM which is capable of calculating molecular coordinates even if several PSFs overlap in any given frame as well as algorithms optimized for accuracy or speed (20–22) (see Note 8). Figure 2a, b show representative TIRF and PALM images respectively.

2. Select a representative region of interest (ROI): Typically the whole acquired field of view is not analyzed. This is because the quantitative cluster analysis assumes the analyzed regions do not contain cell edges etc. A region of interest for further analysis must, therefore, be selected. This is typically a square area 2–5 µm across. The optimal size will depend on the density of molecules within the region. Good statistical cluster analysis typically requires around 100–1,000 localized molecules per square micron. The ROI can be chosen by viewing a scatter plot of the *x*–*y* coordinates (e.g., in Matlab). The data set should be cropped to this ROI and points that lie outside discarded (see Note 9).

3. Calculate the Ripley's $K$-function for the ROI: Ripley's $K$-function measures the degree of clustering of a spatial point-pattern relative to a random distribution by counting the number of encircled points within concentric circles centered on each point. The $K$-function ($K(r)$) at a radius $r$, for an area $A$, containing $n$ points is calculated as:

$$K(r) = A\sum_{i=1}^{n}\sum_{j=1}^{n}\left(\frac{\delta_{ij}}{n^2}\right) \quad \text{where } \delta_{ij} = 1 \text{ if } \delta_{ij} < r\text{, otherwise } 0$$

where $\delta_{ij}$ is the distance between points $i$ and $j$. For a random distribution this scales linearly with area. Linearize the $K$-function to generate the $L$-function using:

$$L(r) = \sqrt{\frac{K(r)}{\pi}}$$

Such that it scales linearly with radius.

4. Plot the value *of* $L(r)$-$r$ vs. $r$. For a spatially random distribution, $L(r)$-$r$=0 at all spatial scales. Positive $L(r)$-$r$ values indicate clustering above what would be expected of a random distribution. The greater the value of $L(r)$-$r$ at any given $r$, the greater the degree of clustering at that spatial scale. Confidence intervals (e.g., 99%) can be generated by simulating randomly distributed points with the same number and region size as the data. Depending on the number of points, 100–500 random distributions should be generated and analyzed to generate reliable statistics. Figure 2f shows a representative Ripley's K curve with 99% confidence intervals.
5. Calculate the Getis and Franklin's (G&F's) $L$-function for local point pattern analysis: While the Ripley's $K$-function produces a mean value for the ROI as a whole, G&F's L generates a value for the clustering level of each point individually. Essentially, this counts the number of points encircled by a circle of a specific radius centered on that point (relative to the value expected of a random distribution). Choose a radius (cluster scale) with which to interrogate the pattern for clustering. Appropriate values might be the size of expected clusters, or the spatial scale at which the Ripley's $K$-function curve peaks. Calculate G&F's $L$-function for every molecule, $i$, at that radius (say, for example, 50 nm) using:

$$L_i(50) = \sqrt{\frac{A\sum_{j=1}^{n}\left(\frac{\delta_{ij}}{n^2}\right)}{\pi}} \quad \text{where } \delta_{ij} = 1 \text{ if } \delta_{ij} < 50\text{, otherwise } 0$$

6. Use the $x$ and $y$ coordinates and the G&F's $L_i(r)$ value for each point to generate a cluster heat map: This can be achieved in

several packages and programming languages but is simple in Matlab. First, generate 2 1D arrays covering the range of the $x$ and $y$ values of the ROI. Typically arrays with 10 nm resolution are sufficient. Use the meshgrid command to convert the arrays into a 2D grid on which to interpolate the surface. Use the griddata function to interpolate the surface from the coordinates, $L_i(r)$ values and the 2D grid. Plot the surface using the contour function (typically with the Jet color scale) and apply the desired cosmetic changes. Be careful to apply the maximum and minimum color values appropriately so as not to saturate the color scale. Save the plot as both a .mat file and as a TIFF file (Fig. 2c) (see Note 10).

7. Generate a binary map from the heat map: Convert the Jet color map to a binary map. Be careful to choose an appropriate level for the black–white transition. Save the plot as both a .mat file and as a TIFF file (Fig. 2d, e).
8. Calculate the fraction of molecules in clusters: Determine the value of $L_i(r)$ that corresponds to the black–white transition in the binary image. Count the number of molecules in the ROI that have a $L_i(r)$ greater than this value. Calculate the percentage of molecules in clusters as this value divided by the total number in the ROI, times 100.
9. Extract clustering parameters from the binary image: Open the binary image in ImageJ and crop out any border around the ROI. Convert the image to a binary image. Use the Analyse Particles function to extract the parameters from the ROI. The exact parameters extracted are determined by the Set Measurements function and can include, among others, the number of clusters, cluster size, and circularity (see Note 11).

## 4. Notes

1. See Invitrogen.com for additional protocols for different cell types for the use of lipofectamine transfection reagent. Other transfection procedures, such as, electroporation can also be used, depending on the cell type.
2. Turning on the microscope several hours before starting to image helps reduce drift caused by thermal changes during the lengthy acquisition. This is especially true if not using gold bead fiducial markers and drift compensation software. Using a full incubation chamber around the system is also advisable. Sample drift will reduce resolution and lead to artificially large clusters being detected.
3. 30 ms integration times are typically appropriate for fluorescent proteins such as PS-CFP2 and mEos2. It is possible to perform

live-cell PALM (on unfixed cells) by increasing the laser power and decreasing the camera integration time. This increases the time resolution, but typically results in a degradation of the localization precision.

4. A good way to determine if you have TIRF illumination is try to focus up onto the equatorial plane of the cells. Under TIRF no other plane other than the basal plane should be visible upon changing focus. If TIRF is not achievable, there may be a problem with the illumination alignment of the microscope or the sample may not be placed flat on the microscope stage.
5. Imaging can be continued until no more blinking molecules are visible in the field of view. Getting an accurate, quantitative measurement of the number of molecules in the field of view is difficult however. While it is possible to compensate for the repeated activation of fluorescent molecules (23, 24), the contribution of noise, the fraction of molecules detected above noise, the fraction of fluorescent proteins that correctly mature and so on are generally not known.
6. Higher levels of 405 nm laser illumination increase the rate of photo-activation but also lead to an increase in background due to cellular autofluorescence. For the highest localization precisions it is typically better to use low levels of 405 nm laser illumination and image for a greater number of frames in order to acquire sufficient molecules. It is also possible to pulse the 405 nm laser and acquire frames only when the laser is off, although this can lead to greater variation in the density of observed molecules and hence overlapping PSFs.
7. 2-channel localization microscopy is possible with sequential acquisitions. These might be 2-color PALM using PS-CFP2 and mEos2 or 2-color PALM and (d)STORM using one fluorescent protein and a red small molecule dye (e.g., Cy5) together with immunostaining.
8. Depending on the method and the specific software package employed, there are typically a number of parameters which must be set by the user in order to optimize the single molecule detection and localization process. One of the most critical is the signal-to-noise ratio required for an event to be counted as a molecule. Too high and the density of detected molecules will be too low. This degrades the final resolution (as resolution needs Nyquist sampling densities of molecules in the ROI) and makes generating good statistical data impossible. Too low and noise is counted as true events leading to artificially low determinations of the degree of molecular clustering.

   Note that most software packages output several other pieces of information on each event in addition to the *x*–*y* coordinates. One of the most useful to obtain is the localization precision of each molecule. If this is available to the user, we recommend "cropping" the data set by precision to exclude badly localized molecules which would otherwise mask clustering. In our experience, 50 nm precision is typically an appropriate level for this process. The frame in which each event was detected may also be saved. This can be used in live-cell PALM data as the time-axis on which to plot various clustering parameters.

9. The cluster analysis methods typically require edge-compensation for the ROIs because molecules located very close to the edge will appear less clustered as they have less nearby neighbors. While the ROI does not have to be square, square regions suffer less from edge effects than rectangles as the area/perimeter ratio is highest. Edge compensation can be achieved by analyzing a slightly larger region than desired and then cropping the generated cluster map, or by weighting edge points appropriately to compensate for this effect. Some point-pattern analysis packages such as SpPack offer a variety of edge correction options (25).
10. Matlab offers a range of interpolation algorithms for use with the griddata function. A simple linear interpolation is computationally quickest, however the "v4" option typically generates the most accurate results. This can, however, result in long processing times so a powerful computer system is recommended for the analysis.
11. Once all the parameters have been obtained they can be combined in various ways to generate new parameters such as "number of clusters per square micron," "average number of molecules per cluster," relative molecular density inside and outside of clusters' and so on.

## Acknowledgment

This work was supported by the Australian Research Council (ARC), National Health and Medical Research Council (NHMRC), and the Human Frontiers Science Program (HFSP).

### References

1. Schermelleh L, Heintzmann R, Leonhardt H (2010) A guide to super-resolution fluorescence microscopy. J Cell Biol 190: 165–175
2. Betzig E, Patterson GH, Sougrat R et al (2006) Imaging intracellular fluorescent proteins at nanometer resolution. Science 313: 1642–1645
3. Rust MJ, Bates M, Zhuang X (2006) Sub-diffraction-limit imaging by stochastic optical reconstruction microscopy (STORM). Nat Meth 3:793–796
4. Heilemann M, van de Linde S, Schüttpelz M et al (2008) Subdiffraction-resolution fluorescence imaging with conventional fluorescent probes. Angew Chem Int Ed 47:6172–6176
5. Axelrod D (2001) Total internal reflection fluorescence microscopy in cell biology. Traffic 2:764–774

6. Heilemann M, van de Linde S, Mukherjee A, Sauer M (2009) Super-resolution imaging with small organic fluorophores. Angew Chem Int Ed 48:6903–6908
7. Thompson RE, Larson DR, Webb WW (2002) Precise nanometer localization analysis for individual fluorescent probes. Biophys J 82: 2775–2783
8. Ripley BD (1977) Modelling spatial patterns. J R Stat Soc Series B Stat Methodol 39:172–192
9. Lillemeier BF, Mortelmaier MA, Forstner MB et al (2010) TCR and Lat are expressed on separate protein islands on T cell membranes and concatenate during activation. Nat Immunol 11:90–96
10. Owen DM, Rentero C, Rossy J et al (2010) PALM imaging and cluster analysis of protein heterogeneity at the cell surface. J Biophoton 3:446–454
11. Williamson DJ, Owen DM, Rossy J et al (2011) Pre-existing clusters of the adaptor Lat do not participate in early T cell signaling events. Nat Immunol 12:655–662
12. Getis A, Franklin J (1987) Second-order neigborhood analysis of mapped point patterns. Ecology 68:473–477
13. Shroff H, Galbraith CG, Galbraith JA, Betzig E (2008) Live-cell photoactivated localization microscopy of nanoscale adhesion dynamics. Nat Meth 5:417–423
14. Juette MF, Gould TJ, Lessard MD et al (2008) Three-dimensional sub-100 nm resolution fluorescence microscopy of thick samples. Nat Meth 5:527–529
15. Huang B, Wang W, Bates M, Zhuang X (2008) Three-dimensional super-resolution imaging by stochastic optical reconstruction microscopy. Science 319:810–813
16. Shtengel G, Galbraith JA, Galbraith CG, Lippincott-Schwartz J et al (2009) Interferometric fluorescent super-resolution microscopy resolves 3D cellular ultrastructure. Proc Natl Acad Sci 106:3125–3130
17. Simons K, Ikonen E (1997) Functional rafts in cell membranes. Nature 387:569–572
18. Pike LJ (2006) Rafts defined: a report on the keystone symposium on lipid rafts and cell function. J Lipid Res 47:1597–1598
19. Henriques R, Lelek M, Fornasiero EF et al (2010) QuickPALM: 3D real-time photoactivation nanoscopy image processing in ImageJ. Nat Meth 7:339–340
20. Holden SJ, Uphoff S, Kapanidis AN (2011) DAOSTORM: an algorithm for high- density super-resolution microscopy. Nat Meth 8: 279–280
21. Mortensen KI, Churchman LS, Spudich JA, Flyvbjerg H (2010) Optimized localization analysis for single-molecule tracking and super-resolution microscopy. Nat Meth 7:377–381
22. Smith CS, Joseph N, Rieger B, Lidke KA (2010) Fast, single-molecule localization that achieves theoretically minimum uncertainty. Nat Meth 7:373–375
23. Annibale P, Vanni S, Scarselli M, Rothlisberger U, Radenovic A (2011) Quantitative photo activated localization microscopy: unraveling the effects of photoblinking. PLoS One 6:e22678
24. Annibale P, Scarselli M, Kodiyan A, Radenovic A (2010) Photoactivatable fluorescent protein mEos2 displays repeated photoactivation after a long-lived dark state in the Red photoconverted form. J Phys Chem Lett 1:1506–1510
25. Perry G (2004) SpPack: spatial point pattern analysis in excel using visual basic for applications (VBA). Environ Modell Softw 19:559–569

# Chapter 7

## High-Content Super-Resolution Imaging of Live Cell by uPAINT

**Grégory Giannone, Eric Hosy, Jean-Baptiste Sibarita, Daniel Choquet, and Laurent Cognet**

### Abstract

In this chapter, we present the uPAINT method (Universal Point Accumulation Imaging in Nanoscale Topography), a simple single-molecule super-resolution method which can be implemented on any wide field fluorescence microscope operating in oblique illumination. The key feature of uPAINT lies in recording high numbers of single molecules at the surface of a cell by constantly labeling while imaging. In addition to generating super-resolved images, uPAINT can provide dynamical information on a single live cell with large statistics revealing localization-specific diffusion properties of membrane biomolecules. Interestingly, any membrane biomolecule that can be labeled with a fluorescent ligand can be studied, making uPAINT an extremely versatile method.

**Key words:** Super-resolution microscopy, Single molecule detection, High-density single particle tracking, Nanoscopy, Membrane biomolecule dynamics

## 1. Introduction

By pushing optical resolutions down to the scale of the molecules themselves, super-resolution optical microscopy techniques have revolutionized biomolecular imaging in cells (1, 2). Super-resolved images obtained from single molecule detections are in fact the collection of the super-localizations of single emitters imaged successively at low density on the specimen of interest. This is usually performed by iteratively activating a small number of photo-activable fluorophores among a dense population of inactivated ones. Such approaches inherently require photo-activable or photo-switchable fluorophores which can be either engineered fluorescent proteins (PALM (3, 4)) or dyes immersed in reducing agents to allow the photo-conversion (STORM (5)). Importantly, these methods have difficulties to access very densely labeled regions due

Alioscka A. Sousa and Michael J. Kruhlak (eds.), *Nanoimaging: Methods and Protocols*, Methods in Molecular Biology, vol. 950, DOI 10.1007/978-1-62703-137-0_7, © Springer Science+Business Media, LLC 2013

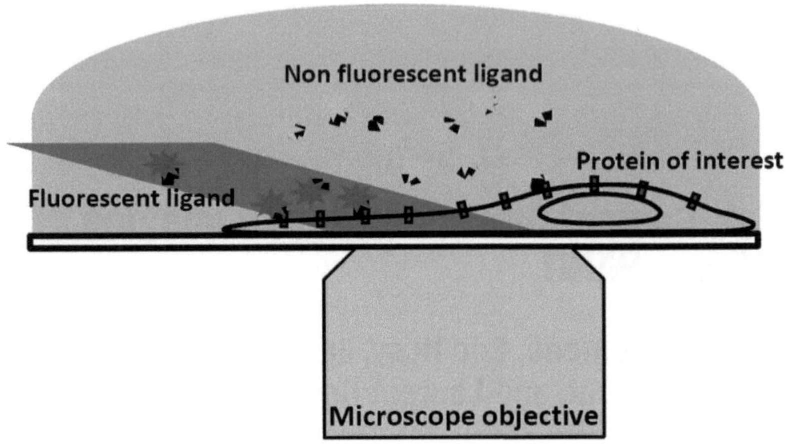


Fig. 1. The principle of uPAINT is based on the real-time imaging and tracking of single fluorescent ligands while they label their membrane biomolecules targets. A low concentration of fluorescent ligands is introduced in the extracellular medium such that a constant rate of membrane molecules is being labeled during the imaging sequence. Oblique illumination of the sample is used to excite predominantly fluorescent ligands, which have bound to the cell surface while not illuminating the molecules in the above solution.

to spontaneous photoswitching, which prevents imaging individual fluorophores in these regions (6). Furthermore, up to now, neither PALM nor STORM allow the versatile study of endogenous molecules at nanometer resolutions on live cells (2). For this purpose, the universal PAINT (uPAINT) method (7), which generalizes the Point Accumulation Imaging in Nanoscale Topography, was developed (8). uPAINT does not require the photoactivation of single molecules and is instead based on the real time imaging and tracking of single fluorescent ligands while they label their membrane biomolecular targets (Fig. 1). In other words, the single emitters are imaged at low density upon labeling as opposed to other single molecule based super-resolution methods where the samples are labeled prior to imaging at high density and stochastic photoactivation is used during the imaging sequence.

One of the key features of uPAINT is to image in real time single fluorescent ligands binding to their target. For this purpose, high affinity luminescent ligands are placed in the recording medium at the appropriate concentration under the microscope so that single molecules can be imaged and their trajectories can be isolated. Oblique illumination is required in order to selectively excite fluorescent ligands that have bound to their target and not those present in solution above the cellular membrane. Akin to any single molecule based super-resolution method, image analysis is an important step. Furthermore, as thousands of long lasting trajectories can be obtained on single cells with uPAINT, the amount of information generated per cell is considerable. It thus requires adequate dynamic analysis and visualization strategies to be handled. The following sections describe each of these steps and discuss the protocols in order to successfully achieve results from uPAINT experiments.

## 2. Materials

There are different types of equipment needed for successful uPAINT experiments. They are used for labeling, acquisition and image processing. Examples of equipment and their suppliers used in our laboratories are listed below.

### 2.1. Reagents

1. Fluorescence free immersion oil (FF Cargille).
2. Fluorescent Tetraspec beads (Invitrogen) for image registration.
3. Protein labeling kit for obtaining fluorescent ligands, e.g., antibodies coupled to NHS ester Atto dyes (ATTO-TEC GmbH) or NHS ester Cy3/Cy5 dyes (Amersham CyDye™ Antibody Labeling Kits, GE Healthcare Life Sciences).
4. Salt based recording medium: 100 mM NaCl, 5 mM KCl, 10 mM Glucose, 2 mM $CaCl_2$, 2 mM $MgCl_2$, at pH 7.4 (Hepes 20 mM/NaOH).
5. Ringer medium: 150 mM NaCl, 5 mM KCl, 2 mM $CaCl_2$, 2 mM $MgCl_2$, 10 mM HEPES, 11 mM Glucose, at pH 7.4 (if performing exogenous labeling; see Subheading 3.3).
6. Trypsin/EDTA (if performing exogenous labeling, see Subheading 3.3).
7. Phosphate Buffer Solution (PBS).
8. Bovine Serum Albumin (BSA).
9. Dimethyl Sulfoxide (DMSO).
10. Gel filtration column filled with G-25 Sephadex (Sigma-Aldrich).
11. Dulbecco's Modified Eagle Medium (DMEM).
12. Fetal Bovine Serum (FBS).
13. Rat collagen I or human fibronectin.
14. Poly-lysine solution.
15. Amaxa Nucleofactor II Device (Lonza).
16. Neurobasal Medium.

### 2.2. Instrumentation

1. Inverted microscope (Olympus IX 71, Nikon TiE or equivalent).
2. High NA (1.45 or 1.49) oil immersion objectives.
3. Continuous Wave (cw) lasers for excitation depending on the fluorophore to be imaged, e.g., 23 mW HeNe laser (Thorlabs), frequency doubled Nd:Yag (Coherent) or solid state lasers.

4. Low noise highly sensitive electron multiplying CCD camera, e.g., QuantEM or Cascade (Photometrics).
5. Several optical and opto-mechanical components including mirror and lenses.
6. Fast shutter (Uniblitz) or AOTF (AA optoelectronic).
7. For each dye, an appropriate set of filters (Chroma or Semrock) is required.
8. Fluorescence free immersion oil (FF Cargille).
9. Ludin opened sample holder (Life Imaging Services).
10. Microscope temperature (37°C) control system (Life Imaging Services).
11. Fluorescent Tetraspec beads (Invitrogen) for image registration.
12. Computer for image acquisition.
13. Software for image acquisition (Metamorph).
14. Computer for image processing and visualization.
15. Software for image processing and visualization (Metamorph, ImageJ, Matlab).

## 3. Methods

### *3.1. Optical Setup*

Image acquisition of single molecules with high signal-to-noise ratio is a critical step for efficient reconstruction of super-resolved images. The final nanometric resolution will directly depend on this signal-to-noise ratio (9, 10).

#### *3.1.1. Microscope*

This protocol was optimized using an inverted microscope equipped with a 1.45 NA 100× oil immersion objective. Highly fluorescent organic dyes (e.g., ATTO647N or Cy3) are detected using a sensitive and rapid EM-CCD camera. Fluorescence excitation sources are CW lasers (e.g., frequency doubled Nd:Yag laser, solid state laser or He:Ne laser). Excitation laser beams enter through the fluorescence epiport of the microscope and illuminate a wide field area of 10–20 μm in diameter of the sample by focusing the beam in the back aperture of the objective. For uPAINT, oblique illumination is required to avoid exciting the out of focus fluorescent ligands in the solution. This is achieved by translating the focused beam with respect to the axis of the objective, on the periphery of the back aperture of the objective (Fig. 2), akin to total internal fluorescence microscopy. Illumination intensities are of a few kW/cm$^2$. For each type of dye used, an appropriate set of fluorescent filters is required. The total detection efficiency of a typical experimental single molecule setup is in the range of 5–10%.

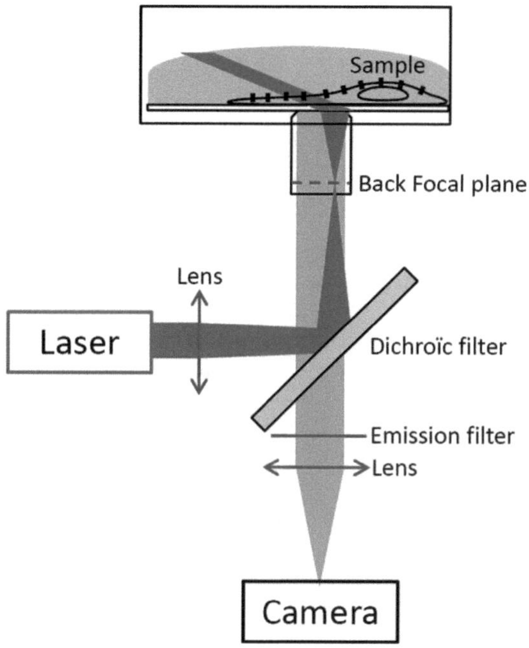


Fig. 2. Schematics of the optical setup.

*3.1.2. Resolution and Trajectory Lengths*

With this detection efficiency, 1,000 counts are commonly detected in 50 ms from a typical single fluorophore (11). This translates in a typical pointing accuracy (or localization precision) of 40–50 nm. We emphasize that different definitions of the pointing accuracy are often used. Here, we use the Full Width at Half Maximum (FWHM) of the distribution of the localizations obtained from a fixed molecule which is repetitively imaged. With this definition and using the Rayleigh criterion to define the resolution at which two molecules can be discriminated, the resolution is equal to the pointing accuracy. Some authors alternatively use the standard deviation of fixed molecule localizations for defining the pointing accuracy or localization precision, which provides values 2.3 times lower than the definition used here. We generally prefer our definition, since, in the latter case, the pointing accuracy gives lower values than the resolution (factor of 2.3), which might induce confusing comparisons of different sets of data.

If longer integration times are used (at the price of fewer data points in a trajectory), improved pointing accuracies are obtained. For a shot noise limited detection, the pointing accuracy will improve with the inverse of the integration time (10) yielding ~20–30 nm resolution for 150 ms integration time. It is worth mentioning that when mobile molecules are imaged, the movements of the molecules might affect the pointing accuracy.

The number of points of a single molecule trajectory depends on the photo-physical properties of the imaged dye. In practice,

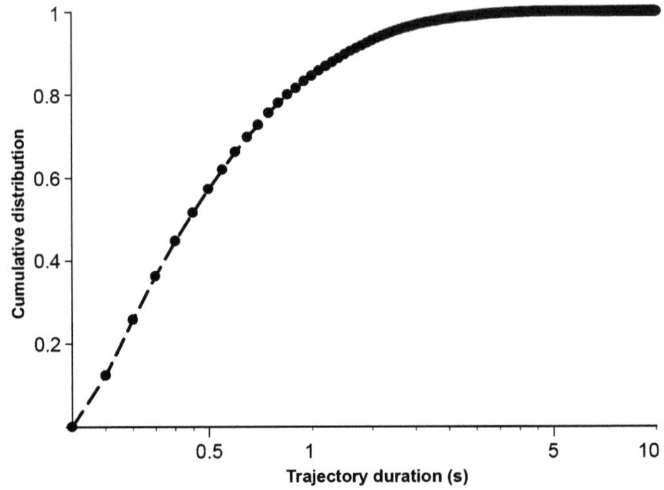


Fig. 3. Distribution of single molecule trajectory lengths measured with ATTO647N on a live cell. Adapted from (7).

and for a given dye, it depends on the excitation intensity and integration time used. Typically, one uses close to video rate imaging (50 ms integration times) at saturation intensities (~1 kW/cm$^2$ for the most common dyes). With these parameters, one can obtain typically 10–15% of the trajectories lasting more than 1 s, in the case of ATTO647N dyes (7) (Fig. 3).

With uPAINT, and unlike standard single molecule methods, single point source emitter detection is ensured by the real time imaging of each individual ligand binding to its target membrane molecule regardless of the number of fluorophores it carries. Thus, one can take advantage of this feature to improve the pointing accuracy by coupling multiple dyes per ligand. For instance, with four fluorophores per ligand instead of one, the average pointing accuracy is improved by a factor of 2 under identical excitation intensity. Optionally, in order to lengthen the obtained trajectories, one can reduce the excitation intensity while keeping the same pointing accuracy as when a single dye molecule is used (7).

#### 3.1.3. Oblique Illumination, Choice of the Angle

By creating a thin sheet of light above the coverslip surface, oblique illumination allows selecting individual fluorescent ligands which have bound to the cell surface while rejecting the great majority of molecules in solution. This improves signal to noise ratios and avoids photobleaching of unbound fluorescent ligands. The choice of the angle depends on the thickness of the cells to be imaged. A typical choice is ~5° such that the resulting sheet of light has a thickness of $\Delta x$ ~2 μm in the center of the field of view. Assuming 3-dimensional isotropic diffusion of the fluorescent ligands freely diffusing in the solution, the diffusion coefficient $D$ is well known and relates the ligand hydrodynamic diameter $d$ and the fluid viscosity $\eta$ by the Stokes-Einstein relation: $D = \frac{k_B T}{3\pi\eta d}$, where $k_B$ is the Boltzmann constant and $T$ is the temperature. For ligand hydrodynamic sizes of $d$ ~5–10 nm, $D$ is of the order of ~40–80 μm²/s and the average time they spend in the 2D sheet of light is $2\Delta x^2/2D$ which lies below ~100 ms. This indicates that such unbound fluorescent molecules should not appear statistically in more than one or two consecutive images if 50 ms integration times are used. Such unwanted events can be rejected in the analysis by omitting the first two molecular detections in each trajectory (see below).

### 3.2. Fluorescent Labels

High affinity fluorescent ligands for the molecule to be studied are needed with uPAINT (see Note 1). We describe here a procedure when an antibody is used as a label, either to study endogenous proteins when a high quality antibody is available or GFP expressed proteins using anti-GFP for instance. Other types of fluorescent labels can also be used depending on the proteins studied, such as $Ni^{2+}$ tris-nitrilotriacetic acid (TrisNTA) when the target protein bears an extracellular poly-histidine tag (7, 12). The protocol for antibody labeling with ATTO647N-NHS-ester (Atto-Tech) is a modified version of the manufacturer's procedure. Keep in mind that coupling a too low quantity of antibodies decreases the efficiency of the reaction. An amount of 200–300 μg is the minimum quantity required to obtain a good coupling.

1. For coupling, antibodies should be in an amine-free buffer. If the buffer contains amines, dialysis should be performed and antibodies should be resuspended in PBS 1×. For many commercial antibodies, addition of BSA is required during the experiments in order to decrease nonspecific labeling. If this is the case, BSA should not be mixed to the antibodies during the antibody/dye labeling procedure to prevent preparing fluorescent BSA proteins.
2. Sodium carbonate (1 M) should be added to the solution to obtain pH 8.3 necessary to protonate the amino group of lysines (about 100 μl of $NaHCO_3$ per ml of antibody solution).
3. Fully dissolve 1 mg of ATTO647N-NHS-ester in 1 ml of anhydrous, amine free DMSO. Then add a threefold molar excess of reactive dye to the antibody solution. Protect the sample from light and agitate for 3 h at room temperature.
4. Use G-25 Sephadex columns to separate the conjugate from the free dye molecules. The columns are used with the gravity protocol. Equilibrate the column with PBS 1×. The first colored and fluorescent zone to elute will contain the conjugated antibodies. The second colored and fluorescent fraction contains the unlabeled free dye (hydrolyzed NHS-ester).

5. It is important to estimate the concentration of the different fractions and the average number of dyes coupled per antibody. After coupling it is also crucial to test if the ATTO647N-NHS-ester labeling does not affect the specificity of the antibody (this might happen if a lysine is present at the antibody binding site). In the latter case, ATTO647N-maleimide coupling can be a good alternative.

### 3.3. Sample Preparation

Depending on the type of proteins and cells to be studied, the following example protocols may be modified. Although it is often more relevant to target endogenous proteins, in some cases the use of transfected proteins is required. In particular, this is the case when highly specific antibody against the protein of interest does not exist or for studies involving protein mutants. A convenient strategy is to fuse the protein of interest with an extracellular tag having known very good ligands (GFP, histidine tags, biotin/avidin, etc.). Here, we describe two example protocols for studying exogenous or endogenous proteins (Subheadings 3.3.1 and 3.3.2, respectively). These protocols can be easily adapted to different kind of adherent cells (COS7, fibroblasts, etc) (see Note 2).

#### 3.3.1. Exogenous Proteins

This protocol gives the user tight control for reducing nonspecific labeling on the surface of the coverslip which is a critical issue in uPAINT experiments.

1. Immortalized mouse embryonic fibroblasts and COS7 cells are cultured in DMEM with 10% FBS.
2. Transient transfection of plasmids encoding membrane proteins fused to GFP or poly-histidine tags (GFP-GPI, 6His-Trans-Membrane) are performed using Fugene 6.
3. The day of the experiment, cells are detached with trypsin/EDTA (0.05% for 2 min), trypsin inactivated with 10% FBS DMEM, washed and suspended in serum free condition in Ringer medium, and incubated for 30 min before plating on the chosen extracellular matrix protein (rat collagen I or human fibronectin).
4. Experiments are performed 3 h after cell plating. Optionally, cells can be co-transfected with a fluorescent protein reporting the subcellular localization of the region of interest (e.g., adhesion sites).

#### 3.3.2. Endogenous Proteins

Studying endogenous proteins depends on the availability of highly specific primary antibodies. Be aware that antibodies can affect the properties of the studied protein, promoting activation, inhibition, or cross-linking. Therefore, it is best to test in a functional assay if the antibody used for uPAINT affects the function of the target protein. In this example, endogenous AMPA receptors could be studied at the surface of live hippocampal neurons.

1. Cultures of hippocampal neurons are prepared from E18 Sprague-Dawley rats following the method previously described (13).
2. Cells are plated at a density of $200 \times 10^3$ cells/ml on poly-lysine pre-coated coverslips. To localize excitatory post-synapses, neurons can be electroporated (Lonza nucleofactor protocol) with Homer1C-GFP just after dissection and processed 2/3 weeks later.
3. Cultures are maintained in serum free neurobasal medium and kept at 37°C in 7.4% $CO_2$ for 10–15 days in vitro.

*3.3.3. Preparing Samples for Imaging*

1. Before acquisition, coverslips are incubated with a solution containing a low concentration of fluorescent beads. Adsorption of a few beads on the glass coverslip will provide immobile reference objects to correct long-term mechanical instabilities of the microscope. The concentration and incubation time should be adjusted so that 2–3 beads are present on average in the imaging field. Gentle rinsing is performed in order to avoid detaching adsorbed beads.
2. Then, a coverslip is mounted on an open chamber and 600 μl of recording medium is added onto the cells. This solution has to be free of vitamins and other cyclic compounds (present at high concentration in the culture medium) because of their intrinsic fluorescence which might produce unwanted background signals.
3. Osmolarity has to be adjusted, by adding NaCl and/or glucose, depending on the osmolarity of culture medium measured just before the experiment.
4. The acquisition is initiated when the area of the cell preparation to be studied is selected. Differential Interference Contrast or fluorescence imaging of a particular fluorescently tagged protein can help to perform this selection (e.g., GFP imaging of excitatory synapses or other subcellular compartment). Once this selection is performed, the sample should not be moved throughout the experiment.

### 3.4. Acquisition

Images are typically acquired at video rates or faster, depending on the mobility of the studied molecule. If the streaming mode of the camera is not used, a fast shutter synchronized with the camera can eventually be added in the excitation path to prevent sample illumination in between frames to limit dye photobleaching.

The acquisition procedure consists in recording several temporal stacks of images. For typical experiments, 3–4 stacks of 4,000 images are recorded. Immediately after identification of the acquisition

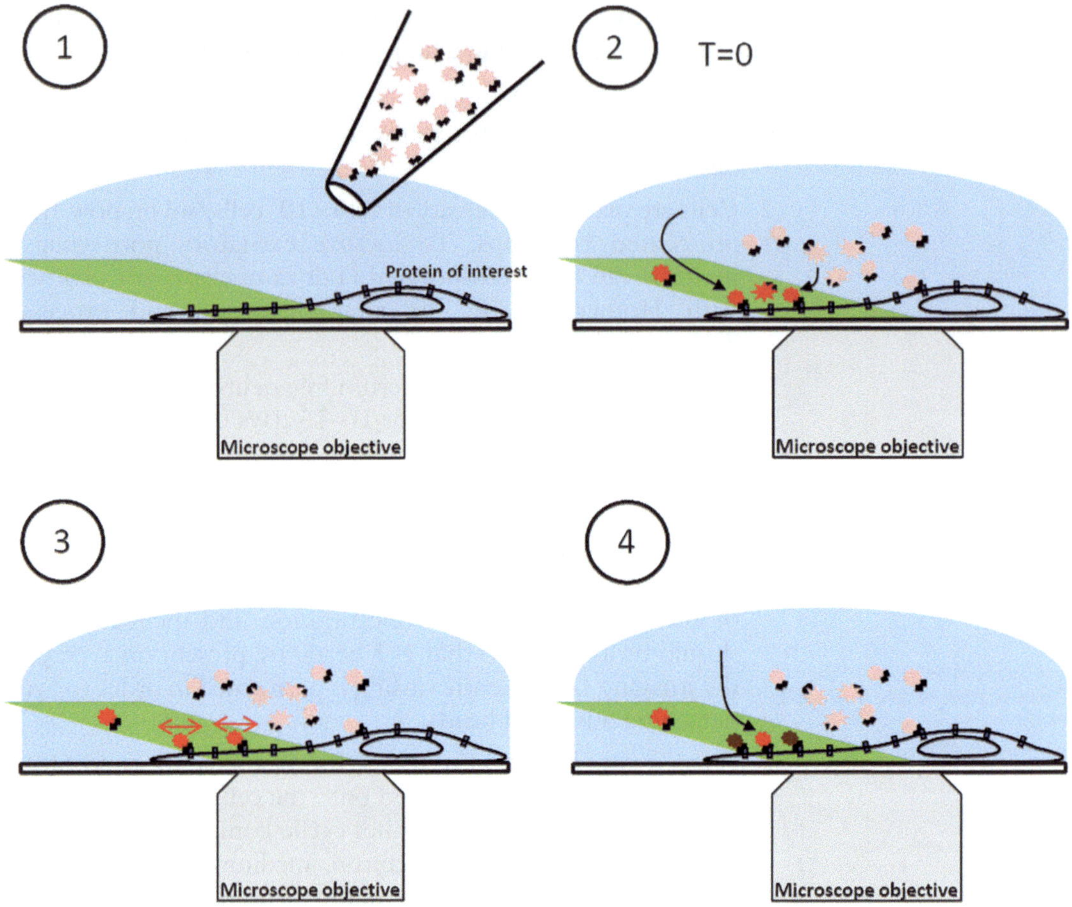


Fig. 4. Cartoon of an imaging sequence. Non-illumintated dye molecules are displayed in *pink*, fluorescent ones in *red* and photobleached ones in *brown*.

area and before the beginning of recording, 1–5 μl of fluorescent ligands are carefully added and the medium homogenized with 200 μl Pipetman to avoid any mechanical drifts of the sample (Fig. 4).

Soon after the addition of the fluorescent ligands, diffraction limited fluorescent spots start to be visible in the field of view (Fig. 5). The quantity of ligands to be added has to be adjusted as a function of its affinity, and the desired labeling density (Fig. 5). For instance, the higher the protein mobility, the lower the spot spatial density should be in order to facilitate trajectory reconstructions and avoid trajectory mixing. To obtain super-resolved images with the highest resolutions, the highest signal-to-noise ratio should be obtained. This can be done by using the longest integration time compatible with molecule movements during each image but at the price of a loss of dynamic information due to fewer number of data points obtained in each trajectory. On the other hand, high content information on the dynamics of the molecules is

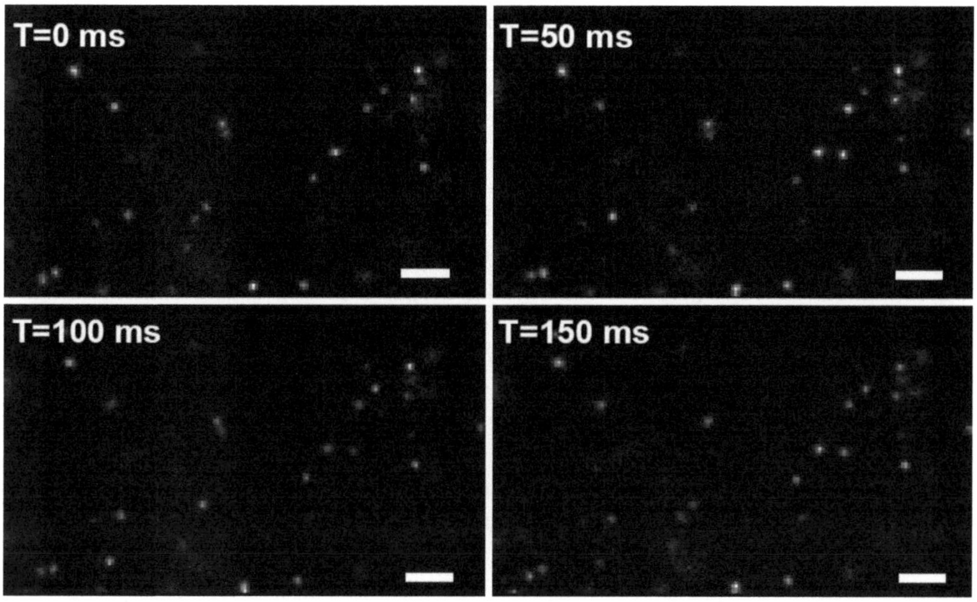


Fig. 5. Typical raw images acquired at different times during the imaging sequence. Fluorescent spots appear stochastically at the surface of the living cell upon fluorescent ligand binding on their target. These spots can be tracked on several images as long as the fluorescent molecules do not photobleach. Scale bar: 5 μm.

better obtained at the highest imaging rates to obtain more data points per trajectories but at the price of slightly lower spatial resolutions. Thus, the choice of the ligand concentration and imaging rates will result from a compromise and should be adjusted for each set of experiments.

It is also important to control the variation of the osmolarity during recording. For instance, in typical experiments at 37°C, where 600 μl of mounted volume is used, the osmolarity increases by 20 mOsmol after 15 min of recording.

### 3.5. Analysis

This ultimate step allows quantifying and visualizing the localization and dynamics of the molecule of interest, at spatial resolution below the diffraction limit of light microscopy. We use custom-made programs written in C/C++ and Visual Basic programming languages. These modules are launched as a DLL inside the Metamorph software environment in order to be able to visualize some preliminary results directly on the acquisition workstation.

#### 3.5.1. Detection and Trajectory Analysis of Single Fluorescent Ligand Molecules

In each image of the recorded sequence, single fluorescent ligands appear as diffraction-limited bright spots. Two different approaches can be used, one based on centroid determination using wavelet transforms, and another one using Gaussian fitting.

Routine single dye localization can be performed using a wavelet transform segmentation process. We use the “à trous” algorithm with a B-Spline of third order (14), since it is very fast and provides very good pointing accuracy, similar to Gaussian fitting methods commonly used in super-resolution microscopy. We keep the second wavelet map, which contains most of the diffraction limited signal, while more than 80% of the noise is filtered in the first map and the background is contained in the larger maps. We then typically apply a fixed threshold of 1–2 times the background noise standard deviation. We use a custom watershed algorithm to separate close molecules. Finally, only the objects of surface superior to 3 pixels (~0.7 μm²) are kept. Figure 6b displays a typical uPAINT image and the corresponding single molecule localization. A stack of dimension 128 × 128 pixels × 5,000 frames takes about 120 s to process on an Intel XEON 2.4 GHz CPU, including the image loading to RAM, leading to about 50,000 detections and 3,000 trajectories

As an alternative to wavelet transform analysis, the positions for each single molecule can be obtained by fitting the fluorescent signal by the following two-dimensional Gaussian function (11, 15, 16):

$$I(x,y) = I_0 \exp\left(-\frac{(x-x_0)^2+(y-y_0)^2}{2\sigma^2}\right)$$

Where $I_0$ is the amplitude of the Gaussian (given as a grayscale value); $\sigma$ is related to the FWHM of the diffraction pattern by FWHM = 2.35σ. Typically, the FWHM of the setup is around 350 nm for 660 nm emission wavelength. The parameters $x_0$ and $y_0$ give the central positions of the spot. The errors in the fitting coefficients vary for each analyzed spot because the fitting accuracy depends on the local signal-to-noise ratios. The FWHM of the distribution of $x_0$ and $y_0$ for fixed molecules represents the pointing accuracy (see definition and remarks above), which only depends on this signal-to-noise ratio and is typically well below the optical resolution limit of about 50 nm in this work.

Nevertheless, since the aim is to localize moving single molecules, the movement occurring during the image acquisition may affect the apparent shape of the signal. This prevents the Gaussian fitting with fixed calibrated size to be achieved accurately and motivates the use of segmentation based on centroid as the wavelet method presented above.

#### *3.5.2. Trajectory Reconstruction*

Trajectories are reconstructed by connecting object positions from one image to the next. Several algorithms can be used. We use a global tracking method based on simulated annealing algorithm for its flexibility and rapid convergence (17). It consists of minimizing a global energy function *E*, taking into account the intensity and velocity of each detected molecule. Since molecules are constantly appearing and disappearing in the field of view, "birth" and "death" events are also taken into account in the energy function. Only trajectories lasted more than several points (ten) are kept for further analysis. This allows the rejection of false single molecule

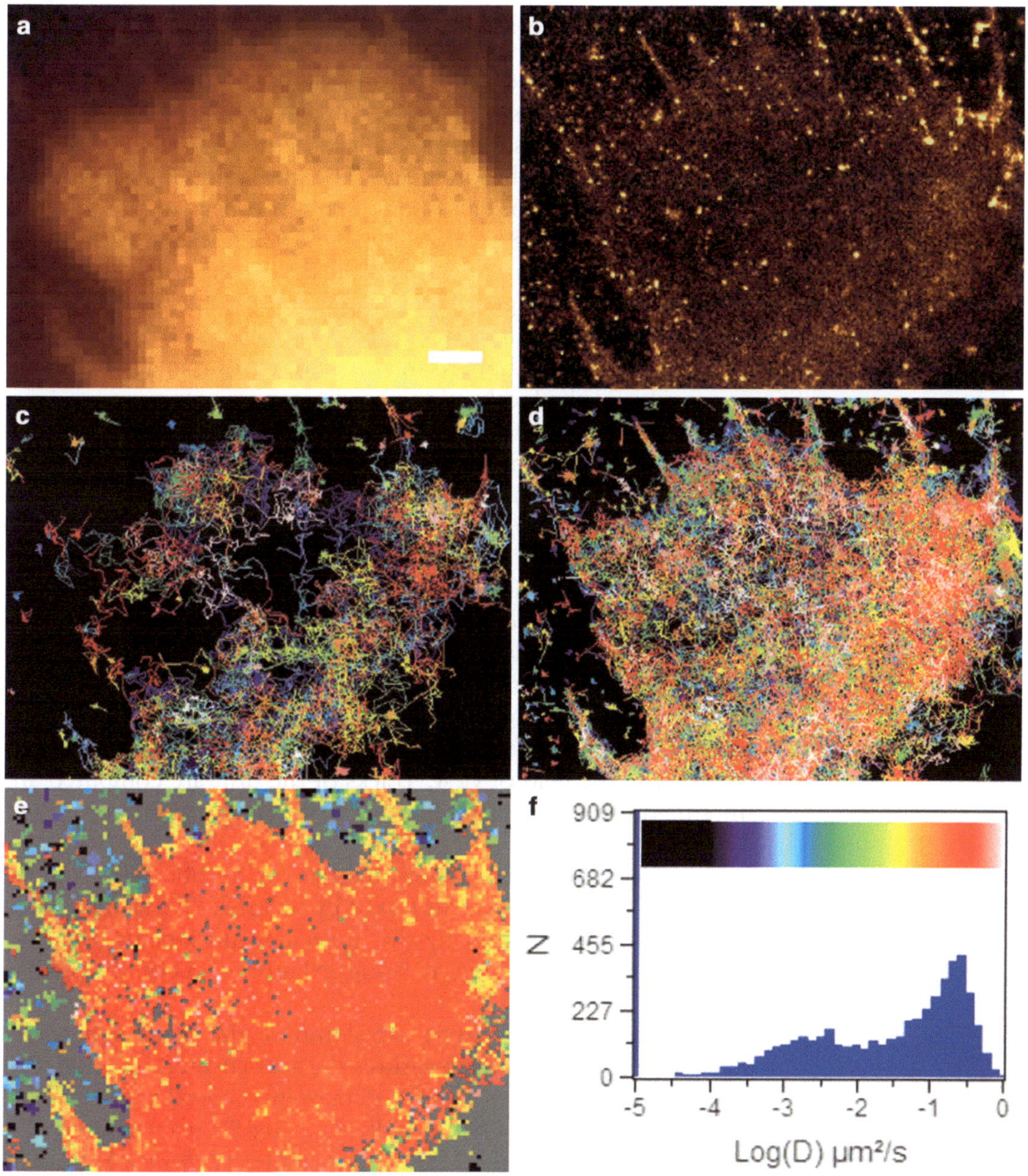


Fig. 6. Illustration of the reconstruction. (**a**) Wide-field fluorescence image of a live fibroblast expressing cytosolic fluorescent GFP and a membrane protein of interest (here the transmembrane domain of the PDGF receptor (7)). Scale bar 1.5 μm. (**b**) Super-resolved image reconstructed from 310,000 localizations of fluorescent ligands (TrisNTA-ATTO647N) which have bound to their target protein at the surface of the cell. Images of the protein trajectories with lengths >50 points (**c**) and >10 points (**d**) ($N$=550 resp. 6,000). (**e**) Diffusion map: spatially resolved color-coded diffusion constants of the proteins shown in (**b–d**) as defined in the histogram presented in (**f**) computed on trajectories with more than 20 points.

detection that could occur due to noise, as well as, to keep trajectories which are long enough for quantification.

Low labeling densities are preferred to facilitate this process and minimize connection errors (see Figs. 5 and 6).

*3.5.3. Super-Resolved Image Reconstruction*

Super-resolution images are reconstructed based on the numerous molecule localizations: all the molecules detected during the acquisition sequence are pooled to form a high-resolution image. Since the resolution of the super-resolved image depends on the localization accuracy of the single molecules, the user can define the pixel size of the super resolution image. The ideal sampling is defined as less than, or equal to, half of the resolution needed to satisfy the Nyquist criterion (18). Typically, a 256×256 image with a pixel size of 160 nm, analyzed with a resolution of 40 nm, will lead to a super-resolution image of dimension 2,048×2,048 with a pixel size of 20 nm (sub-sampling factor of 8 times). Each detected molecule will then contribute to the super-resolution image by adding a grey value at the localization coordinates.

It is possible to represent a super-resolution image of various additional parameters, like the intensities, trajectories, planes of detection, or any quantification performed on the mobility (see below). Thanks to high molecule density provided by uPAINT, it is also possible to perform the cartography of any of the quantification parameters by averaging all the measurements in pixels of dimensions defined by the user.

In order to a take into account the possible drifts occurring during the total duration of the acquisition (which is in the range of few minutes), an automatic registration step is performed using the super localization and tracking of the immobile fluorescent beads recorded in the field of view. Tracks of the drift are computed and smoothed using a median filter of 5 before registration.

*3.5.4. Dynamic Quantification*

Different types of dynamic information can be obtained from the analysis of the trajectories measured by uPAINT. We list here a few standard ways of quantifying the molecules movements occurring on a cell, without aiming at being exhaustive about the analysis strategies.

For each trajectory, one can compute the mean square displacement (MSD), which measures the area covered by a molecule over time. For a trajectory of N data points (coordinates $x(t)$, $y(t)$ at times $t=0$ to $(N-1)\times\Delta t$, with, e.g., $\Delta t=50$ ms, the inverse of the acquisition rate), the MSD for time intervals $\tau = n\times\Delta t$ is calculated using the formulae:

$$\mathrm{MSD}(n\times\Delta t) = \sum_{i=1}^{N-n} \frac{\left[x((i+n)\times\Delta t)-x(i\times\Delta t)\right]^2+\left[y((i+n)\times\Delta t)-y(i\times\Delta t)\right]^2}{N-n}$$

For the time interval $\tau=n\Delta t$, the MSD and its error bar are thus calculated on $N-n$ points.

The MSD is widely used to extract diffusion characteristics from trajectories. For instance, a linear MSD over time is characteristic of a molecule diffusing freely, while the confined movement of molecules in a domain is indicated by a plateau reached by its MSD over time. For each MSD, the instantaneous diffusion coefficient, $D$, can also be calculated from linear fits of the first four points (corresponding to 200 ms) of the MSD using MSD $(\tau) = \langle r^2 \rangle (\tau) = 4D\tau$. Histograms of the molecules diffusion constants measured on a large number of molecules on a given cell can thus be obtained (Fig. 6f).

Two-dimensional maps of molecule mobilities can also be produced by displaying in a color-coded pixel (whose size is chosen by the user, typically 200 nm) the median of the step lengths (corresponding to the distance between two consecutive points of a trajectory) found in this pixel (7) or the mean instantaneous diffusion constant of the molecules detected in this pixel (see Fig. 6e–f for example). These types of maps provide dynamic information of subcellular regions of a given cell (7).

## 4. Notes

1. In uPAINT, the specificity of the ligand used is certainly the most crucial parameter. Indeed, on the contrary to other labeling procedures with external probes, one does not wash the preparation after placing the fluorescent ligand in solution. Thus, highly specific ligands should be used.
2. At the moment, this protocol is mainly dedicated to membrane protein imaging. The method can also be adapted to intracellular protein imaging in fixed permeabilized cells.

## Acknowledgments

We wish to thank B. Lounis for helpful discussions. This research was funded by Centre National de la Recherche Scientifique (CNRS), the Région Aquitaine and the Agence Nationale pour la Recherche (ANR), the Fondation pour la Recherche Médicale, and the European Union's seventh framework program for research and development ERC grant Nano-Dyn-Syn.

### References

1. Hell SW (2007) Far-field optical nanoscopy. Science 316:1153–1158
2. Huang B, Bates M, Zhuang X (2009) Super-resolution fluorescence microscopy. Annu Rev Biochem 78:993–1016
3. Betzig E et al (2006) Imaging intracellular fluorescent proteins at nanometer resolution. Science 313:1642–1645
4. Hess ST, Girirajan TP, Mason MD (2006) Ultra-high resolution imaging by fluorescence photoactivation localization microscopy. Biophys J 91:4258–4272
5. Rust MJ, Bates M, Zhuang XW (2006) Sub-diffraction-limit imaging by stochastic optical reconstruction microscopy (STORM). Nat Methods 3:793–795

6. Geissbuehler S, Dellagiacoma C, Lasser T (2011) Comparison between SOFI and STORM. Biomed Opt Express 2:408–420
7. Giannone G et al (2010) Dynamic superresolution imaging of endogenous proteins on living cells at ultra-high density. Biophys J 99:1303–1310
8. Sharonov A, Hochstrasser RM (2006) Wide-field subdiffraction imaging by accumulated binding of diffusing probes. Proc Natl Acad Sci U S A 103:18911–18916
9. Bobroff N (1986) Position measurements with a resolution and noise-limited instrument. Rev Sci Instrum 57:1152–1157
10. Thompson RE, Larson DR, Webb WW (2002) Precise nanometer localization analysis for individual fluorescent probes. Biophys J 82:2775–2783
11. Tardin C, Cognet L, Bats C, Lounis B, Choquet D (2003) Direct imaging of lateral movements of AMPA receptors inside synapses. EMBO J 22:4656–4665
12. Grunwald C et al (2011) Quantum-yield-optimized fluorophores for site-specific labeling and super-resolution imaging. J Am Chem Soc 133:8090–8093
13. Groc L et al (2004) Differential activity-dependent regulation of the lateral mobilities of AMPA and NMDA receptors. Nat Neurosci 7:695–696
14. Starck J-L, Murtagh F (2006) Astronomical image and data analysis, 2nd edn. Springer, New York
15. Schmidt T, Schuetz GJ, Baumgartner W, Gruber HJ, Schindler H (1996) Imaging of single molecule diffusion. Proc Natl Acad Sci U S A 93:2926–2929
16. Cheezum MK, Walker WF, Guilford WH (2001) Quantitative comparison of algorithms for tracking single fluorescent particles. Biophys J 81:2378–2388
17. Racine V et al (2006) Multiple-target tracking of 3D fluorescent objects based on simulated annealing. 2006 3rd IEEE international symposium biomedical imaging: macro to nano, vols 1–3, IEEE international symposium on biomedical imaging, 1020–3
18. Shannon CE (1949) Communication in the presence of noise. Proc IRE 37:10–21

# Chapter 8

# Super-Resolution Fluorescence Imaging with Blink Microscopy

**Christian Steinhauer, Michelle S. Itano, and Philip Tinnefeld**


## Abstract

Recently, a new approach for super-resolution microscopy has emerged which is based on the successive localization of single molecules. The majority of molecules are prepared to reside in a nonfluorescent dark state, leaving only a few single molecules fluorescing. The single molecules can subsequently be localized on the camera image. Successive localization of all molecules allows reconstruction of a super-resolved image of the labeled structure. A variety of ways for limiting the number of locatable molecules have been developed recently which expand this current field of imaging. Here we describe a super-resolution microscopy method that employs the use of reversible, generic dark states, for example radical ion states. This method requires only a single laser source and can be carried out with many fluorescent dyes, in some cases, even in living cells. We provide a step-by-step procedure for this method, which we have called Blink Microscopy.


**Key words:** Blink Microscopy, Super-resolution fluorescence imaging, Super-resolution standards, Immunostaining

## 1. Introduction

Since the publication of STORM and (F)PALM in 2006 (1–3), several similar approaches have been developed based on the wide-field detection and localization of single molecules (4–14). These approaches mainly differ from each other in the way that single molecule fluorescence events are generated. Generally, this is achieved by affecting how the majority of fluorophores are prepared to reside in a nonfluorescent state while a small subset emits, resulting in point-spread functions originating from single molecules. An overview of different methods can be found in review articles in various journals (15–20).

This chapter focuses on, and describes in detail, the unique features of Blink Microscopy. Blink Microscopy employs single-molecule

Alioscka A. Sousa and Michael J. Kruhlak (eds.), *Nanoimaging: Methods and Protocols*, Methods in Molecular Biology, vol. 950, DOI 10.1007/978-1-62703-137-0_8, © Springer Science+Business Media, LLC 2013

Fig. 1. An example of Blink Microscopy data from filamentous actin on a glass surface. *Left*: At the beginning, all fluorophores are in their ON state. This ensemble of fluorescent molecules provides a diffraction limited image. *Center*: The excitation laser quickly drives all molecules into their OFF-state while individual molecules switch back ON and can be localized as single-molecule events with high precision. This step is at the heart of Blink Microscopy and most experimental parameters described in this chapter aim at optimizing this single-molecule blinking. *Right*: all found localizations are binned into a super-resolved image (reprinted from ref. 36 with kind permission from Springer).

blinking, a well-known feature of organic fluorophores, to cycle dyes between a fluorescent ON-state and a nonfluorescent OFF-state.

With the exception of stimulated emission depletion microscopy (STED) and other realizations of the RESOLFT (REversible Saturatable OpticaL Fluorescence Transitions) concept (17), many emerging super-resolution (SR) imaging methods are based on the successive localization of single molecules (1, 2). In these methods, the majority of molecules are prepared to reside in a non-fluorescent dark state, leaving only a few single molecules fluorescing. These single molecules are subsequently localized by fitting a two-dimensional Gaussian function to the point spread functions (PSFs) of their fluorescence intensities. This temporal separation of fluorescence is the most crucial step in localization-based microscopy; how to achieve this separation using organic fluorophores is a major focus of the following sections. The super-resolved image is then reconstructed from all single-molecule localizations. An example of Blink Microscopy data is shown in Fig. 1.

Interestingly, both STED/RESOLFT and localization-based approaches can use the same types of organic fluorophores, although the photophysical requirements for the fluorophores are considerably different. Both approaches require very photostable fluorophores that can be switched between an ON- and an OFF-state (16, 17). One exemplary method exploits generic dark states (e.g., radical anion states) of the fluorophores in order to switch the major fraction of molecules OFF, so that the remaining subpopulation of molecules can be precisely localized. In this method, fluorescent dyes are switched OFF by transiently accepting an electron from a reductant molecule, which then yields dark states of up to several hundred seconds (Fig. 2a) (6, 21, 22). Such dark states appear as blinking in single-molecule intensity transients (see Fig. 2b). Therefore, this technique has been termed "Blink Microscopy."

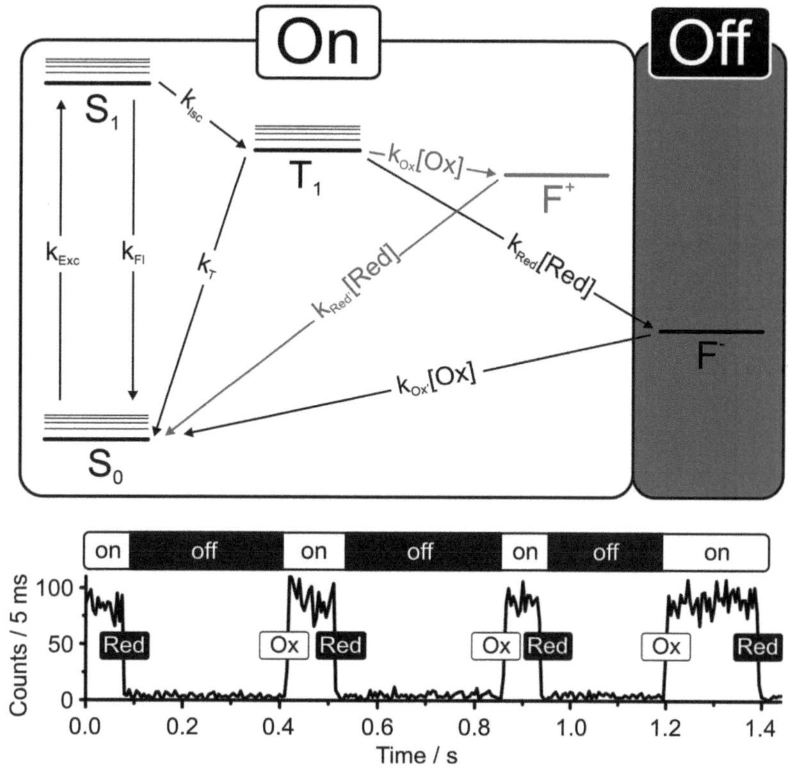


Fig. 2. Dark state reactions used for Blink Microscopy (taken from ref. 16). (**a**) Jablonski diagram showing the photophysical states of a single fluorophore. The driving force behind the OFF-switching reaction is provided by fluorescence excitation ($k_{exc}$), followed by occasional intersystem crossing ($k_{ISC}$) and reduction by a reductant in solution ($k_{Red}$). The fluorophore switches back ON due to oxidation by oxygen or an added oxidant ($k_{Ox}$). Note that there is no need for a separate switching laser and that the reaction is fully reversible, so that each dye can run through many ON-OFF cycles. (**b**) Fluorescence vs. time transient showing several switching cycles of a single fluorophore.

To rationalize the blinking, we take a brief look at the thermodynamics of photoinduced electron transfer reactions. Following excitation, the fluorescent dye is shelved from the ground state, $S_0$, to the first excited single state, $S_1$. In other words, the dye has entered a state of higher energy. In this state, the dye is a better electron acceptor than it was when it was in the ground state because the highest occupied molecular orbital of the molecule is only singly occupied, and can therefore accept an electron quite easily. When a putative electron donor such as ascorbic acid or a thiolate ion collides with a dye in the excited singlet or triplet state, electron transfer to the dye can occur. The lifetime of the radical anion (for a dye molecule with an assumed neutral ground state) depends on the stability of the radical anion and the concentration of potential electron acceptors, such as oxygen. Thus, the lifetime can be controlled in vitro by adapting the oxidant (e.g., the oxygen) concentration. For in vivo measurements, certain dyes that show sufficiently long OFF-state lifetimes in physiological oxygen concentrations have to be used. Due to their high reduction potentials, examples of suitable dyes include the oxazines ATTO655, ATTO680, and ATTO700 (16, 22).

Long OFF-states are of profound importance for super-resolution microscopy methods that utilize successive single-molecule localizations. The OFF-state lifetime and the time it takes to read out the photons emitted during one ON-state are crucial parameters affecting the obtainable resolution. More molecules can be registered for an area of the size of the point-spread function when the ratio of OFF- to ON- times is large. This requires the use of dyes with long OFF-states, comparably strong lasers, and fast cameras to enable fast readouts (16).

Belonging to the class of techniques that employ successive single-molecule localizations, Blink Microscopy distinguishes itself by working with common organic dyes and requiring only one laser (for each color, in multi-color applications). In addition, Blink Microscopy can work in the presence of oxygen, and offers a rather broad range of possible buffer conditions that are close to physiological. Blink Microscopy has been used to study cytoskeletal structures, such as actin filaments in fixed cells, amyloid aggregates, DNA nanostructures, and single-molecule assembled structures (6, 22–25). In parallel, very similar techniques have been developed by the Sauer and the Hell lab, which have been termed dSTORM and GSDIM (5, 7, 26), respectively, showing that single-molecule blinking can even be applied in living cells and might also be based on dark states other than radical ion states (27, 28).

In this article, we provide a detailed protocol for Blink Microscopy, including the microscope setup, chemical and sample preparation, measurement procedures, and issues to consider for data analysis and trouble-shooting. The method is exemplarily demonstrated by imaging DC-SIGN, a transmembrane lectin that forms domains on the membrane of dendritic cells. These domains are commonly too small to be resolved by conventional fluorescence microscopy (29).

## 2. Materials

### *2.1. Chemicals and Consumables*

1. MilliQ water.
2. PBS, pH 7.4, preferentially as tablets to dissolve in water.
3. Ascorbic acid (AA).
4. Methyl viologen (MV).
5. Mercaptoethylamine (MEA).
6. Beta-mercaptoethanol (ME).

7. TCEP (Sigma 646547).
8. Tris.
9. Catalase (Sigma C100-50 mg).
10. Glucose oxidase (Sigma G2133-50KU).
11. Glucose.
12. Glycerol 86%.
13. LabTek chambered cover glasses (Nunc 155411).
14. 96-well Plate sealing sheets.
15. 16% Para-formaldehyde (PFA) stock solution.
16. Primary antibody corresponding to a protein of interest.
17. Secondary antibody, preferably unlabeled $Fab_2$ fragments.
18. Amine reactive dye, e.g., ATTO655 NHS-ester. Dissolve in DMF, make 0.1 mg aliquots and dry in a speed-vac. Store at −20°C.
19. Oxygen scavenging stock solution (20×): 5 ml 100 mM Tris–HCl, 25 mM KCl, pH 7.5; 5 ml Glycerol (86%); 40 μl TCEP (0.5 M); 40 μl Catalase (20 mg/ml as supplied); 10 mg Glucose oxidase. Store in working aliquots at −20°C. Due to its negligible influence on dye photophysics, TCEP is used instead of DTT.
20. NAP-5 Columns (GE Healthcare).
21. NanoRuler calibration standards to test sub-diffraction resolution (www.sts-nano.com).

### 2.2. Microscope

1. Leica SR-GSD, Zeiss Elyra PAL-M or Nikon nSTORM commercial microscope system or comparable custom-built setup.
2. Appropriate excitation laser with at least 100 mW output power.
3. High-quality emission and excitation filters.
4. High numerical aperture (≥1.4) oil-immersion objective with 60× or 100× magnification.
5. Back-illuminated EMCCD-camera.
6. Stage micrometer for pixel size calibration.

## 3. Methods

### 3.1. Sample Preparation

#### 3.1.1. Cells

Cells are grown in standard media and at recommended conditions. Growing the cells on LabTek chambered cover glass slides offers the advantages that they can easily be transferred to the microscope, buffers can be changed easily, and chambers can be sealed for oxygen removal. When using other non-glass cell culture slides, the oxygen permeability of some plastics should be considered.

Detailed conditions for the proper growth and fixation of cells depend strongly on the type of cells used and the molecule being imaged, and are not described here. However, it is important to note that fixation chemicals may have an influence on dye photophysics. In addition, fixation methods, especially those which use hardening media, might alter the shape of the structures of interest. Dehydrating formulas, for example, tend to flatten cells and change the appearance of the nucleus (30). Therefore, if Blink Microscopy is unsuccessful or yields unexpected results, you may want to consider using alternate methods for fixation. Correspondingly, we recommend keeping the samples in buffer, if possible. Another benefit of keeping cells in buffer is that it allows for simple and fast tuning of the ON- and OFF-times by changing the reductant and oxidant concentrations in the buffer solution.

#### 3.1.2. Fluorophores

It is as important to choose an appropriate fluorophore and appropriate imaging and staining conditions, as it is to use a highly sensitive microscope. There are many options available in regards to choosing an appropriate fluorophore to use for imaging, but in general, commercially available dyes (e.g., ATTO dyes) work well for use with Blink Microscopy (5, 6, 22, 26).

#### 3.1.3. Antibody Labeling

As with other immunofluorescence-based labeling methods, a primary antibody must be carefully chosen to ensure both reliable specificity and affinity for the desired target molecule. It is best to directly label the primary antibody with the fluorophore, as this will lead to both less spatial distance between the molecule of interest and the dye, and a simpler estimation of the number of dyes per target protein. Alternatively, if labeling the primary antibody is either technically difficult or results in altered binding to the target protein, sequential immuno-labeling with an unlabeled primary antibody and a labeled secondary antibody can be employed. Keep in mind, however, that this antibody complex can have a spatial extent of 10–20 nm, which may be a limiting factor when imaging very small structures, for example, quantifying nano-domain sizes.

Good protocols for labeling antibodies with fluorescent dyes are provided by the suppliers of reactive dyes, such as NHS-esters and maleimides. Our standard procedure for labeling with amine-reactive NHS-esters is as follows:

1. Dissolve the protein in PBS at 0.5–1 mg/ml.
2. Adjust the pH of a 250 ml antibody solution to be between 8.5 and 9.
3. Dissolve 0.1 mg of dye in 10 ml of DMF, or another water-free solvent.
4. Add 1–5 μl of dye solution to the antibody solution. Whole IgG molecules contain more amine groups and require more dye than, for example, $Fab_2$ fragments.

5. Shake at room temperature for at least 3 h.
6. Remove excess dye using a desalting column (e.g., NAP-5, GE Healthcare).
7. Determine the degree-of-labeling by absorption spectroscopy of the filtrate.

For Blink Microscopy, it is best to have a low fluorophore number per target molecule. In part, this reduces the likelihood that more than one dye will be simultaneously emitting within the area of a point-spread function (PSF) of the dye. If more than one dye is emitting within a single PSF it is likely that the dyes will be rejected during image processing steps and not included in the final SR image, or will lead to a wrong localization. Both will worsen the quality of the final SR image. Additionally, more fluorophores within a small area may lead to physical contact between dyes, which can alter their photophysical properties and compromise Blink Microscopy

#### 3.1.4. Cell Labeling

In order to generate an accurate SR image (without missing information from unlabeled target molecules) or to estimate the number of molecules from a SR image, it is important to label the structure of interest to saturation or as completely as possible. Saturation can be assured by immunostaining samples with increasing dilutions of primary antibody (combined with an excess of labeled secondary antibody, for sequential immunostaining experiments). The concentration of the primary antibody should be increased until the fluorescence intensity of the structure, using identical acquisition settings for wide-field microscopy, saturates. The lowest antibody concentration that achieves the maximum possible fluorescence intensity is then used for the experiments. For very dense structures, it is possible to label with antibody concentrations below saturation levels in order to achieve single-molecule localization during acquisition. However, in this instance, it should be noted that the final SR image will not represent the complete structure, so the staining should occur at concentrations as close to saturation as possible. Of course, this labeling characterization only works with primary antibodies with high specificity for their target molecule. If antibodies bind nonspecifically, the fluorescence intensity will saturate at much higher concentrations, where the nonspecific fluorescence will outshine the actual signal.

Fixation and staining of cells will depend upon the particular cell type and target structure to be studied. Careful attention should be given to the choice and concentration of antibodies. As an example, the procedure for staining transmembrane proteins DC-SIGN and Dectin-1 on human dendritic cells is given:

1. Wash cells three times with PBS. Incubate cells in between washes to maintain cell health.

2. Fix with 4% PFA/PBS, by incubating at room temperature for 20 min. Depending on the structure to be visualized, cell permeabilization may be necessary.
3. Wash cells three times with PBS. At this stage, cells can be stored in PBS w/0.05% sodium azide at 4°C for several weeks
4. Block by incubating 30 min or overnight in PBS + 0.1% BSA.
5. Stain with 100 μl of the primary antibody for the first protein of interest (in this example, mouse monoclonal IgG2b specific for human protein DC-SIGN) for 20 min at room temperature with a concentration of ≈10 μg/ml in PBS + 0.1% BSA.
6. Wash cells three times with PBS.
7. Add 100 μl of the secondary antibody (in this example, goat anti-mouse ATTO655-Fab) at a concentration of ≈2 μg/ml in PBS w/0.1% BSA. Incubate at room temperature for 15 min in the dark (covered with foil or in a drawer).
8. Wash cells five times with PBS.
9. If a second color is required, proceed as described in steps 10–14 below. Otherwise see step 15.
10. Stain with 100 μl of the other primary antibody for the second protein of interest (in this example, goat polyclonal IgG specific for human protein Dectin-1) for 20 min at room temperature with a concentration of ≈10 μg/ml in PBS + 0.1% BSA.
11. Wash cells three times with PBS.
12. Add 100 μl of appropriate secondary antibody for the second protein (in this example, rabbit anti-goat Alexa546). Incubate at room temperature for 15 min in the dark (covered with foil or in a drawer).
13. Wash cells five times with PBS.
14. Store cells in 2 ml PBS, at 4°C, sealed with parafilm and protected from light.
15. Visualize with fluorescence for Blink Microscopy.

### 3.2. Microscope Set-Up

A prerequisite for any super-resolution approach that is based on the localization of single molecules is the ability to detect single fluorophores with a good signal-to-noise ratio (SNR). This imposes an upper limit for the resolution that can be achieved for a given dye. It is also important to use a strong laser, combined with a fast camera, which can drastically reduce the image acquisition time and increase the achievable resolution by optimizing the ratio of ON-times to OFF-times.

The task of breaking the diffraction barrier is not trivial and requires attainment of the highest detection efficiency, signal-to-noise ratio and optical resolution. Any trade-offs made regarding the components of the microscope will certainly compromise SR image quality and resolution.

#### 3.2.1. Microscope Body

Any microscope body that has an entrance for a laser light source and a side-/front- or bottom port to attach an EMCCD camera can be used. An autofocus system is not necessary; however, it may be helpful if longer image acquisition times are required. Manual refocusing during the acquisition of a SR image introduces too many vibrations.

#### 3.2.2. Light Source

Blink Microscopy requires a relatively strong CW laser that has at least 100 mW output power. This intensity is needed in order to drive the fluorophore into its dark radical form within milliseconds of illumination. An even higher-powered laser may be beneficial if the light is coupled into an optical fiber, or if a large sample area is to be illuminated.

#### 3.2.3. Filters and Dichroics

High-end filters and dichroics are essential to obtain a suitable signal-to-noise ratio. For most solid-state laser sources, a clean-up filter is required to block nonpeak wavelength light emitted by the laser. The dichroic should be specific for the chosen laser and be able to reflect >90% of the respective wavelength. The emission filter must have an OD > 6 in the range that the clean-up filter has an OD < 3. When multiple lasers are used, care must be taken that each emission filter can completely block each laser. Some filters tend to have low ODs in regions a few hundred nanometers away from their transmission window.

#### 3.2.4. Objective

For sufficient magnification and collection efficiency, high-NA objectives with 60× or 100× magnification should be used. A numerical aperture of at least 1.4 is required for objective-based total internal reflection (TIR) microscopy, which offers a much better SNR than conventional wide-field microscopy.

#### 3.2.5. CCD-Camera

For the detection of single molecules, each emitted photon is important. So the detection efficiency should be as high as possible. Back-illuminated EMCCD cameras are the most widely used, because they offer a quantum efficiency of >90% between 480 and 700 nm. A new generation of CMOS chips can acquire images much faster, but are currently only available in front-illuminated versions that have quantum efficiencies around 50–60%.

#### 3.2.6. Optical Alignment

Beam path optics are provided by the microscope manufacturers. If possible, the illuminated area should be set to an area as small as possible (yet sufficiently large to illuminate the desired field of view) in order to maximize illumination density. Depending on the camera's physical pixel size, additional magnification in the detection path may be used to achieve an image pixel size between 80 nm and 140 nm.

For later analysis it is essential to measure the effective image pixel size for each configuration (objective, additional magnification, and camera). For this purpose, stage micrometers with 10 μm line distances are commercially available.

1. Mount the stage micrometer on the microscope and align it so that the lines appear vertical on the camera.
2. Count the number of lines ($n_{\text{lines}}$) visible on the camera and determine the number of pixels ($n_{\text{px}}$) between the outermost lines.
3. The pixel size in nanometers is: $\frac{n_{\text{lines}} \times 10{,}000\text{nm}}{n_{\text{px}}}$.

### 3.3. Does the Microscope Detect Single Molecules?

Before starting a SR measurement, make sure the microscope is working properly. This can be done by preparing a simple test slide with single fluorophores.

1. Prepare a 5 mg/ml BSA solution.
2. Incubate the BSA solution on a glass slide for 15 min.
3. Wash the slide, mount it on the microscope, and add the labeled antibody used for imaging at a concentration of $\approx 10^{-8}$ M in the BSA solution.
4. Incubate for 5 min and wash two times with the BSA solution (when imaging with a TIR microscope, the dye solution can remain on the slide, because the solution background should not be visible).
5. Cool the camera to −55°C, set the EM gain to 250, make sure the optics are set for TIRF illumination, and adjust the laser power to approximately 10 μW/μm$^2$.
6. Single molecules should be visible on the glass surface. If too many molecules have bound, repeat with reduced concentration of labeled antibody and/or incubation time. If there are no molecules visible, supplement the BSA solution used in the first step with some primary antibodies that will be bound by the labeled secondary antibody.
7. Test for the presence of single molecules by continuously imaging with high laser intensity and observing single-step photobleaching, or count bleaching steps and compare to the expected dye-to-protein ratio calculated according to the degree-of-labeling of the secondary antibody.
8. At this point an *estimate* for the localization precision can be obtained by recording the raw data and analyzing it as if it was actual Blink Microscopy data (see Subheading 3.8). The FWHM of the resulting spots in the super-resolution image reflects the expected resolution for the Blink experiment.

### 3.4. Testing the Resolution with NanoRulers

To ensure that the microscope is not only able to detect single molecules, but is also able to perform super-resolution measurements, it is favorable to actually test for the ability to resolve sub-diffraction distances. Since 2011, there are commercially available calibration standards specifically for this purpose (www.sts-nano.com).

The resolution stated in many publications is based on experimental evidence from cellular structures. This approach is certainly not inaccurate, but may be hard to reproduce and depends on how long the researcher is willing to search for "the right structure." This structure consists of two spots or lines at a distance large enough to be resolved and small enough to demonstrate the resolution advantage of using the respective method (also see Note 1). A novel approach for creating nanoscale distance standards, based on DNA nanotechnology, can supply the required resolution test sample (24). Based on the DNA origami technique (31), fluorophores are placed at specific positions on a 100 × 70 nm grid. Thousands of identical structures, all exhibiting a distance of, 80 nm, for example, can be immobilized on a glass surface and imaged with Blink Microscopy or other complimentary techniques. These standards, named "NanoRulers," are easy to use without further technical knowledge and provide reliable information on the status of the microscope.

### 3.5. Preparing the Dye to Blink

Depending on the fluorophore used, slightly different procedures are required to make the fluorophores blink in the right way for Blink Microscopy. Some exemplary procedures are listed below. They should be considered examples to act as a starting point for optimizing the respective conditions for each cell type and protein of interest. Details on how to fine-tune these conditions can be found in Subheading 3.6.

#### 3.5.1. Cyanine Dyes (e.g., Cy3b and Cy5 (6))

Cy5 is widely used in fluorescence imaging. Previous SR imaging techniques like STORM (2) and dSTORM (8) were first demonstrated with this fluorophore, but rely on a different mechanism (32, 33). The mechanism of Blink Microscopy requires the fluorophore to be reduced from an excited state, forming a radical anion. As in STORM and dSTORM, oxygen must be removed, as it acts as a triplet quencher and oxidant. This can easily be done enzymatically as described in the literature (34, 35). Also, the absence of oxygen drastically increases the photostability of the dye.

1. Prepare an AA working solution by diluting the 100 mM stock 1:10 in PBS.
2. Fill 95% of the LabTek chamber, containing cells that have been stained (as in Subheading 3.1.4), with PBS containing 1% glucose (→700 μl).
3. Add 7 μl of AA working solution.
4. Add 35 μl of oxygen scavenging solution, which should completely fill the chamber.
5. Mix quickly (by pipetting up and down) and seal chamber using sealing sheets.
6. Wait approximately 5–10 min, until oxygen is removed (see Notes 2 and 3).

*3.5.2. Oxazine Dyes (e.g., ATTO655 (22))*

Due to the relatively low energy level of the radical anion of ATTO655, its reduced state is remarkably stable, even in the presence of oxygen. Because of this, oxygen removal is not required to obtain sufficiently long OFF-times for Blink Microscopy. To prevent photobleaching, however, oxygen can be removed and the OFF-times adjusted by addition of an alternative oxidizing agent, such as methyl viologen.

Another characteristic feature of ATTO655 is that the number of photons emitted can be tuned by varying the concentration of AA. High photon numbers offer the advantage of higher localization precision, but at the cost of imaging speed and vice versa. The ideal parameters for each type of experiment are found empirically (22).

1. Prepare a working solution of AA by diluting the 100 mM stock 1:10 in PBS.
2. Fill 95% of the LabTek chamber with PBS (containing 1% glucose if oxygen removal is required).
3. Add 7 μl of AA working solution. No oxygen removal is required.
4. Optimize Blink parameters as discussed in Subheading 3.6.

*3.5.3. Other Dyes*

Many other dyes can potentially be used for Blink Microscopy. We found the fluorophore Alexa 546 to work with the same parameters as given for ATTO655 in the previous section. It may be helpful to also try other reducing agents, like the thiols mercaptoethylamine (MEA) or beta-mercaptoethanol (ME), if ascorbic acid does not show the desired effect (see dSTORM, ref. 26). In the case of thiols, the reducing species is the thiolate, which means that thiols are used at higher concentrations and reproduced through the acid–base equilibrium. For thiols, however, the pH has to be carefully controlled, since the effective thiolate concentration is strongly pH dependent.

***3.6. Finding the Right Blink Parameters (36)***

In a first step, acquire 2000 frames at the fastest camera integration time (≈4 ms) and highest laser power (≈200 mW). Discard the first 10–200 frames, when too many dye molecules are still in their ON-state. Browse through the raw data looking at individual frames. It is essential that the raw data is of good quality, exhibiting single-molecule blink events with a good SNR. The following sections should help to identify possible error sources and eliminate them.

*3.6.1. Check for Single Molecule Events*

When looking at consecutive frames, diffraction limited spots should be identified. They must be perfectly circular (within the limit of the pixel size), uniform in brightness, and appear and disappear within one or two frames. If all spots appear asymmetric or elliptical in the same way, microscope detection optics should be checked. Otherwise, the dyes are not blinking properly and multiple spatially distributed emitters are forming an asymmetric emission pattern.

*3.6.2. Laser Intensity/ON-Times*

Assuming one is imaging with the highest laser intensity possible and has succeeded in imaging single-molecule events, look at individual fluorescent regions over time. Most Blink events (ON-times) should have a length of 1–3 frames. Consider that ON-times are distributed exponentially.

1. If the majority of events exhibit ON-times of three or more frames, the laser intensity can be increased. Alternatively, the ON-times of dyes like ATTO655 can be reduced by adding higher amounts of a reducing agent. If both these measures are not feasible or sufficient, a longer ON-time is acceptable in the case that the OFF time is sufficiently long and/or the sample is not too densely labeled.
2. If all or almost all fluorescent events occur within just one frame, meaning fluorescence starts in one frame and is no longer detected in the next frame, the laser power is set too high and should be decreased until a distribution of ON-times, as described above, is obtained.

*3.6.3. OFF-Times*

For localization-based super-resolution microscopy to work, Blink events from individual molecules within a diffraction-limited area must be separated in time. In principle, just one dark frame in between two Blink events is sufficient, however, the exponential distribution of OFF-times makes it necessary to have a cumulative average OFF-time of at least a few frames. This cumulative OFF-time is the time while all molecules within the area of one diffraction limited spot are in their dark state, and is the cumulative effect of a stochastic distribution of the OFF-times of all single emitters.

If fluorescence events are not separated by several dark frames, try adjusting (decreasing) the oxidizing agent concentration. Alternatively, if the degree-of-labeling of the secondary antibody is greater than two, relabeling of the antibody with a lower dye concentration is recommended.

*3.6.4. Camera Integration Time and Laser Intensity*

According to ref. 36, the camera integration time is ideally set to equal approximately twice the average ON-time of a single fluorophore. In most cases, the fastest possible recording time (5–10 ms) is desired, as this minimizes drift effects. Depending on whether the camera speed or the laser power is the limiting factor, the other must be adjusted to obtain optimal resolution. To increase the camera frame rate, a subset of the CCD pixels can be read out to achieve faster acquisition. This is especially applicable when using TIRF illumination, which often does not excite the whole field-of-view.

### 3.7. Data Collection

*3.7.1. Software*

The collection of data is relatively uncritical, and therefore there are a large range of feasible possibilities. One of the simplest possibilities is the use of the software supplied by the camera manufacturer, combined with subsequent analysis in a separate software program. More elegantly, one could use highly integrated software that controls the whole microscopy system and integrates live display of the super-resolution image as it is recorded. This software is usually provided with commercial microscopes.

*3.7.2. Camera Settings*

1. Cool the camera to –55°C; this can take a few minutes.
2. Enable EM gain and set it to 250 (see Note 4).
3. Enable frame transfer mode. This is an important setting. Otherwise there will be pauses for CCD readout between the camera frames. Photons emitted during this time will not be detected (up to 80%).
4. Set the camera integration time (see Subheading 3.6.4).

*3.7.3. Recording Raw Data*

1. The laser should run at the full power used for SR imaging. When imaging to find the structure of interest and adjust the focus, the laser should be attenuated to about 5 mW to avoid introducing significant photobleaching.
2. It has proven practical to use a fast remote-controlled shutter to protect the sample from photodamage while removing (or adjusting) the attenuator.
3. Before opening the shutter, image acquisition should be started to collect some dark frames. This has the advantage that when the shutter is opened, an image is acquired where all fluorophores are still in their active state. This resembles a standard wide field image that can subsequently be compared to the SR image.
4. Images can be acquired until no more fluorophores are active (complete photobleaching has occurred).

***3.8. Post-processing***

As the raw data of a SR image is a stack of thousands of diffraction-limited images, the information of each frame must be extracted and combined into a single graph (see Note 5). There are several approaches that differ in speed, accuracy, and their level of corrections. If you are using a custom-built microscope, there are several programs freely available on the internet (9, 37, 38). Usually, in a first step, single-molecule Blink events are identified. In this step, algorithms can be applied that exclude emission from multiple molecules, which are identified by shape or intensity. Second, the identified spots are fit to a two-dimensional Gaussian function. The most straightforward approach utilizes a least-square fit. This yields very accurate results, but is relatively time consuming. Simpler alternatives involve the centroid-finding functionalities included in most programming environments, e.g., LabView, Igor, or Matlab. Faster, more specialized algorithms are developed constantly (39).

As an example, a set of Blink Microscopy data was analyzed with three different software packages that are freely available (see

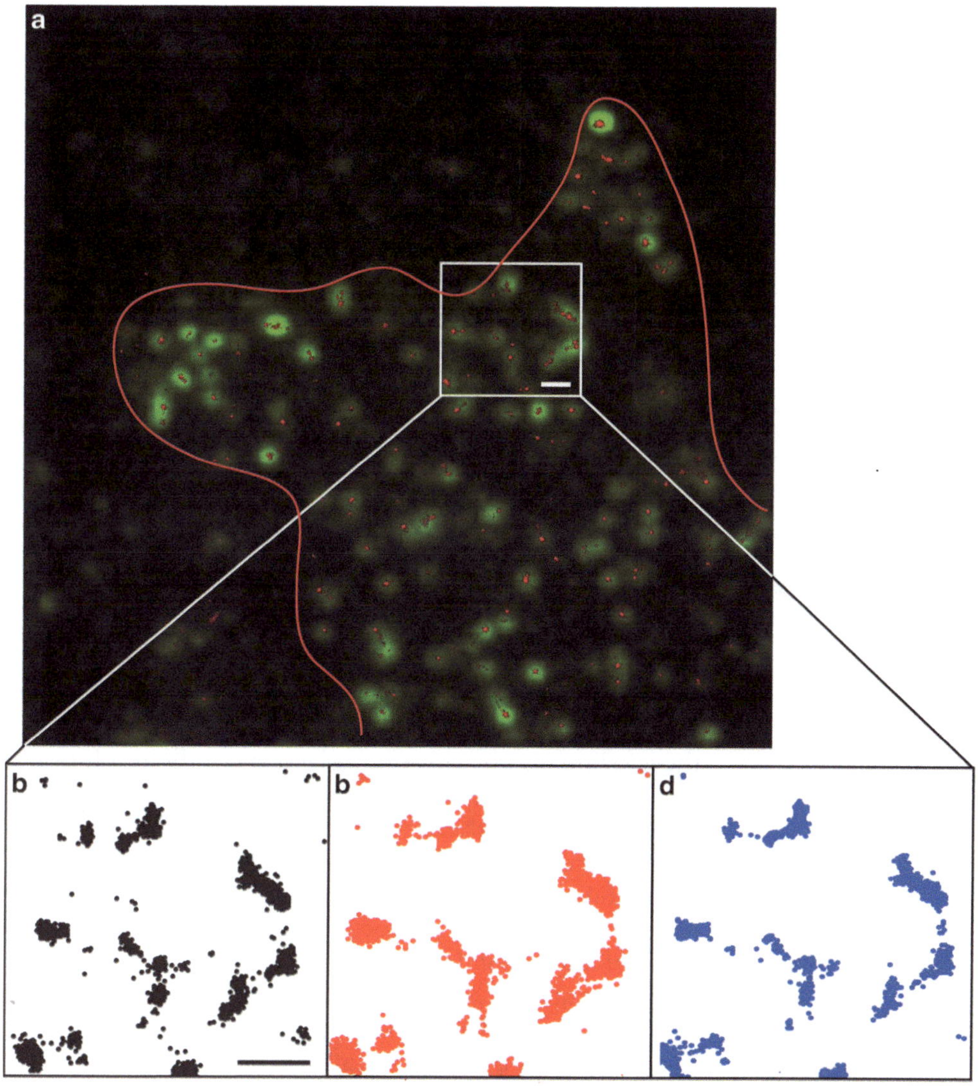


Fig. 3. (**a**) An example of the localization of the protein DC-SIGN on fixed human dendritic cells following sequential immunostaining (with DCN46, Invitrogen, and ATTO655-Fab, ATTO-TEC) and imaging with Blink Microscopy. The standard TIRF image is shown in *green*, overlaid with *red* SR data. The cell boundary is indicated by a *red line*. A subset of SR raw data was analyzed with (**b**) DNA-PAINT software, (**c**) quickPALM software and (**d**) the IGOR standard spot finding algorithm. Each dataset was plotted in Origin to obtain comparable graphical output (unbinned, non-convoluted spot positions). Scale bars are 1 μm.

Fig. 3). The software available from the publication of DNA-PAINT (9), the built-in Igor localization algorithm, and the ImageJ plug-in named quickPALM (38) were employed using standard settings. For better comparison, molecule coordinates were exported and plotted in Origin. This comparison is not meant to identify "the best" software, or to give a comprehensive overview of what is available, but to show that the choice of data analysis software may have an influence on the interpretation of SR data.

## 4. Notes

1. *What is the resolution, actually?* One of the most important and probably most asked questions about super-resolution microscopy is: "what is the best resolution one can get?" Each publication on a new SR method claims to achieve a number as low as possible, and this claim is generally supported by using an experimental example. However, it is hard to predict the resolution limit of a particular experiment in advance (16). In practice, determining the achieved resolution limit will require finding and imaging very small structures that are at or below the resolution limit of the experiment (see discussion in Subheading 3.4). Additionally, there are several aspects relating to the conditions used for labeling that can further limit or enhance the achievable resolution. The first resolution-limiting factor, relating to the density and size of the label, is independent of the imaging method chosen. In general, the higher the degree of labeling, the more detail can be resolved in an image. Therefore, the labeling of the sample must be sufficiently high to show the required amount of detail for the desired resolution. Once again, quoting Nyquist in a different context (40), the labeling density must be at least twice the resolution of the image. In practice, if one wants to discern two objects at a distance of 10 nm, there must be room for a fluorescent label every 5 nm or less. Speaking in a simplified, descriptive way, there must be room for a gap (or label) between the two objects. When using indirect immunostaining methods (using primary and secondary antibodies), the achievable resolution may be limited to >30 nm, because the antibodies are generally 5–15 nm in size. This is one of the reasons why we prefer using $Fab_2$ fragments over whole IgGs.

   Secondly, the resolution of localization-based microscopy indirectly depends on the labeling density, the number of fluorophores per diffraction-limited area. If a structure is densely packed, there is a higher probability that there will be more than one dye in an ON-state within a diffraction-limited area than if the same number of dyes is diffusely spread over a larger area. If a software program attempts to fit a diffraction-limited spot that contains more than one dye, false localizations may be present in the final SR image and the overall resolution of the SR image will be decreased (36). Increasing the OFF-times of the dyes will effectively decrease the labeling density because it will reduce the number of dyes that are emitting in a PSF at a given time. Therefore, imaging a densely packed structure requires longer OFF-times than when imaging a diffusely spread signal. In practice, this means that longer OFF-times are required when imaging a densely packed cluster of proteins than when imaging a cytoskeleton protein that is

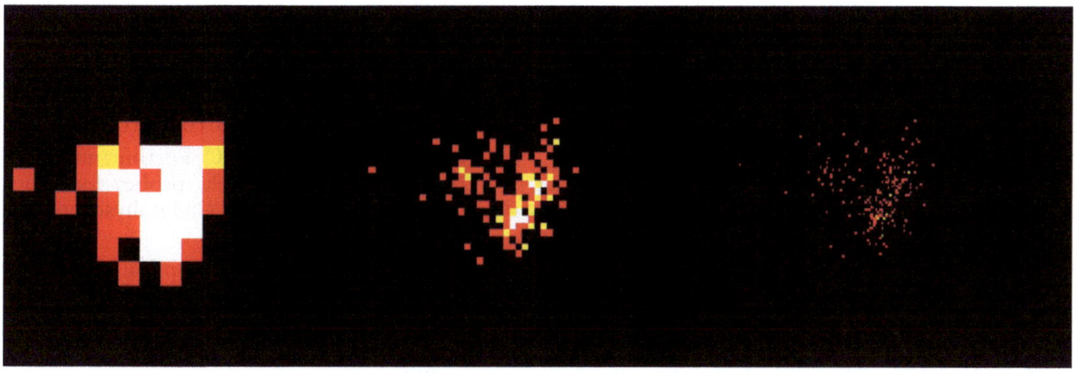

Fig. 4. Effects of bin size on SR image appearance. SR data binned in 30, 10, and 3 nm pixels (*left* to *right*).

diffusely located along a filament (16). In this manner, the resolution obtained in one published experiment is not necessarily what can be achieved in a different context.

2. If there is more than a small bubble of air in the chamber, increase all volumes proportionally.
3. There is no need for an oxidizing agent, as cyanine dyes switch on spontaneously by the remaining oxygen.
4. All current EMCCD cameras offer a linear gain in the range of about 1–1,000. If your camera gain cannot be set to 250, set it to the highest possible value.
5. *It's a graph, not an image.* The output of the software described in Subheading 3.8 reports fluorophore coordinates in a double precision floating-point format (e.g., $x=31.23589062$, $y=9.14215860$). Each coordinate by itself is zero-dimensional, so it does not occupy any space in an image. The pixels in a SR image are created by binning the number of localizations within a certain area (e.g., 10 μm divided into 10 nm pixels). The size of these bins can be chosen arbitrarily and changed in post-processing without loss of information. However, the more important question is: What binning is actually reasonable? According to the sampling theorem of Nyquist (40), the bin size must be at least half the expected resolution of the method. For a resolution of 20 nm, then, the recommended pixel size is somewhere between 5 and 10 nm. Figure 4 shows a particular structure binned in 30, 10 and 3 nm pixels. Choosing the bins too large (left) makes it impossible to determine the substructure in the example. The center binning reveals a structure in the area of interest. When the bins are chosen too small (right panel), mostly single pixels are visible, which do not form an image easily perceived by the eye. The pixel size must therefore be chosen carefully, particularly when automatically analyzing SR images to determine cluster sizes. This should actually be done not on a SR image, but directly on the localization coordinates with specialized algorithms.

## References


1. Betzig E, Patterson GH, Sougrat R et al (2006) Imaging intracellular fluorescent proteins at nanometer resolution. Science 313:1642–1645
2. Rust MJ, Bates M, Zhuang X (2006) Sub-diffraction-limit imaging by stochastic optical reconstruction microscopy (STORM). Nat Methods 3:793–795
3. Hess ST, Girirajan TP, Mason MD (2006) Ultra-high resolution imaging by fluorescence photoactivation localization microscopy. Biophys J 91:4258–4272
4. Sharonov A, Hochstrasser RM (2006) Wide-field subdiffraction imaging by accumulated binding of diffusing probes. Proc Natl Acad Sci USA 103:18911–18916
5. Folling J, Bossi M, Bock H et al (2008) Fluorescence nanoscopy by ground-state depletion and single-molecule return. Nat Methods 5:943–945
6. Steinhauer C, Forthmann C, Vogelsang J, Tinnefeld P (2008) Superresolution microscopy on the basis of engineered dark states. J Am Chem Soc 130:16840–16841
7. van de Linde S, Kasper R, Heilemann M, Sauer M (2008) Photoswitching microscopy with standard fluorophores. Appl Phys B 93: 725–731
8. Heilemann M, van de Linde S, Schuttpelz M et al (2008) Subdiffraction-resolution fluorescence imaging with conventional fluorescent probes. Angew Chem Int Ed Engl 47:6172–6176
9. Jungmann R, Steinhauer C, Scheible M et al (2010) Single-molecule kinetics and super-resolution microscopy by fluorescence imaging of transient binding on DNA origami. Nano Lett 10:4756–4761
10. Gunkel M, Erdel F, Rippe K et al (2009) Dual color localization microscopy of cellular nanostructures. Biotechnol J 4:927–938
11. Flors C, Ravarani CN, Dryden DT (2009) Super-resolution imaging of DNA labelled with intercalating dyes. Chemphyschem 10: 2201–2204
12. Schoen I, Ries J, Klotzsch E et al (2011) Binding-activated localization microscopy of DNA structures. Nano Lett 11:4008–4011
13. Schwering M, Kiel A, Kurz A et al (2011) Far-field nanoscopy with reversible chemical reactions. Angew Chem Int Ed 50:2940–2945
14. Giannone G, Hosy E, Levet F et al (2010) Dynamic superresolution imaging of endogenous proteins on living cells at ultra-high density. Biophys J 99:1303–1310
15. Huang B, Babcock H, Zhuang X (2010) Breaking the diffraction barrier: super-resolution imaging of cells. Cell 143:1047–1058
16. Vogelsang J, Steinhauer C, Forthmann C et al (2010) Make them blink: probes for super-resolution microscopy. Chemphyschem 11: 2475–2490
17. Hell SW (2007) Far-field optical nanoscopy. Science 316:1153–1158
18. Heilemann M, Dedecker P, Hofkens J, Sauer M (2009) Photoswitches: key molecules for subdiffraction-resolution fluorescence imaging and molecular quantification. Laser Photonics Rev 3:180–202
19. Henriques R, Griffiths C, Hesper RE, Mhlanga MM (2011) PALM and STORM: unlocking live-cell super-resolution. Biopolymers 95:322–331
20. Fernandez-Suarez M, Ting AY (2008) Fluorescent probes for super-resolution imaging in living cells. Nat Rev Mol Cell Biol 9: 929–943
21. Vogelsang J, Kasper R, Steinhauer C et al (2008) A reducing and oxidizing system minimizes photobleaching and blinking of fluorescent dyes. Angew Chem Int Ed 47: 5465–5469
22. Vogelsang J, Cordes T, Forthmann C, Steinhauer C, Tinnefeld P (2009) Controlling the fluorescence of ordinary oxazine dyes for single-molecule switching and superresolution microscopy. Proc Natl Acad Sci USA 106: 8107–8112
23. Cordes T, Strackharn M, Stahl SW et al (2010) Resolving single-molecule assembled patterns with superresolution blink-microscopy. Nano Lett 10:645–651
24. Steinhauer C, Jungmann R, Sobey TL et al (2009) DNA origami as a nanoscopic ruler for super-resolution microscopy. Angew Chem Int Ed Engl 48:8870–8873
25. Duim WC, Chen B, Frydman J, Moerner WE (2011) Sub-diffraction imaging of huntingtin protein aggregates by fluorescence blink-microscopy and atomic force microscopy. Chemphyschem 12:2387–2390
26. van de Linde S, Loschberger A, Klein T et al (2011) Direct stochastic optical reconstruction microscopy with standard fluorescent probes. Nat Protoc 6:991–1009
27. Wombacher R, Heidbreder M, van de Linde S et al (2010) Live-cell super-resolution imaging with trimethoprim conjugates. Nat Methods 7:717–719

28. Bierwagen J, Testa I, Folling J, Wenzel D et al (2010) Far-field autofluorescence nanoscopy. Nano Lett 10:4249–4252
29. Cambi A, de Lange F, van Maarseveen NM et al (2004) Microdomains of the C-type lectin DC-SIGN are portals for virus entry into dendritic cells. J Cell Biol 164:145–155
30. Cordes T, Maiser A, Steinhauer C, Schermelleh L, Tinnefeld P (2011) Mechanisms and advancement of antifading agents for fluorescence microscopy and single-molecule spectroscopy. Phys Chem Chem Phys 13: 6699–6709
31. Rothemund PW (2006) Folding DNA to create nanoscale shapes and patterns. Nature 440:297–302
32. Heilemann M, Margeat E, Kasper R, Sauer M, Tinnefeld P (2005) Carbocyanine dyes as efficient reversible single-molecule optical switch. J Am Chem Soc 127:3801–3806
33. Dempsey GT, Bates M, Kowtoniuk WE et al (2009) Photoswitching mechanism of cyanine dyes. J Am Chem Soc 131:18192–18193
34. Aitken CE, Marshall RA, Puglisi JD (2008) An oxygen scavenging system for improvement of dye stability in single-molecule fluorescence experiments. Biophys J 94:1826–1835
35. Rasnik I, McKinney SA, Ha T (2006) Nonblinking and long-lasting single-molecule fluorescence imaging. Nat Methods 3:891–893
36. Steinhauer C, Forthmann C, Jungmann R et al (2011) Nanoscopy using localization and temporal separation of fluorescence from single molecules. In: Bartolo BD, Collins J (eds) Biophotonics: spectroscopy, imaging, sensing, and manipulation. Springer, New York
37. Hedde PN, Fuchs J, Oswald F, Wiedenmann J, Nienhaus GU (2009) Online image analysis software for photoactivation localization microscopy. Nat Methods 6:689–690
38. Henriques R, Lelek M, Fornasiero E et al (2010) QuickPALM: 3D real-time photoactivation nanoscopy image processing in ImageJ. Nat Methods 7:339–340
39. Wolter S, Schuttpelz M, Tscherepanow M et al (2010) Real-time computation of subdiffraction-resolution fluorescence images. J Microsc 237:12–22
40. Nyquist H (1928) Certain topics in telegraph transmission theory. Trans AIEE 47:617–644

# Chapter 9

# Photoswitchable Fluorophores for Single-Molecule Localization Microscopy

**Kieran Finan, Benjamin Flottmann, and Mike Heilemann**

## Abstract

Over the past decade, fluorescence microscopy has been revolutionized by the development of novel techniques that allow near-molecular resolution. Many such methods—collectively referred to as "single-molecule localization microscopy" (SMLM)—are based upon the repeated imaging of sparse stochastic subsets of the fluorophores in a sample. Active fluorophores are localized by finding the centers of their point spread functions, and a super-resolution image is constructed.

Key to this strategy is the use of fluorophores that can be switched "on" and "off" in a controllable manner. Here we review the strengths and weaknesses of the wide variety of SMLM-compatible photoswitchable fluorophores and labeling strategies currently available. We also discuss their suitability for live-cell and multicolor imaging, as well as molecular counting.

**Key words:** PALM, Photoactivated localization microscopy, STORM, *d*STORM, Stochastic optical reconstruction microscopy, Super-resolution, SMLM, Single-molecule localization microscopy

## 1. Introduction

### *1.1. Conventional Fluorescence Microscopy Has a Limited Resolution*

Fluorescence microscopy is a proven and powerful part of the modern biologist's toolbox; the information it provides on the spatial distribution of biomolecules has been essential for the understanding of many fundamental cellular processes.

However, it has—until recently—been unable to reveal the fine structural details of the cell. This limitation arises directly from the wave-nature of light, and the structure of the objectives used to detect it: when a point-source of light is visualized using a conventional microscope, it does not appear as a single point, but rather, as a broad spot, or "point spread function" (PSF). The position of the source can be easily determined by finding the center of the PSF (1). However, when several point sources (e.g., fluorophores) are

Alioscka A. Sousa and Michael J. Kruhlak (eds.), *Nanoimaging: Methods and Protocols*, Methods in Molecular Biology, vol. 950, DOI 10.1007/978-1-62703-137-0_9, © Springer Science+Business Media, LLC 2013

close enough to one another, their PSFs overlap, and their precise locations can no longer be determined.

This fundamental law was first expressed by Ernst Abbe (2), who noted that two parallel lines can only be distinguished by a light microscope if their spacing is greater than $\lambda/2NA$, where $\lambda$ is the detection wavelength, and NA is the numerical aperture of the objective. On a typical microscope, this resolution "limit" is ~200 nm in the imaging plane and >500 nm along the optical axis (for $\lambda = 500$ nm). Unfortunately, many biological structures are smaller than the PSFs of the fluorophores used to visualize them, and so are inaccessible to conventional fluorescence microscopy. Although electron microscopy has—to some extent—been able to fill this void, it has its own limitations related to sample preparation and contrast.

### 1.2. Super-Resolution Fluorescence Microscopy

Several recently developed methods allow the resolution of fluorescence microscopy to be increased >10-fold (i.e., to <20 nm; (3–5)). One such strategy—termed "single-molecule localization microscopy" (SMLM) (6)—prevents the PSFs of individual fluorophores from overlapping by sequentially imaging sparse stochastic subsets of the molecules in a sample (7–10). These fluorophores are localized to a high precision by finding the centers of their PSF (1), and their coordinates used to reconstruct a "super-resolution" image. A wealth of different SMLM-based methods have been developed, together with a list of acronyms that reflect technical differences; some well-known abbreviations are PALM (7), fPALM (11), STORM (8), *d*STORM (9), GSDIM (12), SPDM (13), and BM (14).

In conventional fluorescence microscopy, most of the fluorophores in a sample are active, and individual PSFs are not visible. The key to localization-based methods is the use of fluorophores that, for a particular set of excitation and emission wavelengths, can be switched in a controlled manner from a nonfluorescent "off" state to a fluorescent "on" state, and then back again to an "off" state. Although many such fluorophores exist, all have limitations. Thus, a key decision in the development of an SMLM experiment is choosing which to use. Here, we will review the advantages and disadvantages of currently available photoswitchable fluorophores and labeling strategies, with the aim of aiding investigators new to SMLM.

### 1.3. General Considerations: What Makes a Good Photoswitchable Fluorophore?

#### 1.3.1. Photons Emitted per "Burst"

All fluorophores currently used for SMLM emit a limited number of photons before switching into an "off" state. Those that emit large numbers of photons can be localized more precisely because their PSFs can more clearly be distinguished from background noise, and because the effects of "shot" noise are minimized. When background noise is negligible (as is often the case in SMLM), "shot" noise is dominant, and the accuracy with which the center of a PSF can be determined follows directly from the number of photons that created it (1). In theory, the detection of 1,000 photons in a single PSF yields a localization uncertainty of ±5 nm. For 100 photons, the uncertainty increases to ±15 nm (1). In practice, the signal is spread over several pixels of a detector device (camera), and a minimum of 100 photons per single fluorophore is desirable.

*1.3.2. "Off" State Fluorescence*

Many photoswitchable fluorophores are weakly fluorescent in their "off" states. When fluorophores are densely packed, this "background" fluorescence can become so dominant as to obscure the PSFs of fluorophores in the "on" state. When molecular densities are low, however, such background can become negligible. Thus, the extent to which "off" state fluorescence is a problem depends upon the sample to be imaged.

*1.3.3. "On"/"off" State Ratios*

For virtually every fluorophore that has been used for SMLM, there is a practical limit to the fraction of total fluorophores that can be in the "off" state during imaging. Even fluorophores that are synthesized in an "off" state (e.g., paGFP; (15)) are often activated by the lasers used to visualize them. The importance of this "on"/"off" ratio depends upon the density of fluorophores in a given sample; if the fraction of fluorophores in the "off" state is not large enough, individual PSFs will not be distinguishable, and imaging artifacts can arise (16, 17).

*1.3.4. Labeling Strategies*

The investment of resources necessary to produce a super-resolution image depends greatly upon the labeling strategy employed. Immunofluorescence-based methods using conventional fluorophores can produce an image in as little as a day, while producing a stable-cell line expressing a fluorescently labeled protein can take months. The advantages and disadvantages of different labeling strategies are outlined in Subheading 4.

*1.3.5. Photoswitching Cycles*

Photoswitchable fluorophores differ in the number of photoswitching cycles they can perform. Irreversibly photoactivatable fluorophores, for example, are typically activated and read out only once, yielding a single recorded localization per molecule. Reversibly photoswitchable molecules, conversely, yield many temporally separated photon "bursts" per experiment; the corresponding localizations then form a tight cluster around the position of the fluorophore. Thus different classes of fluorophores yield different types of images which will require different analysis strategies.

## 2. Photoswitchable Fluorescent Proteins

A simple way of determining the distribution of a protein of interest using SMLM is to fuse it to an appropriate fluorescent protein. A wide variety of fluorescent proteins can be switched—using different wavelengths of light—from an "off" state to and "on" state,

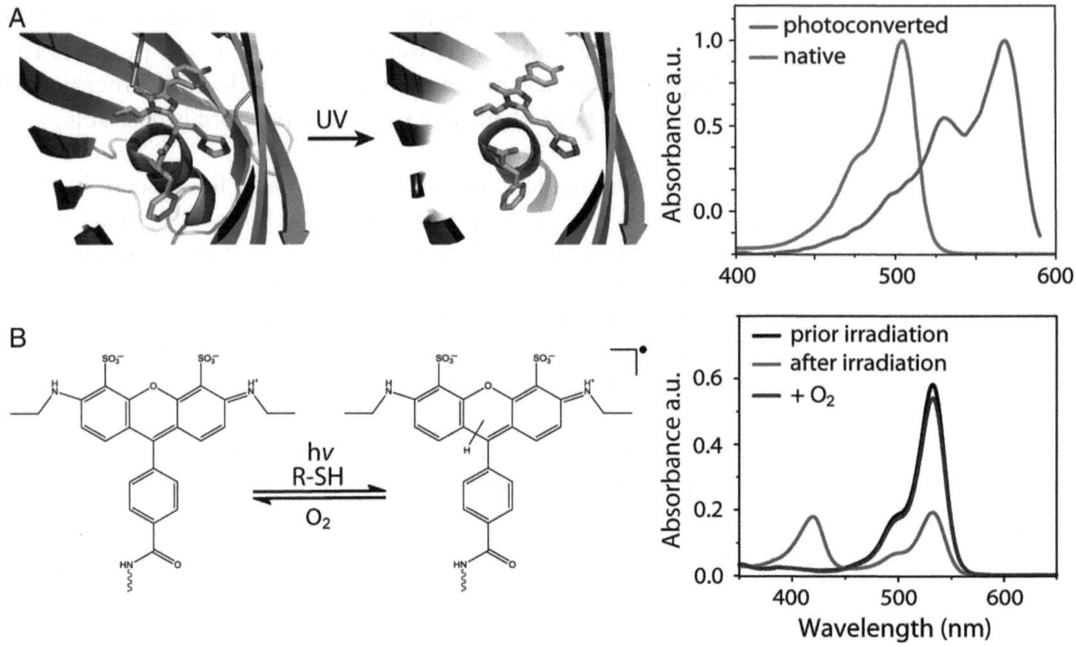


Fig. 1. Photophysical states involved in photoswitching. (**a**) The structure and absorption spectra of mEosFPthermo before (*left*) and after (*right*) activation. UV light induces an intramolecular reaction in which the protein backbone is cleaved and the $\pi$-system of the chromophore is extended. As a result, fluorescence absorption is red-shifted. (**b**) The structure and absorption of Alexa Fluor 532 in its active state (*left*) and dark state (*right*). Under intense illumination with excitation light, the fluorophore is driven into a triplet state and immediately reduced by thiols present in the imaging buffer. The formed radical shows no fluorescence and an increased absorption in the blue spectral range. Irradiation in the new absorption band drives a reaction with oxygen and allows the initial ground state of the fluorophore to be recovered.

and back to an "off" state (18); here, we refer to all as photoswitchable fluorescent proteins (psFPs). Most psFPs are synthesized in their "off" state, and are switched "on" by low doses of near-UV (~405 nm) light. This "on" to "off" transition often involves an irreversible chemical bond breakage (Fig. 1a). They are typically switched "off" again by the laser line used to excite them (often through photobleaching)—but only after they emit a burst of photons (Table 1).

Many of the advantages of using fluorescent proteins are associated with their ability to be genetically fused to their targets (see Subheading 4.2). However, there are also drawbacks associated with this class of fluorophores. psFPs tend to emit less photons per "burst" than fluorescent dyes, and so often produce lower resolution images. "Off" state fluorescence is a reoccurring problem, and can be as high as 1% of "on" state fluorescence. Oligomerization is also a concern, as many psFPs are derived from parents that naturally exist as tetramers. Although protein engineers attempt to abolish this behavior, some residual tendency for aggregation can remain.

In this section, we review a selection of photoactivatable and photoswitchable fluorescent proteins. For a more detailed view on this rapidly growing field, we refer to excellent and recent reviews (18, 19).

**Table 1**
**Photoswitchable fluorescent proteins**

| Name | Ex. Max. | Em. Max. | "On"/"off" contrast | Brightness (×10³) | References |
|---|---|---|---|---|---|
| PAGFP | 504 | 517 | 70 | 3 | (15) |
| PSCFP2 | 490 | 511 | Poor | 11 | (17, 22) |
| Dronpa | 503 | 522 | Good | 85 | (17, 25) |
| BSDronpa | 460 | 504 | 17 | 22 | (25) |
| Padron | 503 | 522 | 143 | 27 | (25) |
| RSFastLime | 496 | 518 | 67 | 27 | (25) |
| IrisFP (green) | 486 | 516 | | 25 | (21) |
| PAmCherry1 | 564 | 595 | 4,000 | 8 | (32) |
| PATagRFP | 562 | 595 | 540 | 25 | (33) |
| Dendra2 | 553 | 573 | | 19 | (29) |
| mKikGR | 580 | 591 | | 33 | (30) |
| tdEosFP | 571 | 581 | 200 | 20 | (26) |
| mEos2 | 573 | 584 | | 30 | (23) |
| GFP* | 500–550 | 560 | | | (36) |
| mIrisFP(red) | 546 | 578 | | 19 | (21) |
| PSmOrange | 636 | 662 | | 9 | (34) |

***2.1. Green Fluorescent Proteins***

Although GFP itself has intrinsic photoswitching properties (20), the first effective photoswitchable fluorescent protein was paGFP (15). It is synthesized in a dark "off" state, but is irreversibly converted by irradiation with 405 nm light into a green-emitting, 488 nm excitable "on" state which can emit ~160 detectable photons before bleaching (21). psCFP2 (22) is another fluorescent protein that emits in the same spectral window (green) as paGFP, but yields more detectable photons (~250 median; (23)). In its "off" state, it emits blue photons when irradiated with 405 nm light; excitation of this blue form irreversibly converts the protein to a green-emitting "on" state excitable by 488 nm light. Unfortunately neither paGFP nor psCFP2 perform well in densely labeled samples; both have substantial "off" state fluorescence (~1% of "on" state for paGFP; (15, 17)), and both are also activated to some extent by the 488 nm laser used to image their "on" states.

Several "reversible" green fluorescent proteins have also been developed. The first to be used for SMLM was Dronpa. This FP emits in the green channel, and has its excitation peak at 503 nm (24). Upon excitation, it emits a median of ~120 detectable photons (17) before entering a stable dark state (with a half life of

840 min; (25)). UV illumination can return it to its "on" state. Dronpa can cycle many times before permanently photobleaching (24). It also appears to have a lower "off" state fluorescence than psCFP2 or paGFP, and so may be more appropriate for high density labeling. Several variants of Dronpa have been created that are slightly brighter or have different spectral properties; some have been used in SMLM experiments (25). One variant, Padron (25), has the opposite photophysical properties to Dronpa; it is switched "on," and excited by 488 nm light, but switched "off" by 405 nm light. As Padron is not deactivated by excitation light, it may have a high photon yield.

IrisFP (21) is a complex photoactivatable FP with multiple switching behaviors. It initially emits in the green band under 488 nm excitation, but after releasing ~170 detectable photons, switches to an inactive state. Exposure to 405 nm light drives the protein back into its original green-emitting state, while further UV exposure converts the green FP to an orange-emitting form excitable by a 568 nm light. This later form emits ~350 detectable photons before photobleaching.

### *2.2. Orange Fluorescent Proteins*

One of the first fluorescent proteins used in SMLM was the orange fluorescent protein EosFP (7). This protein is synthesized as a green emitting FP excitable by 488 nm, but can be—by exposure to UV light—converted to a orange-emitting "on" state excitable by 568 nm (26). The monomeric form of this protein matures poorly at 37°C, and thus seems unsuitable for expression in mammalian cells (26). However, its tandem dimer tdEos is fully functional at 37°C, and has been used extensively in SMLM, emitting a median ~450 detectable photons before photobleaching (23).

An effort to improve EosFP led to the creation of mEos2, a variant that matures correctly at 37°C when expressed as a single fusion protein (i.e., not as a tandem dimer). It is now widely used in SMLM, and has an excellent photon yield (median ~360 detected; (23)). Unfortunately, mEos2 has several properties that can cause problems in SMLM experiments. It has a tendency to aggregate in vitro, and thus may cause artifactual clustering of its fusion partner (23). Activated mEos2 molecules have also been shown to enter long-lived dark states before photobleaching (27, 28); thus in a typical SMLM experiment, individual molecules can be localized several times. For quantitative studies of protein clustering, careful control experiments are needed, as bona fide molecular clusters cannot be easily distinguished from single mEos2 molecules that have been localized multiple times. Similar blinking behavior has also been reported for its EosFP parent (26). Dendra2 and mKikGR are both monomeric green-to-orange photocovertable FPs that lack the tendency of mEos2 to aggregate (29, 30); however, neither yields as many photons as mEos2 (e.g., ~130 median detected for Dendra2; (23, 31)).

Several dark-to-orange photoactivatable FPs have also been developed. PAmCherry1 is reported to yield as many photons as mEosFP (median 413 detected under 568 nm excitation; (32)) and has virtually no "off" state fluorescence. PATagRFP, a second dark-to-red FP, is three times brighter, and roughly twice as photostable as PAmCherry1 (33) in ensemble measurements, and so should also be useful in SMLM experiments. As the fluorescence emission of PATagRFP and PAmCherry1 are spectrally distinct from those of the green fluorescent proteins noted above, they can be used in two-color experiments.

### 2.3. Red Fluorescent Proteins

Recently, a new orange-to-red photoswitchable FP, PSmOrange, has been introduced. This psFP is revolutionary as it is the first to possess a red-emitting "on" state. It is initially orange, with peak emission at 565, but, upon exposure to 488 nm light, switches to a red-emitting FP with excitation and emission spectra comparable to those of Cy5 (34). Unfortunately, the photophysical properties of this psFP do not allow it to be localized as accurately as other psFPs (e.g., PAmCherry1). However, its red excitation/emission spectra may make it useful for in vivo experiments.

### 2.4. Conventional Fluorescent Proteins

In addition to fluorescent proteins that have been designed specifically for photoswitching, a number of conventional FPs have been used as photoswitches. GFP, for example, has long been known to enter stable nonfluorescent off states under constant 488 nm excitation; molecules can be returned to their active states by irradiation with UV light (20). EYFP can be similarly operated as a photoswitch, and has been used for super-resolution imaging in bacteria (35). A different strategy has been presented by the Sedat lab who used "oxidative reddening" to operate EGFP as a photoswitch, and demonstrated super-resolution imaging of histone proteins (36).

### 2.5. Preserving Fluorescent Protein Activity

Although the fluorophore inside a fluorescent protein is somewhat isolated from their environment by the beta barrel surrounding it, their photophysical properties can be easily altered, or destroyed, by exposure to buffers components. Chemical fixation, for example, can inactivate or reduce the activity of psFPs. We and others have found that glutaraldehyde is particularly destructive (18), while formaldehyde has more moderate effects (e.g., formaldehyde does not affect GFP fluorescence; (37)). Other data suggests that fixation can—in some cases—reduce, but not destroy the photon yield of psFPs (32). Mounting media might also have effects on the switching of psFPs. Such media are used to match the refractive index of a sample to that of the coverslip, in order to avoid imaging aberrations. However, most SMLM studies to date have not used mounting media but rather, immersed the samples in innocuous aqueous buffers (e.g., PBS or PHEM; (17, 32, 38)), as a result, little is known about the effects of mounting media on psFP activity.

However, one can speculate that the redox environments and reduced oxygen levels of some mounting media might have strong effects on psFP activity. A recent study, for example, has found that reducing agents and oxygen depletion can both influence psFP behavior (28) (the influence of redox reagents is further discussed in Subheadings 5.1 and 5.2.4). However, we do note that recent SMLM studies that did not use special mounting media have achieved excellent results, probably because they observed only molecules found relatively close to the coverslip (using total internal reflection microscopy, which requires refractive index mismatch). As SMLM is applied to molecules found deeper in samples, we imagine that the issue of mounting media compatibility will become increasingly important.

## 3. Photoswitchable Fluorescent Dyes

Although psFPs are powerful tools for SMLM, it is sometimes desirable to use fluorescent dyes instead. These fluorophores exhibit photon yields that are considerably higher than those of fluorescent proteins, a fact that directly translates to increased localization accuracy and spatial resolution. Under standard imaging conditions (e.g., phosphate buffer saline), synthetic dyes typically spend most of their time in an "on" state. However, several different strategies have been developed that allow dyes to be switched "off," most of which involve redox chemical processes. Novel dyes, which can be activated by the light-induced cleavage of a quenching group (39) have also been developed, and thus can be operated similarly to irreversibly activatable FPs.

### *3.1. Conventional Fluorophores*

Many standard, commercially available dyes can be operated as photoswitches when imaged in buffers that allow excited fluorophores to stochastically enter reversible but long-lived dark states (e.g., containing oxygen scavengers and/or reducing agents). Under these conditions, fluorophores continuously exposed to a moderate intensity of excitation light begin in an "on" state, but transit quickly into a stable "off" state. Individual molecules exit their "off" states stochastically (induced by e.g., "activation" light or oxygen), and emit a burst of photons before being once again driven into an "off" state. The fraction of molecules in their "off" state is determined by the excitation laser intensity and buffer conditions; both should be optimized such that single fluorophores are distinguishable.

One bright and reliable photoswitch is Cy5 (or its derivative, Alexa Fluor 647; Fig. 2). When imaged (using 647 nm excitation) in buffers containing the reducing agent β-mercaptoethylamine (MEA) and lacking oxygen, it emits several thousand detectable

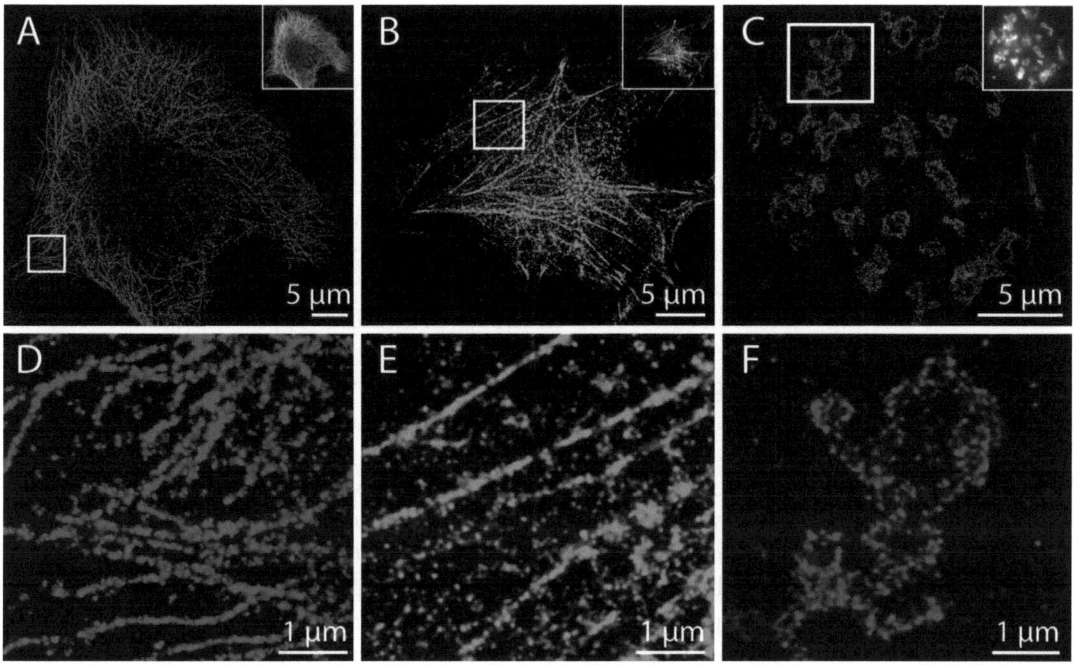


Fig. 2. SMLM imaging of fixed HeLa cells. (a) Alpha-tubulin was immunostained with Alexa Fluor 647 and then imaged in PBS + 50 mM MEA (pH 7.4) using 647 nm excitation light (20,000 frames). Fluorophores in the dark state were controllably reactivated using 405 nm light. (**b**) Cells transfected with actin-mEos2 plasmids were imaged with an excitation wavelength of 568 nm and an activation wavelength of 405 nm (8,000 frames). (**c**) Giantin was immunostained with Alexa Fluor 647 and imaged in the presence of 100 mM MEA (pH 7.4) and an oxygen scavenger (6,000 frames) using 647 nm excitation. (**d**), (**e**), and (**f**) show magnified sections of the images in (**a**), (**b**), and (**c**).

photons before entering a long-lived dark state ($\tau_{1/2}$ = 23 min; (40)). Inactive molecules can be stochastically and controllably brought back into their active states by irradiation with short wavelength light (405 nm, 488 nm, and 532 nm laser lines are all adequate). The structurally similar carbocyanines Cy5.5, Cy7 and Alexa Fluor 700 exhibit similar photoswitching properties (41).

A wide variety of other conventional fluorescent dyes, ranging from the green dye Alexa Fluor 488 to the far red dye ATTO 700, can also be used as reversible photoswitches in the presence of varying concentrations of MEA (Fig. 2; (42)). Such approaches are straightforward, but require freshly prepared MEA solution as well as careful control of buffer pH.

The mechanism through which reducing thiols create dark states is, for many fluorophores, now well understood. In the case of Rhodamine fluorophores such as Alexa Fluor 488, the nonfluorescent state that is generated has been identified as a radical anion using both absorption spectra and electron paramagnetic resonance (EPR) spectroscopy (Fig. 1b; (43)). These radical states have a lifetime of minutes in aqueous buffers, and are thus well suited for stochastic super-resolution imaging of dense samples (where only a small subset of all fluorophores present can be in the fluorescent state; (16)). Oxazine fluorophores, such as ATTO655, exhibit only very weak EPR signals (43); however, Fourier transform infrared (FT-IR) spectroscopy and theoretical calculations have demonstrated that the photoswitched nonfluorescent state is exhibited by fully reduced molecules (44). Photoswitching can often be controlled by adding both reducing and oxidizing reagents to buffers, as in the reducing and oxidizing system (ROXS) approach (45). ROXS allows engineering dark states by adjusting the concentrations of the buffer components that make SMLM possible (14). A number of similar approaches that rely on "on"/"off" switching of conventional synthetic fluorophores through different photophysical routes have been published in the recent years (46–48).

### 3.2. Activator–Reporter Dye Pairs

Pairs of fluorophores held in close spatial proximity can form systems with unique photoswitching properties (49). These photosystems consist of a "reporter" dye (e.g., Cy5), which is imaged directly, and an "activator" dye (e.g., Cy3) placed nearby. When the reporter dye is excited in a buffer containing a reducing agent and lacking oxygen, it quickly enters a dark state (as with direct switching), but can be reactivated by exciting the "activator" dye.

Replacing the "activator" and "reporter" with different dyes (e.g., replacing Cy5 with Cy5.5 or Cy7) yields probes that emit, or are activated, at different wavelengths (41). Such photoswitches are particularly useful for multicolor imaging. Although activator–reporter pairs are a proven tool for SMLM, they have the one drawback that they are not commercially available and so must be synthesized by the user.

### 3.3. Caged Fluorophores

Caged fluorophores are nonfluorescent in their original state and are activated by irradiation with light. Activation involves a chemical bond breakage (often, internal esters or lactones are used) which modifies the chemical structure of the synthetic fluorophore (e.g., establishes a conjugated $\pi$-system), and causes the probes to become fluorescent. Some caged fluorophores have been optimized for super-resolution imaging (39, 50). This strategy allows the number of fluorophores active in a sample to be finely controlled by modulating the intensity of activation light (much like irreversibly activatable psFPs).

In general, SMLM can be performed using any fluorophore that can be driven into a dark state, and a number of approaches have been developed in the past years which are beyond the scope of this article. The average time that the fluorophore resides in this dark state, however, has to be long enough to generate a single-molecule density at the sample (16). Noticeably, an early experimental demonstration of super-resolution imaging through stochastic switching used quantum dots as fluorophores, which exhibit an intrinsic blinking (51).

## 4. Labeling Strategies

### 4.1. Antibodies

One of the easiest ways to label a biomolecule of interest with photoswitchable fluorophores is to use an antibody. This strategy (i.e., immunohistochemistry) has long been a standard method in molecular biology, and has been described in detail elsewhere (52). Typically, cells are fixed, permeabilized, incubated with an antibody that binds the protein of interest (i.e., the primary antibody), and then incubated with a second, fluorescently labeled antibody which binds to the first.

This labeling strategy has multiple advantages. It can be performed on any fixed sample (including primary tissue), and does not require genetic manipulation of target cells. Samples can, theoretically, be labeled and imaged in a single day, although optimization of the labeling procedures is often required in practice. Importantly, labeling can be accomplished using commercially available primary and fluorophore-conjugated secondary antibodies.

However, the technique also has substantial drawbacks. It is difficult, for example, to prove that a primary antibody is binding specifically to its intended target, and not nonspecifically to other antigens, or "sticky" cellular structures. It is also nontrivial to determine what fraction of the target protein is bound; antigens can go unlabeled because they are sterically inaccessible to antibodies, or because they are unrecognizably denatured by fixatives (53). Both problems can be reduced, but not eliminated, by using "light" fixation procedures (e.g., a low concentration of formaldehyde).

Another notable disadvantage of traditional immunohistochemistry is that the complex formed by primary and secondary antibodies leaves fluorophores ~10–20 nm away from their target (Fig. 3). As a result, the resolution achievable using this method will be >20 nm even if individual fluorophores that mark a target can be localized to a much higher accuracy. This problem can be lessened by labeling primary antibodies directly with fluorophores (using e.g., NHS-esters), or by isolating and labeling only the part of the primary antibody that binds the antigen (i.e., the Fab). The use of commercial antibodies heavily labeled with fluorophores can also create problems. When two fluorophores are within close proximity of each other, their photophysical properties can change dramatically. As a result, fluorophores bound to highly labeled antibodies can often fail to switch.

### 4.2. Genetic Tags

Another widely used strategy for labeling a protein of interest with a photoswitchable fluorophore is to alter the target's genetic code so that it is fused (through its N- or C terminus) to a genetically encoded tag. Such tags are usually fluorescent proteins, but can also be "chemical tags," which bind small fluorescent molecules added to the surrounding media.

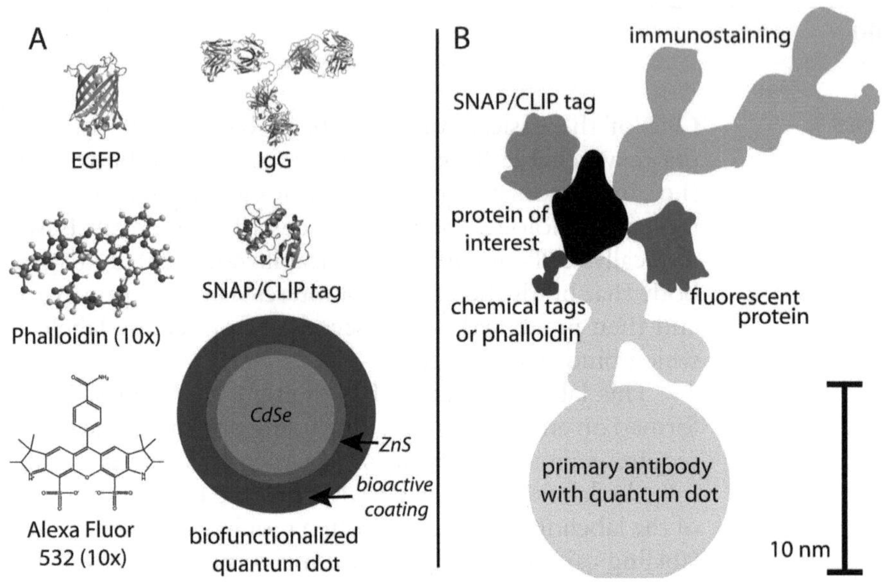


Fig. 3. An overview of the tools currently available for the labeling of biomolecules with photoswitchable fluorophores. (**a**) Comparison of the size of different molecules used for fluorescence labeling. (**b**) An image of a hypothetical protein labeled using different strategies. Antibodies leave their fluorophores ~15 nm away from their target, while fluorescent proteins and SNAP tags bring their fluorophores to within 1–2 nm.

Although the genetic tagging of a protein is a relatively straightforward process, it can also be time-consuming. In bacteria and yeast, tagging can usually be accomplished by directly altering the genomic DNA of the organism and effectively replacing the endogenously expressed protein (54). In mammalian cells, plasmids expressing the tagged protein can be introduced transiently into cells using various transfection methods. However, more reliable biological results are often obtained when "stable" transfectants are selected (55); such cells have incorporated the exogenous DNA directly into their chromosomes, and thus do not have to be exposed to toxic transfection reagents directly before being imaged. Unfortunately mammalian cells usually express endogenous, unlabeled versions of target proteins. This problem can be solved by simultaneously knocking down the native protein with RNAi (targeting 3′UTRs), or by tagging the endogenous protein directly using zinc-finger nucleases.

Several crucial controls must be performed when using genetically encoded tags. As the presence of a bulky tag can reduce or eliminate the activity of a protein of interest, fusion proteins must be shown to be functional. One must also ensure that the tag being visualized in the cell is indeed bound to its target using a western blot (as cleavage at linker regions is not uncommon). Such blots can also reveal the degree to which a protein is overexpressed. Excess amounts of a target protein can cause mislocalization/aggregation in addition to adverse physiological effects.

In spite of these difficulties, genetic tags remain an extremely effective way of localizing cellular proteins. If used correctly, they can allow for the even labeling and visualization of most, if not all, molecules of a given protein. They are also able to position fluorophores very close to target molecules, a feature that should become increasingly important as SMLM-based methods mature (Fig. 3). In addition, they allow for the tagging of proteins for which no effective antibody exists.

#### 4.2.1. Fluorescent Proteins

Fluorescent proteins are the most commonly used genetically encoded tag. When a psFP-fusion protein is expressed, it is very difficult to determine what fraction of psFPs are indeed fluorescent. However, several lines of evidence suggest that, in many cases, the majority of expressed fluorescent proteins are indeed active. The crystal structures of many psFPs expressed in *Escherichia coli* suggest that they form a structurally homogeneous population (e.g., (56)). Moreover, the demonstration of high FRET efficiency between intracellular proteins in vivo suggests that the majority of acceptor FPs (i.e., YFP or RFP) were correctly folded and active (57).

#### 4.2.2. "Chemical" Tags

Although fluorescent proteins are powerful tools for superresolution, they have low photon yields. Thus, protein-based "chemical tags" have been developed that allow labeling with superior synthetic fluorophores while maintaining many of the advantages of genetic tagging.

These systems consist of two components: the first is a small protein tag that has a high affinity for a specific small molecule, and can be genetically fused to a protein of interest. The second component is the "label": the small molecule conjugated to a fluorophore. The label binds the tag with high affinity, thus labeling the protein of interest. The labeling of targets within living cells can be accomplished by adding membrane-permeable labels directly to growth media. For fixed and permeabilized cells, any label can be added to the surrounding solution. In both cases, unbound label can be washed away, leaving (ideally) only fluorophores bound to the target protein.

The "SNAP" tag was the first such tag to be developed (58), and is based upon the affinity of the human DNA repair enzyme O6-alkylguanin-DNA alkyltransferase (hAGT) to its substrate, O6-benzyl guanine. A mutant version of hAGT, the SNAP tag covalently binds benzyl-guanine-conjugated fluorophores with high affinity (59). A second mutant of hAGT, the CLIP tag (60), was engineered to bind to benzyl-cytosine-based labels, and has been used together with the SNAP tag in dual-color imaging. Several other chemical tags have also been developed; the Halo tag (61) covalently binds alkylhalide-modified fluorophores, while the eDHFR tag (62) binds non-covalently to trimethoprim conjugates.

All of these tags have been used to label specific proteins with photoswitchable fluorophores; SNAP has been used with the commercially available and membrane-permeant SNAP-Cell 505 and TMR-Star conjugates (63), although other membrane inpermeant dyes such as Alexa Fluor 647 can also be used in permeabilized fixed cells. The eDHFR and Halo tags have also been used with ATTO 655- and DCDHF-based conjugates (64, 65). Chemical tags can be used in live cell experiments, as many membrane-permeable labels have also been shown to photoswitch in the cellular milieu (47, 63, 64).

The nonspecific binding of fluorophore conjugates is a major problem associated with chemical tags. The resultant increase in background varies depending on the conjugate, cell line, and wash conditions used, but can often be extremely high. We have found nonspecific labeling of mitochondria to be particularly common. Such background can be reduced by using low concentrations of conjugates; however, this strategy risks under-labeling of tags. One promising approach for eliminating background is the use of fluorophore conjugates that are initially quenched, but activated upon reacting with their tags. Such conjugates are being developed and may soon be commercialized (59).

### 4.3. Other Labeling Approaches

A wide variety of other labeling strategies have been developed (66) and will likely prove useful for the labeling of biomolecules with photoswitchable synthetic fluors. In principle, any labeling strategy that can target an appropriate fluorophore without drastically altering its photophysical properties will be useful. Unpublished but promising possibilities include enzyme-based fluorophore targeting, tetra-cysteine tags, phalloidin, fluorophore-conjugated receptor ligands, and click chemistry (67).

## 5. Further Imaging Considerations

### 5.1. Live Cell Imaging

The intracellular environment is complex, spatially heterogeneous, and different in many ways from the simple and well-defined buffers used in the photoswitching strategies described above. These differences can affect the performance and viability of such strategies in vivo.

Although all psFPs are, in principle, suitable for live cell SMLM, they can be influenced by their surroundings. The activated versions of paGFP and psCFP2, for example, lose substantial photostability when the cells expressing them are cultured in media containing riboflavin and other vitamins (68). An additional study of other FPs has found that reducing agents can alter the photophysical behavior of a wide range of psFPs, suggesting that similar effects might occur in vivo (28). The newly developed PSmOrange photoswitches more effectively under cellular redox conditions, although its brightness and photostability seem to be unaffected (34).

Chemical fluorophores are more difficult to use *in vivo*, as many are highly charged and are thus unable to pass through intact cellular membranes. The photoswitching of many synthetic fluorophores also requires the absence of oxygen, and/or the presence of reducing agents such as MEA. Such conditions are toxic to cells, and thus not ideal for *in vivo* experiments.

Certain synthetic dyes, however, are photoswitchable in live cells. ATTO 655, for example, displays photoswitching behavior *in vivo* similar to that observed *in vitro* in 10 mM MEA (42, 64). It is thought that such switching is driven by the intracellular concentration of the reducing agent glutathione, which takes on the role played by MEA *in vitro*. This membrane-permeable dye has been successfully used in combination with the eDHFR tag in the performance of live-cell SMLM. The SNAP tag-compatible dyes TMR-Star and SNAP Cell 505 have been used in a similar manner (63).

### 5.2. "Multicolor" Imaging

One of the most powerful features of SMLM is its ability to determine the relative distributions of different biomolecular species. This is achieved by labeling each species with a photoswitchable fluorophore whose spectral properties (i.e., switching, excitation, and emission wavelengths) allow it to be distinguished from the others in the sample. In practice, switching several different types of fluorophores in the same sample is nontrivial; the laser lines and imaging buffers required for the imaging of one fluorophore species often have adverse effects on others. Nonetheless, various imaging protocols have been established that allow pairs of photoswitchable fluorophores to be simultaneously or sequentially imaged.

#### 5.2.1. Fluorescent Protein Pairs

Using pairs of psFPs is a straightforward way of performing 2-color SMLM. This strategy has been used successfully in both fixed (17) and live (33) cells, and requires no special buffer conditions. One proven psFP pair is tdEos and Dronpa (17), with the tdEos molecules being activated and imaged before the reversible psFP Dronpa is also read out. Green-emitting psFPs (e.g., PA-GFP) can also be imaged in parallel with dark-to-orange photoactivatable FPs (e.g., PAmCherry1). Sparse subsets of both FPs are first turned "on" using 405 nm light before the activated fluorophores are read out (orange before green) and the cycle repeated (32, 33).

#### 5.2.2. Activator–Reporter Systems

One disadvantage of using emission spectra to distinguish between fluorophore species is that chromatic aberrations in detection optics must be corrected for. Using pairs of activator–reporter systems that have identical emission spectra but are distinguishable by their activation wavelengths allows this problem to be sidestepped (41). Unfortunately, some effort to correct for cross-activation and spontaneous recovery of the reporter fluorophore is required (69).

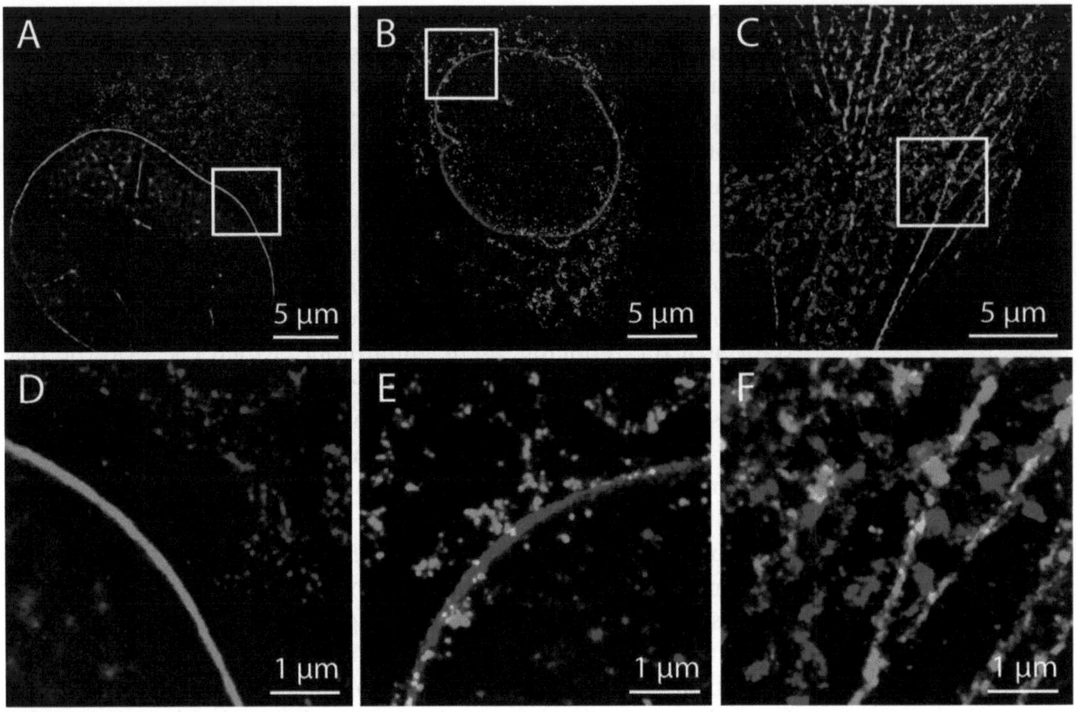


Fig. 4. Dual-color SMLM of fixed HeLa cells. (**a**) Lamin and calnexin were immunolabeled with Alexa Fluor 532 and Alexa Fluor 647 respectively, and imaged in PBS containing 50 mM MEA at pH 7.4. Alexa Fluor 647 was imaged first (using 647 nm light), followed by Alexa Fluor 532 (with 514 nm light). For both fluorophores species, a low intensity of 405 nm light was used to reactivate "off" state fluorophores. (**b**) Identical to (**a**), with labeling scheme reversed. (**c**) Cells expressing actin-mEos2 were immunolabeled with Alexa Fluor 647-conjugated antibodies against calnexin, and imaged under reducing conditions (50 mM MEA, pH 7.4). mEos2 was imaged using 568 nm excitation, followed by Alexa Fluor 647 using 647 nm excitation (8,000 frames per fluorophore species). Chromatic aberration was corrected for using Tetraspecks. (**d**), (**e**), and (**f**) show the enlarged cutouts of (**a**), (**b**), and (**c**), respectively.

*5.2.3. Fluorescent Dyes*

Pairs of synthetic fluorophores can also be used to produce dual-color images, but only if they are spectrally distinguishable, and can be photoswitched in the same buffer. Suitable pairs include ATTO655/ATTO520 (70) and ATTO520/Alexa Fluor 647 (71). We have also found Alexa Fluor 647/Alexa Fluor 532 to be an optimal pair, as both switch in 50 mM MEA at pH 7.4 (Fig. 4). Interestingly, it is often the case that no oxygen scavenging system is needed for Alexa Fluor 647. Since the photoswitching of synthetic dyes consumes oxygen (42), we speculate that high densities of Alexa Fluor 647 molecules (such as those present in samples densely labeled with secondary antibodies which are themselves labeled with many fluorophores) can reduce local oxygen concentrations to levels appropriate for the maintenance of long-lived and stable dark states. For structures which cannot be densely labeled, we recommend the use of oxygen scavenging systems. An alternative strategy is the use of spectrally close carbocyanine fluorophores (e.g., Cy5 and Alexa Fluor 700) in combination with spectral unmixing (72).

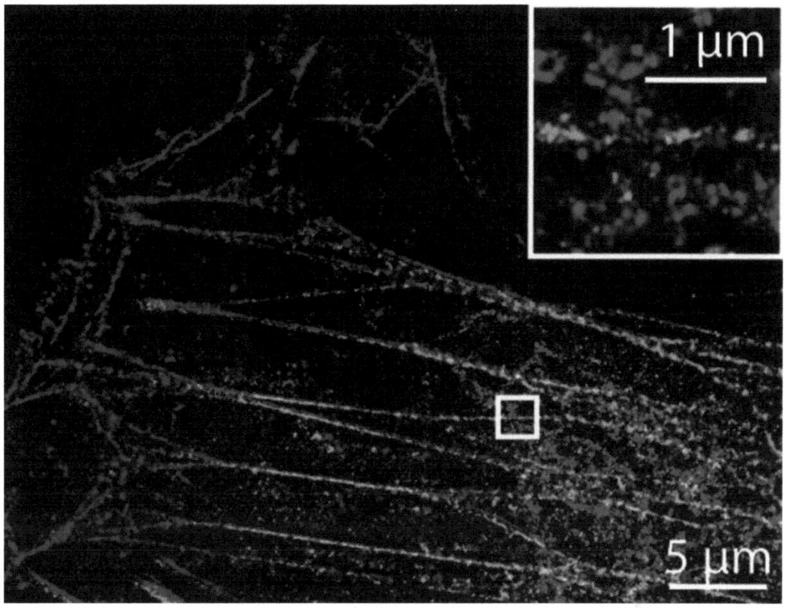


Fig. 5. Three-color SMLM. HeLa cells were transfected with actin-mEos2 plasmids and fixed. Calnexin was then immunolabeled with Alexa Fluor 647 and actin labeled with phalloidin-Alexa Fluor 488. mEos2, Alexa Fluor 488, and then Alexa Fluor 647 were sequentially imaged under reducing conditions (50 mM MEA, pH 7.4; 8,000 images per fluorophores) using 568, 488, and 647 nm laser lines. Fluorophores were recovered/activated using 405 nm light. Inset: a magnified portion of the image.

#### *5.2.4. Combining Fluorescent Proteins and Dyes*

Fluorescent dyes and psFPs can also be combined to yield dual-color images. Although most psFPs function in 10 mM MEA—and so are compatible with dyes such as ATTO655—many function poorly in 100 mM MEA (28). As a result, many psFPs cannot be used in conjunction with most fluorescent dyes. mEos2, however, still shows blinking behavior in 50 mM MEA, and so can be used in conjunction with many synthetic fluorophores (e.g., Alexa Fluor 647; Fig. 4; (28)).

Three-color imaging is more demanding because of the increases in the overlap of excitation and emission spectra, and so requires a careful experimental strategy. However, three-color imaging of Alexa Fluor 488, mEos2, and Alexa Fluor 647 is possible (Fig. 5). Since the unactivated form of mEos2 overlaps with the fluorescence spectrum of Alexa Fluor 488, mEos2 must be imaged first—and completely read out—before Alexa Fluor 488 can be visualized. As Alexa Fluor 488 shows acceptable blinking behavior at 50 mM MEA (pH 7.4), Alexa Fluor 647 can be used as a third color.

### 5.3. Molecular Counting

Because SMLM uses PSFs arising from single fluorophores, it can also—potentially—allow individual molecules to be identified, distinguished from one another, and counted. However, the relationship between the number of localizations recorded in an experiment and the number of fluorophores actually present in a sample can be complex, and depends strongly upon fluorophore, imaging conditions, and analysis method.

The simplest strategy for molecular counting involves using fluorophores that are irreversibly activatable (i.e., irreversibly activatable psFPs). In theory, each molecule can be activated, localized in adjacent frames (which analysis software then counts as a single localization), and then irreversibly photobleached. Unfortunately, such fluorophores often enter long-lived and reversible dark states several times before they are finally photobleached. Such behavior can be easily corrected for if such dark states last only a few frames (i.e., ~100 ms), but become problematic when they last several seconds. Then, determining whether a PSF arose from a newly activated fluorophore, or from a already-localized fluorophore exiting a dark state can be difficult. Studies of the popular psFP mEos2 have shown that the majority of localizations in a typical experiment can arise from such "blinking" (73). Although the blinking behavior of irreversibly activatable photoswitchable fluorophores has not been well studied, it is clear that some fluorophores (e.g., psCFP2) blink less than others (e.g., mEos2; (28, 73)).

Molecular counting is made more difficult by the fact that many fluorophores may never be detected at all. Regardless of the fluorophore used, some fluorophores will photobleach before emitting enough photons to be detected. Fixation can also destroy psFPs by preventing nascent proteins from properly maturing, or by inactivating matured psFPs.

Ultimately, robust positive and negative controls that account for the problems mentioned above must be presented if the claim of molecular counting is to be made. Such experiments will not be trivial and, to our knowledge, have never been convincingly performed.

## 6. Conclusion

We have reviewed photoswitchable fluorescent probes that are suitable for SMLM, discussed the requirements associated with different types of fluorescent probes, and presented various experimental concepts. Single-molecule super-resolution imaging will become an essential tool in biomedical research, and future research will focus on finding various kinds of suitable photoswitchable fluorophores with improved optical properties for both structural imaging and molecular counting.

## Acknowledgements

The authors are grateful to Sebastian Malkusch and Patrick Zessin for critical reading of the manuscript and to Prof. G.U. Nienhaus for providing spectral data for the fluorescent protein mEosFP-thermo. This work was supported by the German Ministry of Education and Research (BMBF; FORSYS initiative, grant nr. 0315262), the German Science Foundation (DFG, grant nr. HE 6166/2-1), and by contract research "Methoden für die Lebenswissenschaften" of the Baden-Württemberg Stiftung (grant nr. P-LS-SPII/11).

## References

1. Thompson RE, Larson DR, Webb WW (2002) Precise nanometer localization analysis for individual fluorescent probes. Biophys J 82: 2775–2783
2. Abbe E (1873) Beitrage zur theorie des mikroskops und der mikroskopischen wahrnehmung. Archive Mikroskop Anat 9:413–420
3. Heilemann M (2010) Fluorescence microscopy beyond the diffraction limit. J Biotechnol 149:243–251
4. Schermelleh L, Heintzmann R, Leonhardt H (2010) A guide to super-resolution fluorescence microscopy. J Cell Biol 190:165–175
5. Hell SW (2007) Far-field optical nanoscopy. Science 316:1153–1158
6. McEvoy AL, Greenfield D, Bates M, Liphardt J (2010) Q&A: single-molecule localization microscopy for biological imaging. BMC Biol 8:106
7. Betzig E, Patterson GH, Sougrat R, Lindwasser OW, Olenych S et al (2006) Imaging intracellular fluorescent proteins at nanometer resolution. Science 313:1642–1645
8. Rust MJ, Bates M, Zhuang XW (2006) Sub-diffraction-limit imaging by stochastic optical reconstruction microscopy (STORM). Nat Methods 3:793–795
9. Heilemann M, van de Linde S, Schuttpelz M, Kasper R, Seefeldt B et al (2008) Subdiffraction-resolution fluorescence imaging with conventional fluorescent probes. Angew Chem Int Ed 47:6172–6176
10. Folling J, Belov V, Kunetsky R, Medda R, Schonle A et al (2007) Photochromic rhodamines provide nanoscopy with optical sectioning. Angew Chem Int Ed 46:6266–6270
11. Hess ST, Girirajan TP, Mason MD (2006) Ultra-high resolution imaging by fluorescence photoactivation localization microscopy. Biophys J 91:4258–4272
12. Folling J, Bossi M, Bock H, Medda R, Wurm CA et al (2008) Fluorescence nanoscopy by ground-state depletion and single-molecule return. Nat Methods 5:943–945
13. Lemmer P, Gunkel M, Baddeley D, Kaufmann R, Urich A et al (2008) SPDM: light microscopy with single-molecule resolution at the nanoscale. Appl Phys B 93:1–12
14. Steinhauer C, Forthmann C, Vogelsang J, Tinnefeld P (2008) Superresolution microscopy on the basis of engineered dark states. J Am Chem Soc 130:16840
15. Patterson GH, Lippincott-Schwartz J (2002) A photoactivatable GFP for selective photolabeling of proteins and cells. Science 297:1873–1877
16. van de Linde S, Wolter S, Heilemann M, Sauer M (2010) The effect of photoswitching kinetics and labeling densities on super-resolution fluorescence imaging. J Biotechnol 149:260–266
17. Shroff H, Galbraith CG, Galbraith JA, White H, Gillette J et al (2007) Dual-color superresolution imaging of genetically expressed probes within individual adhesion complexes. Proc Natl Acad Sci USA 104:20308–20313
18. Lippincott-Schwartz J, Patterson GH (2009) Photoactivatable fluorescent proteins for diffraction-limited and super-resolution imaging. Trends Cell Biol 19:555–565
19. Chudakov DM, Matz MV, Lukyanov S, Lukyanov KA (2010) Fluorescent proteins and their applications in imaging living cells and tissues. Phys Rev 90:1103–1163
20. Dickson RM, Cubitt AB, Tsien RY, Moerner WE (1997) On/off blinking and switching behaviour of single molecules of green fluorescent protein. Nature 388:355–358
21. Fuchs J, Bohme S, Oswald F, Hedde PN, Krause M et al (2010) A photoactivatable marker protein for pulse-chase imaging with superresolution. Nat Methods 7:627–630

22. Chudakov DM, Verkhusha VV, Staroverov DB, Souslova EA, Lukyanov S et al (2004) Photoswitchable cyan fluorescent protein for protein tracking. Nat Biotechnol 22:1435–1439
23. McKinney SA, Murphy CS, Hazelwood KL, Davidson MW, Looger LL (2009) A bright and photostable photoconvertible fluorescent protein. Nat Methods 6:131–133
24. Habuchi S, Ando R, Dedecker P, Verheijen W, Mizuno H et al (2005) Reversible single-molecule photoswitching in the GFP-like fluorescent protein Dronpa. Proc Natl Acad Sci USA 102:9511–9516
25. Andresen M, Stiel AC, Folling J, Wenzel D, Schonle A et al (2008) Photoswitchable fluorescent proteins enable monochromatic multilabel imaging and dual color fluorescence nanoscopy. Nat Biotechnol 26:1035–1040
26. Wiedenmann J, Ivanchenko S, Oswald F, Schmitt F, Rocker C et al (2004) EosFP, a fluorescent marker protein with UV-inducible green-to-red fluorescence conversion. Proc Natl Acad Sci USA 101:15905–15910
27. Annibale P, Scarselli M, Kodiyan A, Radenovic A (2010) Photoactivatable fluorescent protein mEos2 displays repeated photoactivation after a long-lived dark state in the red photoconverted form. J Phys Chem Lett 1:1506–1510
28. Endesfelder U, Malkusch S, Flottmann B, Mondry J, Liguzinski P et al (2011) Chemically induced photoswitching of fluorescent probes-a general concept for super-resolution microscopy. Molecules 16:3106–3118
29. Chudakov DM, Lukyanov S, Lukyanov KA (2007) Tracking intracellular protein movements using photoswitchable fluorescent proteins PS-CFP2 and Dendra2. Nat Protoc 2:2024–2032
30. Habuchi S, Tsutsui H, Kochaniak AB, Miyawaki A, van Oijen AM (2008) mKikGR, a monomeric photoswitchable fluorescent protein. PLoS One 3:e3944
31. Patterson G, Davidson M, Manley S, Lippincott-Schwartz J (2010) Superresolution imaging using single-molecule localization. Annu Rev Phys Chem 61:345–367
32. Subach FV, Patterson GH, Manley S, Gillette JM, Lippincott-Schwartz J et al (2009) Photoactivatable mCherry for high-resolution two-color fluorescence microscopy. Nat Methods 6:153–159
33. Subach FV, Patterson GH, Renz M, Lippincott-Schwartz J, Verkhusha VV (2010) Bright monomeric photoactivatable red fluorescent protein for two-color super-resolution sptPALM of live cells. J Am Chem Soc 132:6481–6491
34. Subach OM, Patterson GH, Ting LM, Wang Y, Condeelis JS et al (2011) A photoswitchable orange-to-far-red fluorescent protein, PSmOrange. Nat Methods 8:771–777
35. Kim SY, Gitai Z, Kinkhabwala A, Shapiro L, Moerner WE (2006) Single molecules of the bacterial actin MreB undergo directed treadmilling motion in *Caulobacter crescentus*. Proc Natl Acad Sci USA 103:10929–10934
36. Matsuda A, Shao L, Boulanger J, Kervrann C, Carlton PM et al (2010) Condensed mitotic chromosome structure at nanometer resolution using PALM and EGFP- histones. PLoS One 5:e12768
37. Li S, Kimura E, Ng R, Fall BM, Meuse L et al (2006) A highly functional mini-dystrophin/GFP fusion gene for cell and gene therapy studies of duchenne muscular dystrophy. Hum Mol Genet 15:1610–1622
38. Greenfield D, McEvoy AL, Shroff H, Crooks GE, Wingreen NS et al (2009) Self-organization of the *Escherichia coli* chemotaxis network imaged with super-resolution light microscopy. PLoS Biol 7:e1000137
39. Belov VN, Wurm CA, Boyarskiy VP, Jakobs S, Hell SW (2010) Rhodamines NN: a novel class of caged fluorescent dyes. Angew Chem Int Ed Engl 49:3520–3523
40. Heilemann M, Sauer M, Margeat E, Kasper R, Tinnefeld P (2005) Carbocyanine dyes as efficient reversible single-molecule optical switch. J Am Chem Soc 127:3801–3806
41. Bates M, Huang B, Dempsey GT, Zhuang X (2007) Multicolor super-resolution imaging with photo-switchable fluorescent probes. Science 317:1749–1753
42. Heilemann M, van de Linde S, Mukherjee A, Sauer M (2009) Super-resolution imaging with small organic fluorophores. Angew Chem Int Ed 48:6903–6908
43. van de Linde S, Krstic I, Prisner T, Doose S, Heilemann M et al (2011) Photoinduced formation of reversible dye radicals and their impact on super-resolution imaging. Photochem Photobiol Sci 10:499–506
44. Kottke T, van de Linde S, Sauer M, Kakorin S, Heilemann M (2010) Identification of the product of photoswitching of an oxazine fluorophore using fourier transform infrared difference spectroscopy. J Phys Chem Lett 1:3156–3159
45. Vogelsang J, Kasper R, Steinhauer C, Person B, Heilemann M et al (2008) A reducing and oxidizing system minimizes photobleaching and blinking of fluorescent dyes. Angew Chem Int Ed 47:5465–5469
46. Egner A, Geisler C, von Middendorff C, Bock H, Wenzel D et al (2007) Fluorescence nanoscopy in whole cells by asynchronous localization of photoswitching emitters. Biophys J 93:3285–3290

47. Testa I, Wurm CA, Medda R, Rothermel E, von Middendorf C et al (2010) Multicolor fluorescence nanoscopy in fixed and living cells by exciting conventional fluorophores with a single wavelength. Biophys J 99:2686–2694
48. Lemmer P, Gunkel M, Weiland Y, Muller P, Baddeley D et al (2009) Using conventional fluorescent markers for far-field fluorescence localization nanoscopy allows resolution in the 10-nm range. J Microsc 235:163–171
49. Bates M, Blosser TR, Zhuang X (2005) Short-range spectroscopic ruler based on a single-molecule optical switch. Phys Rev Lett 94: 108101
50. Belov VN, Bossi ML, Folling J, Boyarskiy VP, Hell SW (2009) Rhodamine spiroamides for multicolor single-molecule switching fluorescent nanoscopy. Chemistry 15:10762–10776
51. Lidke KA, Rieger B, Jovin TM, Heintzmann R (2005) Superresolution by localization of quantum dots using blinking statistics. Opt Express 13:7052–7062
52. Goldstein M, Watkins S (2008) Immunohistochemistry. Curr Protoc Mol Biol 14:Unit 14.6.
53. Puchtler H, Meloan SN (1985) On the chemistry of formaldehyde fixation and its effects on immunohistochemical reactions. Histochemistry 82:201–204
54. Thomason L, Court DL, Bubunenko M, Costantino N, Wilson H et al (2007) Recombineering: genetic engineering in bacteria using homologous recombination. Curr Protoc Mol Biol 1:Unit 1.16
55. Mortensen RM, Kingston RE (2009) Selection of transfected mammalian cells. Curr Protoc Mol Biol 9:Unit 9.5.
56. Subach FV, Malashkevich VN, Zencheck WD, Xiao H, Filonov GS et al (2009) Photoactivation mechanism of PAmCherry based on crystal structures of the protein in the dark and fluorescent states. Proc Natl Acad Sci USA 106:21097–21102
57. You X, Nguyen AW, Jabaiah A, Sheff MA, Thorn KS et al (2006) Intracellular protein interaction mapping with FRET hybrids. Proc Natl Acad Sci USA 103:18458–18463
58. Keppler A, Gendreizig S, Gronemeyer T, Pick H, Vogel H et al (2003) A general method for the covalent labeling of fusion proteins with small molecules in vivo. Nat Biotechnol 21:86–89
59. Sun X, Zhang A, Baker B, Sun L, Howard A et al (2011) Development of SNAP-tag fluorogenic probes for wash-free fluorescence imaging. Chembiochem 12:2217–2226
60. Gautier A, Juillerat A, Heinis C, Correa IR Jr, Kindermann M et al (2008) An engineered protein tag for multiprotein labeling in living cells. Chem Biol 15:128–136
61. Los GV, Encell LP, McDougall MG, Hartzell DD, Karassina N et al (2008) HaloTag: a novel protein labeling technology for cell imaging and protein analysis. ACS Chem Biol 3:373–382
62. Miller LW, Cai Y, Sheetz MP, Cornish VW (2005) In vivo protein labeling with trimethoprim conjugates: a flexible chemical tag. Nat Methods 2:255–257
63. Klein T, Loschberger A, Proppert S, Wolter S, van de Linde S et al (2011) Live-cell dSTORM with SNAP-tag fusion proteins. Nat Methods 8:7–9
64. Wombacher R, Heidbreder M, van de Linde S, Sheetz MP, Heilemann M et al (2010) Live-cell super-resolution imaging with trimethoprim conjugates. Nat Methods 7:717–719
65. Lee HLD, Lord SJ, Iwanaga S, Zhan K, Xie HX et al (2010) Superresolution imaging of targeted proteins in fixed and living cells using photoactivatable organic fluorophores. J Am Chem Soc 132:15099–15101
66. Wombacher R, Cornish VW (2011) Chemical tags: applications in live cell fluorescence imaging. J Biophotonics 4:391–402
67. Henriques R, Griffiths C, Hesper RE, Mhlanga MM (2011) PALM and STORM: unlocking live-cell super-resolution. Biopolymers 95:322–331
68. Bogdanov AM, Bogdanova EA, Chudakov DM, Gorodnicheva TV, Lukyanov S et al (2009) Cell culture medium affects GFP photostability: a solution. Nat Methods 6:859–860
69. Dani A, Huang B, Bergan J, Dulac C, Zhuang X (2010) Superresolution imaging of chemical synapses in the brain. Neuron 68: 843–856
70. van de Linde S, Endesfelder U, Mukherjee A, Schuttpelz M, Wiebusch G et al (2009) Multicolor photoswitching microscopy for subdiffraction-resolution fluorescence imaging. Photochem Photobiol Sci 8:465–469
71. Löschberger A, van de Linde S, Dabauvalle MC, Rieger B, Heilemann M et al (2011) Super-resolution imaging reveals eightfold symmetry of gp210 proteins around the nuclear pore complex and resolves the central channel with nanometer resolution. J Cell Sci 125: 570–575
72. Baddeley D, Crossman D, Rossberger S, Cheyne JE, Montgomery JM et al (2011) 4D super-resolution microscopy with conventional fluorophores and single wavelength excitation in optically thick cells and tissues. PLoS One 6:e20645
73. Annibale P, Vanni S, Scarselli M, Rothlisberger U, Radenovic A (2011) Identification of clustering artifacts in photoactivated localization microscopy. Nat Methods 8:527–528

# Chapter 10

# Single-Molecule Tracking of mRNA in Living Cells

Mai Yamagishi, Yoshitaka Shirasaki, and Takashi Funatsu

## Abstract

Some mRNAs localize to specific regions within eukaryotic cells to express their functions. The movement and localization of mRNA molecules provides valuable information about how they concentrate to particular regions. Recent technical advances in optical microscopy and image analysis algorithms enable real-time tracking of single mRNA molecules in living cells. This chapter presents the methods to visualize and track single β-actin mRNA molecules that localize at the leading edge of chicken embryo fibroblasts. Furthermore, this chapter presents an analysis approach for single-molecule tracking data to extract quantitative information about the microenvironments of the mRNA molecules.

**Key words:** Fluorescence microscopy, Single-molecule imaging, Single-molecule tracking, mRNA localization, Brownian motion

## 1. Introduction

Increasing evidence suggests that some mRNAs are targeted to specific regions within cells, and that the localization of mRNA provides a mechanism to generate cell polarity, a process that is crucial in diverse cellular functions (1). Among such mRNAs, β-actin mRNA represents a well-characterized example (2–6). During the process of understanding the mechanism of β-actin mRNA localization, the MS2 system has been developed as a powerful method to visualize single RNA molecules in living cells and has been applied to mammalian cells (7, 8). This system consists of two components: (i) a fusion protein of the MS2 coat protein and a green fluorescent protein (GFP) or its derivatives, and (ii) a target mRNA tagged with an RNA motif of a unique hairpin that is recognized by the MS2 coat protein.

For visualizing the location and movement of a specific RNA in a living cell in real time, the MS2 system was followed by analogous techniques using particular sets of an RNA binding protein

Alioscka A. Sousa and Michael J. Kruhlak (eds.), *Nanoimaging: Methods and Protocols*, Methods in Molecular Biology, vol. 950, DOI 10.1007/978-1-62703-137-0_10, © Springer Science+Business Media, LLC 2013

and its target RNA motif. Such systems include U1A hairpins and the human U1A protein (9). In particular, the $\lambda_N$ system that uses a 22-amino-acid peptide ligand from bacteriophage $\lambda_N$ and the minimal 15-nt RNA hairpin termed boxB (10) is an alternative to the MS2 system when visualizing mRNA in mammalian cells. Moreover, by combining the two systems, the tracking of two kinds of mRNAs in the same cells with different colored fluorescent proteins is possible. These GFP-based visualization systems for RNA observation have an advantage when tracking mRNA at the single molecule level because a tandem array of multiple copies of the RNA motif is usually tagged to the target RNA. Consequently, this approach enables the accumulation of GFP-fused RNA binding protein on the RNA to yield higher signal intensities.

Single molecule tracking (SMT) of biomolecules in a living cell is becoming an important tool for investigating not only the dynamics of the biomolecules but the microstructures of the cell, as reviewed elsewhere (11). This method offers higher spatial resolution (approximately several dozen nanometers) when compared with other methods to observe lateral mobility such as fluorescence recovery after photobleaching and fluorescence correlation spectroscopy. This is because the spatial resolution of other methods is determined by the optical diffraction limit, which is about a few hundred nanometers. In addition, as SMT permits multiscale analysis, it easily reveals so-called anomalous diffusion, which refers to a diffusion process influenced by interactions with the surroundings such as molecular confinement or molecular crowding.

In this chapter, we describe an SMT approach that uncovers the involvement of the microstructures of cells in maintaining the localization of β-actin mRNA at leading edges in living chicken embryo fibroblasts (CEFs). We describe how to label the reporter mRNA with the MS2-EGFP fusion protein, track its movement at the single molecule level, and analyze the tracking data (12).

## 2. Materials

Examples of the major equipment and their suppliers used in our study are listed. Required specifications are discussed in later sections.

1. Plasmid containing MS2-EGFP with a nuclear localization signal (MS2-EGFP) (see Note 1).
2. Plasmid containing β-actin and the MS2-binding site (β-actin-MS2-bs) (see Note 2).
3. Chicken embryo fibroblasts (CEFs) isolated from the hearts of 9- to 10-day-old chick embryos.

4. Glass-based dishes (27 mm diameter coverslips).
5. Culture medium: minimum essential medium (MEM) containing 10% fetal bovine serum (FBS).
6. Incubator for culturing the cells at 37°C in a humid atmosphere with 5% $CO_2$.
7. Transfection reagent: FuGENE 6 (Roche Diagnostics).
8. Plasma cleaner.
9. Light microscope setup: Inverted microscope equipped with epifluorescence microscopy; objective lens (Plan apochromatic N TIRF 60×, NA = 1.45); stage-top incubator; blue solid-state laser (Sapphire 488-200); dichroic mirror of Q505LP and emission filter of FF01-520/35-25; EM-CCD camera (C9100-12; Hamamatsu photonics).
10. Image processing software: Image J (NIH, Bethesda, USA; http://rsb.info.nih.gov/ij/).
11. Tracking program: Plug-in program on Aquacosmos software (Hamamatsu photonics).

## 3. Methods

### 3.1. Specimen Preparation for Real-Time Observation of Single mRNA Molecules

#### 3.1.1. Cell Preparation

1. Load cells onto glass-based dishes prior to microscopic observation. Depending on the type of cells, the glass surface of the dishes may need to be coated with a coating reagent such as collagen I because cells are not generally adhesive to glass (see Note 3).
2. Culture cells to a density of ~50 cells/mm$^2$ in MEM containing 10% FBS in a humid atmosphere with 5% $CO_2$. The optimum density of cells depends on the cell type and the phenomenon of interest. In the case of tracking β-actin mRNA, it is desirable to have the cells separated from each other to form developed leading edges.

#### 3.1.2. Expression of mRNA with the MS2-Binding Site and the MS2-EGFP Protein

1. Replace the culture medium with serum-free MEM just before DNA transfection.
2. Transfect cells with a 9:1 molar ratio mixture of β-actin-MS2-bs and MS2-EGFP vectors using FuGENE 6 transfection reagent, according to the manufacturer's instructions (see Note 4). The optimal ratio of the two vectors should be predetermined because the optimal conditions to visualize cytoplasmic mRNA are determined by the expression amount and the ratio of products that will be changed with the type of promoters and/or the species of the cells. EGFP-labeled mRNA will begin to appear in the cytoplasm 10 h following transfection, and is abundantly observed during 16–24 h.

3. Replace the serum-free MEM with pre-warmed MEM containing 10% FBS prior to microscopic observation, because the cycle of depletion and re-addition of FBS to CEFs was found to enhance localization of β-actin mRNA (13).

*3.1.3. Verification Experiment for Labeling of the Cytoplasmic mRNA*

Whether or not cytoplasmic mRNA is labeled should be verified through a comparative experiment using two types of vectors: one expressing mRNA containing MS2-bs and the other expressing mRNA without MS2-bs (Fig. 1). If the MS2 system works well, some fluorescence signal can be seen within the cytoplasm only in the cells transfected with the mRNA expression vector containing MS2-bs. For example, co-expression of β-actin-MS2-bs and MS2-EGFP results in bright fluorescence signal at the leading edge of

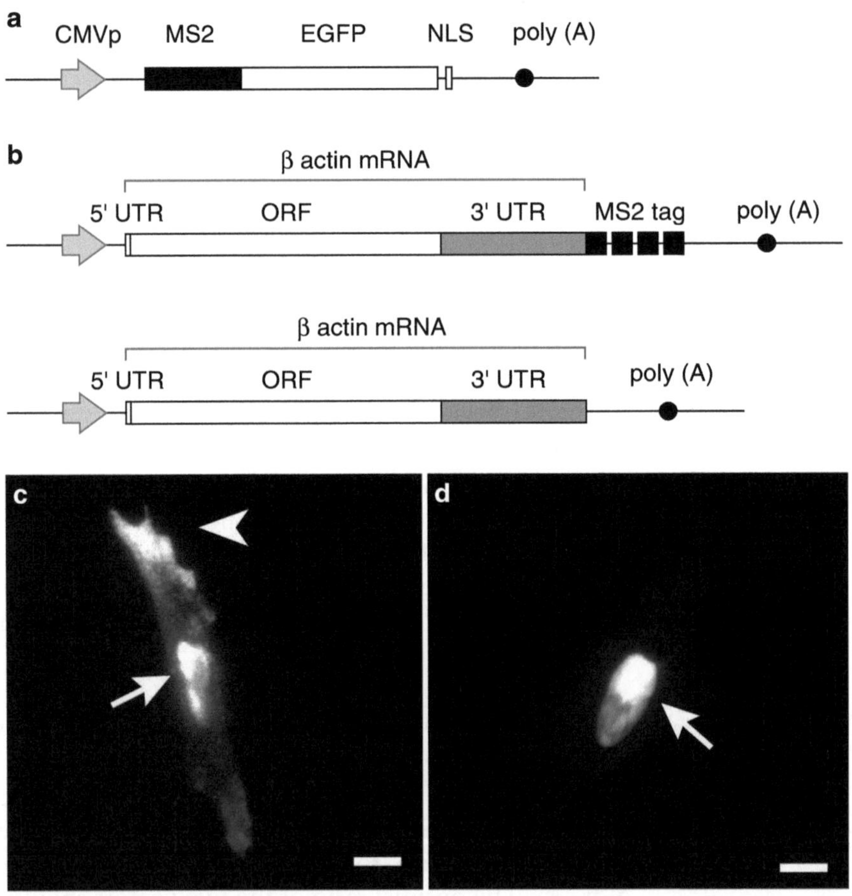


Fig. 1. Verification for labeling of the cytoplasmic mRNA. (**a**) A DNA construct for the expression of the MS2-EGFP protein with an NLS. (**b**) DNA constructs for the expression of β-actin mRNAs with and without the MS2-bs. (**c**) Distribution of MS2-EGFP in a CEF expressing mRNA with an MS2-bs. An arrow and an arrowhead indicate the nucleus and the leading edge of the cell, respectively. β-actin mRNA is localized to the leading edge of the cytoplasm. (**d**) Distribution of MS2-EGFP in CEFs expressing mRNA without an MS2-bs. MS2-EGFP is localized to the nucleus (*arrow*). The scale bars represent 10 μm (adapted from ref. 12 with the permission of Elsevier).

the cytoplasm, as well as in the nucleus (Fig. 1c), whereas MS2-EGFP accumulates only in the nucleus when the MS2-bs is absent (Fig. 1d). These results indicate that β-actin mRNA is successfully labeled and, furthermore, localized at the leading edge, as observed for native β-actin mRNA (3, 4).

### 3.2. Quantification of the Fluorescence Intensity of each Fluorescent Spot

Cytoplasmic mRNAs will be observed as bright spots. To confirm how many mRNA molecules are present in each bright spot, it is recommended that the fluorescence intensities of individual bright spots are compared with those of single EGFP molecules.

#### 3.2.1. Optical Setup for the Measurement of the Fluorescence Intensity

1. For live cell imaging, use an inverted epifluorescence microscope with an incubation system, such as a stage-top incubator, to maintain the temperature of the cells at 37°C (see Note 5).
2. Use a high magnification and high numerical aperture objective lens, such as PlanApo N TIRF (60×, NA = 1.45) oil immersion objective, to identify single spots in the cytoplasm and suitable to ensure the collection of faint fluorescence signals from single EGFP molecules.
3. Illuminate the specimen with blue laser power of a few μW/μm$^2$ at the sample surface. Generally, the intensity distribution of the laser beam is approximated by a Gaussian function, which will result in different efficiencies of excitation by inhomogeneous illumination. Therefore, the beam line should be expanded by a beam expander, and the intensity data should be acquired within a limited area of irradiation around the center to avoid too large a variation in the irradiation intensity (for example, within 10% variation).
4. Place a dichroic mirror (Q505LP) and emission filter (FF01-520/35-25) combination optimal for collecting EGFP fluorescence emission in the microscope (see Note 6). An excitation filter is not required, because the introduction of an excitation filter in the laser line may cause uneven illumination owing to the interference fringe.
5. To observe the quantized photobleaching of single EGFP molecules, acquire images at 20 frames per second, or faster, using a cooled electron-multiplying charge-coupled device (EM-CCD) camera (a cooled EM-CCD camera is best suited for quantification of the fluorescence intensity of single EGFP molecules). The electron-multiplying gain register can increase the signal more than several hundredfold before any readout noise is added, therefore keeping the noise level low. Furthermore, a cooled EM-CCD camera incorporating back illuminated type CCD chips is recommended for its high quantum efficiency (see Note 7).

*3.2.2. Measurement of the Fluorescence Intensity of Single EGFP Molecules*

To accurately measure the fluorescence intensity of single EGFP molecules, a purified solution of EGFP is required. The preparation method for obtaining recombinant EGFP is described elsewhere (14).

1. Load purified EGFP at the concentration of ng/ml level in a saline solution such as PBS onto a pre-cleaned glass coverslip to be absorbed nonspecifically (see Note 8). Excess free EGFP is then washed with the saline solution prior to observation.
2. Acquire series of images of single EGFP molecules to measure changes in fluorescent intensity before and after quantized photobleaching. We typically monitor the intensity for at least 10 s at video rate under conditions where single EGFP molecules photobleach in a few seconds on average.
3. Measure the fluorescence intensities with select regions within the image series using image analysis software (see Note 9). The region of interest (ROI) for specifying a fluorescence molecule is made by the circle with the smallest diameter to cover the Airy disk. The ROI for calculating the background fluorescence is made by a slightly larger diameter circle. The margin between the smaller ROI and the larger ROI, which forms a concentric ring that surrounds the Airy disk, is used to calculate the background fluorescence intensity.

*3.2.3. Measurement of the Fluorescence Intensity of mRNA Molecules in Living CEFs*

1. With the transfected CEFs cells on the microscope, acquire single image snapshots under the same optics and imaging conditions (optics setup, camera frame rate) as in the quantification of the intensities of single EGFP molecules (see Note 10).
2. To analyze the fluorescence intensities of mRNA molecules in the transfected CEFs, draw ROIs the same size as those used in the quantification of the intensities of single EGFP molecules. To extract the position of bright spots in the cytoplasm of CEFs, dynamic thresholding is an effective approach and is described below.

In a previous study, we completed such an analysis from 1,354 bright spots in the cytoplasm and 352 EGFP molecules on the coverslips, and compared the distributions of fluorescence intensities using histograms (12). An average intensity of single EGFP molecules was determined from the histogram fitted with a Gaussian distribution. The histogram of intensities of bright spots was fitted with a single Gaussian distribution with the average equivalent to six molecules of EGFP (Fig. 2a), indicating that each individual spot was bound with an average of six EGFP molecules. Up to eight EGFP molecules can bind to each single β-actin-MS2-bs mRNA molecule. The determined number of EGFP molecules

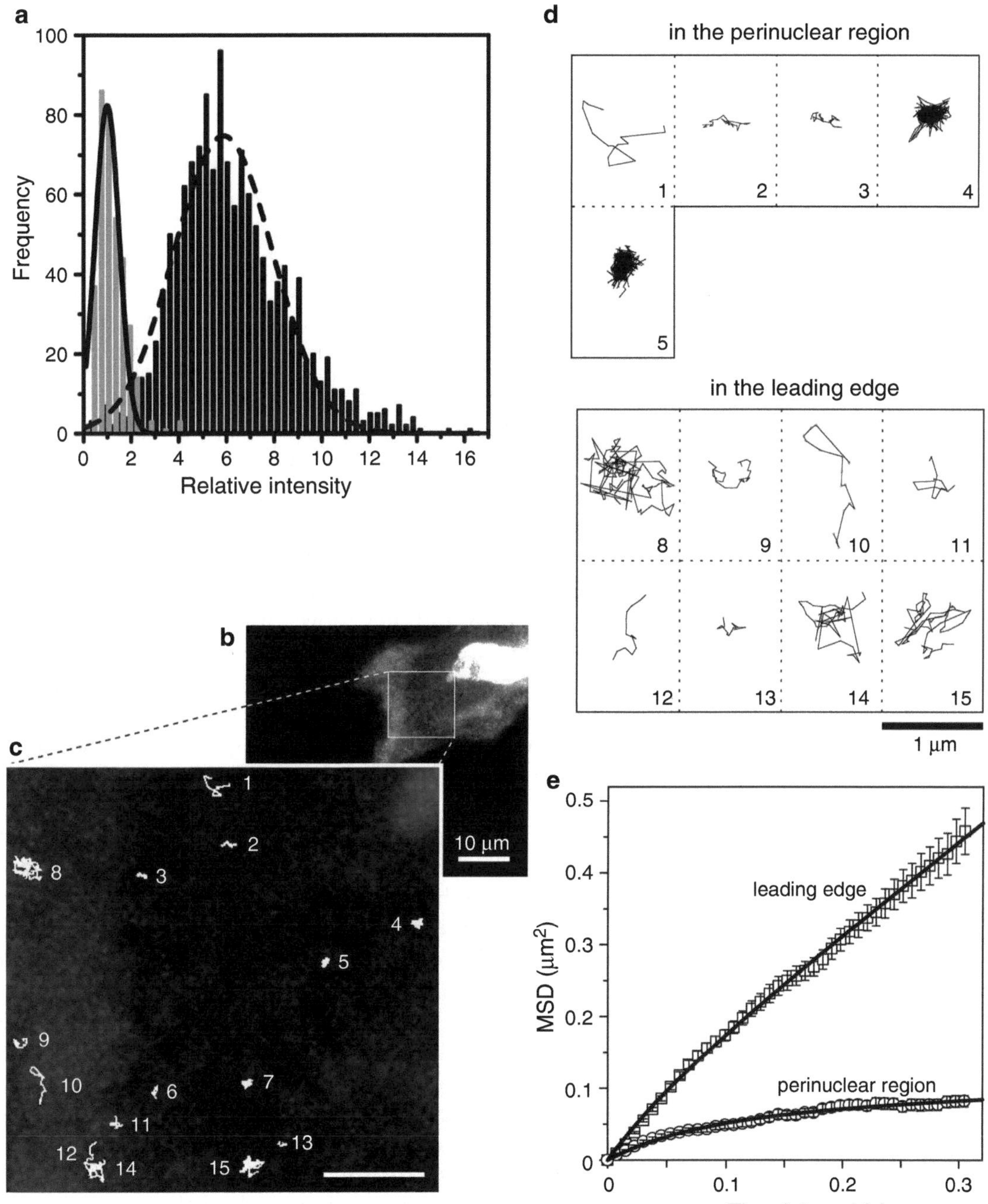


Fig. 2. Single-molecule tracking of β-actin mRNA. (**a**) Histogram of fluorescence intensities of single EGFP molecules (*gray bars*) and β-actin mRNA molecules labeled with EGFP (*black bars*). A *solid line* and a *broken line* indicate Gaussian distributions fitted to the data. The fluorescence intensity is normalized by the average intensity of single EGFP molecules. (**b**) Wide-field epifluorescence micrograph of β-actin mRNA molecules labeled with EGFP in a CEF. (**c**) Magnified image of the rectangular region in (**b**). Trajectories of 15 mRNA molecules are numbered. (**d**) Galleries of trajectories of mRNA molecules in the perinuclear region (No. 1–5) and at the leading edge (No. 8–15) of the cell. Nos. 6 and 7 in (**c**) were excluded from the analysis because they are not located in the perinuclear or leading edge regions. (**e**) Ensemble-averaged MSD–*t* plots for the perinuclear region (*circles*) and the leading edge (*squares*). *Bars* indicate SEM. Data were fitted with Eq. 7 (adapted from ref. 12 with the permission of Elsevier).

in the cytoplasmic spots was about the same as the expected number of GFP molecules for a single mRNA. Therefore, we conclude that the individual bright spots seen in the cytoplasm are single β-actin mRNA molecules.

### 3.3. SMT of β-Actin mRNA Labeled with MS2-EGFP

#### 3.3.1. Epifluorescence Microscopy for SMT

The same inverted microscope with the equipment described in Subheading 3.2.1 can be used for SMT; however, the camera setting should be reconsidered depending on the motility and spatial crowding of the target molecules. Successive images should be captured at a fast enough frequency to track each molecule of interest in the presence of high concentrations of other molecules. Otherwise the molecule of interest mixes with other molecules.

1. Place in the laser path a field stop that corresponds to a tracking area (15 μm in diameter) at the specimen plane, so that EGFP molecules outside of the tracking area are not illuminated.
2. Capture images of 10.6 × 10.6 μm rectangle field in the tracking area by an EM-CCD camera every 8 ms with intense laser illumination (a few $\mu W/\mu m^2$ at the sample surface) because β-actin mRNA molecules at the leading edge of CEFs are present at a high density and they move rapidly. A large number of fluorescent spots are photobleached by the intense laser and the remaining spots decrease in density such that they can be distinguished as discrete spots. Track the remaining fluorescent spots and newly appearing fluorescent spots from outside of the illumination area.

#### 3.3.2. Protocol of SMT

There are many software packages suitable for automated particle tracking. Some require their licenses to be purchased and some are offered by several research groups as freely downloadable files. In this article, we refer to the plug-in program for trajectory analysis running on Aquacosmos software that we have successfully used in the single mRNA molecule tracking studies. This program facilitates the definition of a target molecule position by fitting a Gaussian curve to the fluorescence intensity profile of the molecule (or calculating the centroid coordinate of the bright spot) and also searches for the nearest-positioned bright spot in the following frame as the same molecule after displacement. The plug-in program for trajectory analysis of Aquacosmos determines the position of targets using a Gaussian fitting method that is commonly used in fluorescent nano-imaging microscopy (15). The program on Aquacosmos uses the *x*- and *y*-axis projection profiles of the fluorescence of target spots, and its algorithm is superior in increasing the speed of the calculation; however, the program is susceptible to the effect of background noise (see Note 11). Therefore, the program can work successfully when tracking bright spots on a smooth background. Conversely, rough and noisy images could be processed to give smooth backgrounds prior to SMT.

During the successive Gaussian fitting against serial images, every bright spot on each image should be distinguished from the background. Methods for background smoothing and subtraction are discussed in steps 1 and 2 below, respectively. Examples for SMT in our study are shown in Fig. 2b–d.

1. Background smoothing of noisy images: Several spatial or time averaging methods are commonly used. Spatial averaging methods, which are based on a spatial low-pass filter including a mean filter, a median filter and a Gaussian filter, remove noise as well as details on the targets to give blurred images. The intensity of target spots is sometimes lost by spatial averaging and a faint spot cannot be distinguished. On the other hand, time averaging, also called moving average or rolling average, averages a series of a certain number of time data to smooth short-term fluctuations such as background noises. Time averaging is likely to have little effect on the intensities of the target spots, because relatively constant values of pixels against time are less affected by the time averaging process. However, time averaging intrinsically involves inevitable effects, especially on the observation of diffusing molecules that are permanently or temporarily confined within some compartments. The effect of time averaging upon changing the exposure time and frame rate has been reported (16). The effect of time averaging is that the position of a molecule appears to be more centralized to the compartments especially when the averaging time is comparable to, or longer than, the time for the target molecules to reach the compartment boundary. In order to reduce such an effect, we process the images using a four-frame exponential moving average with a 2/3 smoothing factor and end this process with spatial averaging of neighboring $3 \times 3$ pixels. All the averaging processes are performed with ImageJ.
2. Extraction of bright spots: In order to extract the bright spots, the program on Aquacosmos adopts a binary threshold, although images can be processed using a rolling-ball background subtraction before tracking when the images have an uneven background. The rolling-ball subtraction is particularly effective against uneven but smooth continuous backgrounds often caused by inhomogeneous illumination. If images have more extreme local enhancement in the background level, such as the images of heterogeneously distributed β-actin mRNA in CEFs, another background subtraction may be required because rolling-ball methods are not sufficient to handle such rough images. We have introduced a dynamic thresholding method to distinguish target spots from the rough background (17). This method uses a spatial averaged image as a background of the original image, which enables us to set the background level at each location depending on the undulation of

the real background. In our study, the spatial averaging of neighboring $11 \times 11$ pixels is used as the background.

*3.3.3. Categories of Single Molecule Motion*

1. Statistical and quantitative analyses of SMTs are usually carried out on the basis of the plot of the mean-square displacement (MSD) against the time interval (MSD–$t$ plot). The main purposes of MSD–$t$ analysis are to reveal the type of movement undergone by the molecule and to estimate some physical quantities such as a diffusion coefficient value $D$ or a velocity of translatory motion $v$. The MSD of mRNA molecules for a given time lag of $n\Delta t$ is calculated by the following equation:

$$\mathrm{MSD}(n\Delta t) = \frac{1}{N-n}\sum_{i=1}^{N-n}\left[r(i\Delta t + n\Delta t) - r(i\Delta t)\right]^2, \tag{1}$$

where $\Delta t$ is the time interval and $r(i\Delta t + n\Delta t)$ describes the position of the particle following a time interval of $n\Delta t$ after starting at position $r(i\Delta t)$; $N$ is the total number of frames in the sequence of the particle, and $n$ and $i$ are positive integers (18, 19). Based on the MSD–$t$ plots, different modes of motion can be characterized as shown in Fig. 3 and they can be expressed as one of, or a combination of, the following equations (20, 21):

Stationary mode (δ: precision of position determination):

$$\mathrm{MSD}(t) < \delta^2. \tag{2}$$

Simple diffusion mode ($D$: diffusion coefficient):

$$\mathrm{MSD}(t) = 4Dt. \tag{3}$$

Confined diffusion mode ($L$: size of a compartment, $\tau$: equilibration time):

$$\mathrm{MSD}(t) = \frac{L^2}{6}\left(1 - \mathrm{e}^{-\frac{t}{\tau}}\right). \tag{4}$$

Directed mode ($v$: velocity of translatory motion):

$$\mathrm{MSD}(t) = (vt)^2. \tag{5}$$

2. In the case of our study, the RNAs that move less than 0.03 $\mu m^2$ in its MSD over 60 frames (457.2 ms) are defined as static. Static mRNA molecules are frequently observed at the perinuclear region (28%) but rarely observed at the leading edge (1%).
3. We can focus on how the diffusion of β-actin mRNA in leading edges and in perinuclear regions differ from each other (Fig. 2b–e). In such cases, the ensemble-averaged MSD is helpful in interpreting the differences between the two. The ensemble-averaged MSD for all the molecules undergoing

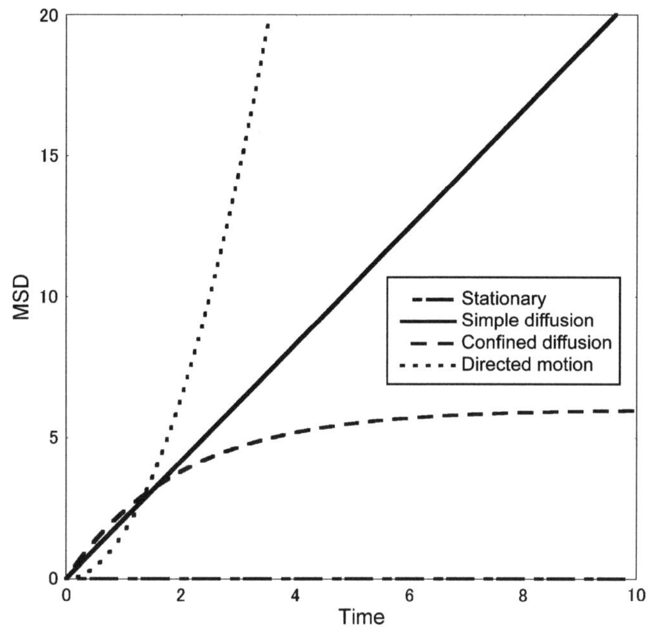


Fig. 3. MSD–$t$ plots for different modes of motion.

Brownian motion in the two regions can be calculated, respectively, by the following equation:

$$\mathrm{MSD}(n\Delta t)=\left\{\sum_{k=1}^{N_{\mathrm{P}}}\left(N_k-n\right)\right\}^{-1}\sum_{k=1}^{N_{\mathrm{P}}}\sum_{i=1}^{N_k-n}\left[r_k(i\Delta t+n\Delta t)-r_k(i\Delta t)\right]^2, \quad (6)$$

where $\Delta t$ is the time interval and $r_k(i\Delta t+\mathrm{n}\Delta t)$ describes the position of the $k$th particle following a time interval of $n\Delta t$ after starting at position $r_k(i\Delta t)$; $N\mathrm{p}$ is the total number of particles in the area and $N_k$ is the total number of frames in the sequence of the $k$th particle, and $n$ and $i$ are positive integers.

4. We can also investigate how the diffusion of mRNA in the leading edges differ from that in the perinuclear regions by comparing the diffusion coefficients for longer time intervals that are derived from ensemble-averaged MSDs. The static mRNAs are excluded from the analysis of the movement of mRNAs. The ensemble average of the MSD is calculated for all molecules undergoing Brownian motion in the leading edges and in the perinuclear regions, and is plotted as a function of time (Fig. 2e). The obtained MSD–$t$ plots are fitted with the following equation:

$$\mathrm{MSD}=\frac{L^2}{6}\left(1-\mathrm{e}^{-\frac{t}{\tau}}\right)+4D_{\mathrm{MACRO}}t \quad (7)$$

that represents a hop diffusion (21, 22), which results from the temporary confinement of the diffusing molecule in some compartments, where the parameter $D_{MACRO}$ is the macroscopic diffusion coefficient characterizing the diffusion passing through the compartments (Fig. 2e). $D_{micro}$, the microscopic diffusion coefficient characterizing the unrestricted diffusion within a compartment, is obtained from Eq. 8:

$$D_{micro} = \frac{L^2}{2\pi^2\tau}. \tag{8}$$

In our case, the values of $D_{micro}$ (0.29 μm²/s) and $D_{MACRO}$ (0.32 μm²/s) at the leading edges are similar, indicating that mRNAs were in nonrestricted Brownian motion rather than anchored to leading edges. On the other hand, the ratio of $D_{micro}$ (0.24 μm²/s) to $D_{MACRO}$ (0.020 μm²/s) is about 12 at the perinuclear regions, indicating that the movement of mRNAs in those regions was restricted by some barrier of the microcompartment.

## 4. Notes

1. The wild-type MS2 cDNA was obtained from pHybLex/Zeo-MS2 (Invitrogen, Japan), and its FG-loop was eliminated (23) using a QuikChange Site-Directed Mutagenesis Kit (Stratagene, La Jolla, USA). This mutation prevents the assembly of coat protein dimers into virus-like particles. The MS2 mutant, EGFP, and a nuclear localization signal (NLS) DNA were fused using PCR and cloned into pcDNA 6/V5-His A (Invitrogen) (Fig. 1a). NLS was included so that only the EGFP protein bound to its target mRNA would be present in the cytoplasm
2. The MS2-binding site (8) was created from pRH3′ (Invitrogen). Four tandemly repeated MS2-binding sites (MS2-bs), which bind eight MS2 proteins, were used. The cDNA of chicken cytoplasmic β-actin was amplified from a chicken fibroblast (secondary) cDNA library (Uni-ZAP XR Premade library; Stratagene), which contained 18-nt of a 5′ UTR, 1,128-nt of the ORF and 591-nt of the 3′ UTR. The β-Actin full sequence was cloned into pcDNA with or without the MS2 tag (Fig. 1b).
3. The coating reagent should be carefully determined because it may affect morphology, motility and viability of the cells as well as the mRNA distribution pattern. In the case of CEFs, the cells can attach to glass surfaces without any coating, and some types of coating drastically change the physiological state of a cell. Additionally, CEFs also show changes in their

morphology and motility after sub-culturing. Therefore, we seed CEFs on glass-based dishes without a coating just after harvesting the cells from chick embryos. They are maintained and undergo every treatment on the same glass-based dishes.

4. The type of transfection reagent used is dependent on the cell type (e.g., established cell lines and primary cells). The reagent used will determine the transfection efficiency and cell viability.
5. We used the IX71 system from Olympus but equivalent microscopes are available from many vendors.
6. Dichroic mirrors and emission filters, which are optimum for EGFP observation, are available from many vendors both as a complete GFP filter set and as single items. In our studies, we used a dichroic mirror of Q505LP (Chroma) and an emission filter of FF01-520/35-25 (Semrock).
7. A C9100-12, deep-cooled, back-thinned EM-CCD camera from Hamamatsu photonics was used in our experiments.
8. The glass coverslips should be cleaned in advance using a plasma cleaner or by immersion in a strong alkaline solution, because they often have fluorescent fine grains on the surface that have fluorescence intensities similar to those of single EGFP molecules. Coverslips can be cleaned using $O_2$ plasma generated by a plasma cleaner for 15 min at maximum power. If a plasma cleaner is not available, coverslips can be alternatively cleaned as follows. They should be sonicated in a table-top sonicator for 30 min in 0.1 M KCH, then rinsed thoroughly in Milli-Q water and dried in a class 1,000 clean cabinet.
9. Subsequent quantification analysis can be performed using commercially available or free image analysis software. If you do not have any software, ImageJ (NIH, Bethesda, MD) is helpful, which is a public domain open source software.
10. For photobleaching, image capture should be initiated prior to excitation and the first or second frame that begins to show a fluorescence image should be used for image analysis. In most cases, the second frame is brighter than the first, so the use of the second frame is recommended.
11. The precision of the position determination using the method largely depends on the level and pattern of the background noise. In order to estimate the precision of the analysis, mRNA expressing cells fixed with some reagent such as 4% paraformaldehyde should be prepared and analyzed in the same way as a control sample with static mRNA. Furthermore, a rectangle ROI that is arbitrarily set around the target bright spot should be small enough not to include any unnecessary spotty noises that will impede Gaussian fitting performed within the ROI.

## Acknowledgments

We thank Yasuhiro Sasuga, Kayoko Terada (Tokyo Metropolitan Institute of Medical Science) and Yoshie Harada (Kyoto University) for constructing the tandemly repeated MS2 tag sequences. This research was partly supported by a Grant-in-Aid for Scientific Research on Innovative Areas (21121004) and by the Japan Society for the Promotion of Science (JSPS) through its "Funding Program for World-Leading Innovative R&D on Science and Technology (FIRST Program)" to T.F.

### References

1. Martin KC, Ephrussi A (2009) mRNA localization: gene expression in the spatial dimension. Cell 136:719–730
2. Lawrence JB, Singer RH (1986) Intracellular localization of messenger RNA for cytoskeletal proteins. Cell 45:407–415
3. Sundell CL, Singer RH (1991) Requirement of microfilaments in sorting of actin messenger RNA. Science 253:1275–1277
4. Kislauskis EH, Zhu X, Singer RH (1994) Sequences responsible for intracellular localization of beta-actin messenger RNA also affect cell phenotype. J Cell Biol 127:441–451
5. Ross AF, Oleynikov Y, Kislauskis EH, Taneja KL, Singer RH (1997) Characterization of a beta-actin mRNA zipcode-binding protein. Mol Cell Biol 17:2158–2165
6. Oleynikov Y, Singer RH (2003) Real-time visualization of ZBP1 association with beta-actin mRNA during transcription and localization. Curr Biol 13:199–207
7. Bertrand E, Chartrand P, Schaefer M et al (1998) Localization of ASH1 mRNA particles in living yeast. Mol Cell 2:437–445
8. Fusco D, Accornero N, Lavoie B et al (2003) Single mRNA molecules demonstrate probabilistic movement in living mammalian cells. Curr Biol 13:161–167
9. Brodsky AS, Silver PA (2000) Pre-mRNA processing factors are required for nuclear export. RNA 6:1737–1749
10. Daigle N, Ellenberg J (2007) $\lambda_N$-GFP: an RNA reporter system for live-cell imaging. Nat Methods 4:633–636
11. Kusumi A, Shirai YM, Koyama-Honda I, Suzuki KGN, Fujiwara TK (2010) Hierarchial organization of the plasma membrane: investigations by single-molecule tracking vs. fluorescence correlation spectroscopy. FEBS Lett 584:1814–1823
12. Yamagishi M, Ishihama Y, Shirasaki Y, Kurama H, Funatsu T (2009) Single-molecule imaging of β-actin mRNAs in the cytoplasm of a living cell. Exp Cell Res 315:1142–1147
13. Latham VM Jr, Kislauskis EH, Singer RH, Ross AF (1994) Beta-actin mRNA localization is regulated by signal transduction mechanisms. J Cell Biol 126:1211–1219
14. Makino Y, Amada K, Taguchi H, Yoshida M (1997) Chaperonin-mediated folding of green fluorescent protein. J Biol Chem 272:12468–12474
15. Yildiz A, Forkey JN, McKinney SA et al (2003) Myosin V walks hand-over-hand: single fluorophore imaging with 1.5-nm localization. Science 300:2061–2065
16. Ritchie K, Shan X-Y, Kondo J et al (2005) Detection of non-Brownian diffusion in the cell membrane in single molecule tracking. Biophys J 88:2266–2277
17. Davies ER (2005) Machine vision: theory, algorithms, practicalities, 3rd edn. Morgan Kaufmann Publishers, San Francisco
18. Lee GM, Ishihara A, Jacobson KA (1991) Direct observation of Brownian motion of lipids in a membrane. Proc Natl Acad Sci USA 88:6274–6278
19. Qian H, Sheetz MP, Elson EL (1991) Single particle tracking. Analysis of diffusion and flow in two-dimensional systems. Biophys J 60:910–921
20. Kusumi A, Sako Y, Yamamoto M (1993) Confined lateral diffusion of membrane receptors as studied by single particle tracking (nanovid microscopy). Effects of calcium-induced differentiation in cultured epithelial cells. Biophys J 65:2021–2040
21. Suzuki K, Ritchie K, Kajikawa E, Fujiwara T, Kusumi A (2005) Rapid hop diffusion of a G-protein-coupled receptor in the plasma

membrane as revealed by single-molecule techniques. Biophys J 88:3659–3680
22. Tomishige M, Sako Y, Kusumi A (1998) Regulation mechanism of the lateral diffusion of band 3 in erythrocyte membranes by the membrane skeleton. J Cell Biol 142: 989–1000
23. Peabody DS, Ely KR (1992) Control of translational repression by protein-protein interactions. Nucleic Acids Res 20:1649–1655

# Part II

## Electron Microscopy

# Chapter 11

# Semiautomatic, High-Throughput, High-Resolution Protocol for Three-Dimensional Reconstruction of Single Particles in Electron Microscopy

**Carlos Oscar Sorzano, J.M. de la Rosa Trevín, J. Otón, J.J. Vega, J. Cuenca, A. Zaldívar-Peraza, J. Gómez-Blanco, J. Vargas, A. Quintana, Roberto Marabini, and José María Carazo**

## Abstract

In this chapter we describe the steps needed for reconstructing the three-dimensional structure of a macromolecular complex starting from its projections collected in electron micrographs. The concepts are shown through the use of Xmipp 3.0, a software suite specifically designed for the image processing of biological structures imaged with electron or X-ray microscopy. We illustrate the image processing workflow by applying it to the images of Bovine Papilloma virus published in Wolf et al. (Proc Natl Acad Sci USA 107:6298–6303, 2010). We show that in the case of high-quality, homogeneous datasets with a priori knowledge about the initial volume, we can have a high-resolution 3D reconstruction in less than 1 day using a computer cluster with only 32 processors.

**Key words:** Single particle analysis, Electron microscopy, Image processing, 3D reconstruction, Workflows

## 1. Introduction

The study of the structure of protein and protein complexes by Transmission Electron Microscopy (TEM) provides key insight into the way that these macromolecules perform their function in the cell (1–3). One of the techniques available to perform these studies is called Single Particle Analysis (SPA). In SPA, thousands of projections of different copies of the same molecule are computationally combined in a single volume. It is assumed that all particles being analyzed correspond to exactly the same conformation. If this is not the case, projection images are separated in different classes of homogeneous populations. The quality of the images

Alioscka A. Sousa and Michael J. Kruhlak (eds.), *Nanoimaging: Methods and Protocols*, Methods in Molecular Biology, vol. 950, DOI 10.1007/978-1-62703-137-0_11, © Springer Science+Business Media, LLC 2013

Fig. 1. *Top-left*: Original image. *Top-right*: Image affected by CTF with no envelope decay. *Bottom-left*: Image affected by CTF with envelope decay (the decay acts as a lowpass filter). *Bottom-right*: Image affected by CTF with envelope decay and noise.

returned by the microscope is rather low: first, the overall image contrast is lost by the TEM microscope (1) (see Fig. 1); second, the structural information is severely degraded at high frequencies by aberrations of the microscope; finally, the macromolecule is not isolated in the vacuum but depending on the sample preparation it is surrounded by a thin layer of amorphous ice (cryo-EM) or a thicker layer of carbon and a heavy metal salt (negative staining), which introduces an important background signal that interferes with the signal from the complex under study. Altogether, the Signal-to-Noise Ratio (SNR) is well below 1/10, i.e., there is 10 times more noise than signal.

In this chapter, we introduce the standard image processing workflow needed to produce a volume from a collection of micrographs. The workflow is presented with the use of Xmipp 3.0 software. Note that other software packages may define different workflows, but in any case, the final structure of a complex must be obtained through a user-defined sequence of the standard steps introduced in this chapter.

In short, the standard image processing workflow starts by screening the micrographs to check that they are not astigmatic or drifted and realize the maximum frequency available. Then particles are selected from the micrographs either manually or semiautomatically. Particles are extracted from the micrographs into a gallery and they are again screened to find possible wrongly picked particles. The set of selected particles is aligned and classified attempting to identify possible 2D inhomogeneities. Possible contaminants or alternative structures are removed from the dataset. Next, an initial model is constructed either from the particles themselves or from a priori knowledge about the particle being reconstructed. Finally, the initial model is further refined using the projection images and its resolution estimated. At this level of model refinement it is still possible to have a mixture of different structural populations, and there are methods to sort them into different homogeneous classes.

## 2. Materials

1. *Software*. SPA image processing is normally performed via software packages like Spider (4), Eman (5), Imagic (6), or Xmipp (7) among others. These software suites allow image processing starting from the raw micrographs and ending at the final reconstructed three-dimensional (3D) structure (see Note 1).
2. *Hardware*. The whole process is rather demanding of computational resources and it is normally performed in computer clusters or supercomputers (cloud computing is an obvious choice for the future but at present it is not in place). The operating system of this kind of computers is always Unix-like, and therefore, Unix is the natural environment for these software packages. An average configuration of Xmipp uses a cluster with 8–16 Gb of RAM memory per node, 8–32 cores per node, and several nodes. Most Xmipp programs scale well up to 128 processors. Beyond this point, inter-process communications and disk access may become a bottleneck, although the optimal performance is rather system dependent and has to be tested on each cluster configuration.

## 3. Methods

In the following, we describe the mainstream protocols needed to perform a 3D reconstruction starting from the electron micrographs.

### 3.1. Micrograph Screening

The first step is to check the quality of the collected micrographs. Only high-quality micrographs should progress to further analysis in a high-resolution analysis. For medium-low resolution analysis, one might include not so good micrographs depending on the

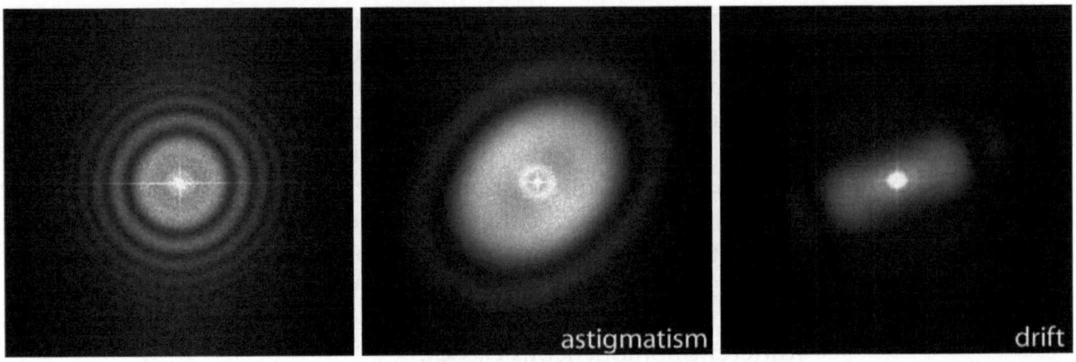


Fig. 2. *Left*: Example of a good micrograph, the PSD has circularly symmetric Thon rings. The presence of many Thon rings is normally associated to the preservation of structural information in high frequency. *Middle*: Example of astigmatic micrograph, Thon rings are elliptical. *Right*: Example of drifted micrograph, Thon rings appear incomplete.

resolution loss that one is willing to tolerate. The Xmipp protocols produce a number of criteria that may help to screen good from bad micrographs.

#### 3.1.1. Identifying Astigmatic and Drifted Micrographs

Good micrographs have a homogeneous background level (without any smooth gradient along the micrograph), are not astigmatic, have no drift, and have high-resolution structural information (visible Thon rings in high frequencies) (see Fig. 2). Astigmatic micrographs could, in principle, be processed. However, in practice they are avoided since most programs cannot track correctly the astigmatism angle through all the sequence of iterative alignments.

Micrographs can be semi-automatically screened by estimating their Power Spectrum Density (PSD) and their Contrast Transfer Function (CTF) (8). The PSD is an estimate of the energy distribution of the micrograph over frequency. Astigmatic micrographs have elliptical rings, while non-astigmatic micrographs have circular rings. Drifted micrographs have masked Thon rings (they do not appear to be complete). The CTF is normally described by a number of parameters, among which the most important are the microscope operating voltage and the defocus (9). However, Xmipp provides a full 2D characterization of the CTF, as well as, the background noise (8), which also plays a role in the accurate determination of the defoci and cannot simply be "filtered out."

In order to correctly estimate the PSD it is important that the digital micrograph does not have empty borders or labels as in Fig. 3.

#### 3.1.2. Determining the Maximum Resolution of the Micrographs

The information content of the micrograph is an important issue to consider. We need to estimate at which resolution the information content of the micrograph fades into the noise (let us call $f_{max}$ to this frequency). This frequency is characterized by a strong

Fig. 3. Example of digitized micrograph with blank areas in the corners and a label for identification. The PSD may not be correctly estimated if these elements are present in the digital micrograph.

decay of the envelope of the CTF (Xmipp reports the frequency at which the CTF envelope drops below 1% of its maximum value). 3D Reconstruction algorithms may recover a few Angstroms more of resolution beyond this frequency, but not many, and the final volume resolution is strongly determined by the maximum resolution visible in the micrographs.

Related to the maximum attainable resolution are the sampling rate and the downsampling factor. The sampling rate relates spatial frequencies measured in 1/Angstroms with digital frequencies measured in cycles/sample (the maximum digital frequency is 0.5 cycles/sample (10)). The relationship is $\omega = fT$, where $\omega$is the digital frequency, $f$is the spatial frequency in 1/Angstroms, and $T$is the sampling rate in Angstroms/pixel. When we downsample the images, we increase the pixel size. Downsampling brings in two important benefits: first, images are obviously smaller with the subsequent gain in disk space and computing time; second, downsampling reduces the noise in the images by cutting out in Fourier space a region where no signal is present.

In principle, we can increase the pixel size by a factor $K$as long as $0.5 \geq f_{max}KT$ (downsampling by a factor larger than this introduces an effect called aliasing in which high-frequency components may be strongly degraded). In practice, it is normally preferred to have a safety region such that $0.2 \leq f_{max} KT \leq 0.3$ (this is so to avoid the aliasing of the signal with the noise). For instance, if $f_{max} = \frac{1}{6} \AA^{-1}$, and $T = 1\AA / \text{pixel}$, then we can downsample by a factor $1.2 \leq K \leq 1.8$. Note that the downsampling factor may not be an integer number. In the Xmipp protocols, downsampling is performed in Fourier space (and that is why it can accept non-integer factors). Downsampling in Fourier space provides the best accuracy even for integer downsampling factors since the micrograph PSD is multiplied by a rectangular window (11). On the other side, downsampling by simply averaging neighboring pixels (an operation often called *binning*) introduces strong distortions in the high-frequency components (11).

Considering $\omega_{max} = f_{max} T$ ($T$ being the effective sampling rate after downsampling), ideal micrographs have $0.2 \leq \omega_{max} \leq 0.3$. Micrographs with $\omega_{max} \leq 0.2$ (also called oversampled) can be safely downsampled since no special gain in resolution will be obtained by so finely sampled images and the noise present may hinder the quality of the final reconstruction. Micrographs with $\omega_{max} \geq 0.3$ may suffer from aliasing resulting in reconstruction artifacts (they are said to be undersampled).

#### *3.1.3. Criteria to Identify Bad Micrographs Based Only on the Micrograph PSD*

1. *PSD correlation at 90°*. The PSD of non-astigmatic micrographs correlates well with itself after rotating the micrograph 90°. This is so because non-astigmatic PSDs are circularly symmetrical, while astigmatic micrographs are elliptically symmetrical. High correlation when rotating 90° is an indicator of non-astigmatism. In Xmipp, this criterion is computed on the enhanced PSD (12).
2. *PSD radial integral.* This criterion reports the integral of the radially symmetrized PSD. This criterion can highlight differences among the background noises of micrographs. This criterion is computed on the enhanced PSD.
3. *PSD variance.* The PSD is actually estimated by averaging different PSD local estimates in small regions of the micrograph. This criterion measures the variance of the different PSD local estimates. Untilted micrographs have equal defoci all over the micrograph, and therefore, the variance is due only to noise. However, tilted micrographs have an increased PSD variance since different regions of the micrograph have different defoci. Low variance of the PSD is indicative of non-tilted micrographs.
4. *PSD Principal Component Variance.* When considering the local PSDs previously defined as vectors in a multidimensional space, we can compute the variance of their projection onto the first principal component axis. Low variance of this projection is indicative of a uniformity of local PSDs, for example, this is another measure of the presence of tilt in the micrograph.

5. *PSD PCA Runs test.* When computing the projections onto the first principal component, as discussed in the previous criterion, one should expect that the sign of the projection is random for untilted micrographs. Micrographs with a marked nonrandom pattern of projections are indicative of tilted micrographs. The larger the value of this criterion, the less random the pattern is.

#### 3.1.4. Criteria to Identify Incorrectly Fitted CTFs

1. *Fitting score.* The CTF is computed by fitting a theoretical model to the experimentally observed PSD. This criterion is the fitting score. Smaller scores correspond to better fits. A complete description of Xmipp fitting score is given at Sorzano, Jonic (8).
2. *Fitting correlation between the first and third zeroes.* The region between the first and third zeroes is particularly important since it is where the Thon rings are most visible. This criterion reports the correlation between the experimental and theoretical PSDs within this region. High correlations indicate good fits.
3. *First zero disagreement.* If the CTF has been estimated by two different methods (normally Xmipp and Ctffind (13)), then this criterion measures the average disagreement in Angstroms between the first zero in the two estimates. Low disagreements are indicative of a correct fit.

#### 3.1.5. Criteria to Identify Bad Micrographs Based on the Fitted CTF

1. *Damping.* This is the envelope value at the border of the PSD. Micrographs with a high envelope value at border are either wrongly estimated or strongly undersampled.
2. *First zero average.* This is the average in Angstroms of the first zero over all possible directions. Normally, this value should be between 4 and 10 times the effective sampling rate in Angstroms/pixel.
3. *First zero ratio.* This measures the astigmatism of the CTF by computing the ratio between the largest and smallest axes of the first zero ellipse. Ratios close to 1 indicate no astigmatism.

### 3.2. Particle Picking

The next step is to pick particles from those micrographs passing the previous screening step. Particle picking can be performed manually or semiautomatically. Manual particle picking simply lets the user to choose those particles he is interested in from the electron micrographs. The ease of picking particles depends on the particle size, image contrast, contamination level, etc. The task can be facilitated by displaying the micrograph with a moderate zoom-out factor.

Image filters can also help to better visualize the particles of interest. In Xmipp we provide the following filters: color dithering, bandpass filter, anisotropic diffusion, mean shift, background subtraction, and contrast/brightness enhancement. Particularly useful are color dithering and bandpass filtering. Color dithering (14) is an algorithm that reduces differences between adjacent gray values. Bandpass filter (15) filters out gray-scale variations that are either too small or too large with respect to the size of the particle we are interested in.

#### *3.2.1. Manual Particle Picking*

Manual picking is sometimes criticized for biasing the dataset towards those shapes that the user better recognize or has in mind as the possible projections of her structure. In order to avoid this bias, automatic or semiautomatic particle picking algorithms are sometimes preferred as an objective way of choosing particles. At the same time, the particle picking process, which in general is a tedious and time-consuming task, is computationally assisted and accelerated. Algorithmically chosen datasets contain a nonnegligible amount of wrongly picked particles (contaminants, true particles on carbon, particle conglomerates, etc.). These datasets have to be carefully scrutinized to remove wrongly picked particles from the onset. This can be done by a manual revision of the automatic picking results, and/or by classifying in 2D the selected particles and eliminating those classes corresponding to wrongly picked particles.

#### *3.2.2. Semiautomatic Particle Picking*

Xmipp allows semiautomatic particle picking (16). The algorithm has been designed to keep a low false-positive rate (FPR), i.e., picking as few wrongly picked particles as possible. In order to keep the FPR as low as possible, the algorithm must be trained by the user in the kind of particles he is interested in. The first step of the training requires the user to manually pick about 100–200 particles. Then, the algorithm learns a set of features describing the particles being picked. In the next step, the algorithm tries to automatically select particles from a micrograph that has not been manually processed. The first attempt will pick a number of true particles, but many other true particles may be left. The user is required to pick those "unseen" particles. He is also required to correct the algorithm by removing those particles that have been wrongly picked automatically. This correction information helps the algorithm to distinguish true particles from other objects looking alike. This process of the algorithm trying to correctly pick particles and the user correcting for errors is repeated on more micrographs until the user is satisfied with the algorithm results. At each step, the algorithm learns from its errors and next time it will try to better distinguish between particles and nonparticles. No automatic particle picking algorithm is absolutely infallible. As a rule of thumb, they can be applied when the imaging conditions are not specially challenging.

### *3.3. Particle Extraction, Screening, and Preprocessing*

The next step is to extract the particles from the micrographs and form a stack of projections of our particle. This stack of projections is further analyzed in 2D or 3D in subsequent steps. Projection screening helps to identify projections that are not typical (in the statistical sense). Nontypical particles may correspond to underrepresented projection directions or states of our particle, but also to wrongly picked particles and outliers. If an automatic particle picking algorithm has been used, nontypical particles normally correspond to wrongly picked particles, such as particles on edges. Finally, particle preprocessing helps to highlight or concentrate specific features of our dataset.

#### 3.3.1. Particle Extraction

When extracting the particle projections, there are a number of actions we can take:

1. *Phase flipping.* Use the information from the estimated CTF to compensate for phase reversals introduced by the microscope at different frequencies. This phase correction is better performed on the micrographs (not the projections), since we have enough information to perform a deconvolution.
2. *Taking logarithm.* Depending on your acquisition system you may have to take the logarithm of the data in order to have a linear relationship between the gray values in the image and those in the volume.
3. *Contrast inversion.* Most image processing algorithms expect to see the particle as a white object over a dark background. However, some imaging conditions produce just the opposite. At this moment, you may invert the contrast if your particles are black over a white background.
4. *Normalization.* The same projection in different micrographs may have different gray values. Even within the same micrograph there might be a light gradient causing the gray values to be different. In order to eliminate a local gradient, a ramp in the gray values is fitted for each projection image and then subtracted from the image. Then, the image values are linearly transformed so that in the background there is zero mean and standard deviation equal to one. Noise statistics should be similar in all projections after this normalization step.
5. *Dust removal.* Sometimes dust, hot or cold spots can be seen in a projection. These pixels are identified by noting that their gray values are normally far from the mean of the rest of the image. You should choose to fill these pixels with a random value from a Gaussian with zero-mean and unity-standard deviation.

#### 3.3.2. Particle Screening

Automatic picking algorithms have a nonnegligible FPR (i.e., they pick locations in the micrograph that do not actually correspond to true particles). Particle screening is a successful way of identifying them.

For each image, we calculate the gray values histogram, square the gray values and compute the radial average of these squared values. The gray histogram and the radial average are stacked into a multivariate vector associated to each projection. Now, we perform a PCA analysis of the whole set of projections and project the multivariate vector onto the PCA space spanned by the first two eigenvectors. Then, we analyze the multivariate normality of these projections. The normality is measured as the Mahalanobis distance of the PCA projection to the space origin. Typical projections have a small distance, while nontypical projections have a larger distance. The dataset is sorted according to their distance (called *z*-score). Normally, wrongly picked particles, particles on edges, contaminants, etc., have a large distance and are sorted to the end of the list. At this point, the user may discard the particles he considers that do not correspond to the structure under study, or that for some reason do not follow the general trend.

#### *3.3.3. Particle Preprocessing*

Sometimes, we may want to apply an image processing filter to our images in order to reduce noise, highlight certain features, mask out the background, etc. In Xmipp protocols the following choices are available.

1. *Scaling*. Change the size of the projection images. Normally a size reduction is performed so that image processing is faster. The amount of noise is also reduced.
2. *Fourier lowpass, highpass, and bandpass filtering*. Low frequencies correspond to slow variations in the image; too high frequencies normally come from noise or very fine details. You may use these filters to remove any frequency band that you are not interested in. Cutoff frequencies are normalized to 0.5 (they are called digital frequencies). The digital frequency of an object whose diameter is $D$ pixels is $\omega = \frac{1}{D}$. The Xmipp protocols offer a wizard that allows interactive selection of the cutoff frequencies so that we may preview the effect of the filter in our images.
3. *Fourier Gaussian filtering*. We may implement multiresolution approaches by a series of Gaussian filters in Fourier space. Coarse representations are obtained by multiplying in Fourier space by a Gaussian of small bandwidth. Finer representations are obtained by enlarging the bandwidth of the Gaussian filter.
4. *Dust removal*. Similar to the dust removal option in the previous section.
5. *Normalization*. Besides the noise normalization with background subtraction described in the previous section (the recommended option), Xmipp also offers to normalize the images to have zero mean and unitary standard deviation, or to normalizing without background subtraction.
6. *Masking*. Apply a mask to the images to concentrate the analysis on a specific region. A wizard helps to choose among a wide variety of masks.

### 3.4. 2D Analysis

Two-dimensional analysis is the most common way of getting acquainted with the structural information contained in the projection images. In a way, it is an Exploratory Data Analysis (EDA) that is very much used in other data analysis contexts. The whole set of images is classified into different groups as homogenous as possible. Inside each group, projections are aligned and (normally) averaged producing a class representative. The class representative has much less noise (thanks to the averaging or equivalent operation) than the raw projections and allows a better visual identification of the structural features. The price for this improved visualization is the possible blurring introduced by averaging nonidentical projection images. If the group is small enough (between 50 and 150 projections), then this blurring effect is minimized.

2D analysis allows identifying wrongly picked particles, contaminants if they are sufficiently dissimilar from the structure under study, particle aggregations, damaged particles, etc. We should select among the classes all those that are likely to come from the structure we are studying in order to choose a projection population as homogeneous as possible in terms of biochemical species and conformational state. The rest of the classes may be interesting by themselves and worthy of further analysis, although they should not be mixed with the "good" classes in order to avoid contamination in the 3D reconstruction process. This class selection process is tricky in the sense that the user might bias the final 3D reconstruction by removing projections that a priori do not fit with his preconceived idea about the structure being reconstructed. The boundary between removal of contaminants, damaged particles, and wrongly picked particles and the removal of valid projection images is a fuzzy, fine line difficult to characterize.

Class representatives can also be used as input images to a common-lines algorithm (17, 18). This kind of algorithms addresses the initial reference problem. Most 3D reconstruction algorithms need an initial volume as starting point. The initial volume must share some general features with the structure under study. When such a starting volume is not available, common-lines algorithms are capable of producing one by identifying common lines in Fourier space. This identification is rather error prone and sensitive to noise. That is why class representatives, with less noise, are particularly well suited as input for these algorithms.

Xmipp offers several possibilities to perform this 2D EDA, which are briefly described below.

#### 3.4.1. CL2D (Clustering in 2D) (19)

The algorithm performs alignment and classification of the input images. It allows the identification of a large number of classes (the larger the number of classes, the fewer particles will take part in each class; there must be a balance between avoiding blurring by averaging dissimilar projections and having sufficient images in each class to significantly reduce the noise and allowing better visualization). First, the algorithm split the data in a small number of classes that are subsequently subdivided into more and more classes until the desired number of classes is reached. Images are allowed to change class at any moment. This divisive structure has been proved to yield more robust results than directly starting with a large number of classes. The algorithm requires relatively few parameters (the most important is the final number of classes). We can also control how to measure the distance between a raw projection and the class representative (correlation or correntropy (20)) and how to decide to which class a raw projection belongs (classical multireference assignment or robust assignment (19)). Two setups are normally used: (1) correlation and classical multireference assignment, which normally works well with good quality images and it is standard in other software packages; (2) correntropy and robust assignment, which works well with good and bad quality images and it is particular to Xmipp. The output classes are sorted so that one image is similar to the next. Additionally, thanks to the hierarchical nature of the algorithm, we can compute the core of any class (we define the core of the class as the subset of images that were always together in the classification hierarchy). We further refine the core by eliminating outliers in the PCA space spanned by the first two eigenvectors (in the same way as in the particle screening). We refer to these refined cores as stable cores.

*3.4.2. ML2D (Maximum Likelihood in 2D) (21)*

Maximum Likelihood is a different classification framework. Instead of assigning each projection image to a single class, it is assigned to all classes with different probabilities. Classes are then updated with the weighted average of all particles (weighted by probability). The algorithm solves at the same time the alignment and classification problems and it is well suited to work with a relatively small number of classes (less than 20–30). The algorithm can be applied in Fourier space instead of real-space. Fourier space has the advantage of letting noise be correlated, which is actually the case in Electron Microscopy due to the Contrast Transfer Function (CTF). Another advantage of Maximum Likelihood in Fourier space is that the CTF needs not be estimated.

*3.4.3. KerDenSOM (Kernel Density Estimation Self-Organizing Map) (22)*

This algorithm projects the input images onto an output space of class representatives. Nearby classes are similar to each other (as in the output of CL2D but in a 2D topology). The algorithm does not align images to classes, so that they have to be previously aligned. The algorithm is useful to perform a careful analysis of a given region in a set of images that may be the output of a previous classification (CL2D or ML2D) or alignment.

*3.4.4. Rotational Spectra (23)*

The rotational spectrum of an image measures its rotational symmetry. A twofold symmetric image can be rotated 180° (=360/2) and we obtain the same image. A threefold symmetric image can be rotated 120° (=360/3) and 240° (=2 × 360/3) and we obtain again the same image. The rotational spectrum measures the strength of each symmetry in a particular image. The rotational spectrum is rotationally invariant (i.e., if we rotate the image by any angle, the rotational spectrum remains the same), but it is not translationally invariant (i.e., if we shift the image, the rotational spectrum changes). For this reason, it is important to have all the images translationally aligned, and compute the rotational spectrum with rotations around the image center. The analysis of the rotational spectrum may reveal heterogeneities in particles whose projections are relatively symmetric (24) in a particular projection direction. The most important parameters of this step are the two radii between which the symmetry is analyzed (for instance, a ring-shaped projection can have symmetry only between the two radii where the ring is inscribed).

### 3.5. 3D Analysis

The two main steps in 3D are the analysis of homogeneous populations (all projections belong to a single structure) and heterogeneous populations (in the dataset there are projections coming from different structures).

The homogenous population analysis is normally referred to as 3D model refinement. Starting from an initial guess of the structure being reconstructed, a workflow of image processing steps is taken to refine this initial guess. The initial guess may be obtained by the same protein in a slightly different conformation, from a similar protein, from an atomic model, from random conical tilt, from a common-lines algorithm, etc. The initial reference chosen may bias the final result, and it is customary to strongly filter the initial volume so that all details are removed, and only the general shape at very low resolution remains.

The heterogeneous population analysis can be seen as the simultaneous refinement of several volumes with homogeneous population. The model refinement problem is coupled to the problem of classifying the input raw projections into different classes (the homogeneous populations).

#### 3.5.1. 3D Model Refinement

The most basic input of a model refinement algorithm is a starting volume (normally at low or very low resolution) and a set of projections supposedly from a single 3D structure and ideally covering all possible projection directions. The starting volume is projected in all possible directions (reference projections) and the experimental images are compared (after alignment) with all of them. The projection direction of the best matching reference projection is assigned to the corresponding experimental image and the in-plane alignment parameters (in-plane rotation and shift) are annotated. These alignment parameters are then used by a 3D reconstruction algorithm to produce a better estimate of the 3D structure of the macromolecular complex. This process is iterated till convergence or a fixed number of iterations have been performed.

The most important parameter of this algorithm is the plan for 3D alignment search. To align each experimental projection with respect to the current estimate of the volume being reconstructed, we need 5 parameters (3 Euler angles defining the projection direction and the in-plane rotation, and 2 shifts defining an in-plane displacement). We can perform a truly 5D search for these parameters (trying all possible combinations of Euler angles and shifts within a bounded range) or perform a 3D+2D parameter search (looking for the best projection direction keeping the shift fixed, and then looking for the best shift using the projection direction just found). A full 5D search provides more accurate results at the cost of a significant increase of computation time. We can use both strategies during the process: it is custom to start with a 5D search for the first few (23, 25) iterations and then switch to a 3D+2D strategy. The bounded search regions for each parameter (angles and shifts) are also normally diminished in size so that the parameter search is performed in a narrower area (this helps the process to converge while saving time by not looking for the parameters in areas that are rather unlikely to produce a good match). An exhaustive search is performed within the bounded region using a sampling step. As the region becomes narrower, the sampling step is also diminished so that the alignment is more accurate. Typical search ranges for the first iterations allow any angle and shift in the first 4–5 iterations with an angular sampling of 10° (the shift sampling is not needed since the shift is searched in Fourier space by making use of its correlation property, which allows for a continuous shift search). From the fourth to fifth iteration, the angular range is reduced to about twice the angular sampling, and this is progressively diminished towards a sensible value (the smallest identifiable angle is given by $\arctan\frac{1}{R}$, where $R$ is the radius of the object being reconstructed in pixels).

During the alignment and reconstruction process we may also correct for the amplitude effects of the CTF. Xmipp does so by grouping the input projections according to their defocus values. Those images belonging to the same defocus group are reconstructed together, and then the different volumes are merged in 3D by using a Wiener filter (26). Alternatively, a B-factor correction can be employed (25).

Another important issue is how to generate the reference volume for the next iteration from the current reconstruction estimate. Using the raw estimate may produce overfitting since the very high frequencies of the volume may be fitting mostly noise components; this is especially true during the first iterations since the angular sampling is still large and the reconstruction estimate is rather rough. The raw estimate can be masked in real space (using a fixed radius or a user-defined mask) and Fourier space (by lowpass filtering the volume to a certain frequency; normally this cutoff frequency is gradually increased so that the alignment uses a progressively improved reference volume).

*3.5.2. 3D Heterogeneity Analysis*

In case we suspect that our population of projections may come from different 3D structures (different conformations, different oligomeric states, different binding states, etc.) we can simultaneously solve the 3D reconstruction and classification problem. The idea is to sort the input projections into different classes so that each class is homogenous, and then a 3D reconstruction is performed within each class. This is the standard idea behind multireference 3D reconstruction. In Xmipp we have extended the Maximum Likelihood framework to 3D (each experimental projection belongs to all classes and all projection directions with different probabilities; then, the volumes representing each class are updated considering the relative weights of each projection) (27). The algorithm is rather time consuming and, because of this, its applicability is limited to relatively large angular samplings (in the order 5–10° depending on the image size and number of images). At the end of the algorithm, we may assign each experimental image to the class with maximum likelihood. Then, within each class we can run a model refinement (see previous Section) in order to improve the resolution of the class model.

### 3.6. Example 3D Reconstruction

We present the protocols results as applied to the Bovine Papilloma virus images published in (28) and kindly provided by Drs. Wolf and Grigorieff for this book chapter. The dataset consists of 49 micrographs of an approximate size of $10{,}000 \times 10{,}000$ pixels. The sampling rate is 1.237 Å/pixel, the microscope voltage 300 kV, and the nominal magnification ×56,588. At this magnification, the projection of a single virus fits in an image of size $512 \times 512$. The execution time for the different protocol steps are reported for the parallel run of Xmipp on Intel Xeon 2.666 GHz processors. The processors belong to a cluster with 28 nodes, 8 processors, and 16 GB of RAM per node.

*3.6.1. Micrograph Screening*

Estimating the CTF and evaluating the quality of all micrographs took 15 min for the BPV dataset in 32 processors. The estimated micrograph defocus ranged between 1.8 and 2.3 μm and no micrograph was removed for having an astigmatic or drifted CTF (Fig. 4 shows the typical output of the Xmipp protocols for this step). The maximum resolution present in the micrographs is about 6 Å (the CTF envelope drops below 1% its maximum value, which implies a loss in image power by a factor $0.01^2$).

*3.6.2. Particle Picking*

Semiautomatic particle picking was used. We manually trained the algorithm with the first 4 micrographs. 238 virus projections were manually picked among the 4 micrographs (the automatic picking

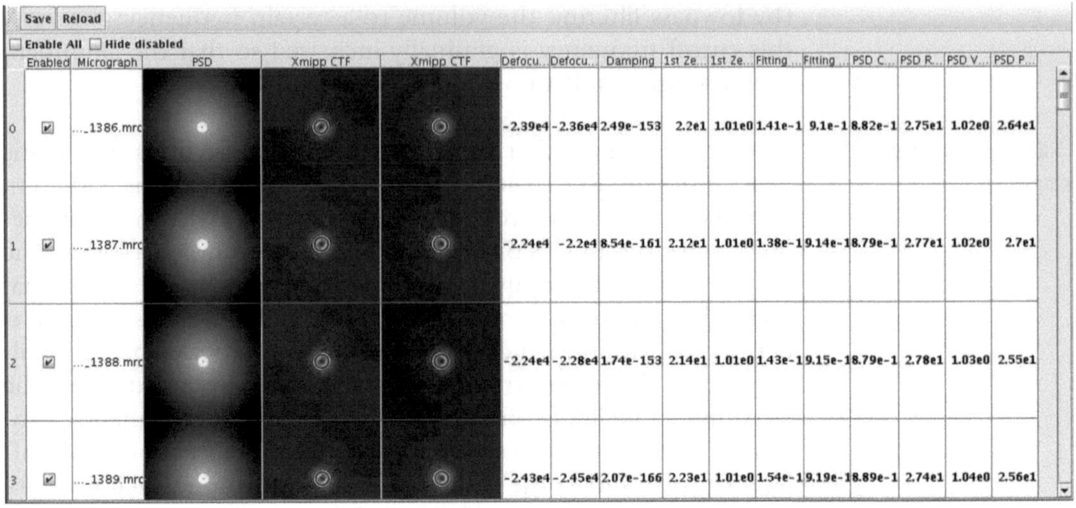


Fig. 4. Display of the micrograph screening results. Each row shows the PSD, fitted CTF, and different quality criteria for a different micrograph. The table can be sorted by any of the quality criteria and bad micrographs can be manually deselected so that they are no longer considered in the analysis.

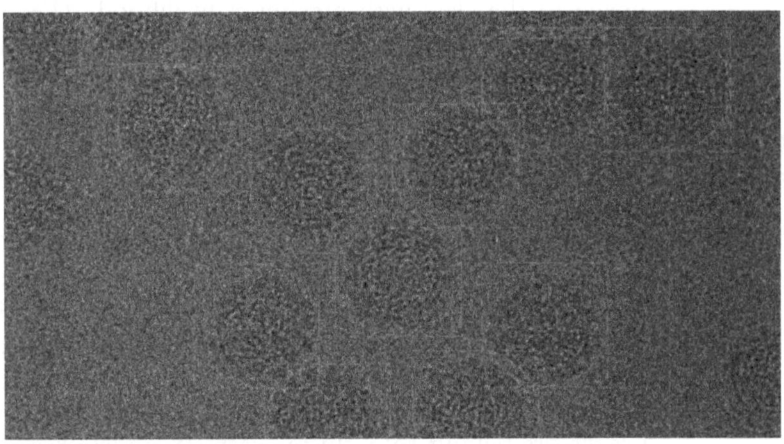

Fig. 5. Sample region of a micrograph with virus particles automatically identified.

algorithm was used after the first 153 manually picked particles). 6,475 particles were automatically selected in 50 min from 45 micrographs (an average of 130 particles/min) (see Fig. 5). Among the 6,475 particles, 1,158 were false-positives (mostly they were true virus particles on the carbon film instead of the grid holes), which amounts to 17.9% of the selected particles. This kind of false-positives was easily removed during the revision of the protocol output. A total of 5,317 (=6,475–1,158) particles were correctly identified. 410 additional particles were manually selected (they corresponded to false-negatives). In total 5,727 (=410 + 5,317) particles were selected during the automatic phase, 92.8% of them were automatically selected. Thus, we see an advantage in the manual workload by using a semisupervised approach to particle picking. Adding the 238 particles manually selected during the training phase then gave a total of 5,965 particles in the dataset.

*3.6.3. Particle Extraction and Screening*

The selected particles were corrected for the CTF phase flip and screened in search of outliers using the multivariate procedure described in Subheading 3.3. The process took 3 min in a single processor. All particles whose PCA distance was larger than 7 were discarded (245 particles out of 5,965, 4.1%; the visual appearance of these particles was not particularly different from the other images in the dataset, but for some reason they did not follow the general trend). Discarded particles mostly corresponded to a couple of micrographs that had a relatively lower correlation between the first and third zero of the CTF. For some reason the CTF in these micrographs was more difficult to estimate, and they also resulted in particles whose projections into the PCA space were outliers. It is worth noting that these two quality measures are completely independent so that both seem to indicate some strange behavior of the discarded micrographs. After the screening, the dataset was composed of 5,720 particle projections.

*3.6.4. 2D Analysis*

In order to explore the acquired projections, we reduced the image size from $512 \times 512$ to $128 \times 128$ (3 min in 8 processors). We then applied a CL2D analysis with 64 classes (6 h in 32 processors). Figure 6 shows the first 24 class averages. The CL2D iterations converged rather clearly, with only 1.7% of the images changing from one class to the other after 15 iterations (this is normally a sign of finding good classes). In the set of class averages, we could not identify any class that did not correspond to true virus projections. Therefore, we assume that the set of experimental projections corresponds to a single, homogenous population.

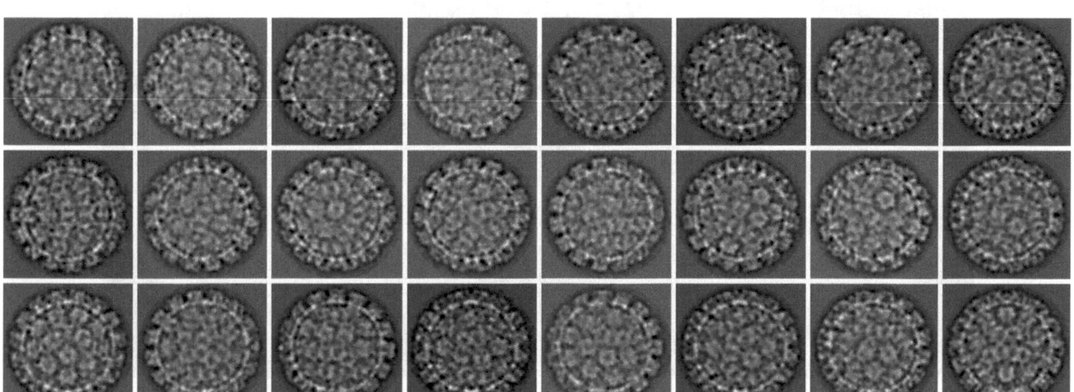

Fig. 6. First 24 classes of the 64 classes calculated by CL2D.

*3.6.5. Initial Volume*

To construct the initial volume, we downloaded the three-dimensional structure of the Bluetongue virus capsid from the Electron Microscopy Data Bank (http://www.ebi.ac.uk/pdbe/emdb, entry code: 5147, Zhang, Jin (29)). This virus is totally unrelated to the Bovine Papilloma virus, and they only share the fact that both have an icosahedral capsid. The volume at the EMDB has an alternate 222 symmetry (twofold symmetry axes on $X$, $Y$ and $Z$). We filtered the Bluetongue capsid to 100 Å, and scaled the capsid to fit in a box of size 128×128×128. We used this lowpass filtered volume as a reference for the alignment and 3D reconstruction of the class averages found by CL2D. We used seven iterations of the 3D Model Refinement pipeline described in Subheading 3.5. The angular sampling was 10° for the first four iterations, 5° in the next two, and 3° in the last one. Images were allowed to freely move in the projection sphere in the first four iterations, and then their Euler angles were restricted to a maximum change of 10° in the next two iterations, and 6° in the last iteration. The first four iterations performed a fully 5D parameter search, while the rest used a 3D+2D search. CTF was not corrected. The whole process took 3 min in 8 processors, at the end of which we had already an initial volume that we rescaled (30 s in a single processor) to a size of 512×512×512. Figure 7 shows the Bluetongue initial structure and its refinement using the CL2D classes. This refined volume will serve as starting volume for the 3D data analysis of the Bovine Papilloma virus.

*3.6.6. 3D Model Refinement*

At this point we entered into a 3D Model refinement process starting from the initial volume constructed in the previous step. We performed 16 iterations of the process. In the first four iterations we performed a fully 5D parameter search allowing the angles to take any value in the projection sphere. The rest of iterations were

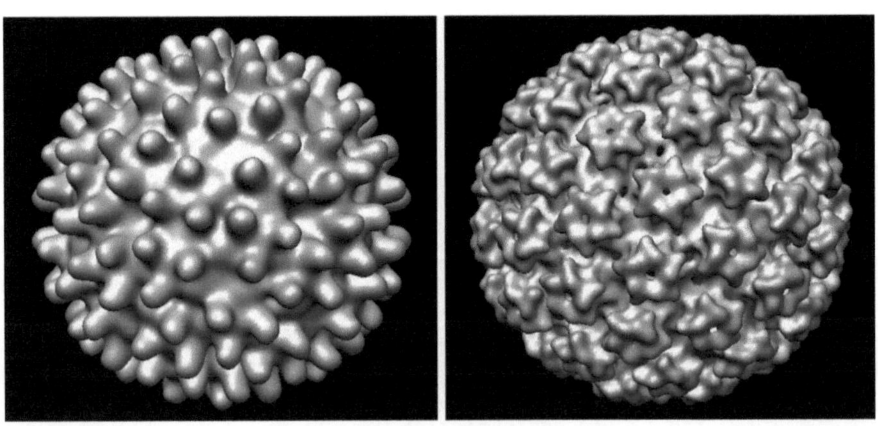

Fig. 7. *Left*: Bluetongue capsid filtered at 100 Å. *Right*: Volume obtained after refining the Bluetongue capsid using the CL2D classes. This example shows the robustness of the 3D reconstruction process in this specific case to the initial volume.

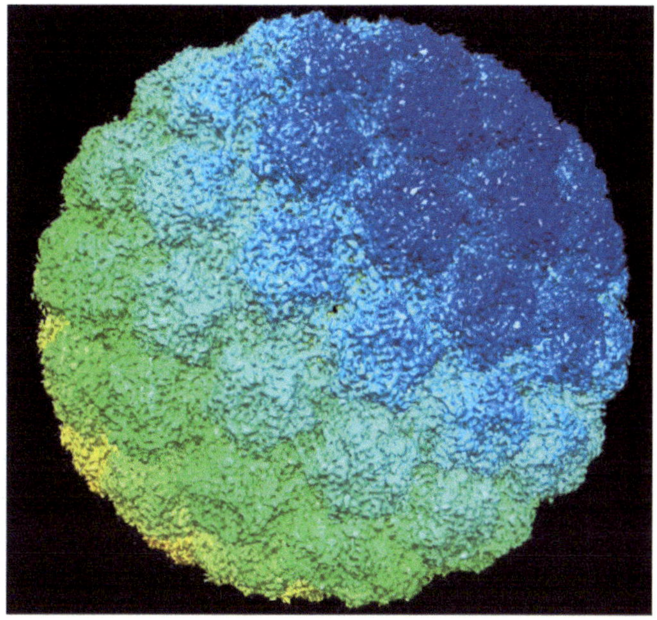

Fig. 8. Final reconstruction of the Bovine Papilloma virus at 5.75 Å.

performed with a 3D + 2D strategy. The angular sampling was 10° in the first four iterations, 5° in the next two, 3° in the following two, 2° in the following two, 1° in the next three, and 0.5° in the last three iterations. In the first four iterations the projection directions could take any value in the projection sphere, in the next two they could move up to a maximum of 30°, in the next two up to a maximum of 15°, in the next two up to a maximum of 10°, and the rest of iterations up to a maximum of 6°.

Starting from iteration 12, we started correcting for the CTF amplitude effects with a Wiener filter (since projections were phase corrected when extracted, the reconstruction up to iteration 11 is phase corrected; the angular step at that level was 2°). The final result is shown in Fig. 8. The resolution achieved in iteration 16 is 5.75 Å (at FSC = 0.5) which is in agreement with the fact that the maximum visible frequency in the micrographs is about 6 Å, and it confirms our hypothesis that the set of images belonged to a single structural population (or at least structural differences are in details finer than those recorded by the microscope). Wolf, Garcea (28) report a resolution of 4.9 Å which is further improved to 3.6 Å by averaging the subunits. The difference from 5.75 to 4.9 Å could be explained by a regularization internal to the 3D reconstruction algorithm used by Wolf, Garcea (28), overfitting or any other reason that increases the consistency of the reconstruction between the two halves of the dataset.

Alternatively we could have performed the amplitude correction by taking the phase corrected volume (Iteration 11), and

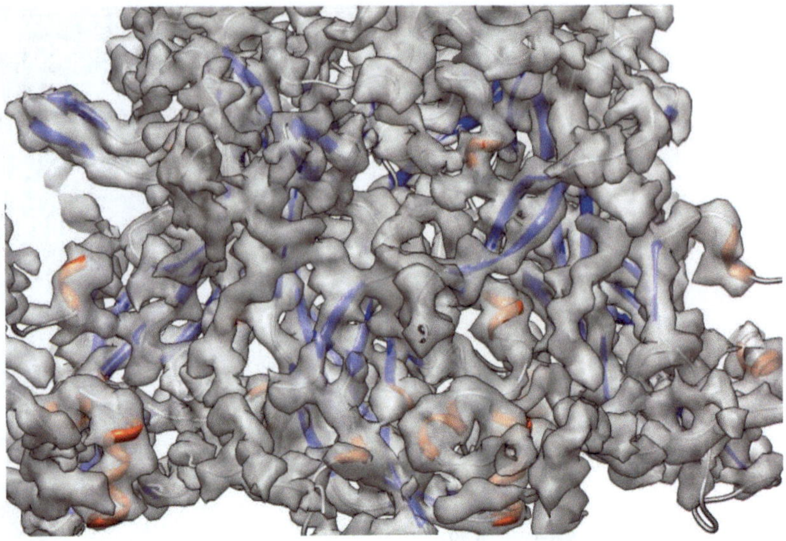

Fig. 9. Overview of the fitting between the EM reconstruction and its atomic model.

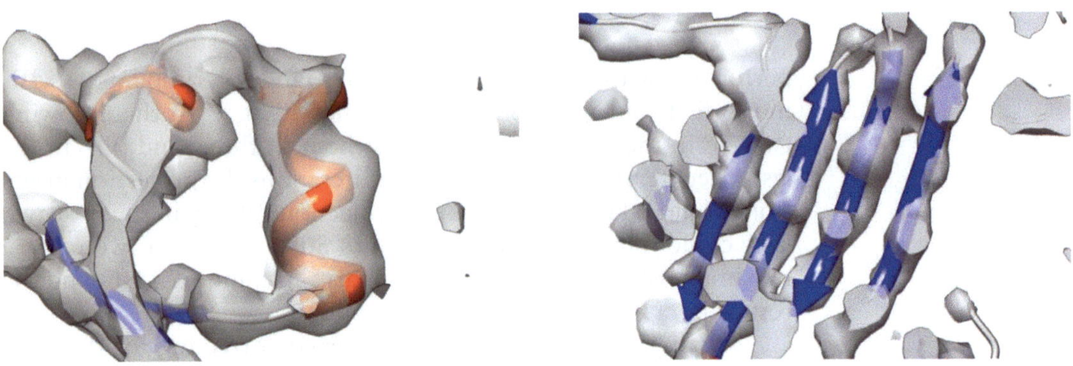

Fig. 10. *Left*: Detail of an α-helix. *Right*: Detail of a β-strand.

applying a B-factor correction (25). This takes 2 min in a single processor, and in this particular case it produced a superior result to the Wiener filter (α-helices were better defined). The resolution of the amplitude corrected volume with the B-factor cannot be measured through the FSC since this measure is insensitive to multiplicative factors.

Wolf, Garcea (28) created an atomic model that was fitted to their EM volume. The atomic model is at the Protein Data Bank (http://www.pdb.org) under the access code 3IYJ. Figures 9, 10, and 11 show some detail of that atomic model fitted to our EM reconstruction. The secondary structure elements (α-helices and β-sheets) can be clearly seen as well as some side chains.

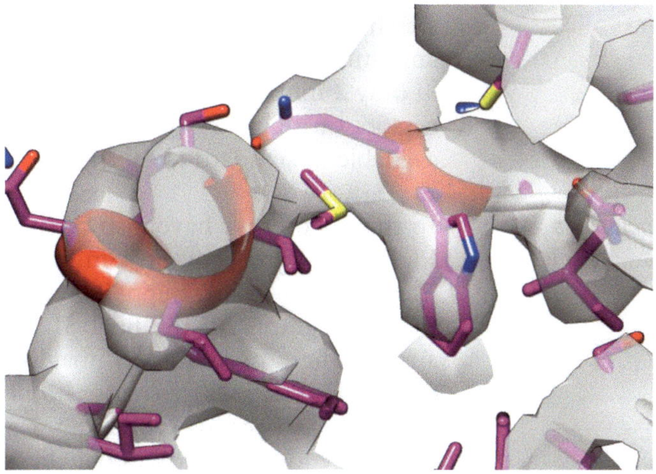

Fig. 11. Detail of the fitting of some side chains.

**Table 1**
**Computing times for each process using a variable number of processors (see text). Each processor is core of an Intel Xeon 2.666 GHz**

| Process | Time |
|---|---|
| Micrograph screening | 15 min |
| Semi-automatic particle picking | 50 min (+1 h manual training and revision) |
| Particle screening | 3 min |
| Particle downscaling | 3 min |
| CL2D | 6 h |
| Initial volume construction | 3 min |
| 3D Model refinement Iterations 1–11 (phase corrected) | 13 h |
| Amplitude correction via B-factor | 2 min |
| 3D Model refinement Iterations 12–16 (amplitude corrected) | 18 h |

*3.6.7. Execution Time*

Table 1 shows the time spent at each of the different steps. In 9 h, we screened the micrographs, selected the particles, screened them, checked that there were no contaminants, and constructed an initial model for the 3D Model refinement. The next 13 h were used to construct a phase corrected volume. Finally, 18 h were needed to construct an amplitude corrected volume with a very fine angular step in the projection matching. In parallel to this last step, a second amplitude corrected volume was computed by B-factor correction in only 2 min. These times were obtained by using at most 32 processors of a cluster, which is a reasonable number of processors available in standard clusters.

### 3.7. Conclusions

In this chapter we have introduced the basic protocols used in 3D Electron Microscopy for Single Particles and illustrated them as they are applied in Xmipp. These protocols cover all the way from the micrographs to the final 3D reconstructed volume. We have applied these protocols to images of the Bovine Papilloma virus capsid. This dataset was rather homogeneous and no heterogeneity analysis was needed. Obtaining a 3D map with a resolution that allows the identification of $\alpha$-helices (below 5.75 Å) took less than 24 h in a cluster with 32 processors available. The presence of heterogeneity in the sample complicates the 3D analysis, specially its validation. However, as shown by this example, 3DEM is advancing towards high-throughput, high-resolution protocols.

## 4. Note

1. All packages have been written with a command-line interface (i.e., the user issue commands from a shell, each one performing a small task). The whole image processing task consists of the concatenation of many small tasks that start at the micrograph level and finish with the 3D reconstruction. However, all packages now offer a graphical interface that provides a more user-friendly interface. In some cases, the different programs have been bundled in a few protocols (30). Protocols reflect the typical image processing pipeline and are suited for most users, especially novice ones. Despite the simplification brought by the protocols, packages still offer their full functionalities so that experienced users may diverge from the most typical image processing path.

## Acknowledgements

C.O.S. Sorzano is a recipient of a Ramón y Cajal fellowship awarded by the Spanish Ministry of Science and innovation (Micinn). J. Vargas and J. Otón are Juan de la Cierva fellows (Micinn). This work has been partly funded by grants CSD2006-0002 (Spanish Ministry of Education), ACI2009-10220 (Micinn), BIO2010-16566 (Micinn), and PIF08-020-2 (CSIC).

## References


1. Frank J (2006) Three-dimensional electron microscopy of macromolecular assemblies: visualization of biological molecules in their native state. Oxford University Press, New York
2. Sorzano COS, Jonic S, Cottevieille M et al (2007) 3D electron microscopy of biological nanomachines: principles and applications. Eur Biophys J 36:995–1013
3. Wang L, Sigworth FJ (2006) Cryo-EM and single particles. Physiology 21:13–18
4. Frank J, Radermacher M, Penczek P et al (1996) SPIDER and WEB: processing and visualization of images in 3D electron microscopy and related fields. J Struct Biol 116: 190–199
5. Ludtke SJ, Baldwin PR, Chiu W (1999) EMAN: semiautomated software for high-resolution single-particle reconstructions. J Struct Biol 128:82–97
6. van Heel M, Harauz G, Orlova EV et al (1996) A new generation of the IMAGIC image processing system. J Struct Biol 116:17–24
7. Sorzano COS, Marabini R, Velázquez-Muriel J et al (2004) XMIPP: a new generation of an open-source image processing package for electron microscopy. J Struct Biol 148:194–204
8. Sorzano COS, Jonic S, Núñez-Ramírez R et al (2007) Fast, robust and accurate determination of transmission electron microscopy contrast transfer function. J Struct Biol 160:249–262
9. Sorzano COS, Otero A, Olmos EM, Carazo JM (2009) Error analysis in the determination of the electron microscopical contrast transfer function parameters from experimental power spectra. BMC Struct Biol 9:18
10. Oppenheim AV, Schafer RW, Buck JR (1999) Discrete-time signal processing. Prentice-Hall, Upper Saddle River
11. Sorzano COS, Iriarte-Ruiz A, Marabini R, Carazo JM (2009) Effects of the downsampling scheme on three-dimensional electron microscopy of single particles. In: Proc. IEEE workshop Intell Signal Proc. Budapest, Hungary
12. Jonic S, Sorzano COS, Cottevieille M et al (2007) A novel method for improvement of visualization of power spectra for sorting cryo-electron micrographs and their local areas. J Struct Biol 157:156–167
13. Mindell JA, Grigorieff N (2003) Accurate determination of local defocus and specimen tilt in electron microscopy. J Struct Biol 142:334–347
14. Akarun L, Yardunci Y, Cetin AE (1997) Adaptive methods for dithering color images. IEEE Trans Image Process 6:950–955
15. Bracewell RN (2006) Fourier analysis and imaging. Springer, New York
16. Sorzano COS, Recarte E, Alcorlo M et al (2009) Automatic particle selection from electron micrographs using machine learning techniques. J Struct Biol 167:252–260
17. Penczek PA, Zhu J, Frank J (1996) A common-lines based method for determining orientations for N>3 particle projections simultaneously. Ultramicroscopy 63:205–218
18. van Heel M (1987) Angular reconstitution: a posteriori assignment of projection directions for 3D reconstruction. Ultramicroscopy 21:111–124
19. Sorzano COS, Bilbao-Castro JR, Shkolnisky Y et al (2010) A clustering approach to multireference alignment of single-particle projections in electron microscopy. J Struct Biol 171:197–206
20. Liu W, Pokharel PP, Príncipe JC (2007) Correntropy: properties and applications in non-Gaussian signal processing. IEEE Trans Signal Process 55:5286–5298
21. Scheres SHW, Valle M, Núñez R et al (2005) Maximum-likelihood multi-reference refinement for electron microscopy images. J Mol Biol 348:139–149
22. Pascual-Montano A, Donate LE, Valle M et al (2001) A novel neural network tecnique for analysis and classification of EM single-particle images. J Struct Biol 133:233–245
23. Crowther RA, Amos LA (1971) Harmonic analysis of electron microscope images with rotational symmetry. J Mol Biol 60:123–130
24. Bárcena M, San Martin MC, Weise F, Ayora S, Alonso JC, Carazo JM (1998) Polymorphic quaternary organization of the Bacillus subtilis bacteriophage SPP1 replicative helicase (G40P). J Mol Biol 283:809–819
25. Fernández JJ, Luque D, Castón JR, Carrascosa JL (2008) Sharpening high resolution information in single particle electron cryomicroscopy. J Struct Biol 164:170–175
26. Frank J, Penczek P (1995) On the correction of the contrast transfer function in biological electron microscopy. Optik 98:125–129
27. Scheres SHW, Gao H, Valle M et al (2007) Disentangling conformational states of macromolecules in 3D-EM through likelihood optimization. Nat Methods 4:27–29
28. Wolf M, Garcea RL, Grigorieff N, Harrison SC (2010) Subunit interactions in bovine papillomavirus. Proc Natl Acad Sci USA 107:6298–6303
29. Zhang X, Jin L, Fang Q et al (2010) A cryo-EM structure of a nonenveloped virus reveals a priming mechanism for cell entry. Cell 141:472–482
30. Scheres SHW, Núñez-Ramírez R, Sorzano COS et al (2008) Image processing for electron microscopy single-particle analysis using XMIPP. Nat Protoc 3:977–990

# Chapter 12

# Mass Mapping of Amyloid Fibrils in the Electron Microscope Using STEM Imaging

**Alioscka A. Sousa and Richard D. Leapman**

## Abstract

Scanning transmission electron microscopy (STEM) in the dark-field mode of operation is a technique regularly used to record high-contrast images from isolated macromolecular assemblies at nanometer resolution. Dark-field STEM images are unique in that they can be readily quantified to provide information on the mass of individual molecular complexes. Importantly, because STEM images contain simultaneous mass and overall molecular shape information, the concept of "mass mapping" can be realized to provide distinctive measurements of the mass per area of planar assemblies or the mass per length of filamentous structures. In this chapter we describe how the STEM technique can be applied to generate characteristic measurements of mass per length from isolated Alzheimer's amyloid fibrils.

**Key words:** Scanning transmission electron microscopy (STEM), Mass mapping, Mass per length, Amyloid fibrils

## 1. Introduction

The electron microscope is a powerful tool with which to obtain high resolution structural information from isolated macromolecular assemblies. Molecules can be visualized by amplitude contrast in preparations of negatively stained samples, or by phase contrast in cryo-transmission electron microscopy (TEM) of frozen-hydrated specimens. In both of these methodologies, however, image intensity does not bear a direct relationship to the mass of the molecule being imaged. The alternative technique of scanning transmission electron microscopy (STEM) enables the acquisition of high-contrast images from unstained isolated molecular assemblies, and a straightforward analysis of such images allows determination of the mass of individual assemblies (1–4).

Importantly, in the electron microscope molecular mass can be directly correlated to molecular shape and, therefore, it becomes

Alioscka A. Sousa and Michael J. Kruhlak (eds.), *Nanoimaging: Methods and Protocols*, Methods in Molecular Biology, vol. 950, DOI 10.1007/978-1-62703-137-0_12, © Springer Science+Business Media, LLC 2013

feasible to measure the mass per area of planar structures or the mass per length of filamentous specimens (5). These types of measurements are unique to the electron microscope and cannot be carried out with other methods for mass determination, such as mass spectrometry and ultracentrifugation. Further advantages include the wide range of masses (from tens of kDa to hundreds of MDa) that can be analyzed, the reasonably high degree of accuracy (between 5 and 10%) that can be typically achieved, and the need for only a very small amount of material on the order of micrograms.

The principle of STEM for molecular mass determination is briefly described as follows. A fine electron probe of sub-nanometer diameter scans across a specimen of unstained molecular complexes deposited onto an ultrathin carbon support film. The charged nuclei of atoms in the specimen elastically scatter a fraction of the incident electrons, which are collected by a high-angle annular dark-field (ADF) detector. The intensity of each pixel in the resultant ADF STEM image is directly proportional to the projected specimen mass contained within the area defined by that pixel. Although the constant of proportionality between image intensity and mass can be calculated from first principles, a much simpler approach makes use of co-deposited tobacco mosaic viruses (TMV) (6). The TMV virus is 300 nm long, 18 nm wide, and its mass per length of 131 kDa/nm serves as a well-defined internal mass calibration standard.

Among the numerous applications of STEM mass measurement, the technique has been most widely applied to analyzing filamentous structures of which amyloid fibrils implicated in Alzheimer's disease are an important and well-studied class (5). In general terms, amyloid fibrils consist of a cross-β sheet structure in which β-strand segments are aligned perpendicular to the fibril long axis and connected by intermolecular hydrogen bonds with a cross-β repeat distance of 0.48 nm. The packing of at least two β-sheets is usually necessary to form a single Alzheimer's amyloid protofilament unit. In turn, variations in protofilament number and arrangement contribute to the substantial polymorphism characteristic of amyloid fibrils. Measurements of mass per length (MPL) in the STEM are unique in that they yield the number of peptide molecules per cross-β repeat on individual fibrils. When combined with information obtained from other structural techniques, this analysis assists in the generation of comprehensive structural models for the assembly of single protofilament subunits into fibril polymorphs (7–11).

In this chapter, we discuss the technique of STEM mass measurement as applied to the important and prevalent amyloid fibril structures, highlighting important aspects of the method related to sample preparation, data collection, data processing and analysis. Although to date most STEM mass measurements have been carried out using specialized, dedicated STEM instruments operating at 40–100 kV acceleration voltage, this approach also works with widely available commercial TEM/STEM microscopes operating at 100–300 kV (12–14). Importantly, the ability to perform molecular mass measurements in modern TEM/STEM instruments brings the technique to a wider community of biologists than has been previously possible. Finally, while this chapter focuses more specifically on MPL measurements of amyloid fibrils, it should also provide useful information related to mass measurement of other biological or synthetic structures.

## 2. Materials

### 2.1. Instrumentation and Software

1. Scanning transmission electron microscope equipped with a field-emission gun (see Note 1). The microscope can be operated at an acceleration voltage in the range of 40–300 kV (see Note 2).
2. High-angle annular dark-field (ADF) STEM detector of high quantum efficiency (see Note 1).
3. Carbon evaporator for making ultrathin (2–3 nm) films of amorphous carbon. We use an Edwards model Auto 306 (Edwards, UK).
4. Glow-discharge unit to render EM grids hydrophilic. We use a home-made unit, but commercial ones are also available.
5. Software for image analysis. Two programs specific for mass measurement are available for automated processing of dark-field STEM images: PCMass (15), developed by Wall at Brookhaven National Laboratory (Upton, NY, USA), and MASDET (16, 17), developed by Krzyzanek and Reichelt (University of Münster, Germany) and Müller and Engel (University of Basel, Switzerland). These programs streamline mass analysis enabling several hundreds or thousands of measurements to be computed. Analysis of STEM images can also be carried out manually using standard image processing software such as ImageJ (National Institutes of Health, Bethesda, USA): http://rsb.info.nih.gov/ij/.

### 2.2. Supplies

1. Solution of amyloid fibrils. The fibrils used in this study were a mutated form of the Alzheimer's Aβ peptide. They were obtained from Oleg Antzutkin (Luleå University of Technology, Sweden; and University of Warwick, UK) and Andrei Filippov (Luleå University of Technology; and Kazan State University, Russia).

2. Ultrapure water.
3. Washing buffer: composition must be the same as the buffer in which the amyloid fibrils are dispersed.
4. 1% Solution of uranyl acetate for negative staining.
5. Lacey carbon-coated 300 mesh Cu grids.
6. Muscovite mica sheets (Electron Microscopy Sciences, USA).
7. Clean tweezers.
8. Filter paper.

## 3. Methods

Although relatively simple to implement, STEM mass measurement of molecular assemblies imposes strict demands on sample quality. First, samples must be free of contaminants of any sort. Second, the complexes must be sufficiently stable not to denature during the steps of sample preparation. A simple visual inspection of STEM images is enough in most cases to identify problems with sample preparation before undergoing any more time consuming quantitative analysis. Provided that a suitable specimen can be prepared, data collection is carried out at low electron doses to keep mass loss of the sample to a minimum. Analysis of the recorded images is then relatively straightforward given that clean specimen areas containing the molecular complexes of interest and TMV are judiciously chosen. In what follows, some relevant issues related to sample preparation (Subheading 3.1), data collection (Subheading 3.2), and data processing and analysis (Subheading 3.3) will be discussed.

### 3.1. Sample Preparation

We find that the somewhat simplified protocol described below works well with amyloid fibrils. Because these fibrils are stable, sample preparation is not as demanding as for other types of samples such as multi-unit protein complexes that can denature and break apart more easily.

#### 3.1.1. Support Film

1. Using a carbon evaporator, deposit a 2–4 nm thick layer of amorphous carbon onto freshly cleaved mica. Carbon films thinner than about 2 nm could break easily under the STEM beam, whereas films thicker than about 4 nm will be unnecessarily thick and reduce image contrast and signal-to-noise ratio.
2. Using scissors, cut the mica in little squares slightly larger than the EM grids onto which the ultrathin carbon films will rest.

3. Float off the pieces of mica in a container with ultrapure water, pick up the carbon films released from the mica surface with lacey carbon-coated 300 mesh Cu grids, blot off the grids with filter paper, let them dry in air, and store for later use.

#### 3.1.2. Negatively Stained Samples for Conventional TEM Imaging

Before attempting to prepare samples of unstained amyloid fibrils for STEM analysis, it is useful to prepare negatively stained specimens for imaging by TEM.

1. Here, use carbon films slightly thicker than the 2–4 nm required for STEM measurements, so that they do not break during the negative stain preparations.
2. Glow discharge the carbon-coated EM grids to render them hydrophilic.
3. Place a 3–5 μL drop of amyloid fibril solution onto one of the TEM grids at a concentration typically between 0.1 and 1 mg/mL.
4. After adsorption for about a minute, blot off the grid and wash it with ultrapure water. This washing step is accomplished by adding a 5 μL drop of water, waiting 10 s, blotting off the grid, and repeating this sequence a few times.
5. After the final blotting, add 3–5 μL of a 1% solution of uranyl acetate, and leave for 60 s.
6. Blot off the uranyl acetate solution completely and let dry.

#### 3.1.3. Unstained Samples for Quantitative STEM Imaging

During sample preparation for mass determination, every step must be carefully executed to produce the cleanest possible specimen.

1. Glow discharge the 2–3 nm thick carbon films to render them hydrophilic.
2. Add a 3–5 μL drop of TMV particles from a 100 μg/mL solution (see Note 3). After adsorption for 1 min, blot off excess TMV solution (never allowing the film to dry) and wash the grid three times with washing buffer. After the final wash, leave a 3 μL drop of washing buffer on the grid for the next step.
3. Inject a 3 μL drop of the fibril solution into the drop left on the grid from the previous step (see Notes 3 and 4). After adsorption for 1 min, blot off the grid and wash a few times with washing buffer (five times) followed by ultrapure water (five times) (see Note 5).
4. After the final step of washing, let the grid dry in air and keep it an evacuated chamber until ready for imaging (see Note 6).

### 3.2. Data Collection

#### 3.2.1. TEM Imaging of Negatively Stained Preparations

Record images of negatively stained amyloid fibrils in standard TEM mode (Fig. 1). This step serves to provide an initial screening of sample quality. Furthermore, the characteristic shape, thickness and twisting of amyloid fibrils can be visualized with higher contrast and resolution in images of negatively stained preparations than in images of unstained samples recorded in the STEM. The different fibril morphologies that are found to coexist in the specimen are then matched to the quantitative MPL data obtained subsequently by STEM.

#### 3.2.2. STEM Imaging of Unstained Preparations: Microscope Set-up

In this section, we assume the reader is familiar with the basic operation of STEM instruments. We also mostly limit our discussion to combined TEM/STEM instruments operating at 100–300 kV, since these are the STEM-capable microscopes most likely to be accessible to biologists.

When setting up the microscope in STEM mode, some of the critical microscope parameters include the extraction voltage, spot size (C1 lens), size of second condenser aperture, and camera length. These parameters, in turn, have direct impact over experimental variables such as the electron dose incident on the specimen, convergence semi-angle of the incident probe, and inner collection semi-angle of the STEM detector.

1. *Beam current*. Choose specific values of extraction voltage and C1 spot size that will give about 8–24 pA of beam current.

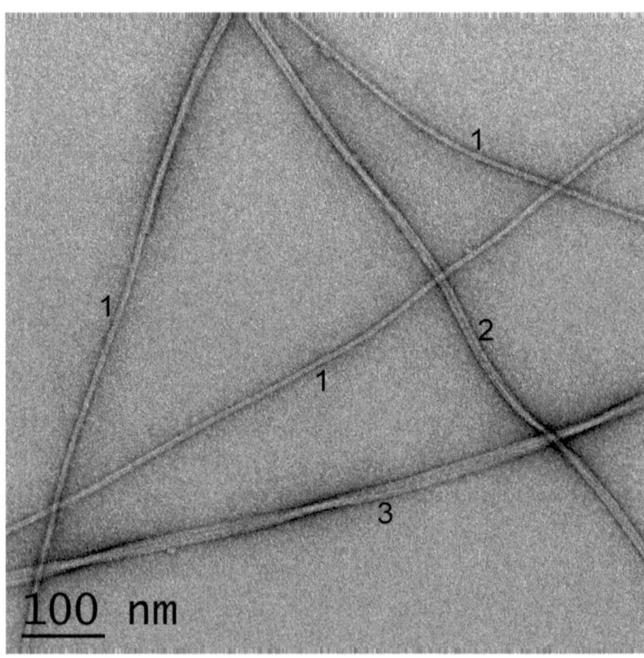


Fig. 1. TEM image of negatively stained amyloid fibrils. Fibrils with three distinct morphologies (numbered *1*, *2*, and *3*) are present in the image.

For usual values of 20 μs pixel dwell time and 1 nm pixel size (see below), this beam current produces a final electron dose of 1,000–3,000 e/nm$^2$ incident on the specimen.

2. *Electron dose.* Keep the total electron dose in the range of about 1,000–3,000 e/nm$^2$, depending on the acceleration voltage of the microscope. Although relatively low, these doses are sufficient to cause some degree of mass loss in the specimen (up to 10%). The rate of specimen mass loss can be reduced by recording images at liquid nitrogen temperature, thus potentially allowing the use of even higher doses to achieve better counting statistics. To improve accuracy especially for data recorded at room temperature, measurements of mass loss as a function of electron dose can be used to extrapolate to the mass of the undamaged complex at zero dose. In general, however, protein complexes and TMV (which is comprised of 95% protein) should have a similar rate of beam-induced mass loss, and so to a first approximation, corrections for mass loss are not strictly necessary.
3. *Convergence semi-angle.* Examine the shadow image of the specimen inside the intense undiffracted electron beam on the fluorescent screen to help select and center the second condenser aperture. For 100–300 kV STEM instruments, the convergence semi-angle angle of the probe is typically around 10 mrad.
4. *Collection semi-angle.* Choose a proper value of camera length, which dictates the inner collection semi-angle of the ADF STEM detector (the smaller the camera length, the larger the inner semi-angle). The minimum possible value of inner collection semi-angle is constrained by the convergence semi-angle of the STEM probe (around 10 mrad). We recommend selecting an intermediate value of inner semi-angle (e.g., 20 mrad) that is neither too close to the convergence angle of the probe (i.e., only 1 or 2 mrad larger) which could introduce a large spurious intensity into the ADF image during scanning, nor too large (i.e., >40 mrad) which would exclude a large fraction of the elastically scattered electrons from the ADF detector resulting in unnecessarily noisy images (see Note 7).

#### *3.2.3. STEM Imaging of Unstained Preparations: Data Acquisition*

1. Once the microscope has been properly set up, search at lower magnification for regions of interest containing fibrils and TMV. This search must be done quickly and at an electron dose much lower (10× lower or more) than the 1,000–3,000 e/nm$^2$ used for data collection.
2. Acquire 1,024 × 1,024-pixel STEM images of fibrils and TMV at a pixel size of approximately 1 nm. Ensure that fibrils and TMV are present in the same image frame.

3. Check overall sample quality. Figure 2a shows a specimen region with clean background, co-adsorbed TMVs as well as amyloid fibrils that are suitable for quantitative mass analysis. High contrast STEM images can often reveal the presence of residual salts, denatured proteins, unwashed monomers, or other contaminants. Figure 2b exemplifies a fibril (arrow) with impurities attached. In the case of amyloid fibrils, which are normally very stable, specimens will typically not contain denatured proteins. On the other hand, samples might often contain small spherical aggregates, which are intermediate products that precede the formation of fibrils. These spherical aggregates can be tolerated in a specimen provided that it is possible to find clean enough regions of background, TMV, and fibrils free of attached aggregates.
4. Check the TMV viruses to confirm that the accumulated electron dose is not too high. TMV rods have one end concave and the other convex, but under conditions of high accumulated electron doses these ends lose their sharpness (Fig. 2d).
5. In addition, check the overall quality of the TMV viruses. First, make sure that the TMV rods appear sharp along their length. TMV with a fuzzy appearance might indicate not only that the electron dose is too high (Fig. 2d), but also that the image is out of focus or that amyloid fibrils are attaching alongside TMV (which we commonly observe in our specimens) (Fig. 2c). Second, check that the central cavity of TMV is not filled with buffer salts or other contaminants, which can be sometimes detected by STEM (Fig. 2e).
6. Repeat steps 1–5 over different regions of the specimen grid. Samples for STEM mass analysis are not necessarily homogeneous across the entire specimen grid, and as a result some sample regions can be better suited for analysis than others.

### 3.3. Data Processing and Analysis

HAADF STEM images can be analyzed "automatically" using specialized software to yield the MPL of amyloid fibrils, including PCMass (15) and MASDET (16, 17). Here we present step-by-step instructions for manually carrying out measurements of MPL using standard image processing software (ImageJ). Nonetheless, most of the points addressed below are of general nature and must also be kept in mind when using specialized software.

1. Select a STEM image containing both sharp and clean TMV and fibrils as explained in Subheading 3.2.3. The image must also be free of contaminants in the background.
2. First draw a box around a TMV particle as shown in Fig. 3a and integrate the total signal intensity inside this box. The length of the box can be around 50 nm and its width slightly larger than that of the 18-nm wide TMV. The box around

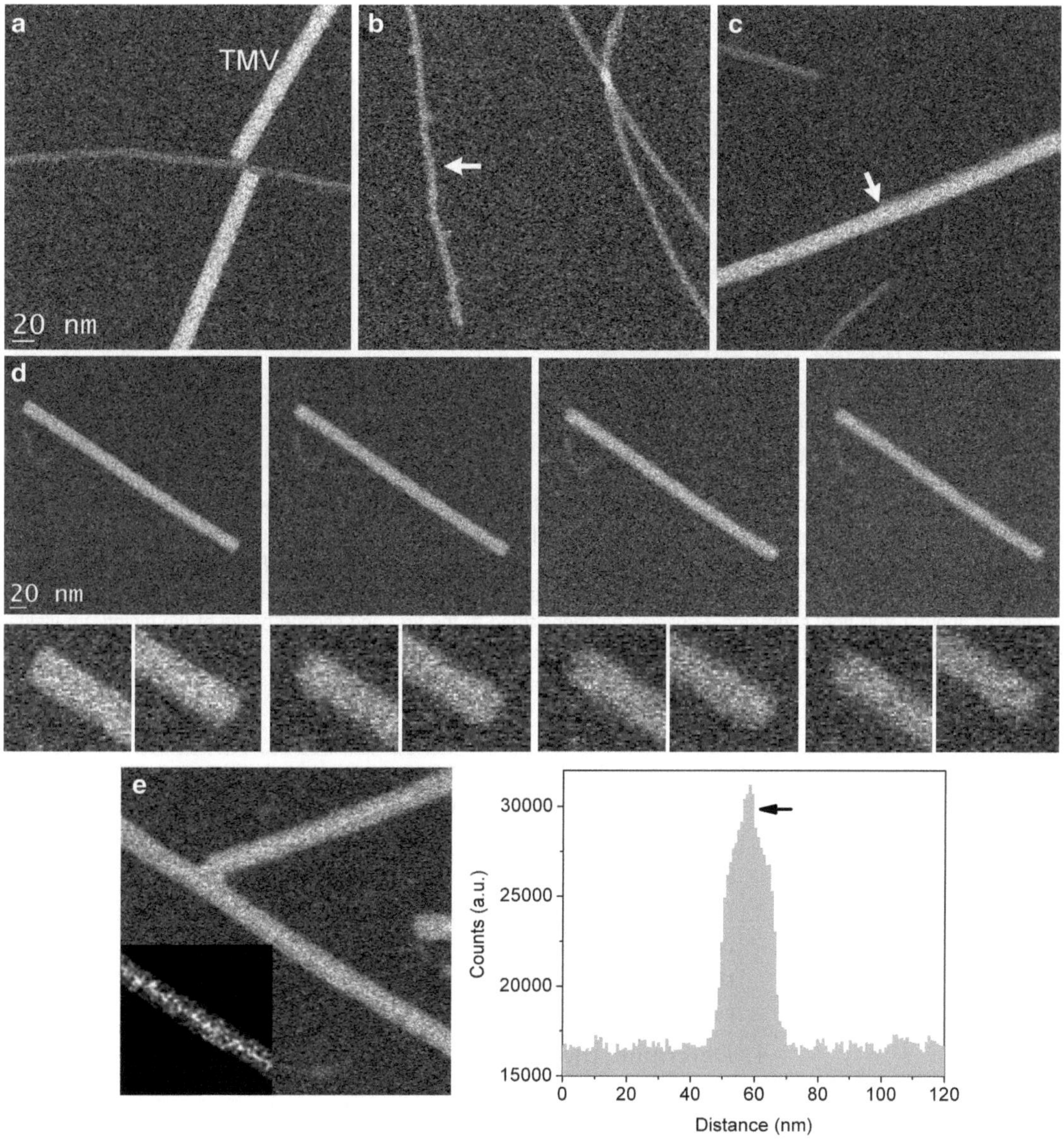


Fig. 2. Quality assessment of dark-field STEM images. (**a**) Image with clean fibril, TMV and background regions suitable for STEM mass analysis. (**b**) Image showing contaminants attached to one of the fibrils (*arrow*). (**c**) Image showing an amyloid fibril attached alongside a TMV virus (see TMV region to the *right* of the *arrow*). The TMV region to the *left* of the *arrow* is clean of any attached fibrils. (**d**) Series of images of TMV at increasing electron dose. Each image was acquired with 3,000 e/nm$^2$, so the last (*rightmost*) image has an accumulated electron dose four times higher than the first. *Bottom* panels show zoomed in views of the concave and convex ends of TMV. Images of TMV become less sharp with increasing electron dose. (**e**) Image illustrating the effect of incomplete washing of specimens. The central cavity of the TMV viruses is filled with buffer salts, which can be seen more clearly in the *inset* image after smoothing and contrast adjustment (notice the brighter contrast in the center and along the length of the TMV). A line profile across the TMV shows increased intensity near the *middle*, thus revealing a filled central cavity.

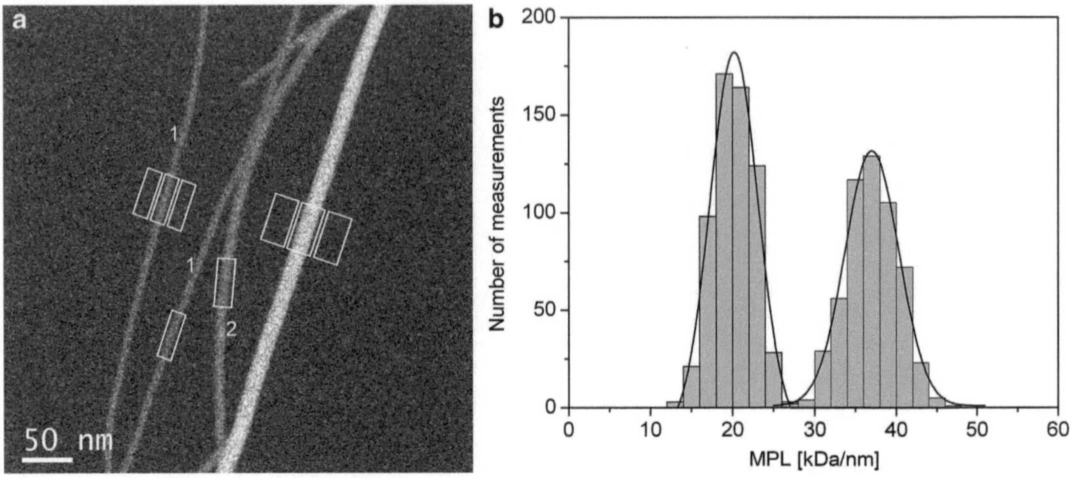


Fig. 3. Mass analysis of STEM images. (**a**) Fibrils with two distinct morphologies (numbered *1* and *2*) are present in the image. These should correspond to fibrils 1 and 2 indicated in Fig. 1. *Rectangular boxes* 50 nm in length delineate the regions from which integrated intensities are measured for TMV, fibrils and background. For simplicity, boxes for background region were not drawn for two of the fibrils. (**b**) Histogram of MPL data (bin size = 2 kDa/nm) obtained from the analysis of about 1,200 fibril segments. Gaussian functions fitted to the first and second peaks of the histogram yielded 20.2 ± 6.9 and 37.0 ± 7 kDa/nm, respectively.

TMV must not be drawn unnecessarily wide, since this introduces random noise in the measurements and decreases their precision.

3. Draw boxes with exactly the same dimensions on either side of TMV, average their integrated intensities, and subtract the result from the integrated TMV intensity from step 2 (see Note 8).

4. Determine the calibration factor $K$:

$$K = \frac{131}{\left(\frac{I_{\text{TMV}}}{L}\right)} \left[\frac{\text{kDa}}{\text{counts}}\right],$$

Where $I_{\text{TMV}}$ is the net intensity value of TMV after background subtraction, $L$ is the box length in nanometers, and 131 kDa/nm is the known MPL value of TMV.

5. Determine the net integrated intensity (i.e., after background subtraction) along several different segments of amyloid fibrils. Figure 3a shows for example a few boxes (about 50 nm in length) along different fibril segments and regions of background.

6. For each fibril segment, divide the net integrated counts $I_{\text{Fibril}}$ by the respective box length $L$ to yield the number of integrated counts per nanometer.

7. Calculate the MPL of each fibril segment as given by:

$$\text{MPL} = K \times \left(\frac{I_{\text{Fibril}}}{L}\right)\left[\frac{\text{kDa}}{\text{nm}}\right].$$

8. Repeat steps 1–7 for several STEM images and calculate the MPL values from at least a couple hundred different segments from fibrils. Preparations that appear "dirty" and/or that contain a large number of different fibril polymorphs may require the analysis of several hundred or a few thousand separate segments.
9. Plot a histogram of MPL data with a bin size of 1–2 kDa/nm, and fit each peak to a Gaussian function. Figure 3b shows for example a histogram obtained from the analysis of 1,200 segments of amyloid fibrils such as those shown in Fig. 3a. Gaussian functions fitted to the first and second peaks of the histogram of Fig. 3b yielded $20.2 \pm 6.9$ and $37.0 \pm 7$ kDa/nm, respectively.
10. Determine the number of peptide molecules per cross-β repeat associated with each fibril polymorph. For this, first calculate the MPL value of a single peptide molecule ($\text{MPL}_p$), which is found by dividing the molecular weight of the peptide by the 0.48 nm cross-β repeat. Then, divide the average MPL values from the Gaussian fits by $\text{MPL}_p$. In the current example, the two populations of MPL in Fig. 3b yielded $2.2 \pm 0.7$ and $4.1 \pm 0.7$ (±standard deviation) peptide molecules per cross-β repeat.

Carefully performed STEM mass measurements can achieve an accuracy of about 5–10% depending on the sample. It is therefore reasonable to assume that the two fibril morphologies in the example of this chapter contain integer numbers (2 and 4) of peptide molecules per cross-β repeat, and that the observed deviations from integer are most probably due to contaminants (e.g., small spherical aggregates) attached alongside some fibrils. The great majority of previous STEM mass measurements on amyloid fibrils in the literature have arrived at integer numbers of peptide molecules per cross-β repeat. A recent study has reported, however, the existence of a non-integer number of about 2.5 peptides per cross-β repeat in single protofilaments of the Alzheimer's Aβ(1–42) amyloid fibril system (11).

## 4. Notes

1. We use a Tecnai TF30 scanning/transmission electron microscope (FEI Company, USA) operating at an acceleration

voltage of 300 kV and equipped with a Schottky field-emission gun. This microscope is fitted with a Model 3000 in-column high angle annular dark-field STEM detector (Fischione Instruments, USA) situated after the projection-lens system and above the viewing screen.

2. STEM mass measurements have been to date successfully performed at several different acceleration voltages, including 40, 80, 100, 300 kV. The use of higher voltage instruments (200–300 kV) could be advantageous for imaging larger structures such as whole organelles as well as polymer nanoparticles for nanomedicine applications, which otherwise might be too thick for imaging at lower voltages.
3. For the most part, the optimal concentration of the TMV and fibril solutions is variable. For the fibrils, some types adhere well to the carbon support film while others do not. It is therefore better to start with a high concentration of fibrils and dilute the solution as needed.
4. Injection of the sample solution minimizes the exposure of the molecules to the air-water interface and reduces the possibility of denaturing. Amyloid fibrils, however, should be fairly stable to conditions that otherwise could cause other molecules to denature.
5. Although not generally required for amyloid fibrils, it is sometimes advantageous to use ammonium acetate buffer (at ~50 mM) instead of pure water when the molecular assemblies break apart during the washes. A chemically inert grid made of titanium should then be used instead of copper. For samples that still break apart, injecting a solution of 0.1% glutaraldehyde into the specimen drop on the grid should solve the problem, although there is a risk of adding some measurable mass to the protein. The glutaraldehyde is left in the drop for about 10 min and the grid subsequently washed as usual (6).
6. Air drying causes some collapse of the specimen's fine structure. For most samples this does not pose problems, since (1) the mass measurements are not affected and (2) the overall shape of the complexes can still be recognized in STEM and matched to that of negatively stained preparations. In cases where the best structure needs to be preserved for STEM imaging, samples should be plunge-frozen in liquid nitrogen after the very final wash and subsequently freeze-dried.
7. For the inexperienced user, a simple approach to help select a reasonable value of detector collection angle would involve recording data at a few different settings of camera length and comparing the results in terms of their mean and standard deviation. Inner semi-angles that are too small might compromise the accuracy of the measurements, whereas unnecessarily

large inner semi-angles will produce noisy images and deteriorate the standard deviation of the measurements. This approach although time consuming would have to be performed just once provided other microscope parameters (mainly the acceleration voltage and size of condenser aperture) were kept unchanged.

8. As an alternative approach, background intensity can be computed by first averaging the pixel values inside larger areas of background close to the TMV segment being analyzed. This average intensity is then multiplied by the area in pixels of the box drawn on the TMV.

## References


1. Muller SA, Engel A (2006) Biological scanning transmission electron microscopy: imaging and single molecule mass determination. Chimia 60:749–753
2. Wall JS, Simon MN, Lin BY, Vinogradov SN (2008) Mass mapping of large globin complexes by scanning transmission electron microscopy. In: Poole RK (ed) Methods in enzymology. Academic, New York, pp 487–501
3. Müller SA, Engel A (2001) Structure and mass analysis by scanning transmission electron microscopy. Micron 32:21–31
4. Engel A (1978) Molecular weight determination by scanning transmission electron microscopy. Ultramicroscopy 3:273–281
5. Steven AC, Hainfeld JF, Trus BL, Wall JS, Steinert PM (1983) The distribution of mass in heteropolymer intermediate filaments assembled *in vitro*. J Biol Chem 258:8323–8329
6. Wall JS, Hainfeld JF (1986) Mass mapping with the scanning transmission electron microscope. Annu Rev Biophys Biophys Chem 15:355–376
7. Goldsbury C, Baxa U, Simon MN, Steven AC et al (2011) Amyloid structure and assembly: insights from scanning transmission electron microscopy. J Struct Biol 173:1–13
8. Antzutkin ON, Leapman RD, Balbach JJ, Tycko R (2002) Supramolecular structural constraints on Alzheimer's β-amyloid fibrils from electron microscopy and solid-state nuclear magnetic resonance. Biochemistry 41:15436–15450
9. Petkova AT, Ishii Y, Balbach JJ, Antzutkin ON et al (2002) A structural model for Alzheimer's β-amyloid fibrils based on experimental constraints from solid state NMR. Proc Natl Acad Sci U S A 99:16742–16747
10. Paravastu AK, Leapman RD, Yau WM, Tycko R (2008) Molecular structural basis for polymorphism in Alzheimer's β-amyloid fibrils. Proc Natl Acad Sci U S A 105:18349–18354
11. Schmidt M, Sachsea C, Richter W et al (2009) Comparison of Alzheimer Aβ (1–40) and Aβ (1–42) amyloid fibrils reveals similar protofilament structures. Proc Natl Acad Sci U S A 106:19813–19818
12. Holmes DF, Graham HK, Trotter JA, Kadler KE (2001) STEM/TEM studies of collagen fibril assembly. Micron 32:273–285
13. Sousa AA, Leapman RD (2007) Quantitative STEM mass measurement of biological macromolecules in a 300 kV TEM. J Microsc 228:25–33
14. Sousa AA, Aronova MA, Wu H, Sarin H, Griffiths GL, Leapman RD (2009) Determining molecular mass distributions and compositions of functionalized dendrimer nanoparticles. Nanomedicine 4:391–399
15. http://www.bnl.gov/biology/stem/Data_Analysis.asp
16. http://campus.uni-muenster.de/masdet.html
17. Krzyzánek V, Müller SA, Engel A, Reichelt R (2009) MASDET-A fast and user-friendly multiplatform software for mass determination by dark-field electron microscopy. J Struct Biol 165:78–87

# Chapter 13

# Elemental Mapping by Electron Energy Loss Spectroscopy in Biology

**Maria A. Aronova and Richard D. Leapman**

## Abstract

Over the past decades there have been significant advances in transmission electron microscopy for biological applications, including in energy filtering and spectrum imaging, which are techniques based on the principles of electron energy loss spectroscopy. These imaging modalities allow quantitative mapping of specific chemical elements with high sensitivity and spatial resolution. This chapter describes the experimental and computational procedures for elemental mapping in two dimensions as well as a more recent extension to three dimensions, which can reveal quantitative distributions of elements in cells on a macromolecular scale.

**Key words:** Electron energy loss spectroscopy, EELS, Elemental mapping, Energy filtered transmission electron microscopy, EFTEM tomography, 3D elemental map, Chromatin structure

## 1. Introduction

Although conventional electron microscopy provides important structural information on macromolecular scale, it is useful to obtain other types of complementary information in order to identify specific types of macromolecules within cellular compartments. One such approach is based on electron energy loss spectroscopy, which uses inelastically scattered electrons to identify specific chemical elements. In biological specimens these elements act as intrinsic stains to localize certain cellular components, for example, the distribution of phosphorus and nitrogen atoms indicate the presence of nucleic acid and protein, respectively (1).

To understand how to measure the distribution of elements in a biological specimen, it is helpful to examine briefly the form of the electron energy loss spectrum (EELS) (2) (see Fig. 1), and how the spectrum is collected. The most dominant feature in the spectrum is the zero-loss peak, which corresponds to incident electrons that are either unscattered and pass undeviated through the

Alioscka A. Sousa and Michael J. Kruhlak (eds.), *Nanoimaging: Methods and Protocols*, Methods in Molecular Biology, vol. 950, DOI 10.1007/978-1-62703-137-0_13, © Springer Science+Business Media, LLC 2013

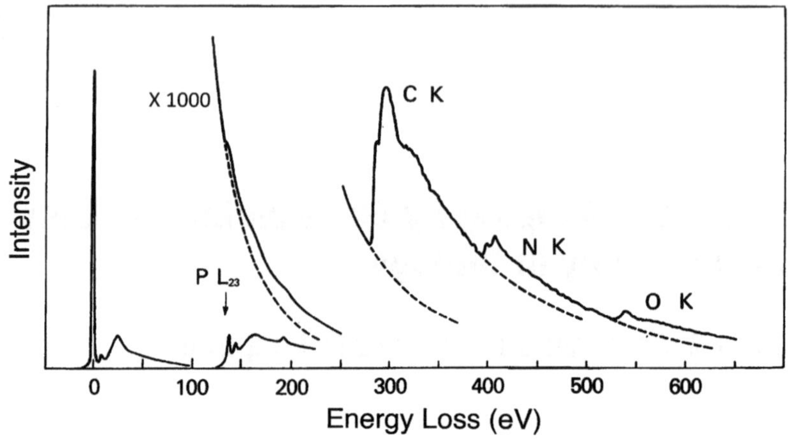


Fig. 1. Electron energy loss spectrum (EELS) from DNA with zero-loss peak, plasmon peak, and superpositioned elemental edges which originate from excitation of inner-shell or core levels of phosphorus, carbon, nitrogen, and oxygen. The *dashed line* indicates the background extrapolation that is used to obtain the raw elemental signal (adapted from ref. 15).

specimen, or are elastically scattered by atomic nuclei with a change in direction but with negligible energy loss. The most probable inelastic scattering events correspond to excitations of atomic outer shell electrons and typically involve energy losses of 10–50 eV. This low-loss spectrum contains a maximum at around 20 eV attributable to a plasmon or collective excitation of the valence electrons; it also includes a tail of decreasing intensity that extends to higher energy losses (range 100–1,000 eV) due to ejection of valence electrons with increasing amounts of kinetic energy. Superposed on this tail due to valence excitations are weak signals or "edges," which originate from excitation of inner-shell or core levels of specific atoms in the specimen and are typically a factor of $10^3$ or $10^4$ weaker than the total inelastic or elastic signals.

One of the approaches for elemental mapping by EELS makes use of the energy filtering in the transmission electron microscope imaging mode (EFTEM) (3, 4). Electrons that are transmitted through a thin specimen enter the magnetic field of the spectrometer and are then dispersed according to their kinetic energy or velocity. A slit placed at the dispersion plane of the spectrometer's magnetic prism selects electrons of a particular energy loss, which then travel through a series of lenses to form a high resolution image of the specimen on a two-dimensional array detector. By inserting an energy-selecting slit, elastically scattered as well as unscattered electrons are allowed to pass, which result in improvements in contrast in some applications.

To obtain element-specific maps from sectioned cells, it is typically necessary to record two images, one at an energy loss just above the atomic core level excitation threshold, with image intensity $I_{post}$, and one at an energy loss just below the threshold, with image intensity $I_{pre}$. By carefully subtracting a fraction of $I_{pre}$ intensity from $I_{post}$ intensity, it is possible not only to determine the net core edge signal but also to quantify the number of atoms of a particular element in the specimen using the theoretical scattering cross section for the core level excitation (2). The number of atoms $n_x$ of an element $x$ in a pixel of dimension $d$ can be determined from the net signal at energies above the core edge $S_x(\Delta, \beta)$ measured in a band of width $\Delta$ defined by the energy-selecting slit and for a collection angle $\beta$, which is defined by the objective aperture:

$$n_x = \left[\frac{S_x(\Delta,\beta)d^2}{I_0(\Delta,\beta)\cdot\sigma_x(\Delta,\beta)}\right] \quad (1)$$

where $I_0(\Delta, \beta)$ is the measured zero-loss intensity in the transmitted electron beam, for the same energy band and objective aperture, and $\sigma_x(\Delta, \beta)$ is the inelastic scattering cross section for the core level excitation for the energy band $\Delta$ and collection semi-angle $\beta$ (5). Usually the number of atoms of element $x$ is much smaller than the total number of atoms in the volume of the specimen, so there is a strong background intensity underlying a relatively weak signal. Therefore, it becomes essential to subtract the background very carefully from the post-edge energy-selected image.

Several methods have been employed to take into account the background intensity (6, 7). The simplest approach is ratio mapping, for which post-edge $I_{post}$ is divided by the pre-edge intensity $I_{pre}$. If the shape of the background remains constant across the entire image (which is rarely the case), the ratio value $k_x$ gives a simple representation of the distribution of element $x$:

$$k_x = \frac{I_{post}}{I_{pre}}. \quad (2)$$

However, it is useful to obtain a modified ratio, by subtracting a constant $k_0$, which corresponds to the measured ratio value in a region of the specimen where the element $x$ is known to be absent.

$$k'_x = \frac{I_{post}}{I_{pre}} - k_0. \quad (3)$$

Multiplying this equation by $I_{pre}$ gives the so-called two-window intensity $S_x$, which is an accurate measure of the core-edge signal, provided that the shape of the background remains constant over the field of interest.

$$S_x = I_{post} - k_0 I_{pre}. \quad (4)$$

When the specimen thickness is nonuniform, which often occurs, the shape of the spectral background varies significantly over the field of interest. Thus, the above mentioned method of calculating the core-edge signal becomes invalid. Previously, this change of spectral shape has been taken into account by collecting additional pre-edge images at different energy losses. For example, in the three-window method, two images are collected below the core-edge (8). By fitting the pre-edge background intensity as an inverse power function, its shape is allowed to vary across the image. However, for beam-sensitive biological specimens the electron dose is restricted and thus the three-window method drastically increases the statistical noise in the extrapolated background, which results in a substantial reduction in analytical sensitivity. For this reason, it is preferable to retain a two-window approach by modifying it to take into account subtle changes in the background shape as a function of the sample thickness $t$. In this case, it is convenient to measure the specimen thickness in units of inelastic mean free path $\lambda$, which can be readily determined from the fraction of transmitted electrons in the zero-loss peak:

$$\frac{t}{\lambda} = \ln\left(\frac{I_{\text{unflt}}}{I_0}\right). \tag{5}$$

Here, $I_{\text{unflt}}$ is the total unfiltered image intensity. The inelastic mean free path for plastic embedding material is typically 150–200 nm for a microscope accelerating voltage of 120 kV and 300–400 nm for a microscope operating at 300 kV.

To incorporate the thickness variation into the two-window method, we compute the spectral intensity as a function of $t/\lambda$ using logarithmic Fourier deconvolution method (7). It is then possible to calculate the ratio $k_0$ as a function of $t/\lambda$, resulting in the $k$-curve or a fitted function, which provides a set of parameters that can be used to adjust the $k$-value at each imaged pixel. It is then straightforward to compute a thickness-corrected or modified two-window elemental map and write the following expression for the number of atoms in a pixel of dimension $d$.

$$n_x = \left[\frac{I_{\text{post}}(\Delta,\beta) - k_0\left(\frac{t}{\lambda}\right)I_{\text{pre}}(\Delta,\beta)}{I_0(\Delta,\beta)\cdot\sigma_x(\Delta,\beta)}\right]d^2 \tag{6}$$

This ability to correct for thickness effects in elemental maps by using a $t/\lambda$ dependent $k$-ratio enables two-window elemental mapping to be performed in specimens that are as thick as one inelastic mean free path, without artifacts due to plural inelastic scattering.

Here we will mainly discuss the application of the modified two-window method on mapping phosphorus (P) in sectioned cells in 2D as well as in 3D. Since DNA contains one phosphate group for every base in the structure, EFTEM imaging of phosphorus provides an intrinsic molecular label for nucleic acid, and can be combined with nitrogen imaging, which provides a means of distinguishing between DNA and protein complexes (see Fig. 2). It is not feasible to map the distribution of phosphorus in frozen

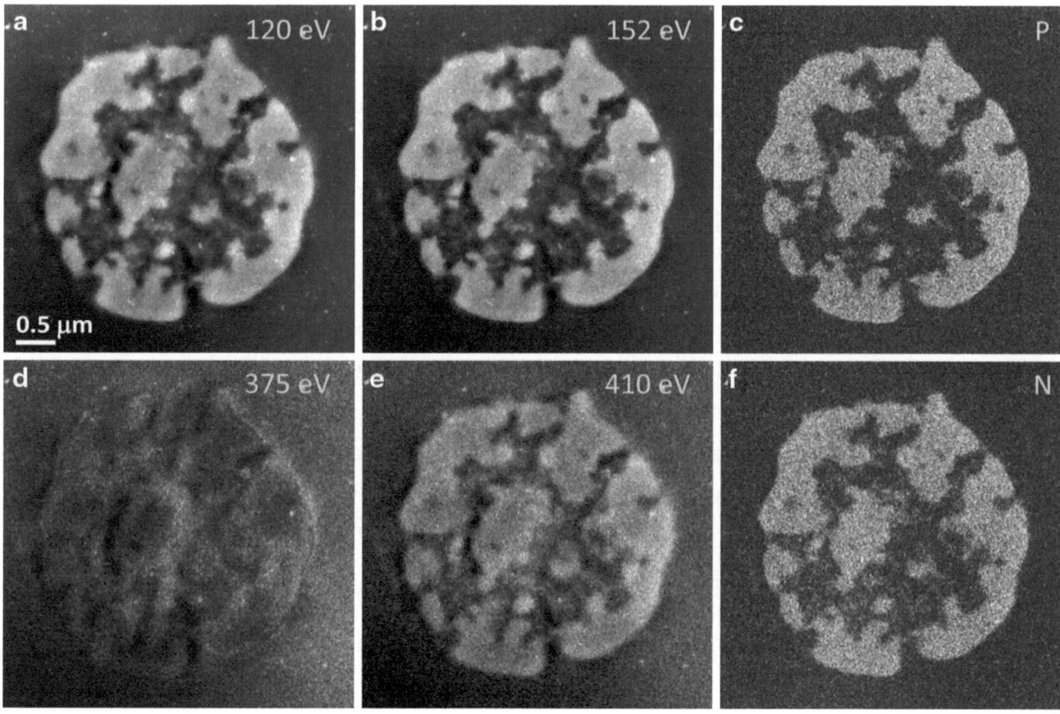


Fig. 2. EFTEM elemental mapping of a thymocyte nucleus. (**a**) and (**b**) show the P $L_{2,3}$ pre-edge image at 120 eV and post-edge image at 152 eV, which are used to calculate the 2D phosphorus map shown in (**c**). (**d**) and (**e**) are the N pre-edge image at 375 eV and N post-edge image at 410 eV, which are used to calculate the 2D nitrogen map shown in (**f**). Major components of the thymocyte nucleus are densely packed DNA containing phosphorus and nitrogen, and protein containing nitrogen but negligible phosphorus. Nitrogen concentrations attributed to protein structures are visible in regions that contain low levels of chromatin regions that appear dark in the phosphorus map (15).

hydrated cells, except at low spatial resolution, because of their high sensitivity to damage. EFTEM can be used to image phosphorus in cells or tissues that have been prepared by high pressure freezing and freeze substitution or conventional chemical fixation, embedding and sectioning. In such preparations, it is important to avoid heavy metal fixatives and stains that could interfere with the phosphorus edge in the energy loss spectrum. Until recently EFTEM phosphorus mapping of cells in unstained plastic sections has involved the collection of two-dimensional maps. By combining EFTEM with tomography, it is now possible to obtain three-dimensional phosphorus distributions of cellular components (9).

## 2. Materials

For performing EFTEM imaging and for processing the acquired data, the following equipment is recommended, with the manufacturer listed. These are suggestions only, since different manufacturers are available for each item:

### 2.1. General Supplies

1. Acetone.
2. Glutaraldehyde.
3. Epon-Aradite (Ted Pella): EMbed-812 and Araldite-502.
4. 10 nm colloidal gold (SPI Inc.).
5. 400 Mesh bare Copper Finder grids (Electron Microscopy Sciences, USA).

### 2.2. Instrumentation

1. Glow discharger.
2. High Pressure Freezer (HPM10, Bal-Tec, Switzerland).
3. Freeze-substitution system (EM-AFS, Leica Microsystems, Germany).
4. Ultramicrotome (EM UC6, Leica Bannockburn, IL).
5. Carbon evaporator (Auto 306, BOC Edwards, UK).
6. Non-tilt and tomography holder (Fischione Instruments, USA).
7. 100–300 kV transmission electron microscope (Tecnai TF30, FEI company, The Netherlands).
8. Tridiem post-column imaging filter (Gatan, USA).
9. 2k×2k Ultrascan cooled CCD detector for direct digital recording (Gatan, USA).

Items 2 and 3 are required only if preparing samples by high-pressure freezing (Subheading 3.1.1).

### 2.3. Software

1. Digital Micrograph (DM) tomography software for automatic data acquisition (Gatan, USA).
2. Filter Control software (Gatan, USA).
3. IMOD software for alignment and reconstruction of tomographic tilt series (http://bio3d.colorado.edu/imod/ University of Colorado at Boulder, USA) (10).
4. Image processing and 3D visualization software: Amira™ (http://www.amiravis.com).

## 3. Methods

Here we describe the steps taken for specimen preparation, general microscope setup, data acquisition and processing obtained using several imaging modes that generate quantitative elemental maps.

### 3.1. Specimen Preparation

To obtain resin-embedded, unstained sectioned cells (see Note 1) for elemental mapping in EFTEM imaging mode, the following is a brief description of the specimen preparation:

*3.1.1. High Pressure Freezing*

1. Specimens are transferred to sample carriers, for example *Drosophila larvae/Caenorhabditis elegans* in 15% sucrose.
2. The samples are then frozen in a high pressure freezer and freeze-substituted in acetone containing 0.2% glutaraldehyde at –90°C for 3 days and then slowly warmed (5°C/h) to 20°C.
3. After rinsing several times in acetone, the samples are infiltrated with Epon-Aradite and polymerized at 60°C for 2 days (see Note 2).
4. The sections are cut to a thickness of 90–150 nm (see Note 3), and picked up on 400 mesh bare copper grid.
5. To ensure the stability of the sections under the electron beam, 5–10 nm carbon films are evaporated directly onto the copper grids containing the specimens using a carbon rod mounted in the evaporator.
6. The grids are glow-discharged in air for 30 s and 10 nm colloidal gold is deposited on top and bottom of the grids (see Note 4). After 5 min, the samples are thoroughly washed with de-ionized water to remove excess reagents.

*3.1.2. Chemical Fixing*

1. Samples are post-fixed with 2.0% glutaraldehyde in PBS. For example wild-type mouse thymocytes are post-fixed for 5 min.
2. After extensive washing in PBS, the specimens are dehydrated in ethanol solutions of 30, 50, 70, 90% for 10 min and 100% for 45 min, before being infiltrated with Epon-Araldite: 30% Epon-Araldite and ethanol for 2 h, 50% Epon-Araldite and ethanol for 4 h, 75% Epon-Araldite for 12 h and 100% for 1 day.
3. The embedded blocks are then polymerized at 60°C for 2 days.
4. The sections are cut to a thickness of 90–150 nm, and picked up on 400 mesh bare copper finder grids.
5. The rest of the steps are the same as mentioned above, for the high-pressure frozen preparation.

### 3.2. General Microscope Setup in EFTEM Imaging Mode

Low noise image acquisition is an important step to ensure generation of nanoscale quantitative elemental maps, whether it is in two or three dimensions. Listed below are the steps needed for data acquisition in the EFTEM imaging mode of the electron microscope:

*3.2.1. 2D EFTEM Imaging*

1. Load the alignment file in the microscope software as well as in Filter Control, which has previously been aligned in the EFTEM mode.
2. Align the largest C2 aperture and an objective aperture of 40-μm diameter.

3. In low magnification mode, find the region on the TEM grid that has no film or specimen on it in order to prepare a gain reference for the CCD camera; the dark current should also be calculated and properly subtracted (see Note 5).
4. Make sure the imaging filter is tuned for several magnifications to provide an isochromaticity of less than 1 eV across the full image field (see Note 6).
5. Align the zero-loss peak relative to the center of the slit. This must be repeated several times during the data acquisition, since the zero loss peak can drift away from the center. Misalignment is demonstrated by shadows in the image. Slit alignment should be close for different magnifications, if step 4 is completed.
6. In the same low magnification, mark the locations of the cellular material of interest.
7. Adjust magnification for data acquisition: 20,000–40,000 are magnifications that we typically use in our experiments.
8. Focus the image at 100 eV energy loss setting.
9. Adjust the intensity of the beam by controlling the spot size (usually smallest) to obtain the maximum brightness.
10. When acquiring filtered (zero-loss) or unfiltered images use car not to saturate the detector. Thus, different binning of the images or acquisition time/frame summation might be used.

*3.2.2. 3D EFTEM Imaging*

Before acquiring tomographic data here are additional items to check for:

1. Gatan EFTEM tomography software in DM (11) offers complete tomographic calibrations including beam tilts, beam and image shifts that need to be completed for each magnification. Once the calibrations are saved in the memory of DM (message will display that the calibrations succeeded) they no longer need to be performed, unless alignments of the microscope are changed. The EFTEM tomography software also offers a set up for controlling the spectrum energy offset, the width of the energy-selecting slit, the tilt of the specimen goniometer, the excitation of the image deflection coils and objective lens, similar to a conventional electron tomography setup. A high-speed shutter allows one to acquire successively as many images as desired at each tilt angle without saturating the CCD detector while maintaining the same illumination of the specimen. All these parameters can be entered into the software and programmed before the data acquisition.
2. Correlation filters that are used for $x$–$y$ and focus tracking need to be checked. Cross correlation is performed on zero-loss

images, enabling correction for specimen drift, as well as correction for focus change based on image shift, induced by tilting the beam.

3. Check the tilt range, making sure other objects on the grid or the holder itself do not obstruct the specimen region of interest. Typically we restrict the tilt range to ±69°, with a tilt increment of 3°, which results in 47 images for a tilt series
4. Check for time/frame acquisition for different images. It is often necessary to bin the images to obtain better. However the acquisition time must be significantly shorter for zero-loss or unfiltered images, otherwise the detector will be saturated
5. Check for the beam tilt that is used to focus and track the images. If the beam tilt is too large, the edge of the objective aperture will be exposed. In that case, adjust the parameters in the software to decrease the beam tilt.

### 3.3. Data Collection

The microscope, as well as, the imaging filter need to be aligned/tuned in order to obtain elemental information from the specimen. It is important to consider the choice of imaging modality, which usually depends on the required information and the type of specimen to be analyzed. In terms of the electron sensitivity and the time to acquire the data, some imaging modes are more practical than others, as described below.

#### 3.3.1. EFTEM 2D

For EFTEM imaging we use a modified two-window approach (see Note 7), for which energy windows are defined by the width of the energy-selecting slit. For each element of interest, one window is chosen below the edge energy with intensity $I_{pre}$ and one above the edge energy with intensity $I_{post}$. Here we focus on imaging phosphorus using the P $L_{2,3}$ edge at around 150 eV (Fig. 2a–c). Follow the instructions from Subheading 3.2.1, for initial setup.

1. Acquire images with intensity $I_{pre}$ at 120 eV and $I_{post}$ at 152 eV (slit width of 20 eV) and bin by 4 to result in 512 × 512 pixel sized images. Depending on the electron dose, acquisition time varies from 16 to 40 s, which can be achieved by summing multiple frames if necessary.
2. Check to see if the images are correlated and make corrections for drift, which can be done in Fourier space. If the section is still shrinking, this data acquisition needs to be repeated (see Note 8).
3. Acquire two images needed for thickness correction, i.e., $I_0$ (zero-loss) and $I_{unflt}$ (unfiltered). These two images will show a uniform intensity, with very few structural details. Do not change the illumination conditions; the electron dose must be the same for all acquired images. The best way to avoid saturatation of the detector is to use different binning parameters (keep the image size 2k × 2k) and reduce the acquisition

time to 0.3–0.5 s, depending on the initial electron dose. If the images don't have enough electrons, increase the acquisition time and adjust binning.

4. The four resulting images are sufficient to calculate a 2D P map.

#### 3.3.2. EFTEM EELS

In this imaging modality one can obtain a complete electron energy loss spectrum averaged over an illuminated region of the specimen, ranging from a fraction of a micron to tens of microns, depending on the electron dose sensitivity of the specimen and the resolution necessary to observe the desired elemental signal (see Fig. 1). Some additional parameters need to be considered when doing this experiment:

Filter Control is used to adjust the spectral parameters for data acquisition. The spectrometer must be properly calibrated to obtain the correct voltage offsets (i.e., drift tube, spectrum/energy offset). The basic procedures are as follows:

1. Align the spectrometer for a particular dispersion or eV/channel (see Note 9) by following the procedures in Filter Control Software.
2. Adjust the focus values for the different lens settings to obtain a zero-loss peak with a narrow width.
3. When the zero-loss peak is on the detector array, collect a low-loss spectrum to calculate the relative thickness of the specimen. Be careful not to saturate the detector; use different illumination conditions from the ones needed to image the core edge signal. If the dark current is not properly calculated, you will see spurious intensities after the plasmon region of the spectrum. In that case re-acquire the dark current reference spectrum.
4. It is good practice to have a reference point in each acquired spectrum (known location of the edge may serve this purpose), so that the calibration of the spectrum can be performed post-acquisiton, and independently of the zero-loss peak position and accuracy of the dispersion.
5. All the other acquisition parameters: time, averaging, dark current and gain corrections can be controlled directly by the Digital Micrograph software (see Note 10).

#### 3.3.3. EFTEM Spectrum imaging

The idea is to combine EFTEM and EELS in one imaging mode and to reproduce the 3D spectral data cube, in which a series of energy-filtered images are taken over a certain energy range with a given slit width. The images exhibit only slight differences in contrast, but a complete set carries detailed spectral information that can be extracted from a region of interest. Figure 3 shows and example, in which a 15-channel spectrum (with 10 eV slit) is obtained from the data cube in order to map a region of the

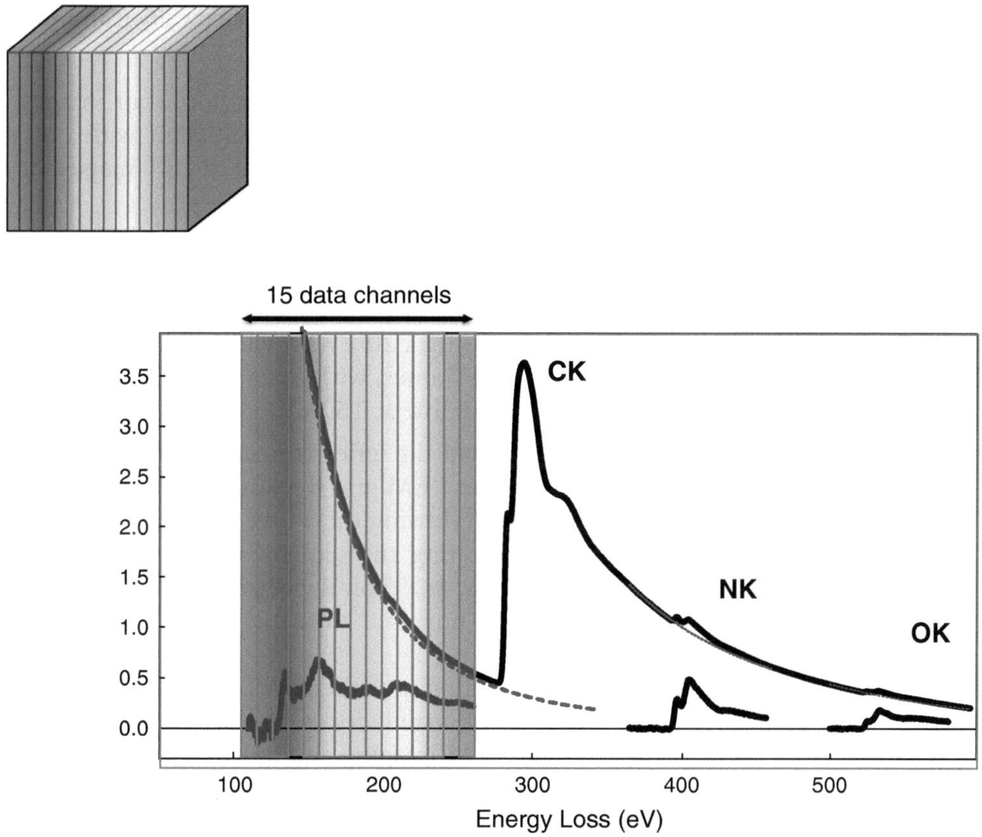


Fig. 3. An illustration of EFTEM spectrum imaging. Each of the 15 images in the cube corresponds to an integrated channel of 10 eV width in the electron energy loss spectrum in the vicinity of a P edge. In this way one obtain complete spectral information from each pixel of the 2D elemental map.

spectrum containing a P edge. In addition to the experimental setup mentioned above for EFTEM (Subheading 3.3.1) and EELS (Subheading 3.3.2), one must consider additional issues:

1. Beam sensitive specimens will be either deformed or develop contamination. Therefore, it is a good idea to check the specimen before data acquisition and, if needed, cool it.
2. As the multichannel images are acquired, it is important to have them aligned for future processing.
3. Once the images are acquired, the spectral information can be extracted and used for quantitative analysis (see Note 11). In principle, this mode of imaging is the most accurate, since one can obtain elemental information even though overlapping edges are present in the spectra. However, the length of acquisition time and the limited area of the specimen are consideration that should be taken into account.

#### 3.3.4. EFTEM 3D

To obtain 3D elemental maps, tomographic data acquisition is necessary. The tomography setup is described in Subheading 3.2.2.

Here we focus on mapping phosphorus, specifically using the P $L_{2,3}$ edge at around 150 eV.

1. Set up the tomography software to acquire four images, with intensities $I_0$, $I_{unflt}$, $I_{pre}$ and $I_{post}$ at each tilt angle with a tilt range of ±69° and tilt increment of 3° (a total of 188 images for mapping P).
2. Acquire the zero-loss image series using a 20-eV slit width, binned 4×, for 0.1 s; unfiltered image series with the same binning and acquisition time; energy-selected series in the vicinity of the P $L_{2,3}$ edge using a 20-eV slit width, binned 4× for 4–8 s, summing over four frames, with a pre-edge window at 120 eV and a post-edge window at 152 eV.
3. When an orthogonal tilt series is needed for dual-axis reconstruction, rotate the TEM grid in plane by 90° and acquire the same 188 images, following the above procedures. A simplified flow diagram indicating the sequence of the acquisition process is shown in Fig. 4a. Typical electron dose on the specimen is about $5 \times 10^7$ $e^-/nm^2$ (see Note 12).

### 3.4. Image Processing and Analysis

Depending on the imaging modality, the acquired data are handled differently. Here we will mainly focus on the data analysis obtained when using the EFTEM imaging mode to map phosphorus. For processing the EELS or EFTEM Spectrum Imaging data, one can refer to ref. 8.

#### 3.4.1. EFTEM 2D

2D elemental maps can be calculated by using a modified two-window method. The goal of the analysis is to produce quantitative elemental maps, in which each pixel corresponds to a number of atoms:

1. Align the acquired four images, with intensities $I_0$, $I_{unflt}$, $I_{pre}$ and $I_{post}$ to correct for specimen drift, even if the shift is a fraction of a pixel. We use a DM script for this purpose.
2. If the $I_0$ and $I_{unflt}$ are not the same size as the other two images, bin them (see Note 13) to the same size as the core images.
3. Since plural scattering can contribute to over/under estimation of the elemental signal, we correct for these effects using a modified *k* ratio to subtract the background, using Eq. 6.
4. Using a DM script (or other software) and Eq. 3, calculate the P map (see Note 14).

#### 3.4.2. EFTEM 3D

Follow the above-mentioned steps in Subheading 3.4.1 to calculate the P maps for each projection in the tilt series from the four acquired series ($I_0$, $I_{unflt}$, $I_{pre}$ and $I_{post}$,). One can achieve this by running the same script (steps 3 and 4) for each acquired image in the tilt series. Here the plural scattering correction is extremely important, since by the time the sample is tilted to 60°, its thickness doubles.

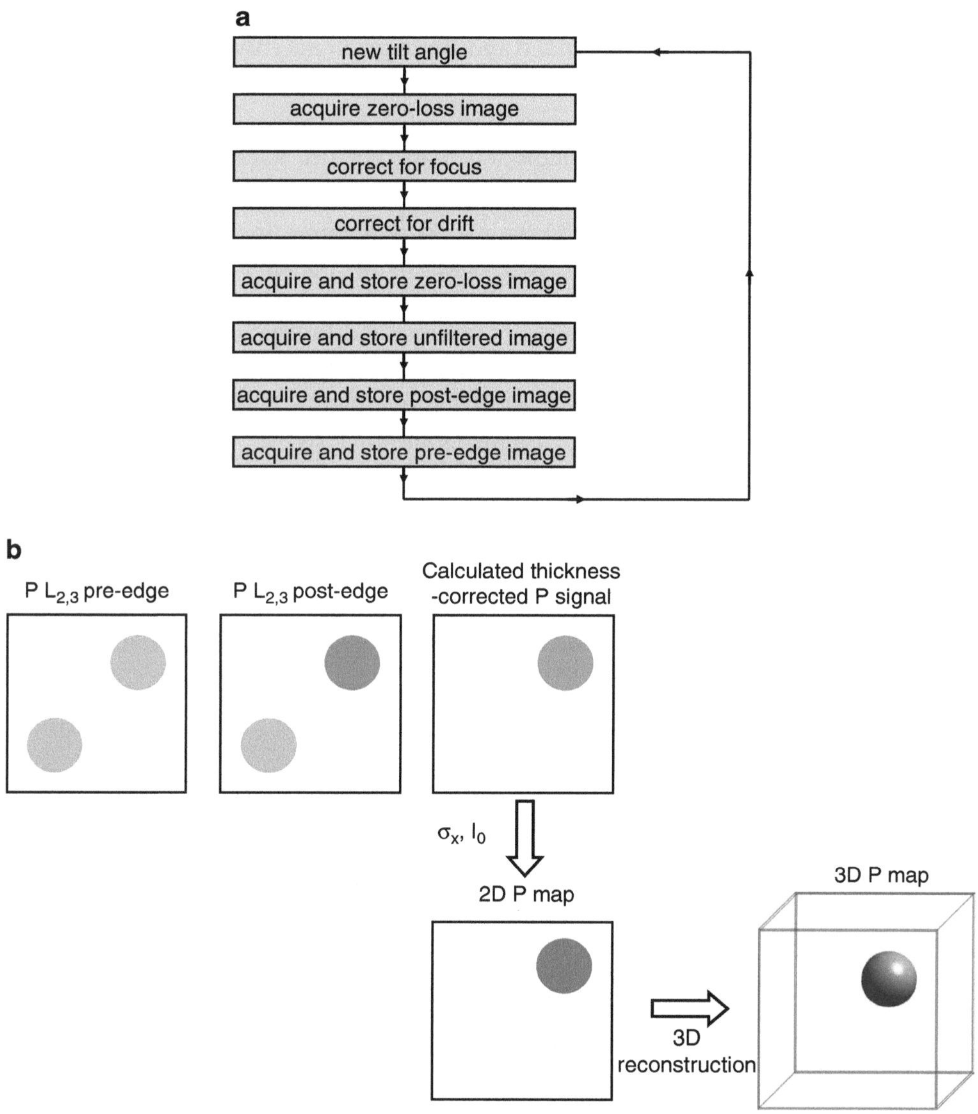


Fig. 4. Steps necessary for obtaining 3D P elemental map. (**a**) Experimental procedure for obtaining tomographic data. (**b**) Steps needed to calculate 3D P map (adapted from ref. 9).

Then follow the procedures below to reconstruct the images into a 3D map. A schematic diagram of this summary is shown in Fig. 4b.

1. Align the $I_0$ or $I_{unflt}$ tilt series using IMOD software (10). The gold fiducials should provide sufficient contrast to perform this task
2. Reconstruct either of the two series to find out the thickness of the tomogram. This information is included later when using simultaneous iterative reconstruction technique (SIRT).

3. Apply the same alignments on the calculated P series. This should be trivial and the alignments should work very well, since all four images were acquired sequentially at each tilt angle.
4. Use an intensity-preserving reconstruction technique to generate a 3D elemental map. We use additive SIRT (12) for this purpose (see Note 15), allowing for negative intensity to exist.
5. Once the 3D data set is generated, in units of atoms/voxel or atoms/nm$^3$, it is ready to be visualized by surface or volume rendering, which helps to distinguish and identify the overlapping structures of interest. Additional image filtering might be required, depending on the noisiness of the data. For visualization we use commercially available software Amira.
6. To perform quantitative analysis on the 3D P distribution, e.g., by estimating the chromatin density distribution (see Note 16), we use DM and Matlab. One way of depicting these quantitative results is by plotting histograms, averaging over a defined area of the specimen. It is helpful to average over a large number of pixels to reduce the effects of different noise sources, due either to data collection or data handling.

Figure 5 shows an example of a 3D mapping of phosphorus in a sectioned *C. elegans* cell. In Fig. 5a, 2D phosphorus distribution is

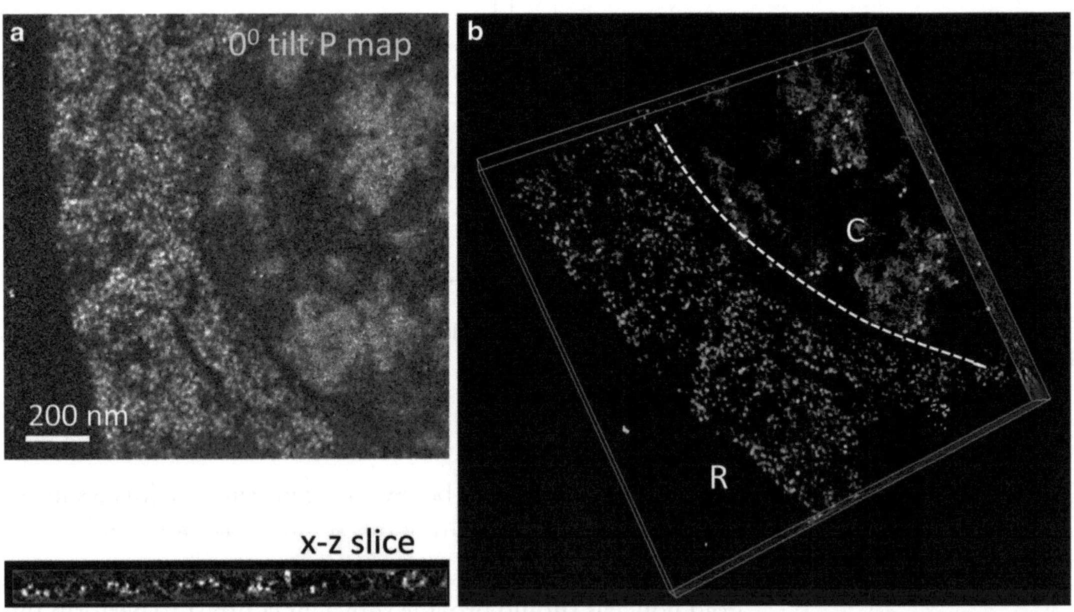


Fig. 5. An example of a 3D mapping of phosphorus in a sectioned *C. elegans* cell. (**a**) 2D phosphorus distribution at 0° tilt. (**b**) Volume rendered tomogram obtained using the SIRT algorithm, in which the value in each voxel is in units of P atoms/nm$^3$. Here two regions of high phosphorus content: cytoplasm containing ribosomes (*R*) (each with about 7,000 phosphorus atoms in its RNA) and nucleus containing areas of densely packed chromatin (*C*). An *x–z* slice extracted from the tomogram shows the roundness and clarity of ribosomes (adapted from ref. 9).

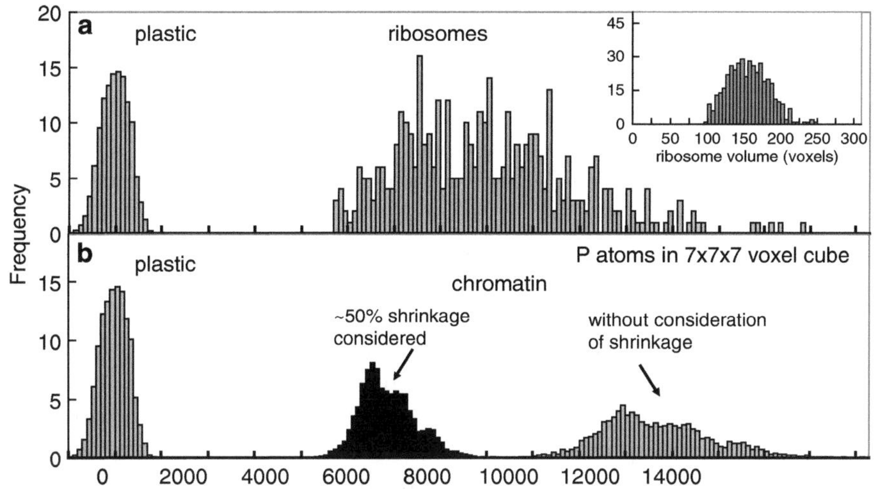


Fig. 6. An example of quantitative analysis of phosphorus content in a specimen of *Drosophila* embryo. Analysis was performed on ribosomes, regions of plastic embedding material, and regions of chromatin. (**a**) In ribosomes, each occupying a volume of 2,990 ± 550 nm$^3$ (152 ± 28 voxels) there were 7,900 ± 1,900 P atoms. In the plastic region, the phosphorous content found in a 7 × 7 × 7 voxel cube, which is similar to a volume that a ribosome would occupy if it were placed in a cube, was −100 ± 300 P atoms (±st. dev). (**b**) The distribution of phosphorus in a region of chromatin contained in a volume of 7 × 7 × 7 voxels, with and without inclusion of the shrinkage factor of the section under electron irradiation. Taking into account the volume shrinkage factor of 0.43, there were 5,800 ± 600 P atoms in a volume containing 7 × 7 × 7 voxels of chromatin, which corresponds to 0.5–0.9 phosphorus atoms per nm$^3$ (adapted from ref. 9).

shown at 0° tilt. It is important to note that if no signal is observable at a single tilt, it will be difficult to map the signal in 3D, although it is known that there will be an improvement in signal-to-noise ratio due to the principle of dose fractionation. Figure 5b shows a volume rendered tomogram obtained using the SIRT algorithm; here the value in each voxel is in units of P atoms/nm$^3$. We depicted here regions of high phosphorus concentration: cytoplasm containing ribosomes (R) (each with about 7,000 phosphorus atoms in its RNA) and nucleus containing areas of densely packed chromatin (C). An *x–z* slice extracted from the tomogram shows the roundness and clarity of ribosomes. From these data, we can also see relatively clear spaces, which could correspond to a sparser density of DNA in regions of euchromatin or to the location of large protein complexes that do not contain high concentrations of phosphorus.

Figure 6 shows an example of quantitative analysis of phosphorus content in a specimen of *Drosophila* embryo. Analysis was performed on ribosomes, regions of plastic embedding material, and regions of chromatin. It was found that ribosomes, each occupying a volume of 2,990 ± 550 nm$^3$ (152 ± 28 voxels) contained 7,900 ± 1,900 P atoms, in agreement with the known phosphorus content and in rough agreement with the ribosome volume of ~4,200 $nm^3$. Although the measured volume of the ribosome is smaller than its known size presumably due to radiation induced shrinkage, the ribosome does maintain its "spherical" shape in the SIRT reconstruction, as demonstrated in Fig. 5. In the plastic region, the phosphorus content found in a $7 \times 7 \times 7$ voxel cube, which is similar to a volume that a ribosome would occupy if it were placed in a cube, was $-100 \pm 300$ P atoms (±st. dev). Figure 6b shows the distribution of phosphorus in a region of chromatin contained in a volume of $7 \times 7 \times 7$ voxels, with and without inclusion of the shrinkage factor of the section under electron irradiation. Taking into account the volume shrinkage factor of 0.43, we obtain $5{,}800 \pm 600$ P atoms in a volume containing $7 \times 7 \times 7$ voxels of chromatin. This corresponds to 0.5–0.9 phosphorus atoms per $nm^3$, which is consistent with a model of 30 nm chromatin fibers that are tightly packed (13, 14). For a more detailed quantitative analysis, refer to ref. 9.

## 4. Notes

1. No heavy elements such as uranium or osmium should be introduced during the sample preparation, since spurious EELS/EFTEM signals can appear in the data.
2. The choice of embedding medium is important in EELS/EFTEM experiments, since the sectioned material needs to withstand high electron doses.
3. It is essential to limit the plural scattering effects, which can prevent accurate extraction of the elemental signal, even for thin specimens. In thicker specimens the elemental signal can be dominated by the spectral background from the embedding material and errors can be greater.
4. Gold particles are used as fiducials for tracking and aligning images if the experiments involve tomographic data acquisition and reconstruction.
5. The proper gain/dark current references are very important for both EFTEM and EELS measurements. If they are not done correctly (after this calibration there should be a uniform electron count distribution across the entire image and spectrum, which is close to zero), there will be spurious intensities either in the images or spectra that may result in false interpretation of the data, i.e., finding an element where none is present.
6. Mis alignment of the imaging filter can cause inaccuracies in determining elemental distributions from EFTEM images.

For spectral acquisition this is also important since the misalignment will make the elemental edges less visible relative to the background.

7. This is referred to as it a modified two-window approach, since the corrections are made for thickness effects at each image pixel. Refer to Subheading 3.4.
8. It often happens that the section continues to shrink during acquisition, which makes calculations difficult and results in shadowing effects that are visible when the images are divided, e.g. in determining elemental ration images. In that case, reacquire the images and check for specimen drift and shrinkage.
9. It is important to avoid dispersion settings that are not focused on the slit (as indicated by Gatan Inc). Also it is necessary to know whether the dispersion is properly calibrated. The choice of the dispersion setting during data acquisition should be made based on the data that is required. A lower dispersion eV/channel corresponds to a smaller total energy loss range, with more detail around the edges of interest but possibly more noise.
10. It might be necessary to adjust several parameters to reduce spectral noise and to provide a smooth background below an elemental edge. Occasionally, reflection peaks or peaks due to improper dark reference and gain normalization are visible in the spectra. Be alert for such artifacts, which can be eliminated by changing the dispersion setting, re-adjusting the illumination, or by re-acquiring reference images, etc.
11. Quantitative elemental maps can be obtained using the same methods as for standard spectral analysis: fit the pre-edge background; subtract the extrapolated spectrum above the edge; and sum the intensity over an appropriate integration window.
12. Be careful not to exceed the maximum tolerable electron. If this occurs, etching of the specimen and thinning of the plastic will be observed, although the elemental signal can still be retained. In these cases, it is probably better to choose another region from which to acquire tomographic data.
13. Make sure that the binning of the zero-loss image corresponds to the binning done in the microscope, for acquiring the core-loss images.
14. The inelastic scattering cross section $\sigma_x$ is a parameter that needs to be calculated in order to obtain elemental maps. This can be done in DM (using the known beam voltage, collection and convergence angles and integration window) or obtained from previously published works.

15. A SIRT reconstruction algorithm is used in this case to preserve the intensity values. Since we are interested in elemental maps, in which each pixel intensity value denotes the number of atoms/pixel or/nm$^2$, it is important to preserve these intensity values in the reconstruction. We use a Matlab program to create separate files from the aligned P series, which are then used in SIRT 3D reconstruction, written and compiled in our laboratory. When SIRT is employed, it is a preferable to use parallel processing to decrease the computing time. Using the computer cluster that is available through the NIH central computing services, a typical time frame for the reconstruction of 47 projections, 512 × 512 pixels each, using 64 processors is ~3 min. If one desktop is used without the help of GPU, it can take 4–5 h.

16. When performing quantitative analysis of 3D P maps, it is helpful to include the shrinkage effects into the calculation. We have previously shown (15) that due to exposure of the plastic section to a high electron dose ($10^7$ e/nm$^2$) the section shrinks ~50% in the *z* direction (parallel to the electron beam direction) and ~10–15% laterally.

## References


1. Bazett-Jones DP, Hendzel MJ, Kruhlak MJ (1999) Stoichiometric analysis of protein- and nucleic acid-based structures in the cell nucleus. Micron 30(2):151–157
2. Egerton RF (1996) Electron energy loss spectroscopy, 2nd edn. Plenum, New York
3. Krivanek OL, Friedman SL, Gubbens AJ, Kraus B (1995) An imaging filter for biological applications. Ultramicroscopy 59(1–4): 267–282
4. Egerton RF (2003) New techniques in electron energy-loss spectroscopy and energy-filtered imaging. Micron 34(3–5):127–139
5. Egerton RF (1979) K-shell ionization cross-sectionsforuseinmicroanalysis. Ultramicroscopy 4(2):169–179
6. Haking A, Troester H, Richter K, Crucifix C, Spring H, Trendelenburg MF (1999) An approach to an objective background subtraction for elemental mapping with core-edges down to 50 eV: description, evaluation and application. Ultramicroscopy 80(3):163–182
7. Aronova MA, Kim YC, Zhang G, Leapman RD (2007) Quantification and thickness correction of EFTEM phosphorus maps. Ultramicroscopy 107(2–3):232–244
8. Leapman RD (1986) Scanning transmission electron microscope (STEM) elemental mapping by electron energy-loss spectroscopy. Ann N Y Acad Sci 483:326–338
9. Aronova MA, Kim YC, Harmon R, Sousa AA, Zhang G, Leapman RD (2007) Three-dimensional elemental mapping of phosphorus by quantitative electron spectroscopic tomography (QuEST). J Struct Biol 160(1):35–48
10. Kremer JR, Mastronarde DN, McIntosh JR (1996) Computer visualization of three-dimensional image data using IMOD. J Struct Biol 116(1):71–76
11. Harmon RT, Meyer S, Hunt JA (2005) Software for EFTEM tomography incorporated into digital micrograph/GMS. Gatan, Inc., Pleasanton
12. Gilbert P (1972) Iterative methods for 3-dimensional reconstruction of an object from projections. J Theor Biol 36(1):105–117
13. Bazett-Jones DP, Hendzel MJ (1999) Electron spectroscopic imaging of chromatin. Methods 17(2):188–200
14. Schalch T, Duda S, Sargent DF, Richmond TJ (2005) X-ray structure of a tetranucleosome and its implications for the chromatin fibre. Nature 436(7047):138–141
15. Aronova MA, Sousa AA, Zhang G, Leapman RD (2010) Limitations of beam damage in electron spectroscopic tomography of embedded cells. J Microsc 239(3):223–232

# Chapter 14

# Cellular Nanoimaging by Cryo Electron Tomography


## Roman I. Koning and Abraham J. Koster



## Abstract

Cryo electron tomography is a technique that allows visualization of biological specimens in three dimensions with nanometer resolution. For cryo immobilized life sciences samples it can reveal cellular morphology, the shape of membranous structures, and depict internal macromolecular arrangements and large proteins. Cryo electron tomography is a unique technique in structural biology research because it is the only tool that enables direct visualization of the cellular space at molecular resolution. Here we present the methods that we apply in our lab to perform cellular cryo electron tomography, which require expertise on cell biology for cell growth, physics for electron microscopy, and image processing for reconstruction and 3D visualization. We define the instrumentation, materials, and protocols for cryo electron tomography of whole cells, including cell growth, specimen vitrification, microscope alignments, data acquisition, tomographic image reconstruction, and 3D visualization techniques.



**Key words:** Cryo electron tomography, Cellular microscopy, Cryo specimen preparation, 3D reconstruction, Visualization


## 1. Introduction

During the last six decades, techniques to chemically fix, stain, and section cells and pieces of tissue have been extensively optimized for imaging by electron microscopy. These advances have brought a wealth of information on cell morphology and on the interaction between cells and their components. Since the late 1990s, 3D imaging based on electron tomography techniques has become more widespread. As a consequence, an overwhelming amount of 3D structural information has emerged to support the 2D data that has been provided by transmission electron microscopy. Also during this time, alternative specimen preparation techniques evolved that are based on physical cryo immobilization techniques and not on chemical fixation procedures. These cryo electron microscopy techniques have been especially important for structure


Alioscka A. Sousa and Michael J. Kruhlak (eds.), *Nanoimaging: Methods and Protocols*, Methods in Molecular Biology, vol. 950, DOI 10.1007/978-1-62703-137-0_14, © Springer Science+Business Media, LLC 2013

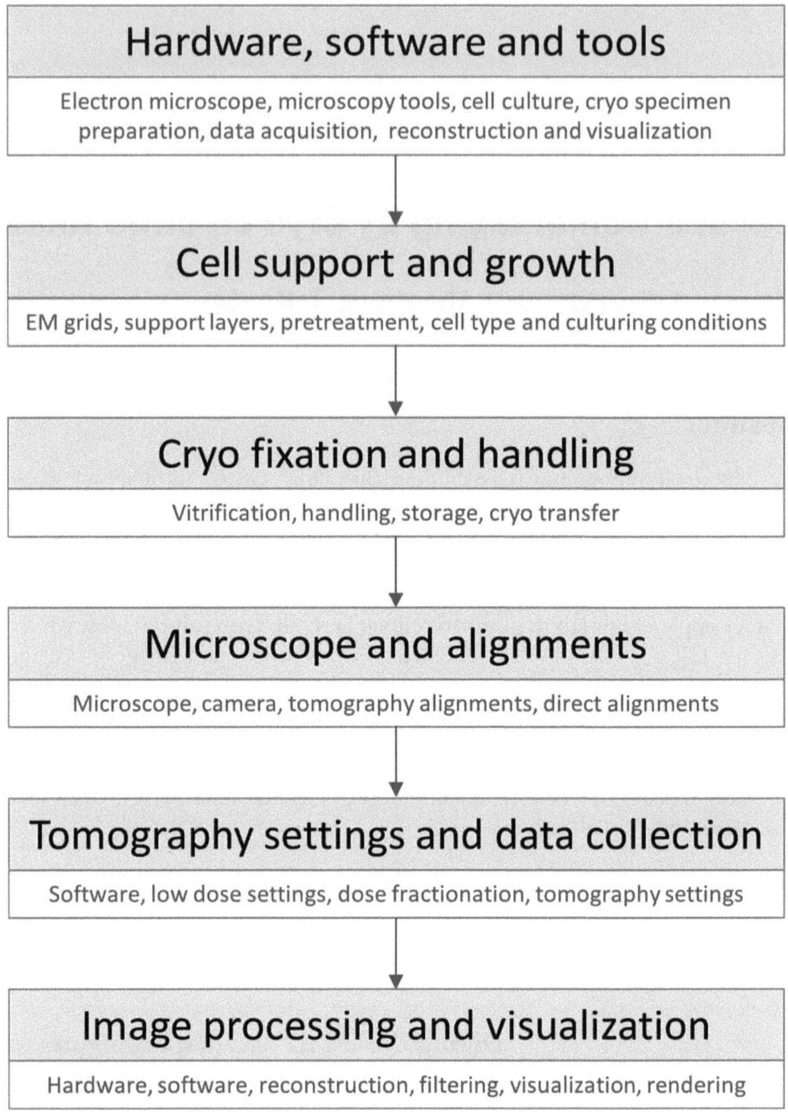


Fig. 1. Overview scheme of steps involved in cryo electron tomographic imaging of cells.

determination of protein complexes and viruses at (sub-) nanometer resolution (1, 2).

Electron tomography can be referred to as a 3D imaging technique that records a series of projection images of a structure from different projection angles and subsequently computes the reconstruction of that structure from these images. When electron tomography is applied to frozen-hydrated specimens, the technique is referred to as cryo electron tomography (or electron cryo tomography). Cryo electron tomography is an imaging technique especially suitable for structural biology applications as it provides a tool to investigate macromolecular assemblies such as microtubules, not only in vitro but also within their native environment in the cell (3, 4).

For reasons of size and uniqueness, structures that are investigated by cryo electron tomography are often not suitable for alternative structural biological techniques such as X-ray crystallography, NMR, or light microscopy. Its resolution range (2–5 nm) actually makes it a suitable technique to bridge the resolution gap between X-ray crystallography and light microscopy.

In the following sections we describe the workflow of cellular cryo electron tomography (a general outline is shown in Fig. 1), give an overview of the hardware, software, and other materials that one needs, describe the methods, and discuss some of the limitations of the technique.

## 2. Materials

Several pieces of equipment are required for performing cryo electron microscopy on cells. The tools can be divided into the following categories: general accessories for (cryo) electron microscopy and cell culture, cryo specimen preparation tools, a suitable electron microscope, and appropriate hardware and software for image processing. A concise overview of the electron microscopy hardware and tools is depicted in Fig. 2. All items are listed below (between brackets the specific type or brand that is used in our lab):

### *2.1. General Accessories*

1. Carbon coater, for extra carbon coating of grids or for the creation of thin carbon films to be put on grids (Emitech KX950, Emitech, The Netherlands).
2. Glow discharger, for making carbon layers hydrophilic (Emitech 350, Emitech, The Netherlands).
3. Tweezers (Ted Pella, USA): (a) Fine tip tweezers, for handling EM grids (Dumont no 5); (b) Bent curved tip or flat tip tweezers, for handling cryo storage boxes during transfer; (c) Long tweezers, for handling cryo storage boxes out of cryo storage tubes.
4. Screw driver, for opening and closing storage boxes.
5. Metal grid mesh, for placing grids during glow discharging.
6. Filter paper, for blotting (Whatman, Kent, UK and Schleicher and Schüll, The Netherlands).
7. 100–300 mesh gold grids with Formvar/carbon film.
8. Cell culture stove and hood.
9. Cell growth chemicals.
10. Upright brightfield light microscope.

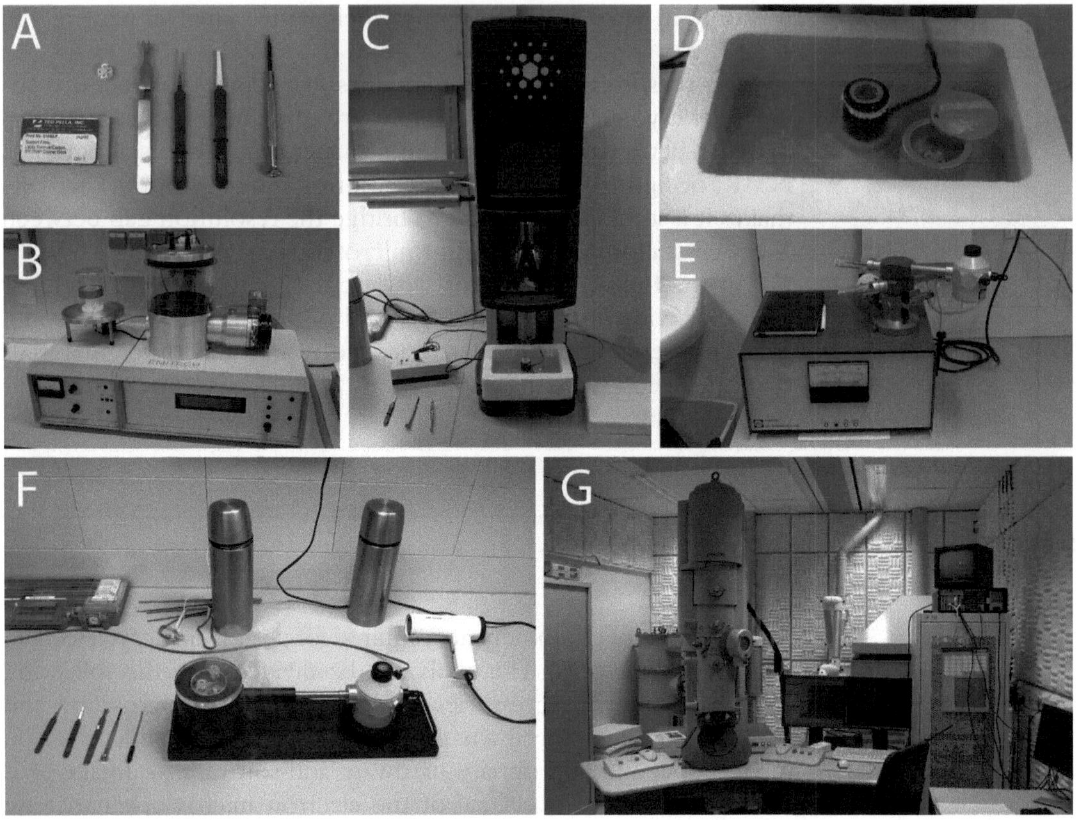


Fig. 2. Overview of electron microscopy hardware and tools used in cryo electron microscopy. (**a**) General tools for cryo electron microscopy: EM grids, storage box for cryo grids, tweezers for grabbing cryo storage box, two tweezers for grid handling, and screwdriver. (**b**) Carbon coater with glow discharging unit. (**c**) Vitrobot with custom-made cryogenic box with heater. (**d**) Cryogenic box for vitrobot. Black cylinder is the ethane container with a heating wire around it. On the right the closable container for cryo EM storage box is depicted. (**e**) Gatan dry pumping station with cryo holder. (**f**) Cryo holder in station during transfer of grids. Cryo-EM grids are transported in stainless steel flasks in which Greiner tubes, holding the cryo storage boxes, are suspended by a rope. (**g**) Tecnai F20 with Gatan GIF 2002.

### *2.2. Cryo Specimen Preparation Tools*

1. 5–15 nm colloidal gold, for fiducial markers in tomography (see Subheading 3.4).
2. Vitrification device with humidity and temperature control, for fast freezing of samples (Vitrobot Mark IV, FEI Company, The Netherlands).
3. Liquid nitrogen, for storing samples and cooling down ethane gas.
4. Ethane or ethane/propane gas, for vitrification.
5. Cryo grid storage boxes (Ted Pella, USA).
6. Cryo storage tubes: Greiner Cellstar 50 ml polypropylene sterile centrifuge tubes (Greiner bio-one, The Netherlands).
7. Stainless steel 1 L liquid flasks for transfer in liquid nitrogen.
8. Liquid nitrogen Dewars for storage.

### 2.3. Electron Microscopy

1. Medium- or high-voltage transmission electron microscope with field emission gun (FEG). We use a 200 kV FEI Tecnai F20 (FEI company, The Netherlands).
2. Specimen stage suitable for tilting a frozen hydrated specimen; e.g., a side entry cryo holder (Gatan 626 or 914 cryo holder, Gatan, Germany).
3. Cryo holder dry pumping station (Gatan 655, Gatan).
4. Energy filter with digital camera (GIF 2002 with 2 k×2 k MultiScan CCD camera, Gatan).
5. Image acquisition software. Various packages are available that can acquire tomographic tilt series: Explore3D (FEI Company); SerialEM (Boulder lab, University of Colorado, USA (5)); TOM toolbox (MPI, Martinsried, Germany (6, 7)); UCSFtomo (UCSF, San Francisco, USA (8, 9)); and others.

### 2.4. Image Processing Computer and Software

1. Reasonably powerful computer workstation with >8 Gb RAM and graphics card or small computer cluster (Hewlett-Packard XC cluster).
2. Tomographic reconstruction and visualization software (IMOD (5); Xplore3D, FEI Company; Amira, Visage Imaging, Germany).

## 3. Methods

In this section all steps are described to get from cultured mammalian cells to a 3D cryo electron tomography reconstruction. The specific choices for the used materials and methods are described in detail, together with their advantages and disadvantages. Here, the settings are described that we used for visualization of microtubules in mouse embryonic fibroblasts (3, 10). However, to make the description more generic, we summarize the rationale for choosing these settings as an aid to adapt them, depending on the lab-specific available instrumentation, biological subject, and final objective of the electron tomography experiment.

### 3.1. Cell Support

1. *Grid material.* Electron microscopy support grids that are used for cell culture are made of gold. Other metals such as copper and nickel are not suitable for cell cultures, as they are cytotoxic.
2. *Mesh size.* Grid mesh sizes between 100 and 300 Mesh are used. The Mesh number should not be too high (>300 Mesh) to keep sufficient transparent space between the grid bars when the specimen is tilted to high angles. On the other hand the metal grid should provide sufficiently sturdy support for the

overlaying film. In our hands, with Mesh numbers equal to and lower than 100 the support films tend to break more frequently upon handling and vitrification of the specimen. A 300 Mesh grid seems to be optimal. Alternatively, 300×75 Mesh grids having different sizes in different directions can be used in order to increase the overall transparency upon tilting. Take care that the latter grids are placed in the specimen holder such that the 75 Mesh spacing is perpendicular to the rotation axis of the holder.

3. *Support layers.* We preferentially use gold grids with a support layer of continuous Formvar, for optimal strength, on both sides coated with evaporated carbon. Formvar support layers are prepared by making a 1% solution in Chloroform or Dicholoromethane, dipping a microscopy glass slide into the solution and, after drying, floating the Formvar layer onto water. Gold grids are placed onto the Formvar layer and picked up again with a glass slide. The Formvar support layer is then carbon coated on both sides. The disadvantage of the use of continuous Formvar/carbon-coated films is that these layers add to the total thickness of the sample. The scattering interactions of the electrons with these layers deteriorate the signal-to-noise ratio of the images and, therefore, the attainable resolution of the 3D reconstruction. To reduce this disadvantage, commercially available holey carbon films, such as Quantifoil, can be used. This type of specimen supports has the advantage that background noise is reduced when these cells are imaged through the holes.
4. *Pretreatment.* Just prior to the use of Formvar/carbon-coated grids, glow discharging is applied in 200 mbar air for 1–2 min at 20–40 mA on a carbon coater with additional glow discharge unit. This is performed to increase hydrophilicity of the grid and improve surface contact with the medium during cell culture. Fiducial gold markers of 5, 10, or 15 nm are applied on the grid before cell culture, their size dependent on the desired magnification, to aid image alignment during the tomographic reconstruction. For this, grids are placed for 1 min on a suspension of 5, 10, or 15 nm protein-A gold, blotted with filter paper and dried in air. To improve biocompatibility and cell adhesion, grids are occasionally coated with Collagen by incubating the grid for several hours with a 1:3 dilution of Collagen in ethanol.

### 3.2. Cell Culture

1. *Cell type.* In previous work, we have used primary mouse embryonic fibroblast cells (MEFs) that were purified from mouse embryos. Briefly, the mouse body was cut into pieces and homogenized with collagenase and trypsin. The homogenate was cultured in a 1:1 mixture of Dulbecco's modified

Eagle's medium and Ham's F10 medium supplemented with 10% fetal calf serum and penicillin and streptomycin antibiotics in a humidified atmosphere with 5% $CO_2$ at 37°C. Other cell types have also proven to be suitable for cryo electron tomography of whole cells, although mainly mammalian cell types that by nature are thin or have the tendency to spread, like fibroblasts and neurons, can be used. The thickness limit that can be meaningfully imaged by transmission electron microscopy depends on the high voltage used. In most cases, ~400 nm is regarded as a maximum thickness to image a frozen hydrated specimen with a transmission electron microscope operating at 200 kV. Consequently, it is not possible to perform cryo electron microscopy of whole cells that have an intrinsically cubic or columnar shape, as well as cellular structures that are positioned near the cell nucleus.

2. *Cell growth*. EM grids are sterilized before cell growth by 15 min UV radiation. Standard cell culture protocols are used for growth on EM grids. MEF cells are cultured in tissue culture flasks and between 10,000 and 30,000 cells are plated into 5 cm diameter tissue culture dishes, in which EM grids are positioned on the bottom. Cell adhesion is allowed to occur overnight and grids are usable up to several days of culture before vitrification. The amount of cells that are plated should not be higher than 30–50% in order to avoid adjacent or overlapping cells. Before vitrification, the cultured cells should be checked by light microscopy for proper cell attachment to the grids, confluency, and spreading. When little to no cells are attached to the grid, one might want to check the toxicity of the grids (by growing cells in an EM grid free culture dish), increase the confluency or the time for cell culture. Keeping the cells in a 100% humid environment and at 37°C during specimen preparation is important to maintain biological and structural integrity. Therefore, after cell culture, which is performed in a specialized cell culture lab, the culture dishes with the EM grids are transferred to a cell culture stove that is located in the same room as the vitrification device. The EM grid with the cells is transferred onto hot plates that are kept at 37°C or, alternatively, transferred as fast as possible into the vitrification device.

### 3.3. Vitrification, Storage, and Transfer

1. *Vitrification*. For the vitrification process, we use the FEI vitrobot Mark IV. This vitrification device has a temperature and environmentally controlled chamber in which the cells are kept before and during blotting (for a detailed protocol see: (11)). Blotting is performed with filter paper (Whatman no. 4 or paper numbers 595 or 597 from Schleicher und Schüll), using the following parameters: temperature, 37°C; humidity, 100%;

blot time, 1–2 s; wait time, 0 s; drain time, 0 s; blot force, 0–25; and blot total, 1.

2. *Custom nitrogen box*. On the vitrobot, we use a custom-made "coolant container." We do not use the standard FEI liquid ethane container for two reasons. In our experience, it is difficult to keep the grids clean from ice contamination (see Note 1) and to keep the liquid ethane at the correct temperature for a prolonged time. Therefore, we use a container that allows heating of the ethane container. In addition, we add a lidded container for a cryo grid storage box in order to prevent contamination (Fig. 2c, d). Additionally, the system has more space for liquid nitrogen and handling. The blotted grid is plunged into liquid ethane (grade PR 3.5 (99.95% pure)) that is cooled by liquid nitrogen and kept in equilibrium with solid ethane (having a matte "milky" appearance). Alternatively a 2:1 mixture of ethane and propane (63.0 mol% propane and 36.98 mol% ethane, custom mixed) can be used. The advantage of the mixture over the pure liquid ethane is that this mixture does not solidify at liquid nitrogen temperatures (12) and no heating of our custom-made container is needed.
3. *Storage*. Vitrified grids are stored in cryo boxes (home-made or Ted Pella, Fig. 2a, top left). The small cryo boxes are closed with a cover that is fixed with a screw and stored in liquid nitrogen filled Greiner tubes that are very well resistant to cryo temperatures and can be used for years without noticeable deterioration. Coded strings are attached to these tubes and they are stored in a large liquid nitrogen Dewar for later use. Also Greiner tubes and grid storage boxes are tagged for later grid retrieval.
4. *Transfer*. Stored boxes with vitrified grids are transferred from the storage container into a Gatan cryo station, holding a Gatan 626 high tilt cryo holder or Gatan 914 cryo electron tomography holder. The cryo holders are baked out and pumped to low pressure using a dry pumping station to retain maximum insulation. Directly after mounting of a grid into the cryo holder, the grid is subsequently transferred from the cryo station into the electron microscope.

### 3.4. Microscope and Alignments

#### 3.4.1. The Electron Microscope

1. We use an FEI Tecnai F20 operated at 200 kV, with an FEG electron source and a 2 k×2 k CCD camera behind an energy filter. Additionally, the electron microscope is equipped with a goniometer that is suitable for tomography, allowing smooth rotation of the stage to angles up to 70°.
2. Data collection is best performed on a medium- or high-voltage electron microscope operating at an acceleration voltage of 200 or 300 kV. With these higher electron voltages thicker samples (up to ~400 nm) can be used. For whole cell imaging

with transmission the sample thickness is one of the most stringent limitations that require a sufficiently high acceleration voltage. As an electron source it is advantageous to use an FEG. An FEG has the advantages over other sources such as $LaB_6$ and Tungsten in that the electron beam is more coherent and more suitable for imaging thicker samples at a relatively large defocus, which is necessary for the generation of phase contrast.

3. Zero loss imaging of the electrons using an energy filter is very advantageous (see Note 2). By using an energy filter in the zero-loss imaging mode (e.g., with an energy slit of 20 eV) the inelastically scattered electrons do not contribute to the final image. Consequently, the signal-to-noise level of images acquired from relatively thick specimen is significantly better compared to unfiltered images. Zero-loss imaging becomes increasingly important at high specimen tilts, since the specimen becomes thicker to the electrons as they traverse it to form a projection image.
4. Imaging should be performed with a digital image detector, which is indispensible for automated image acquisition. In cryo electron tomography, the number of electrons is in general very low due to the necessary dose fractionation: the acceptable total electron dose has to be distributed over as many images as possible. Consequently, the detector should be highly sensitive. To image the specimen in a large field of view with reasonable resolution, the imaging detector size should be at least 2 k×2 k pixels in size but preferable larger (4 k×4 k or even 8 k×8 k). This allows for either a large field of view or binning of pixels to allow increased effective sensitivity. Most cameras that are currently available for electron microscopes are cooled slow-scan CCD cameras that record light that is generated by the interactions of the electrons with a scintillator positioned on top of the detector. Currently, alternative (direct) electron detectors with superior characteristics in terms of image detection quantum efficiency (DQE) have become commercially available.

*3.4.2. Microscope Alignments*

Especially for cryo electron tomography, a proper alignment of the electron microscope is vital to assure optimal imaging conditions and a robust data collection performance. Dedicated care has to be taken with some particular alignments that are important for proper tracking of feature movement and focus variation during the acquisition of the tilt series. Under practical circumstances, an optimal microscope alignment is not always achieved by a microscope supervisor. It is useful when the users have the expertise to tweak and optimize these alignments during the tomography session. Therefore, we think it is useful to describe and discuss some important alignments in this chapter. The necessary alignments are in most cases similar for a wide range of transmission electron microscopes and tomographic acquisition software. We will specifically describe the procedures during a tomography session for a Tecnai (F20) microscope.

1. *Eucentric height.* After insertion of the cryo holder into the microscope, it must be set to the eucentric height. This means that the height of the holder in the goniometer must match the rotation axis of the goniometer. Many microscope alignments are related to this height as it defines the objective lens focus plane. When the stage is not at eucentric height upon applying specimen tilt, the sample shifts orthogonally to the tilt axis. The eucentric height can be set manually or automatically. For the automatic version to function robustly, it is advisable to first roughly set the height manually. A rough manual alignment of the eucentric height can be performed by centering a recognizable feature at low magnification, tilting the stage and then recentering the feature using the *z*-height controls. On our Tecnai the automatic eucentric height function is available in the tomography software suite.
2. *Centering of apertures.* The condenser and objective apertures have to be aligned properly to the optical axis to prevent beam movements during the tomographic data acquisition. A quick way to check if a condenser aperture is aligned is to center a focused beam and observe whether the beam stays centered during spreading of the intensity. If the area does not stay centered, the condenser aperture is not well aligned. The condenser aperture can be centered by moving the aperture using the two knobs on the aperture mounting. Repeat adjustments until the illuminated area stays centered and only changes size. The objective aperture can be centered by switching the microscope to diffraction mode (where, at the appropriate camera length, the objective aperture is visible as a circular shape) and to move the aperture such that the central diffracted beam is centered within the circular shape.
3. *CCD calibration files.* Most of the calibrations and alignments in the tomography software make use of image shift measurements using cross-correlations. Therefore, the quality of the CCD images is an important factor. Unfortunately, in practice, CCD images can suffer from a number of image deteriorating artifacts that have to be corrected for. In most cases, two types of calibration files, often referred to as gain and dark reference images, are needed for the calibration of CCD detectors. Gain references are recorded to correct for the inhomogeneous sensitivity (i.e., gain) of the individual pixels of the detector arising from various sources, e.g., hardware variations of the CCD chip, variations of the scintillator, dust, scratches, and more (13). Dark references are recorded to correct for intensity variations

related to the readout and electronics of the data. Using the Gatan CCD detectors on our systems, the dark references are recorded automatically prior to the recording of the illuminated image. Therefore, only the gain references have to be recorded as a separate step. Gain references are recorded for an even (homogeneous) illumination of the CCD detector. One way to verify whether the gain reference files are fine is to record an image at even illumination (preferably at the same image intensity as is anticipated to record the images during the tilt series) and to compute an auto-cross-correlation of the image. The auto-cross-correlation image should show a sharp intensity peak in the center of the image, while the rest should be noisy without any visible pattern. If a pattern of peaks, or bright lines, appears a new gain reference image should be recorded.

4. *Beam tilt pivot points.* This alignment makes sure that, upon tilting the beam, the illuminated area on the specimen stays centered. When the pivot point of the beam tilt is not correct (i.e., does not correspond to the specimen plane), this can be observed by the movement of the illuminated area when the illumination is tilted. Beam tilting is used as a way to measure the defocus during autofocusing since the amount of shift is directly related to the amount of defocus. The beam tilt pivot points can be corrected in the direct alignment steps of the microscope. Note that for the alignment and focusing it is important that the sample is at eucentric height. When this is not the case, the illuminated area will shift upon beam tilt. In particular, this can induce problems when the illuminated area is relatively small and is only slightly smaller than the CCD.
5. *Rotation center.* This alignment ensures that the illumination traverses through the center of the objective lens. Rotation center alignment is important for minimizing aberrations and image movements during focusing. The rotation center can be set in the direct alignments.
6. *Low dose alignments.* Apart from low dose settings (see below), the low dose calibrations (blanker, image/tilt and peek calibrations) should be performed. Additionally, it is important to set the correct value for the tilt axis rotation in focus mode, since angle 0° is not necessarily set exactly along the tilt axis. Incorrect angle might result in focus deviations. There is an easy way to set the angle of the rotation axis. In low dose focus mode, set the focus angle and distance to 0 and find a feature to set in the center of the camera. Once centered, set the focus distance to the desired value and move the feature along the rotation axis (Y direction only! e.g., by using MF stage Y) and back to the center. When the actual rotation axis is not exactly at 0°, the

feature will not go back to the center. Move the feature as close to the center as possible, and then move it exactly to the center using the focus angle.

Before starting the data collection software, the overall alignments and calibrations can be verified performing the individual auto eucentric height and auto focus routines that are part of the tomography data collection.

Tables 1 and 2 show common microscope and tomography misalignments and possible solutions.

### 3.5. Tomography Settings and Data Collection

Cryo electron tomographic data collection is performed automatically by software that controls the microscope settings. Various packages are available that can acquire tomographic tilt series, and most have the option to record cryo electron tomography tilt series in low dose mode. Next, the parameters and settings specific for the Explore3D package, which we use during data collection, are discussed.

#### 3.5.1. Low Dose Settings

For the acquisition of cryo electron tomographic tilt series a low dose imaging scheme is used. This ensures that during tilt series

**Table 1**
**Typical alignments of a transmission electron microscope**

| Variable | Observed | Possible misalignment/error |
|---|---|---|
| Intensity | Beam/Image moves<br>Beam moves non-concentric | Gun tilt pivot point<br>Condenser aperture |
| Focus | Beam/Image moves | Rotation center |
| Magnification<br>Magnification: LM to Mi<br>Magnification: Mi to SA<br>Magnification: SA to Mh | Beam/Image moves<br>Image moves/Focus changes<br>Image moves/Focus changes<br><br>Image moves/Focus changes | Image shift calibration<br>Image shift LM to Mi (LM images)<br>Image shift Mi to SA (Mi objective lens preset)<br>Image shift SA to Mh (Mh objective lens preset) |
| Spot size | Beam/Image moves | (Spotsize dependent) gun shift |
| Beam shift | Beam moves in slightly wrong direction<br>Beam moves in orthogonal direction | Beam shift<br>EFTEM mode on while imaging is not on GIF |
| Image shift | Beam moves | Beam/Image shift calibration |
| | Blurred Image (FFT thon rings are oval)<br>Beam is oval | Objective astigmatism<br>Condenser astigmatism |

Regularly observed misalignments and errors observed with their possible causes and solutions

**Table 2**
**Tomography software alignments**

| Mode | Observed | Possible cause | Possible solution |
|---|---|---|---|
| Eucentric height | Fails because images do not match | Eucentric height too far-off | Lower magnification<br>Do rough manual alignment<br>Lower tilt angle |
| | Fails because images are noisy | Not enough intensity | Increase beam focus<br>Decrease spot size<br>Increase acquisition time |
| | Fails because filtered images are noisy | Filter settings are not suitable | Try alignment without using filter<br>Try other filter settings |
| | Fails because one image is black | Beam is obstructed by grid bar | Try other position on grid |
| Auto focus | Fails because part of image is black | Objective aperture is not centered<br>Beam is not centered | Center objective aperture<br>Calibrate beam tilt |
| | One of two images is black | Objective aperture is not centered | Center objective aperture |
| Acquisition (focus mode) | Beam is off center | Beam is not centered | Center beam with direct alignments |
| | Auto focus fails | At high tilt angle | Switch to manual focusing<br>Skip focusing |
| Acquisition (tracking mode) | No maximum in cross-correlation | Thick sample<br>Bad filter settings | Try other position<br>Try alignment without filter |
| | Maximum correlation in image center | Little contrast in image<br>Bad CCD reference files | Try other filter settings<br>Redo the CCD gain reference |
| | Wrong correlation peak is selected | Sharp edges in image | Use suppress center peak option<br>Try different position on grid |
| Acquisition (imaging mode) | Samples do not stay centered | Holder drifts at high angles<br>Holder calibrations are unsuitable<br>Large tilt increments | Switch off holder calibration correction<br>Record new holder calibration<br>Use small tilt increments |

Regularly observed misalignments and errors observed during cryo tomographic data collection with their possible causes and solutions acquisition the majority of the electron dose is used for recording the images that will be used for the tomographic reconstruction and a minimal amount of electron dose for other purposes, e.g., keeping the feature of interest within the field of view of the detector. For low dose acquisition, one can set three imaging states: search, focus, and exposure. Each state defines a different setting for the magnification, spot size (C1), intensity (C2), exposure time, and camera binning.

1. Typical settings for the search state during cryo electron microscopy of cells allow the visualization of cellular structures at very low magnifications (see Note 3): magnification of ~5,000×; defocus −100 μm; spot size between 4 and 6; exposure time ~1 s, camera binning 2 or 4 (resulting in a 1 k×1 k image).
2. For the focus state, the illumination is shifted such that the focusing area on the specimen is sufficiently far away from the imaging area. In this manner, the feature of interest is exposed only during the acquisition of the tilt series and not for keeping the specimen in focus. This required distance between the focusing and imaging area depends on the size of the imaged area in imaging mode and on the spreading of the beam in focus mode. Furthermore, the focus position should be located on the tilt axes to prevent a difference in focus between the areas used for focusing and imaging when the specimen is tilted. It is not necessary to minimize the electron dose in the focusing area.

*3.5.2. Xplore3D Tomography Settings*

1. *Calibrations.* When Explore3D tomography is used for the first time, several calibrations have to be performed. These include magnification, image shift, stage shift, beam tilt, focus and stigmator, and eucentric height calibrations. A 463 nm grating grid is necessary for some of the calibrations. Most of the calibration functions can be accessed under the calibrations tab. In addition, the optimized position calibration and the holder calibration have to be accessed under the tomography acquisition tab. The latter is dependent on the actual cryo holder that is used and, therefore, must be performed more often. (1) *Optimized position.* The optimized position is the physical distance between the (vertical) optical axis of the electron beam and the (horizontal) rotation axis of the goniometer. The optimized position can be determined using the Explore3D software during tomography acquisition. For tomography applications the distance should be minimal and ideally be (close to) 0. In practice this distance should not exceed 2 μm to keep it within acceptable limits. When it exceeds 2 μm, the position of the rotation axis of the goniometer has to be mechanically adjusted by a service engineer. (2) *Holder*

*calibration*. In most situations, the Explore3D software makes use of a holder calibration file to anticipate the shifts in XY and Z (defocus) during data collection. Such a calibration file should be obtained for the different type of holders that might be used. During data collection, the differences between the anticipated movements in XYZ of the holder with the actual measured movements will be compensated for by image shifts (electron optics) and changes in focus (objective lens).

2. *Electron dose*. An important parameter in cryo electron tomography is the total electron dose that is used in the tilt series. The total electron dose determines the signal-to-noise ratio of the reconstructed tomogram. The total dose also influences many of the specific tomography settings since it has to be fractionated over many images. In practice an electron dose between 100 and 200 e/Å$^2$ per tilt series is used. An electron dose up to 100 e/Å$^2$ is generally used to ensure that high-resolution information is retained, e.g., for sub-tomogram averaging of structures within the tomogram. The electron dose rate (i.e., the intensity of the beam) can be measured on the microscope prior to the data collection. The measurement can be done using a specific function within the software or determined independently from the counts on the detector using the (calibrated) conversion factor or sensitivity (counts per electron) of that detector.
3. *Magnification*. The magnification defines the pixel size at the level of the specimen (together with the physical pixel size of the detector). An optimal setting of the magnification is determined by the desired resolution and the area to be imaged in one tomogram (field of view). To avoid interpolation artifacts during the subsequent image processing it is advisable that the pixel size is about three to five times smaller than the desired/expected resolution. In practice, the pixel size would be somewhere between 0.5 and 2 nm. On our experimental setup, these requirements require a magnification between ~8,000× and ~30,000×. At lower magnifications, cellular components such as lipid membranes, microtubules, and cytoskeletal filaments are readily visualized. Higher magnifications are advisable when one envisions the use of sub-tomogram averaging techniques.
4. *Defocus*. In cryo electron microscopy defocus is used to generate contrast (i.e., phase contrast). The required defocus depends on the acceleration voltage, the specimen thickness and the magnification. Stronger defocus values are needed to generate acceptable image contrast when working at higher acceleration voltages, with thicker specimens and lower magnifications. In practice, at magnifications less than 20,000× the defocus ranges from 5 to 10 μm, while at higher magnifications a smaller defocus value, e.g., between 3 and 6 μm is a good choice.

5. *Filtering*. The very low dose images that are taken during cryo electron tomography are extremely noisy. Within Explore3D, because of the low signal-to-noise ratio, the robustness of the cross-correlation algorithm to measure shifts between the individual images is sensitive to the settings of a (band-pass) filter that selects the size range of image details that will contribute to the computation of the image shifts. The filter settings are especially important in the absence of sharp contrasting features (e.g., fiducial gold beads or carbon edges) within the field of view. Typical filter settings are 200 nm ± 10 nm for low frequencies (large image details) and 2 nm ± 0.5 nm for high frequencies (small image details). In addition, the option to remove X-rays (i.e., occasional high intensity peaks in the image) should be on, and the option to suppress peak at zero (i.e., to compensate for imperfect image detector quality) should be off. The option to enhance features should be on, as well as the option for 16 pixels tapering. Finally, the option to apply a Hanning window should be on, and the option to apply a Sobel filter off. These options might vary, depending on the magnifications used in search, focus, and exposure mode. Since there is only one filter that should work for different magnifications and settings in all modes, it is necessary to tune the filter settings to work at both high and low magnifications.
6. *Tilt range*. The tilt range is the range of angles over which the images in the tilt series are taken. The maximum tilt angles are generally taken as large as possible to maximize the resolution isotropicity and minimize the wedge of missing data. They are defined by the maximum tilt range that the cryo holder can tilt without blocking the field of view. These maximum angles usually range somewhere between 70° and 80°. Other parameters that define the maximum tilt angles are the Mesh size of the grid, the thickness of the grid bars, and the position of the feature of interest on the grid. Especially at higher specimen tilt angles the (metal) grid bars should not be in the field of view. In practice, the maximum tilt range is most often limited by the sample thickness. At increasing tilt angles the effective thickness of a slab geometry sample (i.e., section) increases with 1/(cosine(tilt angle)). Samples will be twice as thick at 60° tilt, three times at 70.5°, and four times at 75.5°. This increased thickness negatively affects the image quality by decreasing the signal-to-noise ratio (multiple electron scattering events in the specimen) and possibly increasing specimen charging. Often, the maximum tilt angle will range between 60° and 65° to acquire a tilt series of 200–400 nm thick samples, while in samples that are thinner than 100 nm a maximum tilt range of 70° could be applicable.
7. *Acquisition scheme*. Within Explore3D, one can choose between using a continuous scheme and a discontinuous scheme to

collect a tilt series. In a discontinuous scheme, the acquisition starts at 0° (untilted) and the tilt angle increases to the maximum (positive) tilt angle to acquire the first half of a data set. The second half is consecutively taken from 0° going to the maximum negative tilt angle (0° to +60°, then 0° to −60°). In the continuous scheme the acquisition starts at the maximum negative angle and the tilt angle is changed towards the maximum positive angle (−60° to +60°). One disadvantage of the continuous scheme is that for relatively thick specimens it might be hard to find the desired specimen region at high tilt angles. One disadvantage of the discontinuous scheme is that a repositioning of the specimen at 0° to link the first half of the series to the second half will be necessary, a procedure that sometimes fails. Also, it can occur that noticeable differences in image quality between the two halves of the tilt series, for instance due to the continuous increase of radiation damage and a difference in the automatically set defocus, will introduce visible artifacts in the resulting tomogram when the two halves are combined.

8. *Tilt scheme and increment.* The use of a single value for the tilt increment (tilt angle between two successive images in a tilt series) during the whole reconstruction results in a linear tilt scheme. This is the most applied scheme that allows straightforward computation of the total number of images in a tilt series (maximum tilt angle/2+1) and the dose per image (total dose/total amount of images). Typically, the tilt image series ranges from +60° to −60° with a tilt increment of 2°, resulting in 61 images. The linear tilt scheme has the disadvantage that the projections of the tilt series are not uniformly distributed in Fourier space. When a tangential spacing of the angles is implemented (often referred to as the Saxton scheme) the projections are evenly distributed in Fourier space (14). In most data acquisition software packages, both tilt schemes are available as options. The Saxton scheme starts off with relatively large tilt increments at low tilt angles and increasingly smaller tilt increments at high angles. The Saxton scheme can be advantageous if it is likely that high quality images will be recorded at high tilt angles (e.g., with thin specimen).

9. *Exposure time adjustments.* One can choose to adjust the exposure time at higher tilt angles. The rationale behind this is that at higher tilt angles the (slab geometry) samples effectively get thicker and, for a constant exposure time, the image intensity decreases resulting in a worse signal-to-noise ratio. Consequently, because of the low signal-to-noise ratio at these high tilt angles, the images will contribute less to the reconstruction quality while they are especially important to improve the 3D information in the tomogram. Therefore, options are

available to (automatically) increase the exposure time at higher tilt angles. One option is to change the exposure time depending on the tilt angle, in such a way that the image intensity stays constant. However, to keep the required dose within limits, a factor that limits the exposure time to a maximum can be set, e.g., to 1.6. Setting the "distribute dose"option ensures that the total dose used for data collection will be constant by reducing all exposure times with a constant value.

10. *Focusing*. Prior to image acquisition, the defocus should be measured and compensated for as the mechanical accuracy and stability of the stage, the cryo holder and the holder calibration are not perfect. The application of the automatic focusing procedure that is available for collecting the tilts series does require time and is not always as critical. There are several factors that play a role in deciding how often and at what angles focus measurement and compensation would be important. For instance, when a specimen is tilted to 45 degrees, the variability in focus within one image is as large as the field of view, which might be in the order of microns. Therefore, unless one either works at low defocus values (less than 5 μm) or plans to do tomographic CTF correction, and the variation in focus is not changing more than a few microns, focusing is not as critical as in single particle cryo electron microscopy. However, with the chance of losing accuracy, the only thing one wins in not using focusing is time. In many cases, autofocusing is carried out every other acquisition angle.
11. *Tracking*. Tracking consecutive tilt series images is done to minimize the XY shifts. This can be performed by tracking after and by tracking before the acquisition. Using tracking after, the shift of the next image is predicted by cross-correlation of the two preceding ones. Using tracking before, additional images are taken after stage tilting and the shifts are accurately determined. The advantage of tracking after is that no additional images have to be recorded and therefore it is faster. Tracking before is more accurate but at the cost of additional images and electron dose.
12. *Other settings*. In the tomography software, the options start at eucentric height, start with eucentric focus, and apply (holder) corrections are typically all on. Images are saved in mrc format. For an overview of the settings window see Fig. 3.

### 3.6. Image Processing

#### 3.6.1. Hardware

Image reconstruction of tomographic tilt series can be performed on any modern PC, Mac, or computer cluster. Image processing software is available for the operating systems Windows, Mac OS, and Linux. It is often necessary to have sufficient RAM memory available to load the tilt series and the reconstructed tomograms. For visualization it is advisable to have a modern

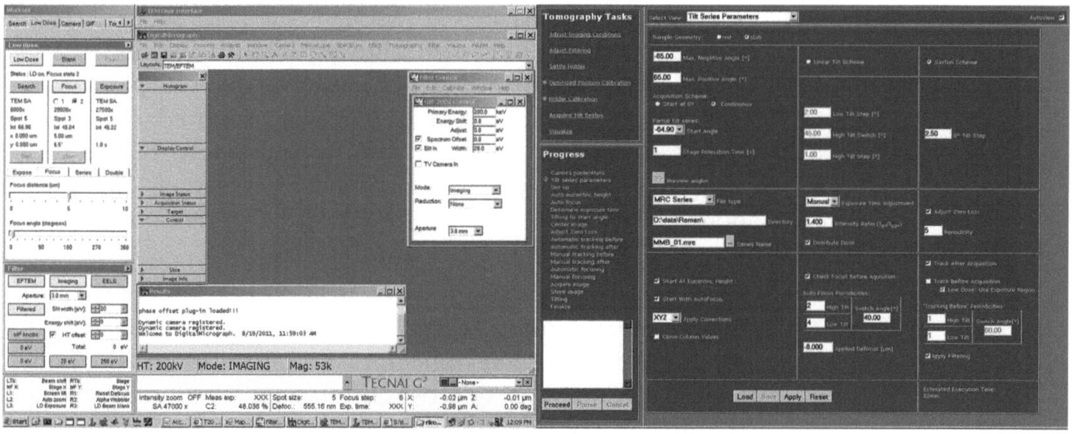

Fig. 3. Screenshot overview of the tomography acquisition settings page in Explore 3D.

graphics card for 3D visualization of rendered images. Typical for image reconstruction and visualization would be a Mac Book pro or a workstation with multicore 64 bits processors running Microsoft Windows 7 or Linux, at least 8 Gb RAM (suitable for 2 k×2 k images), and a Nvidia graphics card with 1 Gb memory. A recent tendency for image processing in electron microscopy has been to use the GPU for processing, instead of the CPU, which makes having a video card which supports CUDA more important.

#### 3.6.2. Software

Several image processing software packages are available for reconstruction of electron tomographic tilt series. The currently most prominent packages are IMOD (5), FEI inspect3D, TOM toolbox (6, 7), and UCSF tomography (8, 9). It is useful to have more than one of these packages available, since each package has specific features that can be useful during image processing. Surface rendering and 3D visualization can be performed using a variety of packages, e.g., Amira or UCSF Chimera (15). Currently, in our laboratory we use mostly IMOD and some features of Inspect3D for tomographic reconstruction, while for visualization we use Amira.

#### 3.6.3. Reconstruction

Tomographic tilt series can be saved in different image formats, depending on the reconstruction software. Prior to reconstruction it might be necessary to change the image format to the one required for the specific program. The tilt series that are recorded with Explore3D are compatible with IMOD.

1. *Preparation.* Examine the raw tilt series and possibly exclude bad images from the tilt series, e.g., images that are partly obstructed, have evidently different defocus values, or clear signs of image deteriorating charging effects. Bad images can also be excluded (ignored) during processing in IMOD or totally excluded using Inspect3D.

2. *Preprocessing.* A first step in IMOD is the preprocessing. In this step hot pixels—pixels with outlying intensity values—that are usually generated by X-rays in the electron microscope or by outliers in the gain or dark field references are removed (i.e., replaced by mean intensity values of surrounding pixels) by the CCD-eraser command using standard values in several cycles, until none or only a few hot pixels are detected by the software.
3. *Course alignment.* The alignment of the images is the most important step of the reconstruction. The quality of the alignment directly impacts the quality of the reconstruction. Alignment is generally done in two steps starting with a course image alignment using cross-correlation. When outer parts of the images are dark, e.g., by obstruction of thick ice, some pixel trimming is applied to enforce a proper alignment. Parameters controlling the course alignment can be adapted in the advanced tab in course alignment, where it is also possible to apply filtering and padding.
4. *Fiducial model generation.* The fine alignment using fiducial markers is performed in two steps. In the first step the gold fiducials are selected. This is performed by manual selection in one image of the tilt series, and the subsequent automated tracking of the position of these fiducial in all other images. As many fiducial markers as possible are added to maximize the alignment accuracy and, therefore, the reconstruction quality. Typically, a limited number of fiducials, maximal 10, are selected. These initially selected gold fiducials are ideally positioned in a thin part of the sample and without any surrounding obscuring features that might result in tracking errors. This initial seed model is then tracked, and the fiducials that are somehow not detected can be included by manual intervention and/or iterative tracking. (In the latest versions of IMOD more advanced options for automated tracking of gold particles and patches are present). The option to use local alignments (separate alignment of image sub-patches) is generally applied to improve tracking. The exact parameters are dependent on the number of fiducial markers within the field of view. It must be noted that the tilt series are aligned solely on the positions of the fiducial gold markers. The actual densities from the biological samples are not used. Since variations in positioning of gold fiducials exist it is good practice to choose the fiducials close to and surrounding the area of interest, since these are best reconstructed.
5. *Fine alignment.* In the second step of the fine alignment, the positioning of the selected fiducials is refined to improve the quality of the alignment. The fiducials that have the largest deviation between their position within the measured image compared to their predicted position are visualized and can be manually recentered. This is the most labor-intensive and

crucial step during the alignment. Typical final mean values for the standard deviation are between 1 and 2. The global variables solve for all rotations, all magnifications and all tilt angles are typically enabled, while distortion correction is disabled.

6. *Tomogram positioning*. Tomogram positioning is performed to fit the reconstructed object into a volume of minimal size (for computational reasons). The thickness of cellular tomograms varies between 400 and 1,200 pixels, and three or four times binned whole tomograms are mostly used to create the boundary model, since this produces more contrast.
7. *Final alignment*. After the tomogram positioning the final aligned stack is calculated. During this step there are also options for CTF correction and 2D filtering, which are not typically performed. Gold beads are erased using the following options "use the existing fiducial model" and "use mean of surrounding points," while in general the used diameter to erase the pixels is ~20% larger than the actual fiducial diameter.
8. *Tomogram generation*. The actual reconstruction of the aligned file is performed using the implemented Weighted-Back Projection (WPB) with standard settings.
9. *Post-processing*. Post-processing includes converting (from 16 bit data) to bytes, swapping Y and Z dimensions, and subsequent squeezing, especially for use in visualization.

### 3.7. Visualization

Cryo electron tomograms are characterized by very low signal-to-noise levels. This is especially the case with cryo tomograms of cells, since the samples are usually thick. The visualization of tomograms, in particular segmentation of substructures of interest and their 3D visualization is, therefore, not straightforward. Image processing and filtering is necessary for interpretation and visualization.

1. *Filtering*. Nonlinear anisotropic diffusion filtering (16) is performed to improve the SNR in cryo tomograms, as implemented in IMOD. Suitable *k*-value parameters for nonlinear anisotropic diffusion should be chosen around the value of the standard deviation of the tomogram, while using between 5 and 40 iterations. Nonlinear anisotropic filtering is computationally intensive and is performed on a Hewlett-Packard XC cluster with 56 nodes and a total of 64 Gb RAM running Linux or using the computers GPU (when available). Few iterations already give rise to clear contrast improvements with the preservation of higher resolution structures. Increasing the number of iterations result in loss of higher resolution features, which might be usable for segmentation and for overview images.
2. *Two-dimensional visualization*. Two dimensional visualization of tomogram slices are inspected in 3dmod (the display tool in

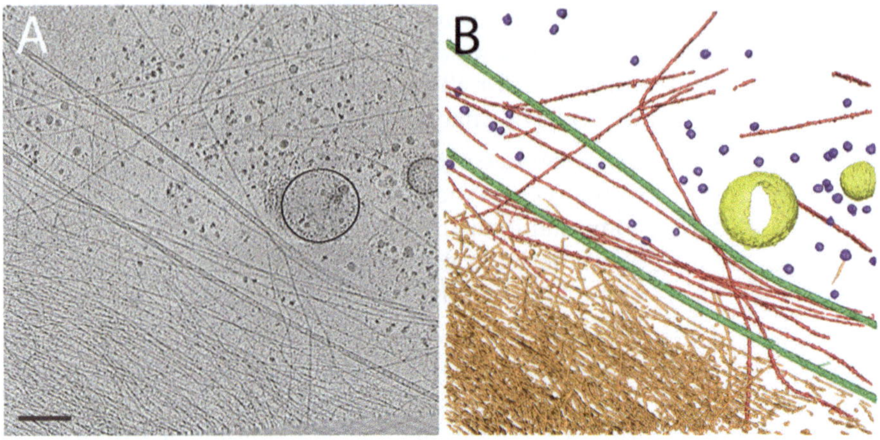


Fig. 4. (**a**) Electron tomographic slice through a whole mount vitrified mouse fibroblast cell. (**b**) Surface rendered image of whole tomogram. Visible are actin (*orange*), microtubules (*green*), intermediate filaments (*red*), (*clathrin coated*) lipid vesicles (*yellow*), and glycosome storage granules (*purple*). Scale bar, 250 nm. Images reproduced from ref. 22 with permission from Wiley-VCH Verlag GmbH & Co.

IMOD). Signal-to-noise levels can be improved by averaging several slices from the tomogram using the slicer window in IMOD. The slicer window also allows change of the slice orientation through the tomogram. Here, also two-dimensional median and anisotropic diffusion filters can be applied (Fig. 4a).

3. *Segmentation and rendering*. Filtered tomograms are used for generation of images and three-dimensional models. Using Amira, these are hand segmented by outlining the regions of interest in the segmentation editor. Microtubules are outlined in the XZ or YZ planes by drawing circles or ovals around them every 10th to 20th slice along the microtubule, followed by interpolation. Lipid vesicles are selected in a similar way along the XY plane. Intermediate filaments, actin filaments and irregular lipid structures are hand tracked (using a drawing tablet as input) throughout the tomogram. Ribosomes and other globular structures are selected by marking with a single point. All selections are improved using a combination of growing, shrinking, and smoothing selections in 3D. In the final segmented structures highest density pixels are selected. These selections are cleaned and smoothened by removing islands and smoothing of labels and surface rendered. This masking approach is time-consuming but produces satisfactory results for surface visualization of cellular cryo electron tomograms (Fig. 4b).

### 3.8. Conclusions and Outlook

By means of cryo electron tomography it is possible to visualize in three dimensions the cellular architecture, networks of macromolecular complexes and the overall morphology of membrane structures inside cells. Using a combination of cryo immobilization techniques and cryo electron tomography, cellular structures can be directly visualized in their native environment at molecular resolution. To achieve nanoscale 3D cellular imaging, high-end microscope equipment and specialized image processing tools are essential. Several technical improvements for investigating cells using cryo electron tomography are under development. These include improving specimen preparation protocols, developing new techniques to produce thin slices from vitrified cells, developing new approaches for the localization of specific structures using electron dense markers (17, 18), and combining high-resolution light microscopy with cryo electron tomography (19, 20).

## 4. Notes

1. *Quality of the sample and contamination.* One aspect that is essential in cryo electron microscopy in general is having good quality samples. Deterioration of the ice quality can occur in different ways at different stages. During vitrification, the temperature of the vitrification liquid has to be low enough to ensure fast freezing to allow vitrification of water without the formation of crystalline ice. When using liquid ethane this is usually done by keeping ethane at its melting temperature (−182°C) by making sure that it has a "milky" appearance, by having a mixture of solid and liquid ethane. A mixture of ethane and propane (2:1) has a lower melting point and does not solidify at liquid nitrogen temperatures and, therefore, is easier to keep at a suitable temperature for vitrification. To prevent recrystallization of the vitrified water after vitrification, the grids have to be kept below −160°C at all times during storage, transfer, and inside the microscope, to prevent the formation of crystalline ice (21), which will destroy the specimen and makes it unusable for imaging. It is important to keep liquid nitrogen shielded from air as much as possible and use shielded systems, specifically in Dewars for long-time storage. Liquid nitrogen is very susceptible to contamination by vapor from moist air and will readily be contaminated with small crystalline ice particles. These specifically stick to surfaces and result in contaminated specimens. Particularly during transfer using a cryo holder, temperatures can rise when not taking proper care of the temperature of the cryo holder. The temperature during transfer can be monitored using a controlling unit. Beware that the temperature sensor is located near the tip of the holder and that there might be a difference between measured temperature and the actual temperature on the grid. When initial cooling is performed in the cryo station, note that

the displayed temperature is always showing −192°C, since the tip is submerged in liquid nitrogen. Enough time should be allowed for proper cooling of the tip by the liquid nitrogen Dewar of the holder.

2. *Energy filtered imaging using the GIF.* It is important to use the energy filter since in thicker cellular samples it gives a huge improvement in contrast. The EFTEM alignments in our setup are separate alignments and, when using these, the above-mentioned TEM alignments are best performed after this EFTEM mode is activated. One difficulty in working with the GIF is that there is a 20× demagnification of the image on the fluorescent screen. One positive feature from this demagnification is that a quick overview at low magnification on the fluorescent screen can be obtained by extensive spreading of the electron beam and switching to lower (LM) magnifications is not necessary. One common problem while working in EFTEM mode is the proper alignment of the GIF entrance aperture at low magnifications (LM and M modes) using the crossover correction.
3. *Recognizing structures and localizing positions for data collection.* When starting cellular cryo electron microscopy on whole mount cells it is not always easy to directly recognize cellular structures at low magnification and localize suitable positions for tomographic data collection. Generally suitable areas for cellular tomography are thin and hydrated. Dehydrated or even fully dried cells can be recognized by high contrast at low magnification (<4,000×). Hydrated cells generally have a high translucency with very low contrast and subcellular structures become apparent only above 8,000× at defocus values around 10–25 μm. Good locations are usually found at the rim of large dark areas (the cells) and at low magnification are often recognized by the presence of round (vesicles) and elongated (cytoskeletal) structures and sometimes mitochondria. Huge amount of large and empty vesicles can be a sign of disintegration of the cell and/or cell death.

## Acknowledgments

R.I.K was financially supported by a Netherlands SmartMix grant and the NIMIC partner organizations (www.realnano.nl) through NIMIC, a public–private program.

## References


1. Chiu W, Baker ML, Jiang W, Dougherty M, Schmid MF (2005) Electron cryomicroscopy of biological machines at subnanometer resolution. Structure 13:363–372
2. Zhou ZH (2008) Towards atomic resolution structural determination by single-particle cryo-electron microscopy. Curr Opin Struct Biol 18:218–228
3. Koning RI, Zovko S, Barcena M et al (2008) Cryo electron tomography of vitrified fibroblasts: microtubule plus ends in situ. J Struct Biol 161:459–468
4. Patla I, Volberg T, Elad N et al (2010) Dissecting the molecular architecture of integrin adhesion sites by cryo-electron tomography. Nat Cell Biol 12:909–915
5. Kremer JR, Mastronarde DN, McIntosh JR (1996) Computer visualization of three-dimensional image data using IMOD. J Struct Biol 116:71–76
6. Korinek A, Beck F, Baumeister W, Nickell S, Plitzko JM (2011) Computer controlled cryo-electron microscopy—TOM(2) a software package for high-throughput applications. J Struct Biol 175:394–405
7. Nickell S, Forster F, Linaroudis A et al (2005) TOM software toolbox: acquisition and analysis for electron tomography. J Struct Biol 149:227–234
8. Zheng SQ, Keszthelyi B, Branlund E et al (2007) UCSF tomography: an integrated software suite for real-time electron microscopic tomographic data collection, alignment, and reconstruction. J Struct Biol 157:138–147
9. Zheng SQ, Sedat JW, Agard DA (2010) Automated data collection for electron microscopic tomography. Methods Enzymol 481:283–315
10. Koning RI, Koster AJ (2009) Cryo-electron tomography in biology and medicine. Ann Anat 191:427–445
11. Iancu CV, Tivol WF, Schooler JB et al (2006) Electron cryotomography sample preparation using the Vitrobot. Nat Protoc 1:2813–2819
12. Tivol WF, Briegel A, Jensen GJ (2008) An improved cryogen for plunge freezing. Microsc Microanal 14:375–379
13. Vulovic M, Rieger B, van Vliet LJ, Koster AJ, Ravelli RB (2010) A toolkit for the characterization of CCD cameras for transmission electron microscopy. Acta Crystallogr D Biol Crystallogr 66:97–109
14. Saxton WO, Baumeister W, Hahn M (1984) Three-dimensional reconstruction of imperfect two-dimensional crystals. Ultramicroscopy 13:57–70
15. Pettersen EF, Goddard TD, Huang CC et al (2004) UCSF Chimera—a visualization system for exploratory research and analysis. J Comput Chem 25:1605–1612
16. Frangakis AS, Hegerl R (2001) Noise reduction in electron tomographic reconstructions using nonlinear anisotropic diffusion. J Struct Biol 135:239–250
17. Mercogliano CP, DeRosier DJ (2007) Concatenated metallothionein as a clonable gold label for electron microscopy. J Struct Biol 160:70–82
18. Wang Q, Mercogliano CP, Lowe J (2011) A ferritin-based label for cellular electron cryotomography. Structure 19:147–154
19. Plitzko JM, Rigort A, Leis A (2009) Correlative cryo-light microscopy and cryo-electron tomography: from cellular territories to molecular landscapes. Curr Opin Biotechnol 20:83–89
20. van Driel LF, Valentijn JA, Valentijn KM, Koning RI, Koster AJ (2009) Tools for correlative cryo-fluorescence microscopy and cryo-electron tomography applied to whole mitochondria in human endothelial cells. Eur J Cell Biol 88:669–684
21. Dubochet J, Adrian M, Chang JJ et al (1988) Cryo-electron microscopy of vitrified specimens. Q Rev Biophys 21:129–228
22. van Tendeloo G, van Dyke D, Pennycook SJ (2011) Handbook of microscopy. Wiley-VCH, Weinheim

# Chapter 15

# Large-Volume Reconstruction of Brain Tissue from High-Resolution Serial Section Images Acquired by SEM-Based Scanning Transmission Electron Microscopy

**Masaaki Kuwajima, John M. Mendenhall, and Kristen M. Harris**


## Abstract

With recent improvements in instrumentation and computational tools, serial section electron microscopy has become increasingly straightforward. A new method for imaging ultrathin serial sections is developed based on a field emission scanning electron microscope fitted with a transmitted electron detector. This method is capable of automatically acquiring high-resolution serial images with a large field size and very little optical and physical distortions. In this chapter, we describe the procedures leading to the generation and analyses of a large-volume stack of high-resolution images (64 μm × 64 μm × 10 μm, or larger, at 2 nm pixel size), including how to obtain large-area serial sections of uniform thickness from well-preserved brain tissue that is rapidly perfusion-fixed with mixed aldehydes, processed with a microwave-enhanced method, and embedded into epoxy resin.


**Key words:** 3D reconstruction, Hippocampus, Microwave-enhanced processing, Neuropil, Perfusion, Scanning electron microscopy, Scanning transmission electron microscopy, Serial sections, Structural plasticity, Ultramicrotomy

## 1. Introduction

Ever since the presence of synaptic junctions was demonstrated by transmission electron microscopy (TEM) in the 1950s (1, 2), serial thin sections of <100 nm thickness were used to visualize them in the three-dimensional context of the surrounding dendrites, axons, and astroglial process, namely, the "neuropil" (3–6). Recently, serial section EM has gained renewed interest in neuroscience as a high-resolution 3D tool (7). It has been used for 3D reconstruction of functionally identified local circuitry (8, 9), as well as for attempts to map all synaptic connections in the brain (termed "connectomics") (10–12). Our laboratory and others use serial section EM to understand how the structure of synapses is modified

Alioscka A. Sousa and Michael J. Kruhlak (eds.), *Nanoimaging: Methods and Protocols*, Methods in Molecular Biology, vol. 950, DOI 10.1007/978-1-62703-137-0_15, © Springer Science+Business Media, LLC 2013

by experience (or in models of learning and memory), which can ultimately alter the connection profile for a neuron (13–15). In this case, a pixel size of ≤2 nm is required for reliable identification and analyses of subcellular structures (e.g., polyribosomes, microtubules, and smooth endoplasmic reticulum) that may play crucial roles in structural plasticity of synapses.

In order to implement the appropriate serial section EM system to address a particular set of scientific questions, one must consider the features that ultimately determine the system's limitations. These include (but are not limited to): platform (e.g., scanning vs. transmission EM), imaging mode (e.g., detection of transmitted vs. back-scattered electrons), serial sectioning method (e.g., diamond knife vs. focused ion beam), specimen presentation (e.g., block face vs. ultrathin sections), and imaging field size. Two serial section EM systems are based on scanning electron microscopy (SEM) with the detection of back-scattered electrons from the block face that is serially removed either by focused ion beam (FIB-SEM) (16–19) or diamond knife inside the SEM chamber (SBFSEM) (20). These approaches can generate thinner sections (~25 nm for SBFSEM and ~5 nm for FIB-SEM) than the sections obtained for traditional serial EM (~45 nm). Thus, the voxels obtained through FIB-SEM imaging can be more isotropic (several nanometers per side). The FIB-SEM is limited by the size of the field that can be milled and imaged, usually less than 100 μm × 100 μm. The destructive nature of SBFSEM and FIB-SEM precludes post-embedding immunolabeling, reexamination of areas of interest at a higher magnification, or use of EM tomography, which can generate virtual sections of less than 3 nm thickness. In other serial section EM approaches, the images are acquired from ultrathin sections. They can be collected automatically on electron-opaque plastic tapes (automatic tape-collecting lathe ultramicrotome, or ATLUM) (21) or manually on glass microscope slides (array tomography) (22) that can be archived for repeated SEM and light microscopy imaging. Array tomography provides a relatively straightforward platform for immunolabeling and correlative fluorescence and electron microscopy. Serial sections collected on traditional TEM grids can be imaged with a TEM custom-fitted with an array of CCD cameras (TEMCA), which rapidly acquires images with larger field size (8). However, TEMCA is cumbersome and expensive, and the resulting images can still suffer from the specimen drifting, optical distortion, and beam damage that are inherent in TEM imaging.

We have recently developed a new method for imaging serial sections using scanning transmission electron microscopy (STEM) based on a field emission (FE) SEM fitted with a multi-mode transmitted electron detector (23). This system allows for semi-automated acquisition of large-field images (single field size of 64 μm × 64 μm square, or more with mosaic) at a high resolution

(1 or 2 nm pixel size) from multiple grids spanning an entire series of ≥200 serial sections loaded on a single specimen holder (with multiple holders, the total section number is only limited by how many can be sectioned). Images taken on this system have less optical distortion and specimen drift, and the resolution is comparable to those taken on a conventional TEM with a bottom-mount CCD camera (pixel dimension = 4,080 × 4,080) at a magnification of about 5,000×. Archived serial sections on conventional TEM grids can be imaged on this SEM system without modification. In addition, less operator involvement is required, and the instrument can be obtained for about half the cost of a conventional TEM with a large-format CCD camera. Recent improvements in computational tools are also making it easier to handle large-field serial images for alignment and analysis, as well as to generate realistic three-dimensional reconstructions of segmented objects (24–32).

In order to take advantage of these advanced tools for studying structural plasticity, however, one must start with well-preserved brain tissue, which can be achieved by rapid perfusion-fixation with mixed aldehydes while the animal is under artificial ventilation to minimize hypoxic damages (33). Once well-fixed tissue undergoes microwave-enhanced processing and is embedded into epoxy resin (34), serial thin sections of uniform thickness (~45 nm) must be obtained from a precisely targeted brain region of interest while minimizing flaws, such as folds, knife marks, and debris from stain or dust (35). The following describes our routine methods to generate series of high-quality EM images from neuropil in the hippocampal dentate gyrus of perfusion-fixed brain tissue embedded in epoxy resin. The resulting image series is then used for quantitative analyses and three-dimensional reconstruction of neuronal and glial structures.

## 2. Materials

General reagents, equipment, and supplies include: purified water (double-distilled or 18 MΩ), 1 N aqueous NaOH, 1 N aqueous HCl, balance, forceps, glassware (e.g., beakers, cylinders, flasks, petri dish, and vials), hot plate stirrer, osmometer, small paint brush, pH meter, sonicator, cyanoacrylate glue (e.g., Super Glue®), desiccant, filter paper, glass microscope slides and cover glass (#1 thickness), Kimwipes®, Parafilm M®, pipettes, razor blades, syringes, syringe filters (pore size = 0.22 μm).

### 2.1. Perfusion-Fixation

1. Animals: Adult rats (typically 4–6 months old; see Note 1). They can be naïve, or ones that underwent in vivo experiments such as electrophysiological recordings or behavioral learning.

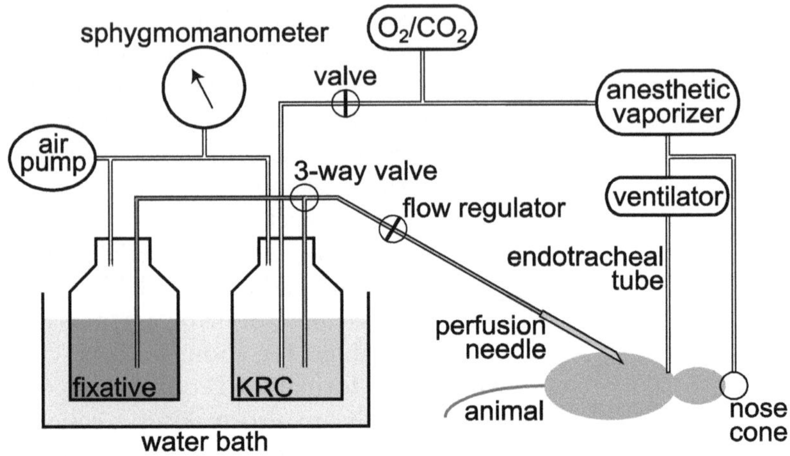


Fig. 1. A simplified diagram of the perfusion apparatus as described in Subheadings 2.1 and 3.1.

2. Anesthetics (halothane or isoflurane; see Note 2), vaporizer (Midmark), and a large desiccator jar.
3. Surgical tools such as scalpels, scissors, suture, forceps, hemostats, bone rongeurs, and spatulas.
4. Perfusion apparatus (Fig. 1): dissection tray, Styrofoam board, pins, $O_2/CO_2$ (95%/5%) tank and regulator, small animal ventilator, nose cone, sphygmomanometer (with the cuff cut off), water bath, 2-L glass bottles, hypodermic needles. These components are assembled with an assortment of stoppers, tubing, connectors, clamps, and valves (see Note 3).
5. Krebs-Ringer Carbicarb buffer (KRC): 2.0 mM $CaCl_2$, 11.0 mM D-glucose, 4.7 mM KCl, 4.0 mM $MgSO_4$, 118 mM NaCl, 12.5 mM $Na_2CO_3$, 12.5 mM $NaHCO_3$, pH = 7.35–7.4, osmolality = 300–330 mmol/kg (see Note 4).
6. Fixative: 2.0% formaldehyde, 2.5% glutaraldehyde, 2 mM $CaCl_2$, 4 mM $MgSO_4$, 100 mM sodium cacodylate, pH = 7.35 (aldehydes and sodium cacodylate are obtained from Ladd Research; see Note 5).
7. Vacuum filter system (pore size = 0.22 μm) to filter KRC and fixative.

### 2.2. Tissue Processing

1. Vibrating blade microtome.
2. 24-well plates.
3. 0.1 M phosphate buffer, pH 7.4.
4. Dissecting microscope.
5. Small dissecting knife, ~4 mm cutting edge (Fine Science Tools).

6. 0.1 M sodium cacodylate buffer, pH 7.4.
7. Agarose: 7% in purified water.
8. Laboratory microwave oven with vacuum chamber and cooling stage (Ted Pella): prior to tissue processing, the magnetron needs to be warmed up by running it for 2 min at 175 W.
9. Tissue processing basket (Ted Pella).
10. Reduced osmium: 1% osmium tetroxide (see Note 6) and 1.5% potassium ferrocyanide [$K_4Fe(CN)_6$ $3H_2O$] in sodium cacodylate buffer.
11. 1% Osmium tetroxide in sodium cacodylate buffer (see Note 6).
12. Uranyl acetate (ACS reagent grade).
13. Ethanol (see Note 7).
14. Propylene oxide.
15. LX-112 epoxy resin embedding kit (Ladd Research).
16. Rotator.
17. Chien embedding mold (Electron Microscopy Sciences, Polysciences, or Ted Pella).
18. Laboratory oven.

### 2.3. Ultramicrotomy and Post-section Staining

1. SynapTek™ beryllium-copper slot grids, slot size = 1 mm × 2 mm or 0.5 mm × 2 mm (Ted Pella or Electron Microscopy Sciences).
2. Pioloform® F (Ted Pella): 0.7% solution in high-purity chloroform.
3. Supplies for casting Pioloform films, including a film casting device (Electron Microscopy Sciences), Ross optical lens tissue paper, 70% ethanol, Alconox® (saturated aqueous solution filtered and diluted to ~20%), and a large glass container (to contain water on which Pioloform films are floated).
4. Ultramicrotome (Leica Microsystems), placed in a Plexiglas enclosure on an air table in a dedicated room (Fig. 2g).
5. Diamond trim knife (DiATOME cryotrim 20).
6. Diamond thin-sectioning knife (DiATOME ultra 35).
7. Supplies for manipulating and picking up serial sections such as eyelash, forceps, and grid box.
8. Grid staining dish (with lid): 55 mm glass Petri dish filled with melted dental wax. Indentations are made on the wax surface with a razor blade so that grids can be inserted vertically.
9. Uranyl acetate: Saturated aqueous solution. Add 6.25 g of uranyl acetate to 100 ml of purified water and sonicate for up to 1 h. Store at 4°C.
10. Reynolds' lead citrate stain (36): In purified water boiled to expel $CO_2$, completely dissolve lead nitrate (Ladd Research)

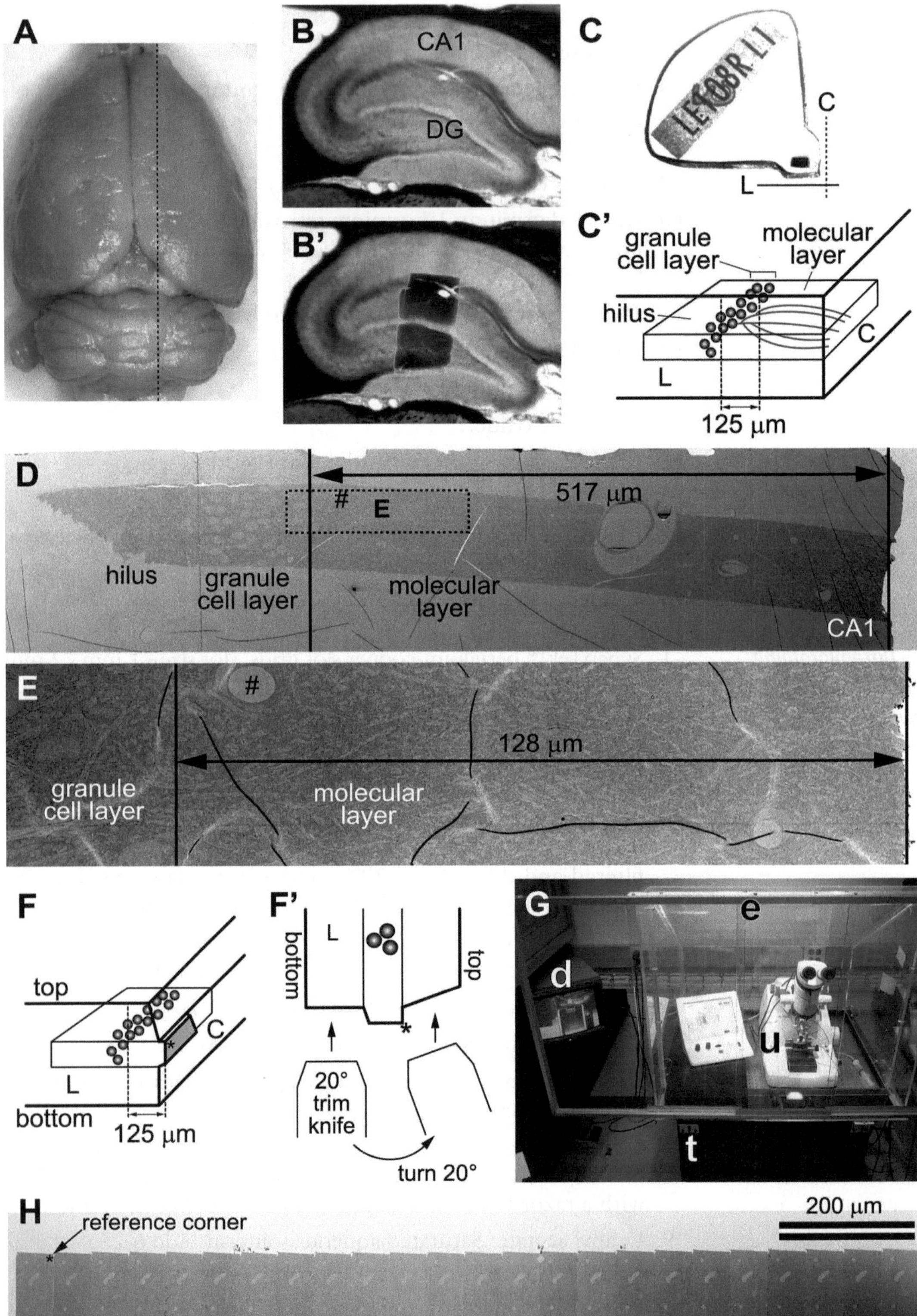
A
B
CA1
DG
B'
C
C
L
C'
granule cell layer
molecular layer
hilus
C
L
125 μm
D
517 μm
# E
hilus
granule cell layer
molecular layer
CA1
E
#
128 μm
granule cell layer
molecular layer
F
top
C
L
bottom
125 μm
F'
L
bottom
top
20° trim knife
turn 20°
G
e
d
u
t
H
reference corner
200 μm

for the final concentration of 2.66%, then sodium citrate (3.52% final). The solution will turn milky white. After letting stand for 30 min with occasional mixing, add fresh aqueous sodium hydroxide (0.16 N final) (see Note 8). To avoid exposure to air, store in 10 ml syringes with syringe filer and hypodermic needle stuck on a rubber stopper at 4°C.

11. Sodium hydroxide: some pellets and 0.02 N aqueous solution.

### 2.4. Serial Section Imaging and Analysis

1. Zeiss FE SEM with a retractable transmitted electron detector and a specimen holder for TEM grids: a SUPRA® 40 in this configuration is used in our laboratory, and the following descriptions of the microscope operation will be based on this model. Other current models of Zeiss FE SEM can also be configured in the same manner, and their operations are essentially identical to ours. At the time of this writing, Zeiss is the sole source of the FE SEM system designed for high-resolution large-field STEM imaging of biological samples as described in this chapter.
2. Zeiss ATLAS™ system integrated with the SEM for semi-automated large field imaging.

Fig. 2. (**a**) Dorsal view of a perfusion-fixed adult rat brain. *Dotted black line* indicates an approximate location from which the slice shown in **b** and **b′** was taken on a vibrating blade microtome. (**b**) A parasagittal slice of the hippocampus. Area CA1 and the dentate gyrus (DG) are indicated. (**b′**) The same slice as in **b**, with a superimposed image of the dentate tissue embedded in the epoxy block shown in **c**. (**c**) An epoxy block containing the dentate tissue dissected from the slice shown in **b** and **b′**. The Chien embedding mold was used so that the tissue can be cut from two faces of the block (C with *dotted line*= "cross-section face"; L with *solid line*= "longitudinal face"). (**c′**) A diagram indicating the orientation of the tissue in the block shown in **c**. The "cross-section face" (C) is parallel to the granule cell layer (*gray spheres*), and dendrites (*gray curved lines* in the molecular layer) of these cells are cut in cross section in this plane. The "longitudinal face" (L) is perpendicular to the cell layer and the cross-section face, and therefore the dendrites are cut longitudinally in this plane. In this example, serial sections are cut from the cross-section face at 125 μm from the top of the granule cell layer (i.e., middle molecular layer). (**d**) An electron micrograph of a test-thin section cut from the longitudinal face of the block shown in **c**. This image is used to determine how much of the tissue must be trimmed to reach the target location on the cross-section face (i.e., 517 − 125 μm = 392 μm to trim). The *rectangle* with *dotted line* indicates the area imaged in E after trimming. "Hash" corresponds to the location of capillary seen in E. Note that flaws in the thin section or Pioloform film (e.g., folds) cannot be avoided in sections of this size (also see **e**), but they are minimized in serial sections cut from a much smaller block face (see **h**). (**e**) An electron micrograph of another test-thin section, taken after trimming to the target location. The target (125 μm) will be well within a series of 200 serial sections (spanning about 9 μm), so the final block face is trimmed at this point. (**f**) A diagram showing the final block face (*gray trapezoid* area) trimmed on the cross-section face of the block. The *asterisk* (also in **f′** and **h**) indicates the "reference corner": the edge extending into tissue from this corner is perpendicular to the block face, and therefore remains constant throughout the series. (**f′**): The *top-down* view of the longitudinal face of the block in **f**, depicting the method for trimming the top and bottom edges of the block face. After the bottom edge is trimmed with a 20° slope, the knife is turned 20° to the right, so that the top edge is trimmed to become perpendicular to the block face. (**g**) In a dedicated room, the ultramicrotome (u) is placed in a Plexiglas enclosure (e) on an air table (t). A video camera is fitted to the ultramicrotome so that serial sectioning can be monitored with a video display (d) after closing the glass front window of the enclosure. (**h**) A low-magnification SEM image of a segment of serial sections on a single Pioloform-coated grid. Large capillaries are evident in each section. The reference corner (indicated by *asterisk*) is used to navigate to the ROI during imaging (see Fig. 3d).

3. Grating replica grid for calibration (Electron Microscopy Sciences).
4. Workstation running FIJI for image alignment and other preprocessing (freely available at: http://fiji.sc/) (see Note 9). A workstation with multiple CPU cores (e.g., ours has 16 cores at 2 GHz) with 64 GB of RAM and 2 TB of hard drives set up in RAID should be sufficiently powerful to handle series of large serial images (~500 MB per image).
5. Workstation running RECONSTRUCT™ (version 1.1.0.0; 32-bit) for image analysis and 3D reconstruction (freely available at: http://synapses.clm.utexas.edu/) (26).

## 3. Methods

### *3.1. Perfusion-Fixation (See Notes 10 and 11)*

This perfusion-fixation procedure is performed best with two persons, one acting as the surgeon (i.e., performing tracheotomy and inserting the perfusion needle) and the other as assistant (i.e., adjusting the delivery pressure, switching from KRC to fixative, etc.). A simplified schematic of our perfusion apparatus is illustrated in Fig. 1. A hand air pump (a modified sphygmomanometer) is used to apply pressure to drive the perfusates (KRC and fixative), which are warmed in a water bath so that they are at 37°C at the tip of perfusion needle (13-gauge). KRC is bubbled with $O_2/CO_2$ (95%/5%) for at least 30 min prior to start of perfusion. $O_2/CO_2$ is also used to drive an anesthetic vaporizer, which is connected to a nose cone (used until tracheotomy is done; step 3 below) and a ventilator (for artificial respiration). The ventilator is in turn connected to an endotracheal tube, which is made with tubing attached to a blunt 16-gauge needle (1 cm long).

1. The vaporizer is set at 5% anesthetic with the flow at 400 ccm. The ventilator should be running at 120 strokes/min with a tidal volume of 1.5 cc.
2. Anesthesia is induced in a large desiccator jar with a wad of several small Kimwipes. Five minutes prior to placing the animal, add ~1.5 ml of halothane or isoflurane to the Kimwipes below the perforated plate and put the lid on the desiccator. Make sure the animal does not come in contact with liquid halothane or isoflurane. Test the animal's anesthesia level with a toe pinch. A sufficiently anesthetized animal will not respond. After attaching the nose cone, the animal is secured with pins onto a Styrofoam board in the dissection tray.
3. To establish artificial respiration, surgically expose the trachea via a chin-to-sternum incision, partially cut across the trachea, and insert the endotracheal tube into the trachea. A piece of

suture is tied around the trachea to secure the inserted tube. The nose cone can be removed at this point.

4. After opening the thoracic cavity and exposing the heart, the right atrium is cut and perfusion needle is inserted into the left ventricle to start perfusion with warm oxygenated KRC delivered at 80 mmHg for a few seconds.
5. The perfusate is switched to the fixative solution, and the pressure is increased to 180 mmHg. The chin is cut to see if the fixative drips out, which confirms that it is delivered to the head (see Note 12). After 10 min of perfusion with fixative, the pressure is reduced to 40–80 mmHg, or to the point where fixative is very slowly dripping from the cut chin of the animal. At the end of perfusion, each animal should have received 1,800–2,000 ml of fixative over the course of about 60 min (see Note 13).
6. The brain is allowed to remain undisturbed in the cranium for at least 1 h (but not more than a few hours). Then it is carefully dissected from the cranium (Fig. 2a) and soaked in the same fixative solution overnight at RT (see Notes 14 and 15).

### 3.2. Tissue Processing (See Notes 10 and 11)

1. The fixed brain is transferred to phosphate buffer and sliced with a vibrating blade microtome. We typically take parasagittal slices at the thickness of 70 μm. The slices are sequentially collected and stored in 24-well plates containing the fixative at RT until EM processing (see Notes 16 and 17).
2. The 70 μm thick slice is transferred to sodium cacodylate buffer, and a piece of tissue containing the area of interest (the hippocampal dentate gyrus in this example; Fig. 2b and b′) is dissected out with a small knife under a dissecting microscope.
3. The tissue is embedded into 7% agarose between glass microscope slides (with thickness of one #1 cover glass) to protect it during the subsequent processing. A block of agarose containing the tissue is cut out and placed in a processing basket filled with sodium cacodylate buffer.
4. After rinses in the buffer, the tissue is osmicated in two steps: (1) Reduced osmium solution for 5 min, followed by the buffer rinses; and (2) 1% osmium tetroxide solution for two 3-min cycles (1 min ON 1 min OFF 1 min ON) at 175 W under vacuum in a laboratory microwave oven. This is followed by rinses in the buffer and then in purified water.
5. The osmicated tissue is then dehydrated through ascending concentrations of ethanol and stained en bloc with 1% uranyl acetate. After it is placed in 50% ethanol for 5 min, transfer the tissue into 50% ethanol containing 1% uranyl acetate, then apply 40 s of microwave (at 250 W without vacuum). Repeat the same microwave procedure for the 70, 90, and 100%

ethanol solutions (each containing 1% uranyl acetate) and again for 100% ethanol (without uranyl acetate).

6. The tissue is placed on a rotator in mixtures of ethanol and propylene oxide (1:1 and then 1:2, 10 min each, followed by two exchanges of 100% propylene oxide over 30 min) and infiltrated with LX-112 epoxy resin (1 h with a mixture of 1:1 = propylene oxide:resin, then overnight with 1:2 = propylene oxide:resin, followed by three exchanges of 100% resin over 3 h).
7. The tissue is arranged into the Chien mold such that one edge is parallel to the granule cell layer ("cross-section face" or C in Fig. 2c and c′) and the other edge perpendicular to it ("longitudinal face" or L in Fig. 2c and c′). This arrangement allows for subsequent ultrathin sectioning of the tissue from two directions: one with longitudinally sectioned dendrites and the other at precise locations in the field with cross-sectioned dendrites (see Subheading 3.3).
8. The epoxy-embedded tissue is cured in a laboratory oven at 60°C for 48 h (see Note 18).

### 3.3. Ultramicrotomy and Post-section Staining (see Notes 10 and 11)

We typically cut 200–300 serial sections at 45 nm thickness from a target location (35). However, there is no particular limit to the number of serial sections that can be obtained and imaged other than the skill of the operator and time (see Notes 19 and 20). Here, an example illustrated in Fig. 2 shows a strategy to target the middle molecular layer of the hippocampal dentate gyrus at 125 μm from the top of the granule cell layer, such that the granule cell dendrites are cut in the cross-section plane. This targeting strategy can easily be adopted to collect serial sections from other brain areas.

1. On an ultramicrotome, the block of epoxy-embedded tissue is secured on a vise-style chuck and trimmed perpendicular to the granule cell layer to expose the tissue on the "longitudinal face" (i.e., along or parallel to the dendritic arbor traversing the molecular layer; L in Fig. 2c and c′). An ultrathin section is collected and used to locate the target area for serial-sectioning (Fig. 2d). Tissue quality is also evaluated in this test-thin section in the electron microscope.
2. Turn the block 90° in the chuck (i.e., parallel to the granule cell layer), and trim it on the "cross-section face" (C in Fig. 2c and c′) to the target region of interest. Cut an ultrathin section of the "cross-section face," which is used to place the final block face (step 4; Fig. 2f). Also cut another ultrathin section from the "longitudinal face" to confirm that you have reached the target location (Fig. 2e). It may be necessary to repeat steps 1 and 2 before the block face reaches the target.

3. Pioloform-coated slot grids are prepared the day before cutting serial sections (see Note 21). Glass microscope slides are cleaned with 70% ethanol and purified water, then coated with 20% saturated aqueous solution of Alconox detergent. The slides are scrubbed dry with Ross lens tissue after each step of cleaning and coating. Using a clean film casting device, Pioloform film is casted onto the slides from a 0.7% chloroform solution of Pioloform F (see Note 22). After being dried in a jar with desiccant, the film is floated on a surface of still water, and slot grids are placed on the film showing a uniform silver interference color (see Note 23). Secure the grids onto the film by tapping them on the edges with forceps, pick up onto Parafilm-wrapped glass slides, and air-dry overnight in a petri dish (see Note 20). The quality of the Pioloform film should be checked under EM.
4. The block face (shaded area in Fig. 2f) is trimmed into a trapezoid shape with a DiATOME cryotrim 20 knife so that the resulting size of the ultrathin sections would be about 70 μm × 200 μm (see Note 24). The block face should protrude from the rest of the block by 20–30 μm (see Note 25). One corner of the block face is trimmed with the knife turned 20° laterally, such that the resulting edge becomes perpendicular to the block face (top-left corner of the block face, marked by asterisk in Fig. 2f, f′, and h). Because the position of this corner relative to the tissue remains constant throughout the series, it later serves as a reference point during imaging (Figs. 2h and 3d).
5. A DiATOME ultra 35 knife is used to cut a series of serial sections with the thickness of 45 nm on an ultramicrotome (see Note 26). The ultramicrotome (u in Fig. 2g) is placed in a Plexiglas enclosure (e in Fig. 2g) on an air table (t in Fig. 2g) in a dedicated room to minimize variations in section thickness caused by vibration, air flow, or humidity and temperature changes. As the serial sections are cut, they should begin to form an unbroken straight ribbon on the water surface. Once this is confirmed, the front window of the enclosure is shut, and a video display (d in Fig. 2g) is used to monitor the rest of the serial sectioning process.
6. When a sufficient number of sections are cut and the ultramicrotome is stopped, the ribbon is broken by tapping it with a fine eyelash into segments small enough to fit onto Pioloform-coated slot grids (see Notes 20 and 27).
7. After being collected on slot grids, the serial sections are air-dried for a couple of hours by placing each grid vertically into a notch on the wax surface of the grid staining dish. Droplets of saturated aqueous solution of uranyl acetate are applied to the section side of the grids with a disposable syringe and

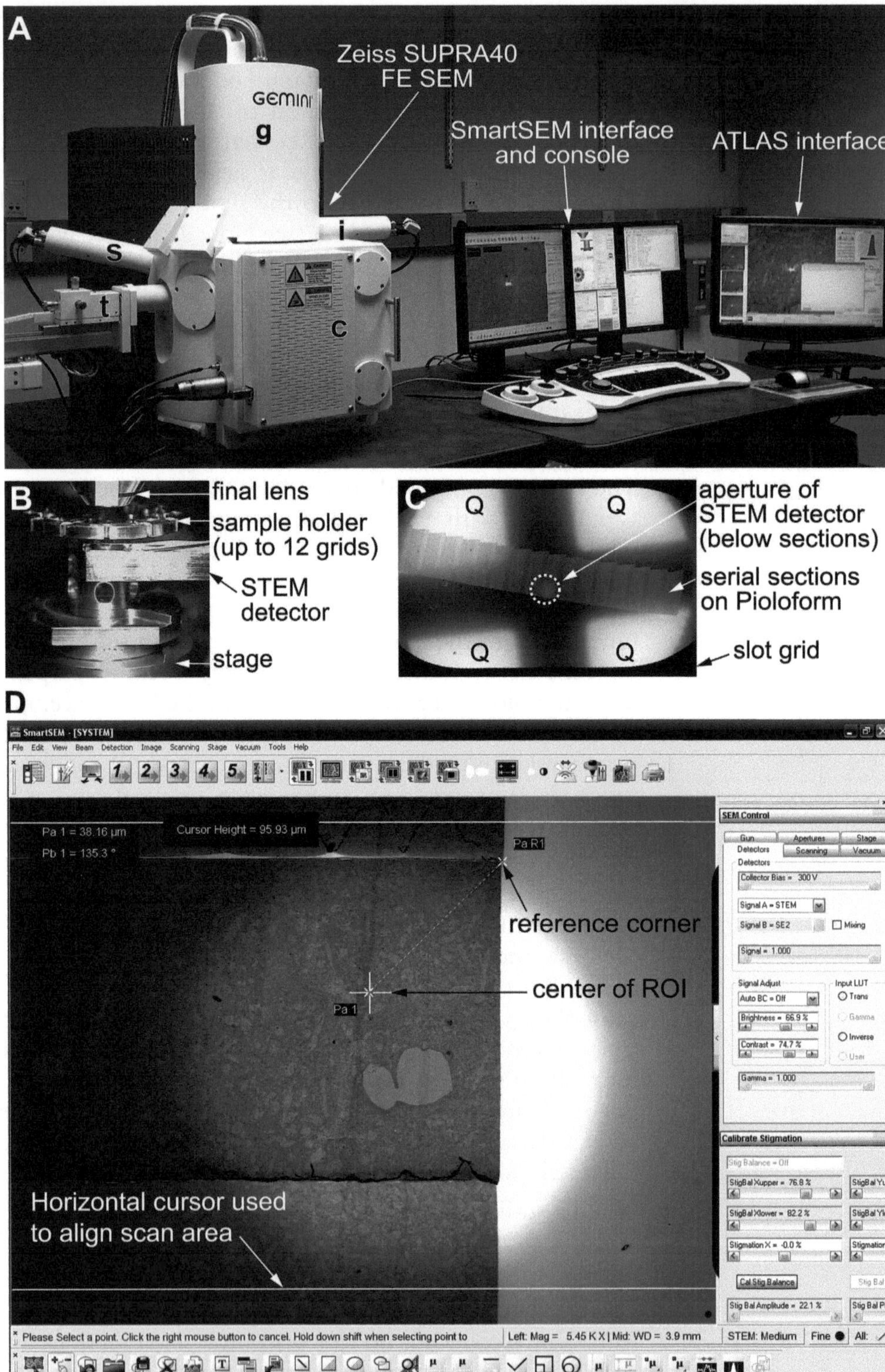
A
Zeiss SUPRA40
FE SEM
SmartSEM interface
and console
ATLAS interface
g
s
t
i
c
B
final lens
sample holder
(up to 12 grids)
STEM
detector
stage
C
Q
aperture of
STEM detector
(below sections)
serial sections
on Pioloform
slot grid
D
reference corner
center of ROI
Horizontal cursor used
to align scan area

syringe filter, and the lid is closed to stain the sections for 5 min. The grids (still kept on the wax surface of the staining dish) are then rinsed at least five times by pouring purified water into the dish. After the last rinse, use small pieces of filter paper to remove excess water carefully from the grids so as not to touch the Pioloform film (see Note 20).

8. The serial sections are then stained with Reynolds' lead citrate (36). This must be done in a $CO_2$-free environment to prevent lead precipitation artifacts: place a few NaOH pellets in the dish (see Note 28) along with a piece of 55 mm diameter filter paper soaked with 0.02 N NaOH solution so that it will stick inside of the lid. Droplets of the lead citrate solution are applied to the section side of each grid, and the lid is closed to stain for 5 min. The NaOH pellets are then removed, and the grids are rinsed first with 0.02 N NaOH solution followed by at least five exchanges of purified water. As in the previous step, filter paper is used to remove excess water from the grids, which are then completely air-dried (preferably overnight) before EM imaging (see Note 20). For a long-term storage, the grids containing serial sections should be placed in grid boxes in a dry environment.

### 3.4. Serial Section Imaging and Analysis

1. Grids containing the series, along with a grating replica grid, are loaded in the specimen chamber of the SEM (c in Fig. 3a) on a 12-grid holder (Fig. 3b).
2. Set up the SEM and the STEM detector for final imaging on the SmartSEM interface: While viewing through the TV camera, adjust working distance between the specimen and final lens to 4–5 mm. Insert the detector and adjust working distance between the detector and the specimen to 4–5 mm. Set the accelerating voltage to 28 kV in high-current mode with a 20 or 30 μm aperture. Switch to viewing with the STEM detector, and align its aperture to the center of field (Fig. 3c, d).

Fig. 3. (**a**) Zeiss SUPRA 40 field emission scanning electron microscope is equipped with a secondary electron detector (s), an in-lens detector (i), and a retractable detector for transmitted electrons (STEM detector; t). The column (g) contains the gun assembly and objective lenses. The specimen chamber door (c) slides open outward with the stage. The SEM is controlled through the SmartSEM interface and console (keyboard and joysticks can be seen in front of the monitors), or the integrated ATLAS system for large-field imaging. (**b**) TV camera view of the specimen chamber showing the arrangement of the final lens, STEM detector, and sample holder clipped onto the stage. Working distance is 4–5 mm between the final lens and the specimen, as well as between the specimen and the detector. (**c**) Low-magnification STEM image of an entire slot grid containing serial sections. Below the sections, the aperture of the STEM detector can be seen (*circle* with *dotted line*), which must to be aligned to the center of the field. Four quadrants (Q) of the detector element are also used for imaging. Imaging mode (normal or inverted) can be set for each detector element on the SmartSEM interface. (**d**) A screenshot image of the SmartSEM interface, showing a low-magnification (×5,450) STEM image of serial sections. The interface has a variety of useful measurement and annotation tools, which are used to navigate to the ROI from the reference corner of each section. The bright circular area seen in the center of the field is the aperture of the STEM detector. The distance between the two horizontal cursor lines is about 96 μm.

In the "STEM control" window, set the central bright-field detector to "normal" and the dark-field quadrants to "inverted" (Fig. 3c). Perform source, aperture, and column alignments, and astigmatism compensation. Set backlash options for rotation and *X*–*Y*, and lock-out beam shift function.

3. Evaluate serial section grids: With the in-lens secondary electron detector at 5 kV, take low-magnification inverted-histogram images of the entire serial sections in each grid (Fig. 2h). These images reveal defects in the sections and Pioloform film (e.g., large folds, tears, holes, and staining artifacts) that would affect the specimen drift rate, as well as alignment and quantitative analyses of the series images. Large tissue structures such as capillaries and cell bodies can also be identified in these low-magnification images.
4. Select regions of interest (ROI) to be imaged: Take images at a higher magnification (covering anticipated size of final image area) of several sections in the center of the series to select potential regions of interest based on experimental criteria. Survey the low-magnification images from the previous step to minimize flaws in the ROI.
5. Define imaging parameters in the "Mosaic Setup and Execution" window of ATLAS interface. This includes: field size (e.g., a single field at 32,000 × 32,000 pixels), pixel size (typically 2 nm), dwell time of the scanning probe (usually 1,200–1,400 ns), auto focus frequency (every tile) and area (FOV ratio = 350% of pixel area), auto brightness frequency (every tile) and area (FOV ratio = 1,200% of pixel area), power options (EHT off when series is complete), and file storage location (this can be on a network).
6. Brightness and contrast settings: When imaging locations are marked in the next step, ATLAS will record not only the location coordinates, but also settings of scan rotation, focus, astigmatism, brightness, and contrast of the image at the time of marking. When serial imaging is initiated, the stage is translated to bring the ROI to the center of the field, and an auto focus routine is carried out starting at the brightness and contrast levels recorded during location marking. During the auto focus routine (as of version 3.5.2.385 of ATLAS), the SEM repeatedly scans the center of the ROI (area size specified in the "Mosaic Setup and Execution" window in step 5), which causes brightening of the scanned area relative to the rest of ROI. Therefore, if the image histogram was already equalized at the time of location marking, the auto focus scan area might become completely bleached. This causes the auto focus to fail (because there would be no contrast to focus on) and/or renders a small area at the center of final images unusable. In order to avoid this, the initial brightness setting needs to be estimated

for the focus scan area (i.e., the entire image will be darkened), such that the area would retain gray pixels during the auto focus routine. Also determine the detector contrast setting for the series by imaging the first section somewhere outside of the ROI at final dwell time, while monitoring the live histogram on ATLAS (histogram seen in Fig. 4a).

7. Mark stage locations for serial imaging. In the "Mosaic Setup and Execution" window, click "Do Stage Backlash" to perform full backlash before beginning, and make sure to select "Auto-backlash on add." In the SmartSEM interface, use angle and distance from the reference corner to navigate to the center of the ROI (Fig. 3d). Use scan rotation to keep each ROI in the same orientation. Then use the ATLAS interface to add the location in the "Mosaic Setup and Execution" window. Sequentially navigate to the same ROI in the following sections to add all imaging locations. Make sure to include the grating replica grid in the series so that the pixel size can be verified for the subsequent image analyses in RECONSTRUCT software (see Notes 29–32).
8. Click "Execute" in the "Mosaic Setup and Execution" window to start imaging.
9. Initial evaluation of serial images: Import the images into FIJI software, and use the TrakEM2 plug-in to apply rigid alignment on the series (32). Check for defects such as stage translation errors and soft focus, and determine sections to be reimaged. Figure 4b illustrates one such section as viewed in TrakEM2.
10. Retake images (typically 5% of all images) and insert them into the series in FIJI/TrakEM2. Realign the new images manually and propagate the new rigid alignment to the rest of the series (or simply rerun the rigid alignment on the entire series).
11. Project the entire series (down-sampled to a larger pixel size) in FIJI (Fig. 4c), or flip through the serial images in TrakEM2, to locate and determine the size of the area to be cropped. The middle part of the series (e.g., 100 middle sections in a 200-section series) should contain very few or no defects (e.g., folds and tears).
12. Stack-crop the series in FIJI/TrakEM2 to desired size, shape, and location.
13. Use TrakEM2 to apply affine and elastic alignment on the cropped series and equalize the histograms.
14. Import the images in the RECONSTRUCT software and manually fine-tune alignment. Calculate pixel size using the grating replica image. Also estimate the section thickness using the cylindrical mitochondria method (37) (see Note 33).

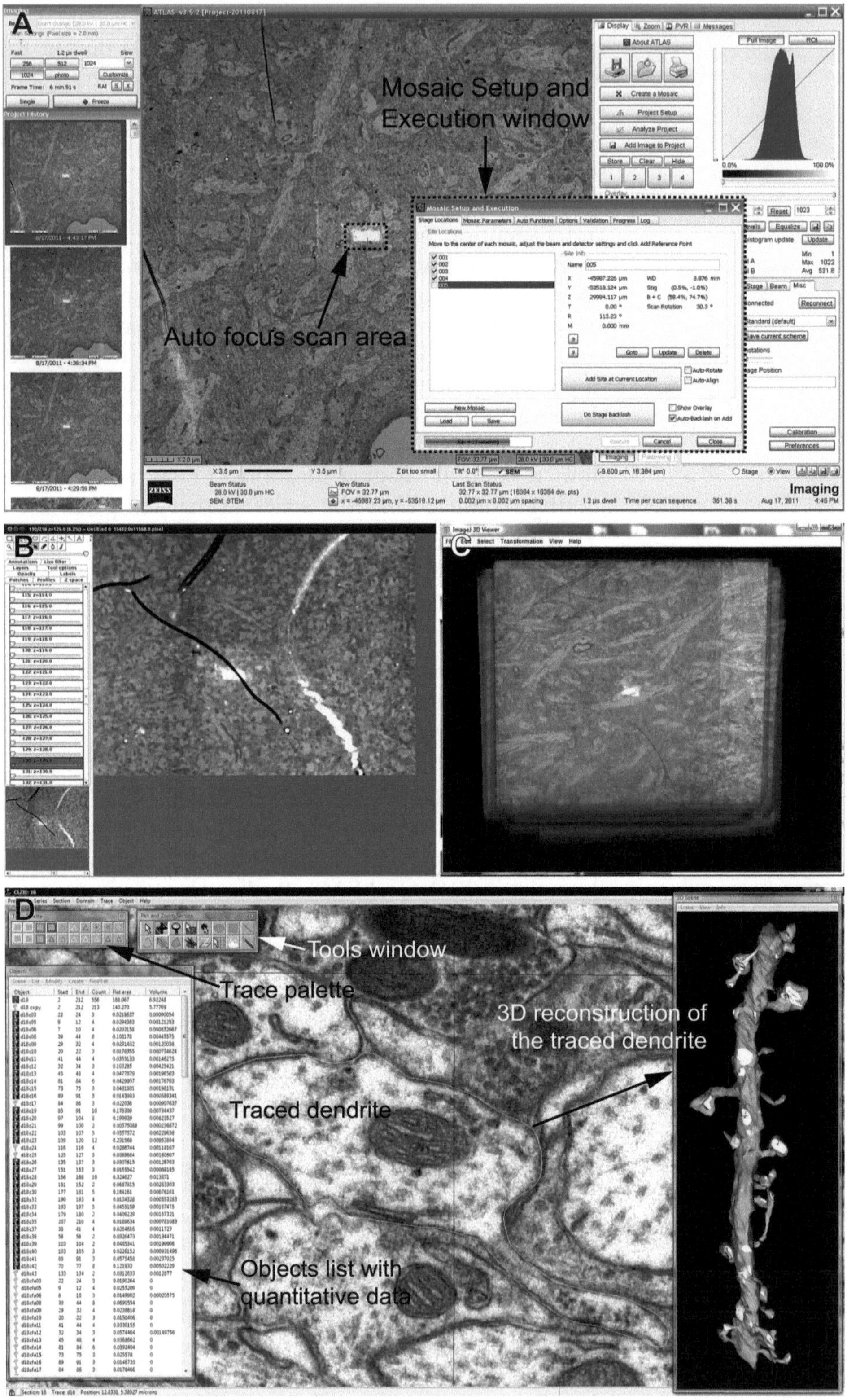
A
Mosaic Setup and Execution window
Auto focus scan area
B
C
D
Tools window
Trace palette
3D reconstruction of the traced dendrite
Traced dendrite
Objects list with quantitative data

15. Use the RECONSTRUCT software to trace structures of interest (e.g., dendrites, postsynaptic density, synaptic vesicles, etc.) and generate 3D reconstructions and quantitative data of traced structures (e.g., area of the PSD) (Fig. 4d). The quantitative data are exported as CSV files for further statistical analysis. 3D reconstructions of the segmented structures can be exported as VRML or XDF files for further rendering and modeling.

## 4. Notes

1. This perfusion method can be adopted for juvenile rats or adult mice. Due to their small size, tracheotomy and artificial ventilation are not performed. Therefore, time between cutting the diaphragm (i.e., loss of breathing) and perfusion with fixative must be as short as possible (30 s or less) to minimize hypoxic damage to the brain tissue (33). The maximum pressure of 120 mmHg should be sufficient, instead of 180 mmHg. Use a 16-gauge needle for perfusion.
2. The use of halothane has been discontinued in the U.S., in favor of isoflurane.
3. If a vaporizer and ventilator are not available, prepare a nose cone by placing a gauze pad into a 50 ml conical tube, to which ~0.5 ml of halothane or isoflurane is added. This is used to maintain anesthesia during the perfusion procedure. As in Note 1, time between cutting the diaphragm and fixative perfusion must be as short as possible to minimize hypoxic damage.
4. To avoid precipitation during KRC preparation, $Na_2CO_3$ must first be dissolved in ~100 ml of purified water and added to the buffer slowly (~2 ml at a time) while maintaining pH between 7.5 and 8 with 1 N aqueous solution of HCl.
5. When stored at 4°C, glutaraldehyde in the stock solution forms large polymers over time, which reduces its ability to penetrate the tissue during perfusion. These polymers also react with

Fig. 4. (**a**) A screenshot image of the ATLAS interface during serial imaging. The Mosaic Setup and Execution window is used to set up a run of serial imaging (see step 5 of Subheading 3.4), as well as to monitor the progress. Auto focus scan area can also be seen as a *bright rectangle* in center of the ROI, which is about 32 μm × 32 μm for this image. (**b**) A screenshot image of TrakEM2 after rigid alignment of a serial section series. Each section is represented as a "Layer" on the left side of the screen. The main window displays an image of the current layer, which in this case is one with soft focus because the auto focus scan fell onto the fold (see Note 29). The sections from which such suboptimal images were taken are then reimaged for replacements. (**c**) An entire serial section series is viewed in FIJI (3D Viewer) software after rigid alignment with the TrakEM2 plug-in. FIJI is also used to project the stack in order to identify and crop areas with minimal flaws. (**d**) A screenshot image of RECONSTRUCT software, which is used to fine-tune alignment, trace structures of interest in the serial images, and generate quantitative data and 3D reconstructions.

osmium tetroxide to generate pepper-like precipitates. To prevent these artifacts, the stock solution should be placed at RT for 3 days prior to preparing the fixative solution.

6. For each run of tissue processing, prepare fresh working solutions of osmium tetroxide from a stock solution (4% aqueous) sealed in glass ampoules under dry nitrogen gas. The stock solution has a pale yellow color. Discard if it appears purple.
7. Because ethanol readily absorbs atmospheric water, a fresh bottle of 100% ethanol should be used for each run of tissue processing that requires dry ethanol.
8. All glassware for lead citrate preparation must be as clean as possible to prevent precipitation. 10% aqueous solution of nitric acid can be used for this purpose, followed by extensive rinses with purified water.
9. The FIJI software could also be used for segmentation and 3D reconstruction of neuropil. Tutorials on FIJI/TrakEM2 are available on the FIJI website (http://fiji.sc/).
10. For all procedures before EM imaging (except for ultramicrotomy), wear appropriate protective clothing. All hazardous waste is contained and disposed of according to local regulations.
11. Perfusion-fixation and tissue processing procedures are performed in a well-ventilated area. For perfusion-fixation, an intake for the air ventilation system can be extended to the work surface so as to avoid unnecessary exposure to fumes from the anesthetic and aldehydes. A necropsy table with down-draft would be optimal for perfusion-fixation. Alternatively, all procedures can be performed in a chemical fume hood.
12. Aside from the cut chin, other signs of a good perfusion are: (1) the feet and tail are pale white, and (2) the tail is stiff down to its tip.
13. Do not allow the fixative bottle to become empty because that could cause air to be perfused into the brain.
14. A well-perfused brain will appear light-gold and have no hematomas or blood in the vessels. A poorly perfused one will have a pink tinge and/or blood apparent in the vessels or hematomas.
15. Time between the end of perfusion and epoxy embedding should be as short as possible to achieve good ultrastructural preservation.
16. When slicing the fixed brain with a vibrating blade microtome, the thickness should not exceed 100 μm in order to achieve optimal penetration of osmium tetroxide solutions during the subsequent processing.
17. Although perfusion-fixed brain tissue is used as an example in this chapter, the same tissue processing method can also be

used for brain slice preparations after microwave-enhanced chemical fixation (13, 34, 38).

18. Hardness of epoxy resin and, therefore, the cutting property of blocks can be adjusted by changing the amounts of monomer components. This is described in the manufacturer's instructions, as well as by Luft (39). Blocks that are "hard", as described in the instructions, seem to produce serial sections of good quality.
19. A 35° diamond knife with a larger water vessel (DiATOME ultra 35 Jumbo) can be used to collect a longer series of serial sections.
20. A detailed description of serial sectioning (including video) is also available in Harris et al. (35) and on our website (http://synapses.clm.utexas.edu/).
21. Select a day with low humidity or use a room with a dehumidifier to coat grids. High humidity seems to diminish film quality.
22. Rinse the film casting device with chloroform and air-dry overnight before use. Also clean the grids by sonicating them in ethanol, then in three exchanges of chloroform (5–10 min each). Spread the grids in a single layer on a petri dish lined with filter paper to air-dry completely.
23. Concentration or drain rate of the Pioloform solution can be adjusted to optimize the film thickness.
24. The wider the block face is, the more likely it is to introduce wrinkles and tears in the serial sections when they are picked up on grids.
25. The block face with its height in this range (20–30 μm) is more stable and provides more uniform sections. The block can be re-trimmed to acquire longer series.
26. Using a DiATOME ultra 35 knife reduces tissue compression compared to a ultra 45 knife (35).
27. Typically, a series of 200 serial sections of this size (70 μm × 200 μm) will require about 10 grids.
28. A small piece of Parafilm can be used to keep the NaOH pellets from coming into contact with the grids in the staining dish.
29. It is useful to check on serial image acquisition after a day or overnight to catch and rectify problems, so calculate imaging time and number of images accordingly. These problems include changes in astigmatism and objective axis alignment.
30. Avoid a fold, tear, or artifact in the center of ROI because auto focus may fail if these flaws are included in the focus area (Fig. 4b).
31. The marked location can be saved on the computer running ATLAS. This can be useful for navigating back to sections that need to be reimaged later (see step 10).

32. Imaging at 16-bit extends the gray levels so that areas of tissue obscured by auto focus (i.e., brightened) or folds in Pioloform film (i.e., darkened) may still be recovered by image processing. This also increases the file size of each image, but the rest of image processing (e.g., alignment) and analysis can be performed after converting the images to 8-bit.
33. This method uses the ratio of the maximum diameter of longitudinally sectioned mitochondria (or other cylindrical objects) to the number of serial sections they span (37). The section thickness can also be estimated directly with a color 3D laser confocal microscope (40). Never use the nominal section thickness (i.e., the feed setting on the ultramicrotome) to calibrate a serial section series in RECONSTRUCT because the actual thickness almost always shows slight variations from it.

## Acknowledgment

We thank Drs. Cliff Abraham and Jared Bowden for tissue samples from which images in the Figs. 2, 3, and 4 were taken. We also thank Laurence Lindsey for help with computational tools and Patrick Parker for help in preparing the manuscript. This work was supported by the US National Institutes of Health grants NS021184, NS033574, and EB002170 to K.M.H. and the Texas Emerging Technology Fund.

### References

1. Gray EG (1959) Electron microscopy of synaptic contacts on dendrite spines of the cerebral cortex. Nature 183:1592–1593
2. Palay SL, Palade GE (1955) The fine structure of neurons. J Biophys Biochem Cytol 1:69–88
3. Novikoff PM, Novikoff AB, Quintana N, Hauw JJ (1971) Golgi apparatus, GERL, and lysosomes of neurons in rat dorsal root ganglia, studied by thick section and thin section cytochemistry. J Cell Biol 50:859–886
4. Spacek J, Lieberman AR (1974) Ultrastructure and three-dimensional organization of synaptic glomeruli in rat somatosensory thalamus. J Anat 117:487–516
5. Thaemert JC (1966) Ultrastructural interrelationships of nerve processes and smooth muscle cells in three dimensions. J Cell Biol 28:37–49
6. Williams RC, Kallman F (1955) Interpretations of electron micrographs of single and serial sections. J Biophys Biochem Cytol 1:301–314
7. Stevens JK, Davis TL, Friedman N, Sterling P (1980) A systematic approach to reconstructing microcircuitry by electron microscopy of serial sections. Brain Res 2:265–293
8. Bock DD, Lee WCA, Kerlin AM et al (2011) Network anatomy and in vivo physiology of visual cortical neurons. Nature 471:177–182
9. Briggman KL, Helmstaedter M, Denk W (2011) Wiring specificity in the direction-selectivity circuit of the retina. Nature 471:183–188
10. Anderson JR, Jones BW, Watt CB et al (2011) Exploring the retinal connectome. Mol Vis 17:355–379
11. Helmstaedter M, Briggman KL, Denk W (2008) 3D structural imaging of the brain with photons and electrons. Curr Opin Neurobiol 18:633–641
12. Mishchenko Y, Hu T, Spacek J, Mendenhall J, Harris KM, Chklovskii DB (2010) Ultrastructural analysis of hippocampal neuropil from the connectomics perspective. Neuron 67:1009–1020

13. Bourne JN, Harris KM (2011) Coordination of size and number of excitatory and inhibitory synapses results in a balanced structural plasticity along mature hippocampal CA1 dendrites during LTP. Hippocampus 21:354–373
14. Knott GW, Holtmaat A, Wilbrecht L, Welker E, Svoboda K (2006) Spine growth precedes synapse formation in the adult neocortex in vivo. Nat Neurosci 9:1117–1124
15. Ostroff LE, Fiala JC, Allwardt B, Harris KM (2002) Polyribosomes redistribute from dendritic shafts into spines with enlarged synapses during LTP in developing rat hippocampal slices. Neuron 35:535–545
16. Cantoni M, Knott G, Hébert C (2010) FIB-SEM nanotomography in materials and life science at EPFL. Microsc Microanal 16:226–227
17. Cantoni M, Genoud C, Hébert C, Knott G (2010) Large volume, isotropic, 3D imaging of cell structure on the nanometer scale. Microsc Anal 24:13–16
18. Knott G, Marchman H, Wall D, Lich B (2008) Serial section scanning electron microscopy of adult brain tissue using focused ion beam milling. J Neurosci 28:2959–2964
19. Knott G, Rosset SP, Cantoni M (2011) Focussed ion beam milling and scanning electron microscopy of brain tissue. J Vis Exp 53:e2588. DOI: 10.3791/2588
20. Denk W, Horstmann H (2004) Serial block-face scanning electron microscopy to reconstruct three-dimensional tissue nanostructure. PLoS Biol 2:e329
21. Hayworth KJ, Kasthuri N, Schalek R, Lichtman JW (2006) Automating the collection of ultrathin serial sections for large volume TEM reconstructions. Microsc Microanal 12:86–87
22. Micheva KD, Smith SJ (2007) Array tomography: a new tool for imaging the molecular architecture and ultrastructure of neural circuits. Neuron 55:25–36
23. Mendenhall JM, Yorston J, Lagarec KG, Bowden J, Harris KM (2009) Large volume high resolution imaging of brain neuropil using SEM-based scanning electron microscopy. Program No. 484.17. 2009 Neuroscience Meeting Planner. Society for Neuroscience, Chicago, IL (Online)
24. Anderson JR, Mohammed S, Grimm B et al (2011) The Viking viewer for connectomics: scalable multi-user annotation and summarization of large volume data sets. J Microsc 241:13–28
25. Chklovskii DB, Vitaladevuni S, Scheffer LK (2010) Semi-automated reconstruction of neural circuits using electron microscopy. Curr Opin Neurobiol 20:667–675
26. Fiala JC (2005) Reconstruct: a free editor for serial section microscopy. J Microsc 218:52–61
27. Helmstaedter M, Briggman KL, Denk W (2011) High-accuracy neurite reconstruction for high-throughput neuroanatomy. Nat Neurosci 14:1081–1088
28. Jain V, Seung HS, Turaga SC (2010) Machines that learn to segment images: a crucial technology for connectomics. Curr Opin Neurobiol 20:653–666
29. Knowles-Barley S, Butcher NJ, Meinertzhagen IA, Armstrong JD (2011) Biologically inspired EM image alignment and neural reconstruction. Bioinformatics 27:2216–2223
30. Lang S, Drouvelis P, Tafaj E, Bastian P, Sakmann B (2011) Fast extraction of neuron morphologies from large-scale SBFSEM image stacks. J Comput Neurosci 31(3):533–545
31. Morales J, Alonso-Nanclares L, Rodriguez JR et al (2011) Espina: a tool for the automated segmentation and counting of synapses in large stacks of electron microscopy images. Front Neuroanat 5:18
32. Saalfeld S, Cardona A, Hartenstein V, Tomancak P (2010) As-rigid-as-possible mosaicking and serial section registration of large ssTEM datasets. Bioinformatics 26:i57–i63
33. Tao-Cheng JH, Gallant PE, Brightman MW, Dosemeci A, Reese TS (2007) Structural changes at synapses after delayed perfusion fixation in different regions of the mouse brain. J Comp Neurol 501:731–740
34. Feinberg MD, Szumowski KM, Harris KM (2001) Microwave fixation of rat hippocampal slices. In: Giberson RT, DeMaree RSJ (eds) Microwave techniques and protocols. Humana Press, Totowa, pp 75–88
35. Harris KM, Perry E, Bourne J, Feinberg M, Ostroff L, Hurlburt J (2006) Uniform serial sectioning for transmission electron microscopy. J Neurosci 26:12101–12103
36. Reynolds ES (1963) The use of lead citrate at high pH as an electron-opaque stain in electron microscopy. J Cell Biol 17:208–212
37. FialaJ C, Harris KM (2001) Cylindrical diameters method for calibrating section thickness in serial electron microscopy. J Microsc 202: 468–472
38. Jensen FE, Harris KM (1989) Preservation of neuronal ultrastructure in hippocampal slices using rapid microwave-enhanced fixation. J Neurosci Methods 29:217–230
39. Luft JH (1961) Improvements in epoxy resin embedding methods. J Biophys Biochem Cytol 9:409–414
40. Kubota Y, Hatada S, Kawaguchi Y (2009) Important factors for the three-dimensional reconstruction of neuronal structures from serial ultrathin sections. Front Neural Circuits 3:4

# Chapter 16

# 3D Imaging of Cells and Tissues by Focused Ion Beam/Scanning Electron Microscopy (FIB/SEM)

**Damjana Drobne**

## Abstract

Integration of a scanning electron microscope (SEM) and focused ion beam (FIB) technology into a single FIB/SEM system permits use of the FIB as a nano-scalpel to reveal site-specific subsurface microstructures which can be examined in great detail by SEM. The FIB/SEM technology is widely used in the semiconductor industry and material sciences, and recently its use in the life sciences has been initiated. Samples for FIB/SEM investigation can be either embedded in a plastic matrix, the traditional means of preparation of transmission electron microscopy (TEM) specimens, or simply dried as in samples prepared for SEM imaging. Currently, FIB/SEM is used in the life sciences for (a) preparation by the lift-out technique of lamella for TEM analysis, (b) tomography of samples embedded in a matrix, and (c) *in situ* site-specific FIB milling and SEM imaging using a wide range of magnifications. Site-specific milling and imaging has attracted wide interest as a technique in structural research of single eukaryotic and prokaryotic cells, small animals, and different animal tissue, but it still remains to be explored more thoroughly. In the past, preparation of samples for site-specific milling and imaging by FIB/SEM has typically adopted the embedding techniques used for TEM samples, and which have been very well described in the literature. Sample preparation protocols for the use of dried samples in FIB/SEM have been less well investigated. The aim of this chapter is to encourage application of FIB/SEM on dried biological samples. A detailed description of conventional dried sample preparation and FIB/SEM investigation of dried biological samples is presented. The important steps are described and illustrated, and direct comparison between embedded and dried samples of same tissues is provided. The ability to discover links between gross morphology of the tissue or organ, surface characteristics of any selected region, and intracellular structural details on the nanometer scale is an appealing application of electron microscopy in the life sciences and merits further exploration.

**Key words:** Scanning electron microscopy, FIB-milling, Sample preparation for electron microscopy

## 1. Introduction

A most compelling feature of FIB/SEM instrumentation has been the integration of the scanning electron microscope and focused ion beam technology into a single FIB/SEM instrument. Focused

Alioscka A. Sousa and Michael J. Kruhlak (eds.), *Nanoimaging: Methods and Protocols*, Methods in Molecular Biology, vol. 950, DOI 10.1007/978-1-62703-137-0_16, © Springer Science+Business Media, LLC 2013

ion beam (FIB) techniques are very important tools for nanostructuring of surfaces and microstructure characterization in the semiconductor industry and material sciences (1). As early as the 1960s FIB instruments were also used to erode biological material and expose subsurface structures (2–5), but difficulties in the interpretation of the results and uncertainties about the side effects of ion manipulation placed significant limitations on the broad application of ion etching in biology.

In the past several years, however, FIB/SEM systems have become uniquely attractive for the nanoscale investigation of biological structures. FIB/SEM is currently used in the life sciences for (a) TEM sample preparation of lamella by a lift-out technique, (b) FIB–SEM tomography of embedded samples, and (c) *in situ* site-specific exposure of subsurface structures by FIB milling and imaging in a wide range of magnifications. The FIB/SEM technology enables investigation of biological samples embedded in plastic, such as those used routinely for transmission electron microscopy (TEM) imaging (6). FIB/SEM can also be performed on dried samples that are typically prepared for conventional SEM imaging (7, 8).

In a FIB device, samples are irradiated with $Ga^{3+}$ ions produced by a liquid metal ion source (LMIS) and accelerated to an energy of 30–50 keV. The ion beam is governed by different parameters including the beam profile, the angle of incidence, ion species, ion dose, and ion energy. The beam profile has a core with the Gaussian type of current-density distribution and the beam diameter may be as small as 5–7 nm (9). The ion beam is focused and rastered across a sample surface by a set of electrostatic lenses, and ions striking the surface with a very high momentum cause some atoms and ions to be sputtered away (9, 10).

FIB milling of a selected region is followed by SEM imaging. The emitted signal in conventional SEM is primarily dependent upon surface topography and on the atomic number of elements on the surface. Topographic contrast is proportional to the intensity of secondary electrons ejected from the sample. In samples embedded in plastic, a backscattered electron (BSE) detector is used to produce an image of the surface. Plastic embedded samples have to be stained with heavy metals, a process known as *en bloc* staining, because the intensity of the BSE signal is highly dependent upon the atomic number (Z) of the specimen.

One of the most important applications of a FIB/SEM system is preparation of samples for transmission electron microscopy by an *in situ* lift-out technique, which involves cutting out electron-transparent slices from specific parts of a resin embedded sample and lifting them out for TEM study. The FIB enables the preparation of large, site specific samples of uniform thickness. This FIB application for biomaterials has been reviewed by Grandfield and Engqvist (*in press*).

Serial FIB and SEM imaging of resin-embedded samples for 3D tissue reconstruction is the more common application of FIB/SEM in the life sciences. By sequential milling of embedded samples and BSE imaging, a large number of images can be generated and combined into a 3D image of a biological tissue. A comprehensive description of sample preparation, FIB/SEM operation performed on a resin-embedded biological sample and 3D image reconstruction has been provided recently by Bushby et al. (11), reporting that with such an approach it is possible, in an automated fashion, to acquire several hundred tomographic SEM images of slices a few nanometers in thickness in the span of less than 1 h. The process begins by defining the volume that is to be FIB sectioned. Volumes are typically 200–500 $\mu m^3$, which translates typically into dimensions of 10 μm wide, 5 μm deep, and 10 μm in height. Suitable FIB milling currents range from 10 to 100 pA. Once milling of the volume is started, the entire process is recorded continuously by real-time capture of SEM images. For example, frames are captured every 10–20 s with scan conditions that result in a very good signal-to-noise ratio (11). The ability to acquire high resolution tomographic SEM slices rapidly at the nanoscale contributes significantly to the practical implementation of FIB/SEM in ultratomography.

The *in situ* site-specific FIB milling of a specimen prepared for conventional SEM and simultaneous imaging of subsurface structures in a wide range of magnifications offers another possibility for the nanoscale investigation of biological structures. Simultaneous imaging of intracellular structures with gross morphology of tissues offers an attractive approach with which to expand sample surface investigations by subsurface structural research at locations of interest. There are many published reports on applications of FIB/SEM on disparate biological samples including single cells (prokaryotic, eukaryotic), multicellular clumps, small arthropods and animals, and different types of tissue.

The aim of this chapter is to describe in detail the proper sample preparation protocols for the visualization of subsurface structures when investigated by FIB/SEM. This will help potential FIB/SEM users to confidently adapt the conventional protocols for each new biological sample type. Examples of FIB/SEM investigations of biological samples will be presented and discussed, with samples differing in chemical fixation, conductive staining, method of drying, and sputter coating.

## 2. Materials

### *2.1. Chemicals and Supplies (See Note 1)*

1. Sodium Cacodylate Buffer, pH 7.2.
2. Phosphate Buffered Saline (PBS), pH 7.4.
3. Aldehydes for fixation (glutaraldehyde, paraformaldehyde).
4. Postfixation: Osmium tetroxide.

5. Dehydration: Ethanol, Acetone.
6. Contrast enhancement: Uranyl acetate.
7. Conductive staining: Osmium tetroxide and thiocarbohydrazide.
8. Air drying: Hexamethyldisilazane (HMDS).
9. Dehydrating and transition fluids: ethanol, acetonitrile, propylene oxide.
10. Embedment: Plastic embedding media (Spurr's resin or similar).
11. Conventional laboratory glass and plastic equipment.
12. Standard SEM microscope stubs.
13. Double-stick adhesive conductive tape (for example carbon tape).
14. Conductive silver paint.

### 2.2. Instrumentation

1. Binocular dissecting microscope, a stereo microscope, with which to view dissected specimens and for three-dimensional observation and depth perception when handling samples during preparation, such as, mounting samples on SEM microscope stubs.
2. Critical point drier. This is not obligatory because samples can be air dried using hexamethyldisilazane.
3. Oven or UV light for polymerization of resin.
4. Sputter Coating System that provides uniform and controllable thickness of the coated layer; here we have used a gold–palladium sputtered (Sputter coater SCD 050, BAL-TEC, Germany).
5. Field emission gun (FEG) SEM equipped with FIB (Fig. 1). In our investigations we have used either a Strata DB235 (FEI Company, Modena, Italy) or Quanta 3D FEG (FEI Company, Eindhoven, the Netherlands) FIB/SEM systems (see Note 2).

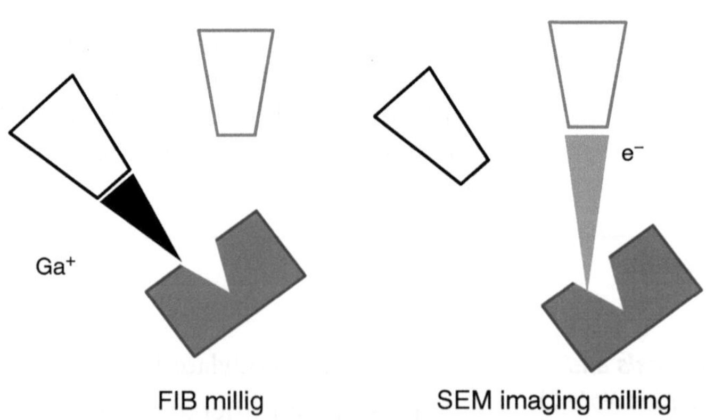


Fig. 1. Schematic illustration of a FIB/SEM system. The scheme shows a sample hit by both ions ($Ga^+$) and electrons ($e^-$) during FIB/SEM operation.

## 3. Methods

### *3.1. Preparation of Dry Samples for FIB Milling and SEM Imaging*

The aim of biological sample preparation procedures is to stabilize and fix the sample to enable imaging. Preparation of dry samples for FIB/SEM operation consists of the following general steps: fixation, contrast enhancement or conductive staining, dehydration with an organic solvent, drying of the solvent, mounting onto specimen stubs, and sputter coating. These steps are described in detail below.

#### *3.1.1. Fixation*

Glutaraldehyde, which is used as a fixative, is efficient in cross-linking proteins and maintaining cell ultrastructure, but penetrates tissues relatively slowly. Formaldehyde penetrates tissues rapidly, apparently due to its low molecular weight. Osmium tetroxide, most commonly used for postfixation, acts both as a fixative and as an electron stain. It preserves many lipids and it is able to stabilize some proteins by transforming them into clear gels without destroying many of their structural features (12).

1. Tissue specimens should be cut into blocks of at least 1 mm in one dimension. Larger specimens will result in incomplete penetration of chemicals, thus reducing the final quality of the results.
2. Fix a tissue block in 0.4% paraformaldehyde and 1% glutaraldehyde in 0.1 M sodium cacodylate buffer (pH 7.2) for 2.5 h at room temperature or overnight at 4°C (see Note 3).
3. Wash 3 times for 10 min in 0.1 M sodium cacodylate buffer (pH 7.2).
4. Post-fix for 1 h at room temperature in 1% osmium tetroxide in distilled water.

#### *3.1.2. Contrast Enhancement or Conductive Staining*

Sample fixation is followed either by *en bloc* staining to enhance contrast or by OTOTO (osmium tetroxide/thiocarbohydrazide/osmium—tetroxide/thiocarbohydrazide/osmium—tetroxide) staining for conductivity. In the OTOTO conductive staining technique, thiocarbohydrazide (T) is used as a ligand. It is bidentate, binding the osmium (O) tetroxide molecules together giving T–O–T. In this manner the natural affinity of osmium for unsaturated lipids is enhanced or amplified. The electrical conductivity imparted to the tissue by this increased deposition of osmium reduces specimen charging during the SEM operation (13).

1. Stain the tissue block for 30 min in a saturated solution of thiocarbohydrazide in distilled water (saturated means the maximum amount of thiocarbohydrazide that can be dissolved in a given volume of water without forming insoluble crystals).
2. Wash three times in distilled water for 10 min each.

3. Stain for 1 h in 1% aqueous osmium tetroxide.
4. Wash three times in distilled water for 10 min each.
5. Stain for 30 min in saturated thiocarbohydrazide in distilled water.
6. Wash three times in distilled water for 10 min each.
7. Stain for 1 h in 1% aqueous osmium tetroxide.
8. Wash three times in distilled water for 10 min each and proceed with dehydration.

#### 3.1.3. Dehydration

1. Wash the tissue in 50% ethanol for 10–15 min.
2. Wash in 70% ethanol for 10–15 min (may be stored overnight at this stage if necessary).
3. Wash in 95% ethanol for 10–15 min.
4. Wash three times in 100% ethanol for 15 min each.
5. Wash three times in 100% acetone for 15 min each.
6. Shake specimen from time to time during the process to ensure full penetration and infiltration of chemicals.

#### 3.1.4. Drying

Critical point drying (CPD) is an automated process that requires approximately 40 min to be completed. It relies on the following physical principle: the liquid in the biological tissue is replaced with a suitable inert fluid whose critical temperature at a realizable pressure is just above the ambient temperature. The choice of fluids is severely limited and despite early work with Freon 13 and $N_2O$, $CO_2$ is now commonly used. With $CO_2$, a critical point of approximately 35°C can be achieved at a pressure of about 1,200 psi. Samples are transferred in 100% acetone to the pressure chamber of a critical point dryer. When the acetone is replaced with liquid $CO_2$ and the temperature is raised to above the critical temperature, the liquid $CO_2$ vaporizes with no change in density. As a consequence, there are no surface tension effects that can distort morphology and ultrastructure. However, another method using hexamethyldisilazane can be used for drying. In our experience, the two drying methods are found to produce a similar appearance of cellular ultrastructure. When the two protocols are compared, we find that HMDS drying takes less effort, has lower costs for chemicals, requires no specialized equipment and has a high rate of success. In our laboratory and elsewhere (14), CPD- and HMDS-dried samples have typically shown equally good results (Fig. 2) (15). The high magnifications in Fig. 2a, b are unusual for SEM investigation of biological samples prepared by drying. This is evidence of satisfactory sample stabilization and conductivity, as well as, of successful FIB operation.

For drying with Hexamethyldisilazane:

1. Incubate tissue block in acetone–HMDS (1:1) for 10 min.
2. Remove the HMDS–acetone and replace with HMDS.

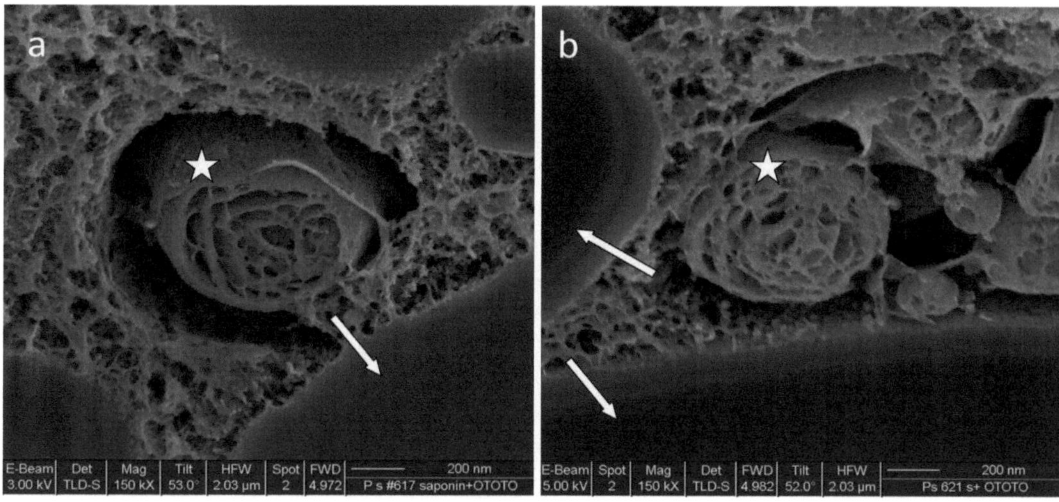


Fig. 2. (**a**) CPD-dried specimen followed by FIB exposure and SEM imaging of the cell interior. Sample was aldehyde-fixed and osmium-postfixed (0.4% paraformaldehyde and 1% glutaraldehyde in 0.1 M sodium cacodylate buffer; 1% osmium tetroxide in distilled water). Homogenous regions correspond to lipid droplets (*arrows*). (**b**) HMDS-dried sample subjected to FIB exposure and SEM imaging of cell interior. Sample was aldehyde-fixed and osmium-postfixed (0.4% paraformaldehyde and 1% glutaraldehyde in 0.1 M sodium cacodylate buffer; 1% osmium tetroxide in distilled water). Note different electron beam voltages (*left* image 3 kV, *right* image 5 kV). *Arrows*—lipid droplets; *asterisk*—lamellar body. Image (**b**) reprinted from ref. 15 with permission.

3. Incubate in HMDS for 10 min.
4. Replace HMDS with fresh HMDS.
5. Incubate in HMDS from 2 to 24 h to allow HMDS to evaporate.

#### 3.1.5. Mounting Dried Samples onto Specimen Stubs

Dried samples are mounted on standard SEM microscope stubs prior to FIB/SEM investigation. Double-stick adhesive tape, conductive silver paint, or both can be used to attach specimens to stubs. Due to the small size of specimens and because sometimes samples are mechanically opened in order to expose their subsurface structure, a stereo microscope with magnification up to 10× is indispensible.

#### 3.1.6. Sputter Coating of Dry Samples

Samples are sputter coated with a layer of carbon or platinum/palladium, or other metals (thickness of ~10–30 nm). Metal atoms ejected from the target by the ionized gas cross the plasma and are deposited onto any surface within the coating unit, including the specimen. A low vacuum is used (0.1–0.05 mbar), which with a modern low voltage sputter coater, enables metal to be deposited at a rate of 1 nm/s. For SEM imaging, when high magnifications are required, sputtering is a very important step that is needed to ensure the thickness and evenness of the cover layer. In Fig. 3a, b we provide examples of different qualities of sample coating.

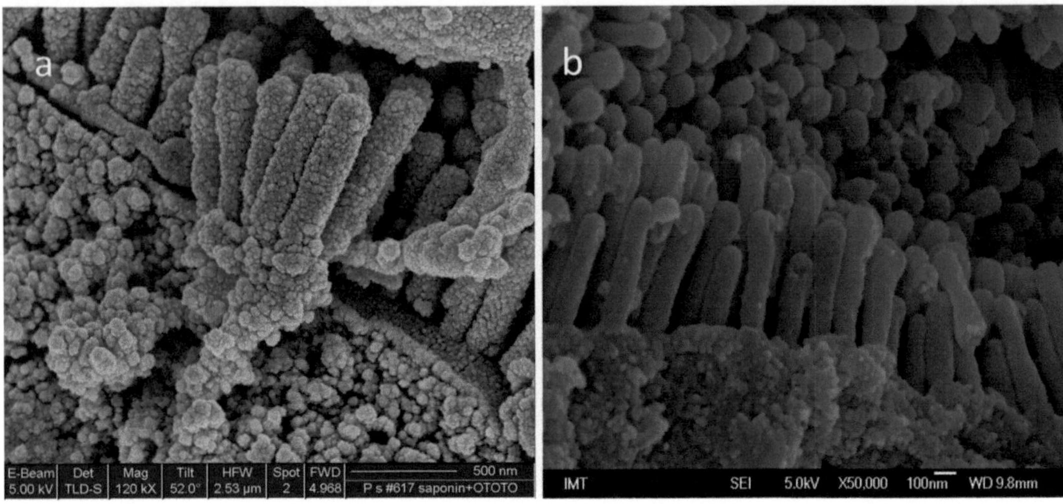


Fig. 3. SEM micrograph of the digestive gland cells prepared for conventional SEM. (**a**) Unsuitable sputter coating when high magnifications are required. Surface deposits are sputtering artifacts. (**b**) Properly coated samples allowing imaging at high magnifications. Image (**b**) taken by M. Hočevar; IMT, Ljubljana, Slovenia.

### *3.2. Sample Preparation for FIB Milling and SEM Imaging by Resin Embedding*

#### *3.2.1. Fixation*

Fixation is the same as described above (Subheading 3.1.1), using a mix of paraformaldehyde and glutaraldehyde in sodium cacodylate buffer.

#### *3.2.2. Staining*

Following fixation, the samples are stained *en bloc* for enhancement of contrast. Uranyl acetate, usually used for staining thin sections for transmission electron microscopy, is used here for *en bloc* staining before dehydration. It also acts as both an electron stain and as a fixative. As a fixative, uranyl acetate stabilizes membranous and nucleic acid-containing structures but also reacts with proteins (12).

1. Wash tissue block at least 3–5 times for 10 min each in distilled water to remove all excess phosphate ions, therefore preventing precipitation of uranyl acetate (UA).
2. *En bloc* stain with 3% aqueous uranyl acetate for 4.5 h at room temperature in the dark to prevent uranyl acetate from being precipitated. This is necessary because UA is photo-reductive.
3. Wash three times in distilled water for 10 min each.

#### *3.2.3. Dehydration*

The dehydration steps are the same as described above (Subheading 3.1.3), using an increasing series of ethanol concentration.

#### *3.2.4. Embedding*

The embedding steps are similar to those commonly used for preparing samples for TEM. To embed our samples we use Spurr's-based resin.

1. Infiltrate for 1 h in 50/50 acetonitrile/Spurr's resin.
2. Remove the 50/50 acetonitrile/Spurr's resin and infiltrate twice in 100% Spurr's resin for 1.5 h each.
3. Place samples in mold and add resin.
4. Embed in plastic using beam capsules or embedding molds, and label each sample.
5. To polymerize, place the samples in a 60–70°C (~65°C) oven for 24–48 h.

#### 3.2.5. Mounting of Resin Embedded Samples onto the Specimen Stubs

Embedded samples are mounted on standard SEM microscope stubs prior to FIB/SEM investigation. The plastic block is cut by a diamond knife in an ultramicrotome until the selected region of a sample is reached. Then the entire block is mounted on the SEM microscope stub. The mounting steps are identical to those used for mounting dried samples (Subheading 3.1.5).

#### 3.2.6. Sputter Coating of Resin Embedded Samples

The sputter coating procedure is not applied to embedded samples. Here, the investigated surface is flat and the sample is conductively stained, therefore no significant charging is typically observed during imaging. In addition, sputtering would cover the region of interest, which is embedded in resin block.

### 3.3. FIB Milling and SEM Imaging

#### 3.3.1. FIB Milling and SEM Imaging Operation

In the FIB/SEM instrument, both beams are aimed at the same point on the specimen surface. A focused high current ion beam of gallium ions ($Ga^{3+}$) is used for site-specific *in situ* sputtering or milling (Fig. 4a). The same beam at low beam currents can be used for polishing (Fig. 4b).

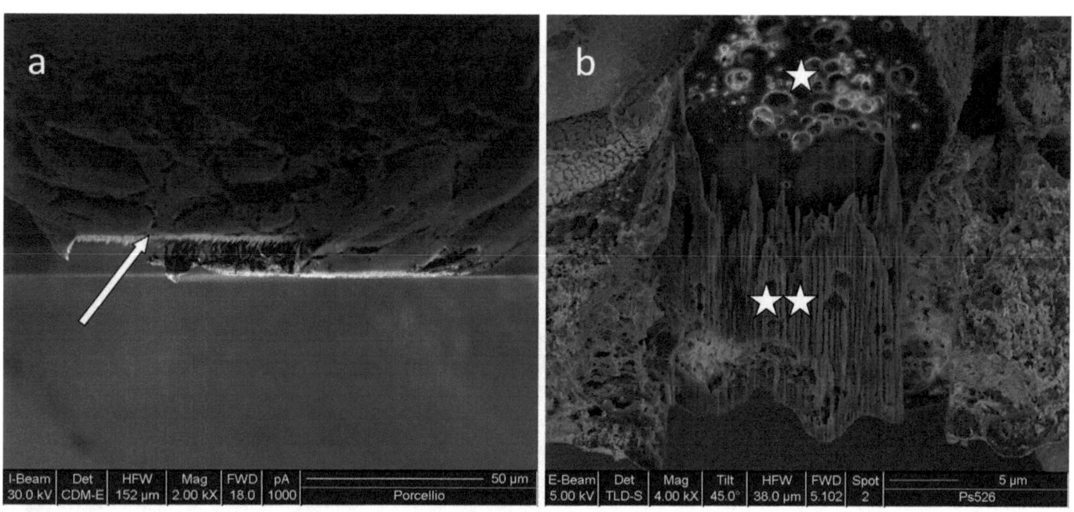


Fig. 4. FIB milling and polishing of biological samples prepared for conventional SEM. (**a**) Top down view from the same direction as the $Ga^{3+}$ ion beam; the region where the $Ga^{3+}$ ions mill the surface is marked with an arrow. (**b**) Polished part of the FIB-milled trench is indicated by one asterisk; the unpolished part is indicated by two asterisks.

In our investigations, dried samples were imaged using a FIB/SEM Strata DB235. For FIB operations (milling and polishing) a gallium ion source was used (7, 15). The ion currents for milling were in the range of 5–7 nA, and 0.3–1.0 nA were used for cleaning. The ion beam accelerating voltage was 30 kV. SEM imaging was performed with the FEG electron column available in the same system, with a specified resolution of 1 nm at 30 kV. The SEM signal was collected by an Everhardt Thornely Detector and processed by a Continuous Dynode Electron Multiplier. The spot size of the SEM probe was in the nanometer range.

Plastic embedded samples that had been previously edge cut with a diamond knife were fixed on brass holders with silver paint (Subheading 3.1.5) and examined by a Quanta 3D FEG. To mill the primary trench, a current of 50 nA was used, and subsequent steps for cleaning across the section were carried out with currents in the range of 1–15 nA. A beam current of 300 pA was used for the final polishing. The backscattered electron (BSE) signal used to image the cross section was collected by mean of a solid state detector placed below the final lens pole piece. A typical cross section was obtained by a sample surface oriented perpendicular to the ion beam (sample surface tilted at 52°); in this condition the take off angle for the BSE signal collection from the cross section (90° to the sample surface—38° with respect to the E-beam) is not optimal. To increase the collection yield the cross section was milled with the sample surface at 0°, allowing for a better orientation of the sidewall (now the cross section surface is at 52° with respect to the E-beam) towards the solid state back scattered electron detector (SSBSED).

The signals from the sputtered secondary ions (Figs. 4a and 5a) or secondary electrons (Figs. 4b and 5b for comparison) are collected

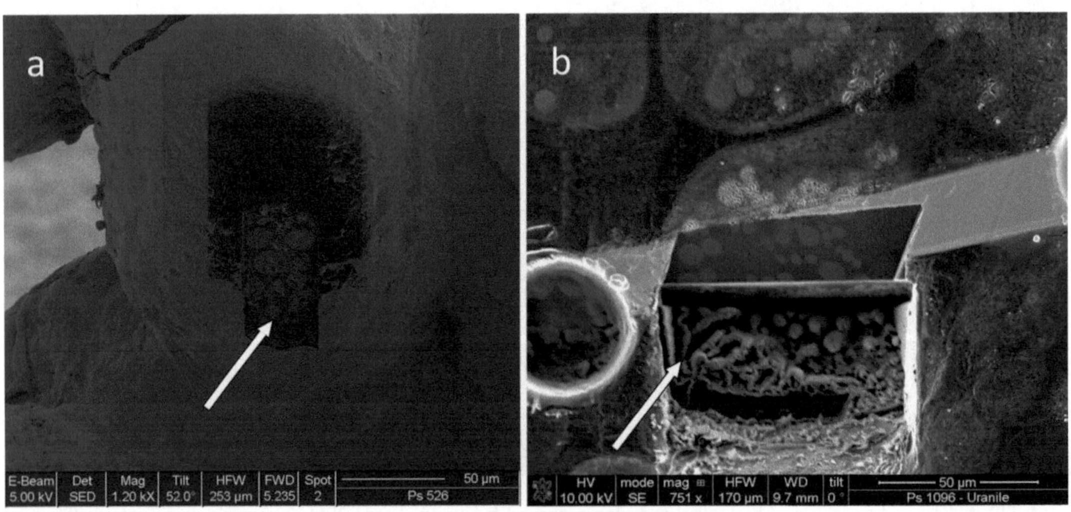


Fig. 5. (**a**) FIB milling and SEM imaging of dry sample prepared for conventional SEM. (**b**) FIB milling and SEM imaging of resin embedded sample as prepared for TEM. *Arrows* indicate FIB milled regions. Image (**b**) is reprinted from ref. 15 with permission.

to form an image. The secondary electrons generated by the primary ion beam provide an image (Fig. 5b), which is used to control the machining process. Ions generated by primary ions can also be collected to form an image (Fig. 4a).

The machining and imaging process for both dry and embedded samples includes the following general steps (see Note 4):

1. Before milling of either dry or embedded samples, a platinum evaporation source is used to make a 1 μm thick coat on the sample surface for its protection during ion milling.
2. Mill the sample with ion beam current ranging from 5 to 7 nA when operation is conducted with Strata DB235 or 50 nA ion beam current when it is done with Quanta 3D FEG.
3. For cleaning mill, an ion beam current between 0.3 and 1.0 nA (Strata DB235; Fig. 5a) or 1–15 nA (Quanta 3D FEG; Fig. 5b) is used.
4. The final step of polishing with Quanta 3D FEG is done using a current of 300 pA.
5. SEM imaging with the Strata DB235 system can be performed with an Everhardt Thornely etector, a Continuous Dynode Electron Multiplier, or a BSE Detector. Imaging with Quanta 3D FEG is performed using a BSE Detector.

#### 3.3.2. Examples of FIB Milling and SEM Imaging of Biological Samples

FIB milling and conventional SEM imaging of dried samples provides structural information on tissue (Fig. 6a, b) and reveals some morphological characteristics of the cell interior (Fig. 6c, d).

In dried samples, SEM contrast is based on topographical contrast. The emitted signal is related primarily to surface topography, but there are also other interactions that increase or minimize contrast from the image. Traditional SEM sample preparation protocols were not designed for subcellular investigation and, consequently, little is known of the appearance of the cellular elements when the cells are opened by FIB milling and imaged by SEM (see Note 5).

A certain similarity is observed between dried and embedded samples that are processed for FIB/SEM. This is true, however, only for particular structures such as lamellar bodies, which exhibit a characteristic shape (Fig. 7a, b). It seems lamellar cellular structures can be, in fact, investigated more effectively with FIB/SEM of dried samples than by TEM or by FIB/SEM of embedded samples (17). Generally, FIB/SEM samples prepared for conventional scanning electron microscopy are suited for the characterization of intracellular morphological features with membranous or lamellar appearance, or structures whose homogeneity or density are different than the remainder of the cell (8). With our present state of the knowledge, however, FIB/SEM of dried samples does not allow unambiguous recognition of cellular organelles, unless they are prepared in parallel for conventional TEM, embedded in plastic,

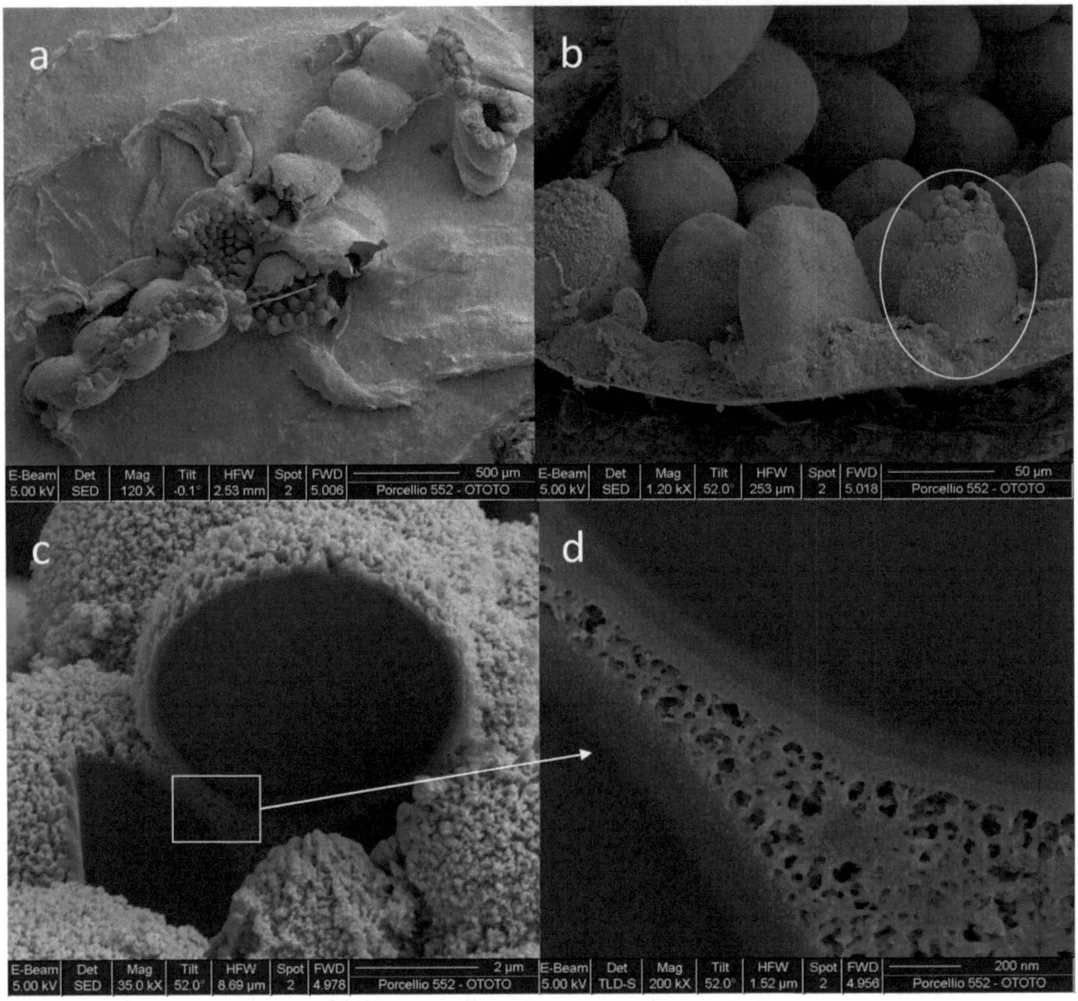


Fig. 6. (**a**) SEM micrograph of tissue gross morphology digestive gland tubes (terrestrial isopod, *Porcellio scaber*). (**b**) Detailed view of digestive gland cells; one cell is encircled. (**c**) FIB milled lipid droplets on the cell surface. (**d**) Detail from the cell interior from the region where two lipid droplets touch.

FIB milled and imaged by backscattered electrons or investigated by TEM. A task for future FIB/SEM investigations of dried samples will be to acquire knowledge and expertise on the recognition of intracellular elements when investigated by FIB/SEM.

The best preparation for FIB/SEM investigation of cell ultrastructure has proved to be aldehyde fixation, uranyl acetate staining and plastic embedment. Here, the backscattering contrast is used to visualize the heavy-metal staining of tissue, prepared using techniques that are routine for TEM. In plastic-embedded samples a compositional contrast is recorded while in dried OTOTO-processed samples a topographical contrast is useful. The conventional TEM sample preparation procedure is not the best choice when samples are dried. FIB/SEM is, therefore, much more

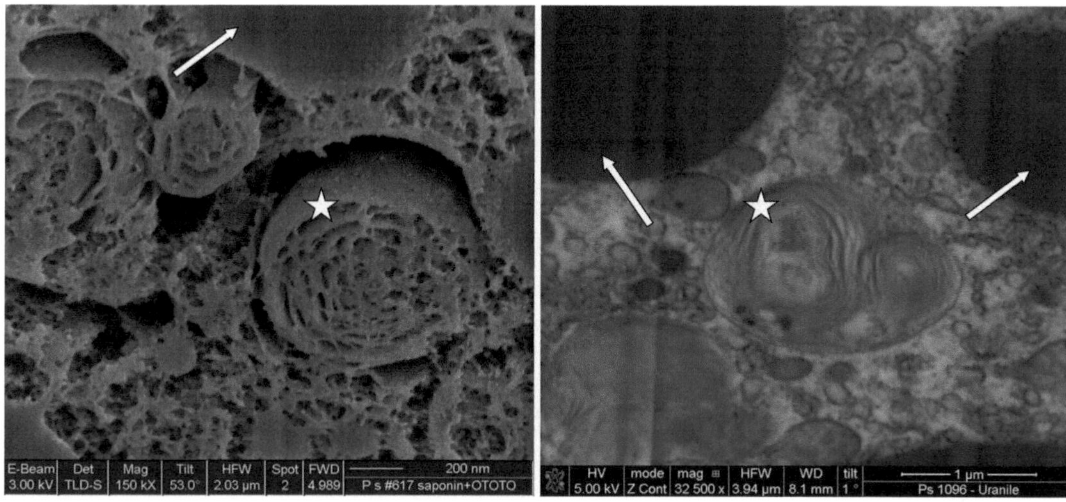


Fig. 7. Comparison between (**a**) SEM micrograph of FIB milled epithelial cells exhibiting lamellar structure prepared for conventional SEM and (**b**) lamellar structure as seen in embedded FIB-milled and BSE-imaged tissue. *Arrows*—lipid droplets; *asterisk*—lamellar body.

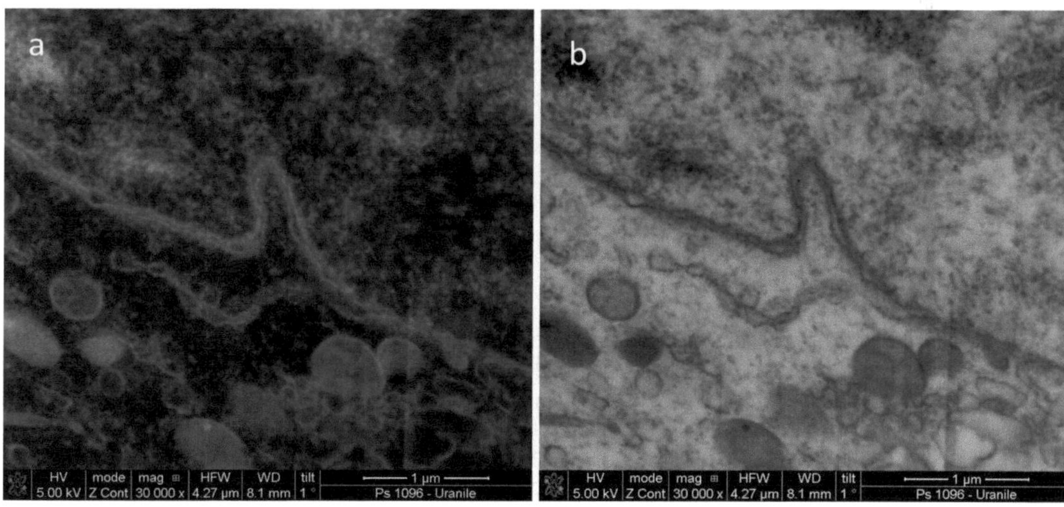


Fig. 8. Electron micrograph of FIB milled area of embedded sample. (**a**) Reverse contrast of back-scattered image. (**b**) Same image as (**a**) in positive contrast. Reprinted from ref. 15 with permission.

popular with, and more frequently applied to, plastic embedded biological samples than to dried samples (17, 18). SEM imaging of plastic-embedded samples enables recognition of cellular organelles and a straightforward comparison with TEM images (Fig. 8).

#### *3.3.3. FIB Milling and SEM Imaging of Biological Samples for 3D Tomography*

For 3D tomography of embedded samples, serial FIB milling (sectioning) operations are followed by SEM imaging. FIB sectioning is performed perpendicular to the sample surface.

The FIB milling and SEM imaging of embedded samples can be used for 3D tomography (18) of larger scale structures like

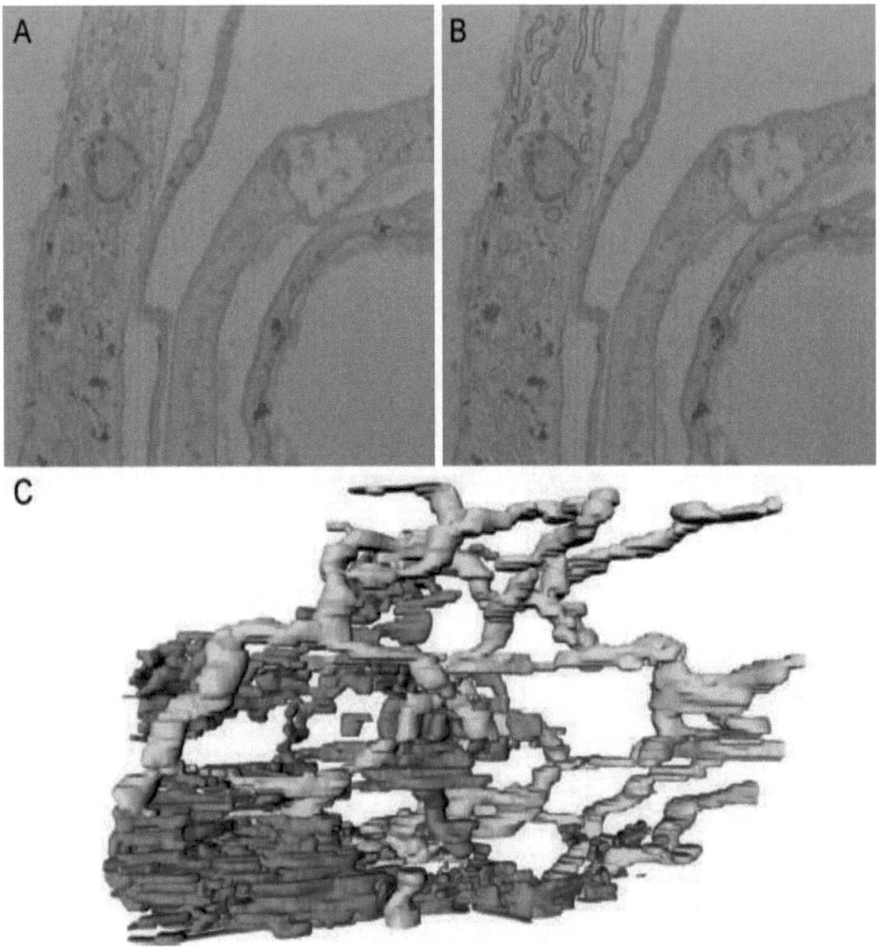


Fig. 9. 3D reconstruction of mitochondria and endoplasmic reticulum from endothelial cells from human umbilical cord. (**a**) Epon-embedded cells were repeatedly FIB milled and SEM imaged. As reported by the authors (20), block face sections of $5.2 \times 4.6\ \mu m^2$ size were imaged with a pixel size of 3.1 nm. (**b**) The endoplasmic reticulum (*green*) and the mitochondria (*orange*) were segmented in individual sections. (**c**) The reconstructed volume ($1.2 \times 4.6 \times 5.4\ \mu m^3$) reveals a very dense and complex network of the endoplasmic reticulum. Also, the mitochondria seem to be connected, forming a large complex. Reproduced with permission from ref. 19 with permission.

entire cells or parts of tissue (Fig. 9) (11, 19). When FIB/SEM is used for 3D tomography, serial sections (FIB cuts) have a thickness in the range of tens of nanometers, and volumes up to 4,000 $\mu m^3$ can be sectioned in a few hours, to tens of hours (18).

Sequential sectioning can also be performed on dried samples. This is illustrated in Fig. 10, which shows the contact between the lamellar body and a small lipid droplet. However, actual 3D structural reconstructions have been carried out only for embedded samples so far (Fig. 9). Further work is needed on sample preparation procedures before dried samples can be more generally used for 3D FIB/SEM imaging.

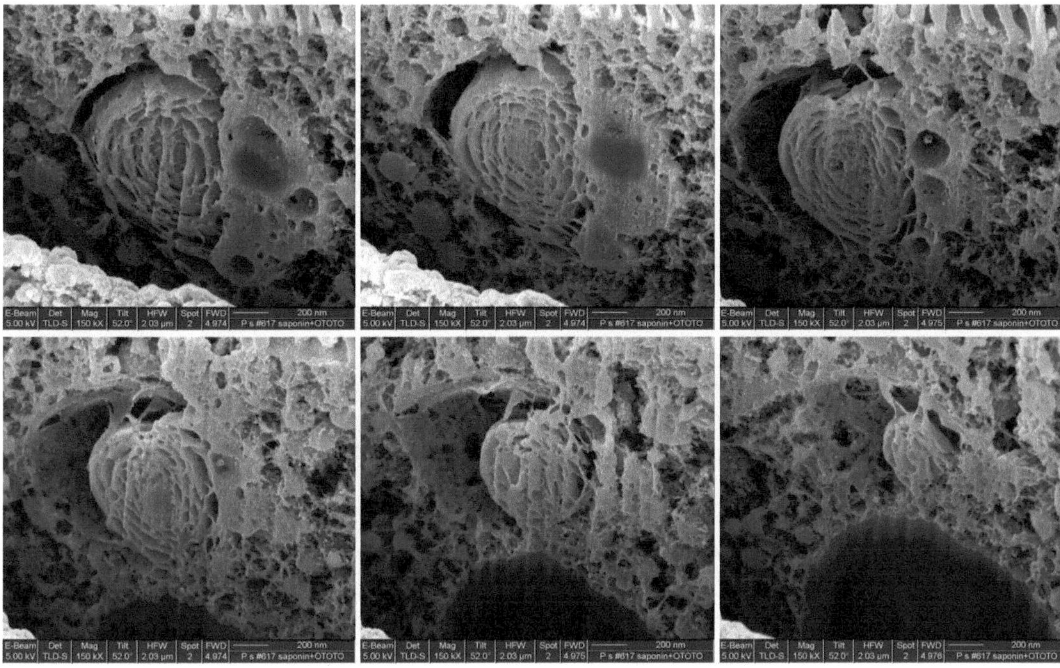

Fig. 10. Serial FIB milling followed by SEM imaging of a lamellar body in a digestive gland epithelium cell of a terrestrial isopod. Reprinted from ref. 16 with permission.

## 4. Conclusion

FIB/SEM on dried samples has the unique capability to link simultaneous investigations of sample gross morphology, cell surface and subsurface characteristics. *In situ* sample manipulation allows selection of any region for FIB manipulation. Aldehyde-fixed and OTOTO-processed samples provide the best topographical contrast for FIB/SEM investigation. For quality FIB/SEM imaging of dried samples, sample preparation protocols must be optimized on a case by case basis, taking into consideration the fixation, dehydration, and drying conditions, depending on the unique properties of the sample of interest. Nevertheless, based on the published literature, and results from our own laboratory, we conclude that FIB milling can reveal new insights on the investigated samples.

Sample preparation for FIB/SEM of plastic-embedded samples follows the same procedure as for TEM. Multiple automated repetitions of FIB milling and SEM imaging, followed by automated collection of 2D information may be performed on both, dried and plastic embedded samples. In the case of embedded samples further steps are performed with the aim of 3D object reconstruction. For this purpose, the 2D images are aligned and computationally processed into a 3D rendering of the specimen structures.

One of the core advantages of the use of FIB/SEM for 3D tissue and cell architecture studies is that the image acquisition process can be completely automated. Ultramicrotoming can be avoided and the process is therefore less time-consuming and labor-intensive (11). Thus, FIB/SEM is a valuable nanoimaging technique suitable for investigating biological samples.

## 5. Notes

1. Safety Note: Most of chemicals used for processing specimens for electron microscopy are hazardous. This is especially true for glutaraldehyde, formaldehyde, osmium tetroxide, cacodylate buffer, embedding medium in liquid form and uranyl acetate. Extreme care should be taken when handling these chemicals. All steps must be performed in a fume hood and gloves should be worn at all times. Excess osmium tetroxide, aldehydes, propylene oxide or resin should be collected in bottles for safe disposal. Fume hoods should always be used when working with a substance that has a threshold limit value (TLV), available on the appropriate Material Safety Data Sheet (MSDS), of less than 50 ppm (for gases). The TLV of a chemical is the level to which it is believed a worker can be exposed on a daily basis without adverse health effects.
2. The electron source in SEMs may use either a thermionic cathode or utilize field emission. SEMs with thermionic sources have lower resolution and are not as suitable for investigation of subcellular structures, while field emission machines are employed for ultrahigh resolution work. These microscopes are usually equipped with in-lens or out lens SE (Secondary Electrons), BSE (Backscattered Electrons) detectors as well as with X-ray detector.
3. Different authors report the use of various concentrations of paraformaldehyde and glutaraldehyde and distinct buffers. In all the examples shown in this chapter, 0.4% paraformaldehyde and 1% glutaraldehyde in 0.1 M sodium cacodylate buffer (pH 7.2) were used.
4. Milling of embedded samples may create problems such as a brightness gradient from the top to the bottom of the section, shadowing from the sidewalls, and a decrease of signal with depth (20). The problem of shadowing can be avoided by milling a sufficiently large, U-shaped trench that surrounds the volume to be sectioned (21). De Winter et al. (20) have developed ion milling conditions and electron imaging conditions suitable for FIB/SEM tomography of insulating biological or geological samples which is based on altered sectioning geometry.

5. Lešer et al. (8) tested different preparation methods for FIB/SEM investigation of dried biological samples. They documented the effect of fixative on subcellular structure and imaging quality in parallel with FIB/SEM of plastic embedded samples and with TEM. They confirmed that FIB/SEM operations on conventionally prepared biological samples require specially modified sample preparation protocols. Conventional SEM protocols could not be used unchanged.

## Acknowledgments

I would like to thank F. Tatti, FEI Co., Milan, Italy who performed all FIB operations. Samples were prepared by my former or present PhD Students, Vladka Lešer, Marjetka Kralj Kunčič, and Živa Tkalec Pipan, whom I would also like to thank. Finally, I would like to thank Bill Milne for English editing and Ana Fortič for technical editing and critical reading of the draft of this chapter.

### References

1. Li J (2006) The focused-ion-beam microscope—more than a precision ion milling machine. JOM 58:27–31
2. Claugher D (1986) A summary of ion beam etching of biological material with special reference to the saddle field source. Scan Electron Microsc 1:139–149
3. Fulker MJ, Holland L, Hurley RE (1973) Ion etching of organic materials. Scan Electron Microsc III:379–386
4. Lewis SM, Osborn JS, Stuart PR (1968) Demonstration of an internal structure within the red blood cell by ion etching and scanning electron microscopy. Nature 220:614–616
5. Spector M, Burns LC, Kimzey SL (1974) Ion beam etching of red blood cells and latex spheres. Nature 247:61–62
6. Knott G et al (2008) Serial section scanning electron microscopy of adult brain tissue using focused ion beam milling. J Neurosci 19:2959–2964
7. Drobne D et al (2007) Surface damage induced by FIB milling and imaging of biological samples is controllable. Microsc Res Tech 70:895–903
8. Lešer V et al (2009) Comparison of different preparation methods of biological samples for FIB milling and SEM investigation. J Microsc 233:309–319
9. Frey L, Lehrer C, Ryssel H (2003) Nanoscale effects in focused ion beam processing. Appl Phys A 76:1017–1023
10. Haswell R, McComb DW, Smith W (2003) Preparation of site-specific cross-sections of heterogeneous catalysts prepared by focused ion beam milling. J Microsc 211:161–166
11. Bushby AJ et al (2011) Imaging three-dimensional tissue architectures by focused ion beam scanning electron microscopy. Nat Protoc 6:845–858
12. Hayat MA (2000) Principles and techniques of electron microscopy: biological applications. Cambridge University Press, Cambridge
13. Friedman PL, Ellisman MH (1981) Enhanced visualization of peripheral nerve and sensory receptors in the scanning electron microscope using cryofracture and osmium-thiocarbohydrazide-osmium impregnation. J Neurocytol 10(1):111–131
14. Braet F, De Zanger R, Wisse E (1997) Drying cells for SEM, AFM and TEM by hexamethyldisinazane: a study on hepatic endothelian cells. J Microsc 186:84–87
15. Lešer V, Milani M, Tatti F, Tkalec ZP, Štrus J, Drobne D (2010) Focused ion beam (FIB)/scanning electron microscopy. Protoplasma 246:41–48
16. Drobne D et al (2005) Electron and ion imaging of gland cells using the FIB/SEM system. J Microsc 219:29–35
17. Drobne D et al (2008) Imaging of intracellular spherical lamellar structures and tissue gross morphology by a focused ion beam/scanning

electron microscope (FIB/SEM). Ultramicroscopy 7:663–670

18. Heymann JAW et al (2006) Site-specific 3D imaging of cells and tissues with a dual beam microscope. J Struct Biol 155:63–73
19. Hekking LHP et al (2009) Focused ion beam-scanning electron microscope: exploring large volumes of atherosclerotic tissue. J Microsc 235(3):336–347
20. De Winter MDA et al (2009) Tomography of insulating biological and geological materials using focused ion beam (FIB) sectioning and low-kV BSE imaging. J Microsc 233(3): 372–383
21. Holzer L et al (2004) Three-dimensional analysis of porous $BaTiO_3$ ceramics using FIB nanotomography. J Microsc 216:84–95

# Chapter 17

# Preparation of Gold Nanocluster Bioconjugates for Electron Microscopy

**Christine L. Heinecke and Christopher J. Ackerson**


## Abstract

In this chapter, we describe types of gold nanoparticle–biomolecule conjugates and their use in electron microscopy. Included are two detailed protocols for labeling an IgG antibody with gold monolayer protected clusters. The first approach is a direct bonding approach that utilizes the ligand place exchange reaction. The second approach describes NHS–EDC coupling of $Au_{144}(pMBA)_{60}$ with IgG. Also included are various characterization techniques for determining labeling efficiency.


**Key words:** Gold nanoparticle, Monolayer protected clusters, Biolabel, Electron microscopy, Cell imaging

## 1. Introduction

The contrast problem in biological microscopy arises from physical similarities, especially in terms of density, among biological molecules. The contrast problem has been addressed in optical microscopy with chemical stains (gram stain, etc.), immunohistochemical techniques, and fluorescent proteins such as Green Fluorescent Protein (GFP). In electron microscopy (EM), the contrast problem is much more severe than in optical microscopy because the palette of available specific electron contrasting (i.e., electron dense) stains is much smaller, and because there are no robust analogs to GFP, despite much work in this area (1, 2). Since gold nanoparticle/protein bioconjugates are currently the most well developed method for imparting contrast in the electron microscope, the objective of this chapter is to give practical guidance for the synthesis of gold nanoparticle/biomolecule conjugates, especially for application in biological electron microscopy.

Bioconjugates of electron contrasting materials—usually gold nanoparticles—with antibodies, nucleic acids, (3), small molecules

Alioscka A. Sousa and Michael J. Kruhlak (eds.), *Nanoimaging: Methods and Protocols*, Methods in Molecular Biology, vol. 950, DOI 10.1007/978-1-62703-137-0_17, © Springer Science+Business Media, LLC 2013

(4), and other proteins (5) are the only well developed tools for imparting special contrast to specific, targeted biological structures in electron microscopy. These nanoparticle conjugates have found application across all biological length scales at which electron microscopy is applied, allowing identification of subunits within multiprotein complexes, the orientation of DNA within DNA/ protein complexes, and the localization of specific proteins or epitopes within sectioned cells and tissues.

In principle, a nanoparticle comprised of any sufficiently heavy element could be conjugated to a biological molecule for imparting electron contrast in biological EM. In practice, gold is most widely used because it is not only exceptionally electron dense among (radio)stable elements but also easier to work with due to its comparative inertness (i.e., nobility). The nobility of gold likely contributes to its comparatively low toxicity as well as its controllable reactivity and conjugation with biological molecules.

There are nearly 1,500 papers published annually that study fundamental or applied properties of gold nanoparticles, as gold has become widely used as a component element in nanoscience studies. Many of the findings of this research, especially with regard to nanoparticle toxicity, the molecular nano-bio interface, mechanisms of particle stability, and novel bioconjugation strategies are of interest in the application of gold nanoparticles to electron microscopy (6–11).

Gold nanoparticle synthetic methods have produced a remarkable array of morphologies, including nanorods and nanoprisms. Here we focus exclusively on conjugation strategies for spherical particles, which are more robustly developed. Spherical gold nanoparticles can be synthesized with nominal inorganic diameter between 0.9 and 250 nm, generally determined by their passivating, protective ligand layers, which are composed of either citrate (colloidal gold, 5–250 nm), phosphine (~0.9 nm) or thiolate (1.5–3 nm) "headgroups." Any of these types of gold nanoparticles can be further "functionalized" through a direct bonding approach using the ligand place exchange reaction where the native protective monolayer can be replaced with incoming thiolate ligands with desirable chemical functional groups distal to the thiolate moiety (see Fig. 1a). This is one of the most well studied (12–16) and exploited reactions on gold nanoparticles and is used for nearly all applications of gold nanoparticles. Additionally, labeling reactions between a biomolecule and gold nanoparticle can be achieved through other strategies, including non-covalent linking via electrostatics as well as covalent interactions by exploiting the native chemical functionality and reactivity of the terminal end of the nanoparticle monolayer and the inherent functionality of various biomolecules. Each of these strategies is discussed in more detail below. A schematic of the most common approaches toward biolabeling gold nanoparticles is depicted in Fig. 1.

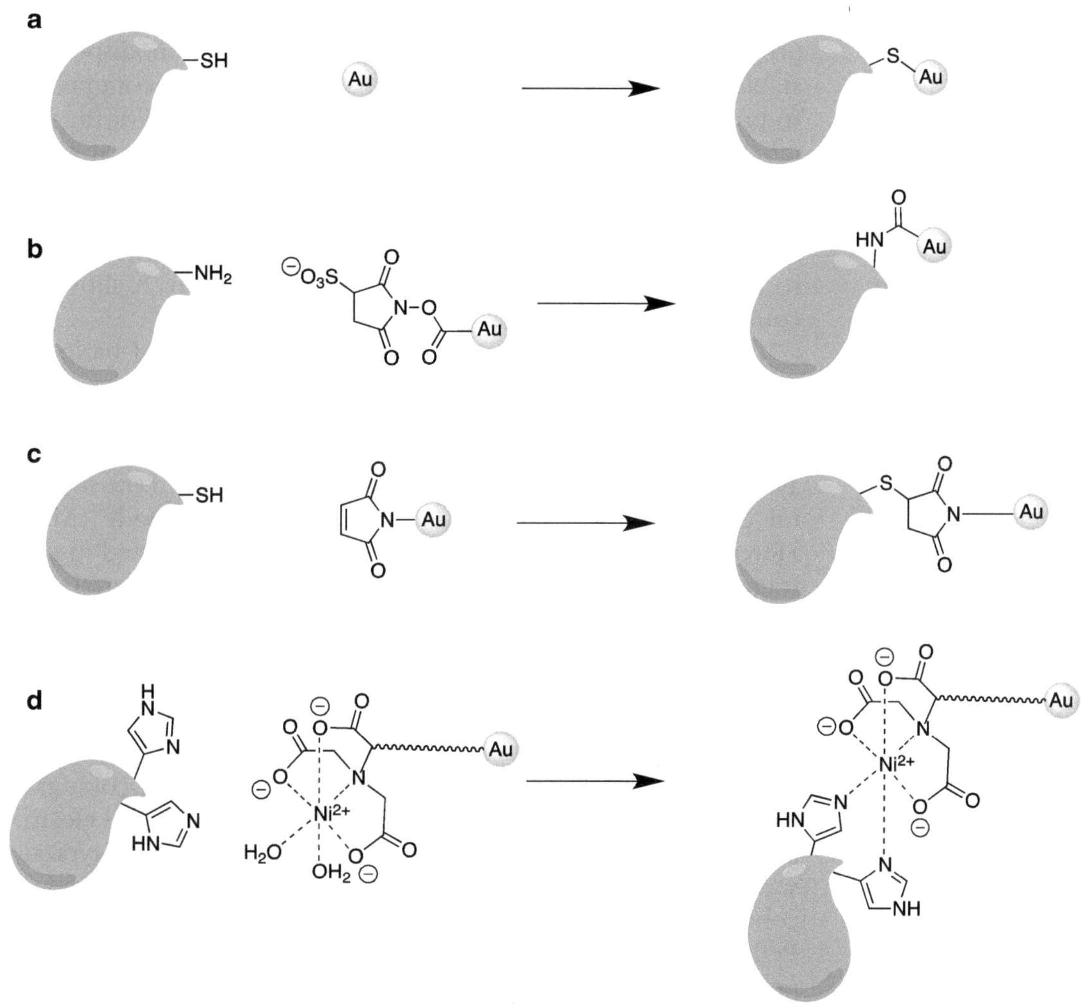


Fig. 1. The four most common strategies for bioconjugation to gold nanoparticles. The *tear-shaped cartoon* represents the biomolecule to be linked to the gold nanoparticle. Only the important functional groups for linking chemistry are shown, additionally the monolayer on the gold nanoparticle is also not depicted. (**a**) Direct bonding approach via ligand place exchange. (**b**) Sulfo-NHS–Ester Amine coupling approach. (**c**) Thiol–Maleimide coupling approach. (**d**) Ni–NTA Chelation approach.

### *1.1. Colloidal Gold: Immunogold*

Colloidal gold has found use in EM imaging of biological tissues since at least 1971, when Faulk and Taylor demonstrated that by simply adsorbing antibodies on the surface of the particle they could be rendered immunoactive (17). Such commercially available immunogold probes have been successfully used in many electron microscopy experiments (18–21). Additionally, selective bio-targeting of immunogold derivatives can be observed by measuring the surface plasmon resonance (SPR) of conjugates (22).

Immunogold probes are typically prepared by direct physisorption of antibodies or proteins to the colloidal gold surface. The nanoparticles used for immunogold preparations are generally

<30 nm in diameter; in some cases 1–3 nm colloidal gold has been prepared, however the samples are typically not monodisperse or stable over long periods of time (23). Generally, it is recommended to prepare protein adsorbed colloidal gold near the p*I* of the protein that will be used. Although labeling is more efficient at higher pH ranges, protein unfolding can occur at higher pH values, limiting biological function and thus utility in an EM experiment. Additionally, it is important to limit the amount of protein adsorbed to colloidal gold probes to prevent desorption, which allows for competition in binding to the target (24).

As seen above, the use of colloidal gold in EM of biological tissues generally takes advantage of noncovalent strategies. Aside from antibody adsorption, another favorable electrostatic interaction used to conjugate biomolecules to colloidal gold includes the avidin–biotin interaction. Copland et al. demonstrated the successful conjugation of a biotinylated monoclonal antibody (Mab), Herceptin® (Mab for HER2/neu receptor over-expressed in early stage breast cancers) (25), to 40 nm streptavidin-coated gold nanoparticles (26).

Colloidal gold conjugates can be purified through repeated centrifugation, where the particle conjugates are collected as a "pellet" and then resuspended in a buffer of choice following supernatant removal. Typically this will be repeated a few times as "wash steps" before final resuspension of the purified conjugate. Colloidal gold particles are generally not amenable to gel electrophoresis purification and visualization methods that are typically successful for smaller gold nanoparticles. Because immunogold is a commercially available material (Nanopartz, Nanoprobes, Ted Pella, BBI, Electron Microscopy Sciences), with well-established protocols available for their production, we do not give detailed protocols in this chapter for electrostatic conjugation to colloidal gold nanoparticles.

### 1.2. Triphenyl Phosphine Protected Clusters: Undecagold and Nanogold®

Bioconjugation of gold nanoparticles, especially the commercially available Nanogold® (1.4 nm) and undecagold [$Au_{11}(P(C_6H_5)_3)_7$, 0.8 nm] derivatives, has found widespread use in electron microscopy for biological imaging (27). Nanogold® in particular has been demonstrated to successfully penetrate 40 μm deep into tissue (28). The smaller size of these clusters in comparison to colloidal gold allows for enhanced stability because linking chemistry takes advantage of covalent bonding strategies. Therefore, the conjugates do not exhibit desorption over time as colloidal gold conjugates do (24). Additionally, unlike colloidal gold, undecagold and Nanogold® do not aggregate over time. These smaller probes are usually visible through TEM alone in very thin specimens; however, silver enhancement can be utilized for increased visibility if necessary (29). Silver enhancement of these probes also allows them to be visible using bright field microscopy (17) as well as Scanning Electron Microscopy (30).

Nanogold® has emerged as a more popular choice than undecagold for biolabeling EM markers presumably due to its larger size, making it visible in unprocessed micrographs. Additionally, it possesses greater stability in the electron beam than undecagold. Nanogold® is sold in two reactive forms, baring either a monomaleimide or monosulfo-NHS ester functionalities which allow conjugation to antibodies via cysteine residues, and lysine residues and terminal amines respectively (see Fig. 1b, c). These reactive strategies have been extended to multiple other biomolecules including oligonucleotides (31), peptides (32), carbohydrates (33), etc. Additionally, Nanogold® conjugates can be prepared by taking advantage of a common protein tag—polyhistadine. Most proteins are engineered to contain $His_6$ tags as a purification handle when bound to a Ni(II)–NTA (nitriloacetic acid) column. This chemistry can also be applied to Nanogold® by coupling $His_6$ tagged proteins with commercially available particles modified to contain Ni(II)-NTA on the surface (see Fig. 1d). Additionally, Kogot et al. demonstrated that 1.5 nm triphenylphosphine capped nanoclusters were capable of conjugation to $His_6$ without the use of Ni-NTA modified particles (34).

Nanogold® conjugates are separated readily from unreacted starting materials using gel filtration, gel electrophoresis, ion-exchange chromatography or high performance liquid chromatography. Because Nanogold® is a commercially available material (Nanoprobes.com, Yaphank, NY), with available protocols once purchased, detailed instructions for labeling will not be provided in this chapter.

### *1.3. Thiolate Monolayer Protected Clusters*

Because we are most experienced working with thiolate monolayer protected gold clusters (MPCs), the rest of this chapter and protocols concern labeling chemistries of thiolate MPCs. MPCs ranging from 1 to 3 nm in diameter fit into what is considered a "magic number" regime, where particles have filled superatomic electron orbitals (35), yielding enhanced particle stability (analogous to the stability noble gases exhibit from a filled outermost orbital of electrons, e.g., the octet rule). These MPCs are composed of a crystalline gold core, where the outermost gold atoms are protected by bridging organothiolate ligands and are of the general formula $Au_n(SR)_m$. Magic number MPCs exist in molecularly defined sizes (general formulae: $Au_{25}(SR)_{18}$, $Au_{38}(SR)_{22}$, $Au_{68}(SR)_{34}$, $Au_{102}(SR)_{44}$, $Au_{144}(SR)_{60}$) and in some cases crystallographically defined sizes ($Au_{25}(SR)_{18}$, $Au_{38}(SR)_{22}$, $Au_{102}(SR)_{44}$) (36–38). In some cases, synthetic attempts will yield multiple cluster core sizes within the magic number regime, which can be separated readily from each other by fractional precipitation methods (see Subheading 3.2) or liquid chromatography.

The nanoclusters we work with preferentially ($Au_{25}$, $Au_{102}$, $Au_{144}$) fit into the magic number regime of molecule-like clusters. Both $Au_{102}(p\text{MBA})_{44}$ and $Au_{144}(p\text{MBA})_{60}$ are protected by *p*MBA

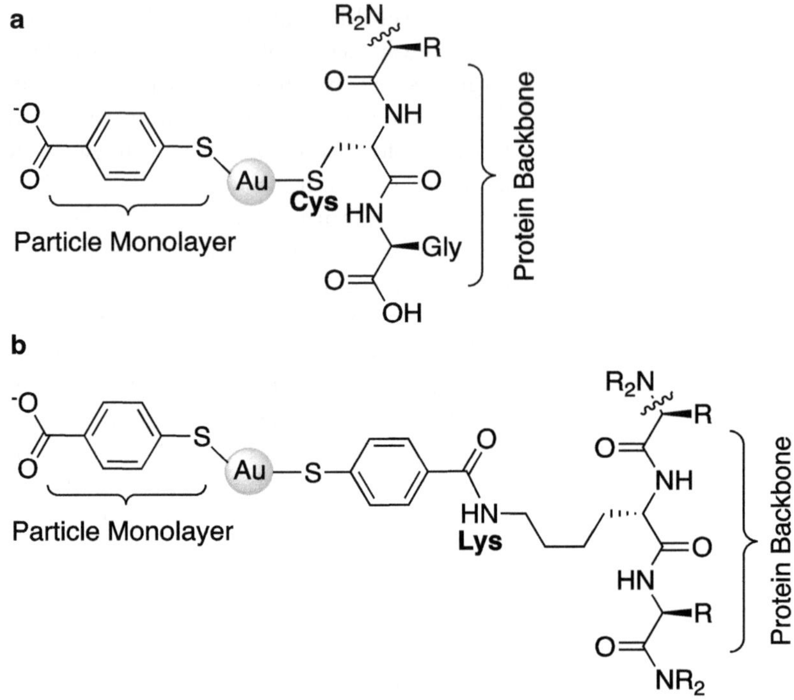


Fig. 2. Schematic of protein conjugation strategies presented in the Subheading 3: IgG@ $Au_{144}(pMBA)_{60}$. (**a**) Direct bonding via ligand place exchange between a near terminal Cysteine residue and the Au MPC. (**b**) NHS–EDC coupling between a carboxylic acid of the *p*MBA ligand layer and a primary amine on a Lysine residue within the protein. Note that for clarity, only one intact *p*MBA of the ligand layer (total 60 ligands) is represented in these depictions.

(*para*-mercaptobenzoic acid) ligands, which provide the clusters with both water solubility as well as a reactive terminal functional group (COOH). The carboxylic acid ligands on $Au_{102}(pMBA)_{44}$ and $Au_{144}(pMBA)_{60}$ allow for NHS–EDC amide bond coupling with any commercially available terminal amines and proteins that display lysine residues, in addition to direct bonding by the place exchange reaction with thiolated ligands and proteins displaying cysteine residues. We have had success labeling $Au_{144}(pMBA)_{44}$ with IgG using both of the aforementioned strategies, depicted in Fig. 2. Detailed protocols for each of these strategies are discussed in Subheadings 3.4 and 3.5 along with helpful tips that should be considered when preparing any bioconjugate.

Water-soluble gold clusters are amenable to functionalization with organic thiolates and proteins exposing cysteine residues via ligand place exchange (13). The ligand place exchange reaction allows for a stoichiometric exchange of incoming thiolate molecules for native thiolate ligands that protect the gold cluster. For example: $Au_{144}(pMBA)_{60} + 4\,SR \rightarrow Au_{144}(pMBA)_{56}(SR)_4 + 4\,pMBA$. The ligand place exchange reaction is employed for nearly all applications of gold particles and has been studied in both kinetic and mechanistic detail since its discovery. The rate and extent of reaction depend on the nature of the incoming and outgoing ligands as well as the oxidation state of the cluster core (13, 14). It has been demonstrated that incoming ligands possessing electron-withdrawing character increase the rate of the reaction. Additionally, oxidatively charging the cluster core increases the reaction rate as well. Any incoming thiolate molecule can be used for this reaction including small molecule organothiolates, proteins displaying cysteine residues, as well as oligonucleotides modified with a 5′ or 3′ thiol. Gold conjugates have been prepared using this strategy for a variety of biomolecules including antibodies (39), proteins (40–43), peptides (44, 45), oligonucleotides (46) and small biologically active molecules (4).

Depending on the functionality of the terminal pendant group on the thiolate ligand layer protecting the gold cluster, various chemical opportunities arise to complete a conjugation or labeling reaction. The MPCs we work with display carboxylic acids, which allow for reactivity with primary amines via NHS–EDC coupling. Similarly, MPCs displaying primary amines can theoretically undergo the same reaction with commercially available carboxylic acids or esters. Nearly any functionalized thiolate can undergo place exchange with MPCs providing additional reactive possibilities. The breadth of standard organic chemistry reactions that could be completed on appropriately functionalized MPCs with variously functionalized small molecules has many possibilities that have not even been discovered yet.

## 2. Materials

### 2.1. Reagents

1. Hydrogen tetrachloroaurate.
2. *Para*-mercaptobenzoic acid (must be sourced from TCI America).
3. Nanopure $H_2O$.
4. Methanol.
5. Ammonium Acetate.
6. Sodium Hydroxide.
7. Sodium Borohydride.
8. *N*-Hydroxysuccinimide (NHS).
9. 1-Ethyl-3-(3-dimethylaminopropyl) carbodiimide (EDC).
10. *Tris*(2-carboxyethyl)phosphine (TCEP).
11. Thiolated biomolecule for labeling (IgG is used in this protocol).
12. Aminated biomolecule for labeling (IgG is used in this protocol).

13. 50 mM Phosphate Buffer, pH 7.5.
14. 0.5 M Phosphate Buffer, pH 7.5.
15. 10× Tris–Boro–EDTA (TBE): 890 mM Tris base, 890 mM Boric acid, 20 mM EDTA, pH 8.3.
16. 1× TBE: 89 mM Tris base, 89 mM Boric acid, 2 mM EDTA, pH 8.3.
17. 2× Nonreducing Loading Dye: 10 mL $H_2O$, 5 mL glycerol, 4 mL 1 M Tris–HCl pH 6.8, 1 g SDS, 0.6 mL 1% bromophenol blue solution (see Note 1).
18. Acrylamide (40%, 19:1).
19. 10× Tris–Glycine SDS Running Buffer: 250 mM Tris, 1.92 M glycine, 1% SDS, pH 8.3.
20. 1× Tris–Glycine SDS Running Buffer: 25 mM Tris, 192 mM glycine, 0.1% SDS, pH 8.3.
21. Coomassie Staining Solution: 40% methanol, 10% acetic acid, 50% $H_2O$, 0.1% (w/v) Coomassie Brilliant Blue-R250.
22. Destaining Solution: 40% methanol, 10% acetic acid, 50% $H_2O$.
23. $Au_{102}(p\text{MBA})_{44}$ or $Au_{144}(p\text{MBA})_{60}$.

### 2.2. Equipment

1. Lyophilizer.
2. Centrifuge(s) capable of spinning 50 mL conicals and 1.5 mL Eppendorf® tubes.
3. Vortexer.
4. Polyacrylamide Gel Electrophoresis Box.
5. Power Supply (at least 200 V).
6. White Light Box.
7. Scalpel.
8. Chromatography Equipment.
9. 1.5 mL Eppendorf® tubes.
10. 50 mL conicals.
11. Rocker or shaking rotator.

## 3. Methods

Herein, we have included a protocol for the preparation of the negatively charged MPC $Au_{144}(p\text{MBA})_{60}$ (see Note 2), as well as two conjugation strategies: direct bonding via ligand place exchange and NHS–EDC coupling. Isolation, purification and characterization methods are also included.

### 3.1. Synthesis of $Au_{144}$(pMBA)$_{60}$ MPC

1. Prepare 95.2 mM *p*MBA in 0.3 M NaOH. Allow stirring overnight (see Note 1).
2. The following day, prepare 28 mM $HAuCl_4 \cdot 3H_2O$ in water.
3. The Au(I)-*p*MBA polymer is then generated in a methanol: water solvent (47% methanol, 53% water) with a 3:1 molar excess of *p*MBA (0.432 mmol, 9 mM final volume) in relation to $HAuCl_4$ (0.144 mmol, 3 mM final volume). This can be prepared in a 50 mL conical by combining in the following order (see Note 3): methanol, nanopure $H_2O$, $HAuCl_4$ solution, and finally *p*MBA solution. We have had success preparing this polymer on a scale of 48 mL final volume (see Note 4). The polymer is then allowed to stir overnight prior to reduction.
4. The following day, prepare a 150 mM solution of $NaBH_4$ in nanopure $H_2O$ (see Note 5) and immediately add 0.5 molar equivalents (relative to Au) to the polymer. We have had success preparing $Au_{144}$(*p*MBA)$_{60}$ at volume scales between 3 and 48 mL (see Note 4). The reduction is allowed to proceed at room temperature with vigorous stirring or vortexing for a *maximum* of 2 h (see Note 6).
5. The product is then precipitated with the addition of ammonium acetate (80 mM final volume) and methanol (equivalent volume to reaction scale volume). Centrifugation at 3220 rcf for 10 min at 4°C will be sufficient to "pellet" the MPC out of solution. The supernatant is then decanted, the MPCs are resuspended in nanopure $H_2O$ and then precipitated again as described above to remove any remaining residual thiolate or $NaBH_4$ impurities.
6. The product is then resuspended in nanopure $H_2O$. For a 25 mL scale, 0.5 mL is an appropriate volume to add.
7. Product purity is quickly assessed by polyacrylamide gel electrophoresis (PAGE). Run the MPC product on a 15% polyacrylamide gel (19:1, 5% glycerol final concentration) in 1× TBE at 110 V for 1–1.5 h. The sample can be loaded into the gel by combining 5 μL of 50% glycerol with 1 μL of MPC sample. If the gel illustrates a large abundance of one band, but a smear reaching towards the well, a fractional precipitation will be required to remove larger material. This protocol is described in Subheading 3.2.

### 3.2. Fractional Precipitation Protocol: Purification of $Au_{144}$(pMBA)$_{60}$ MPC

Typically, the synthesis described above will generate an abundance of $Au_{144}$(*p*MBA)$_{60}$; however, sometimes larger MPC impurities will also be generated, requiring their removal from the desired cluster core size. This is accomplished using a fractional precipitation method, where increasing amounts of precipitating agent are added to the MPC suspension to remove larger products from the desired product: $Au_{144}$(*p*MBA)$_{60}$. This is an effective purification method because larger MPCs require less precipitant (in this case methanol) to crash out of solution than smaller MPCs (likewise smaller MPCs require more precipitant). If fractional precipitation purification is necessary, the following steps should be taken.

1. The MPCs in need of purification should be suspended in approximately 9 mL of nanopure $H_2O$ in a 50 mL conical.
2. Add ammonium acetate to a final concentration of 80 mM.
3. Add methanol to a final concentration of 20%.
4. Centrifuge the sample at 4,000 rpm for 10 min at 4°C. This should remove the majority of really large products if they are present.
5. Decant the supernatant into a fresh 50 mL conical. Save pellet to check by PAGE.
6. Add methanol to the supernatant to a final concentration of 40%.
7. Centrifuge the sample again at 4,000 rpm for 10 min at 4°C. This should remove products that are slightly larger than $Au_{144}(pMBA)_{60}$ if they are present.
8. Decant the supernatant into a fresh 50 mL conical. Save pellet to check by PAGE.
9. Add methanol to a final concentration of 60%.
10. Centrifuge the sample again at 4,000 rpm for 10 min at 4°C. This will precipitate $Au_{144}(pMBA)_{60}$.
11. The supernatant is then discarded and the pellet is dried in vacuo and then resuspended in nanopure $H_2O$.
12. The success of the precipitation is then assessed by running all pellets on a PAGE gel as described above (see Note 7).

### 3.3. Characterization of $Au_{144}(pMBA)_{60}$

1. The MPCs can be first characterized by UV–Visible spectroscopy. Unlike colloidal gold which has a prominent SPR band in UV–Vis, $Au_{144}(pMBA)_{60}$ has a very broad absorption which decreases in intensity from the 300 to 600 nm window of the visible spectrum. At 510 nm $Au_{144}(pMBA)_{60}$ has a calculated molar extinction coefficient of $4.24 \times 10^5$ $M^{-1}$/cm. Beer's Law can be used in conjunction with UV–Vis to calculate the concentration of MPCs.
2. TEM of $Au_{144}(pMBA)_{60}$ indicates these particles are 2.0 nm in diameter.
3. MALDI-MS should reveal a material with a molecular weight of approximately 37,615 Da for the intact particle and 28,363 Da for the gold core. We have had success with MALDI determination of $Au_{144}(pMBA)_{60}$ by including DHB (2,5-dihydroxybenzoic acid) in the matrix and tetramethylammonium (bromide) as the counter-ion.

#### *3.4. NHS/EDC Coupling of IgG with $Au_{144}(pMBA)_{60}$*

This coupling strategy takes advantage of standard amide bond formation chemistry by conjugating biomolecules containing primary amines (lysine residues, terminal amine, etc.) and the carboxylic acid moiety at the terminal end of the $Au_{144}(pMBA)_{60}$ monolayer.

1. Prepare working stock solutions of NHS (0.04 M) and EDC (0.0116 M) in water. These can be aliquoted into 1 mL Eppendorf® tubes and stored at −20°C for future use.
2. Dispense 5 μL of 0.5 M Phosphate Buffer into a 1.5 mL Eppendorf® tube.
3. Add 50 μL of NHS stock solution (16.7 mM final concentration) (see Note 8).
4. Add 5 μL EDC stock solution (480 μM final concentration).
5. Add $Au_{144}(pMBA)_{60}$ to a final concentration of ~130 μM (100 μL of ~300 μM solution in Phosphate Buffer) (see Subheading 3.3, step 1).
6. Allow the NHS and EDC to pre-activate the carboxylic acid monolayer on the MPCs for at least 10 min at room temperature prior to the addition of amine (see Note 8).
7. While the MPC is being activated, prepare a 5 mg/mL IgG stock solution in 50 mM Phosphate Buffer.
8. Add 50 μg of IgG solution (10 μL) to the activated MPC and incubate with mild shaking on a vortexer at room temperature or 37°C overnight (see Note 9).
9. Concentrate sample by freezing and lyophilizing.
10. Resuspend sample in ~50 μL 50 mM Phosphate Buffer.
11. Check labeling efficiency by running polyacrylamide gel electrophoresis (PAGE).

#### *3.5. Direct Bonding of IgG to $Au_{144}(pMBA)_{60}$ via Ligand Place Exchange*

1. Prepare a solution of 5 mg/mL IgG (see Note 10) in 50 mM Phosphate Buffer.
2. Dispense 100 μg (20 μL) of IgG solution into a 1.5 mL Eppendorf® tube.
3. Add TCEP to a final concentration of 100 μM (2 μL of 1 mM stock solution works well for this purpose). Allow reduction for 1 h at room temperature (see Note 11).
4. Add $Au_{144}(pMBA)_{60}$ to a final concentration of ~250 μM (100 μL of ~300 μM in Phosphate Buffer) (see Subheading 3.3, step 1).
5. Allow to incubate by vortexing at room temperature over night (~18 h), this step can also be done at 37°C for a few hours if the biomolecule will retain stability at this temperature for extended periods of time (see Note 9).

6. Concentrate the sample by first freezing and then lyophilizing.
7. Resuspend the sample in ~50 μL of 50 mM Phosphate Buffer, pH 7.5.
8. Assess reaction efficacy using PAGE.

### *3.6. Monitor Labeling Yield with PAGE*

Both labeling strategies described in this chapter can be monitored using a polyacrylamide gel electrophoresis gel-shift assay. The protocols here entail antibody (IgG) labeling to $Au_{144}(pMBA)_{60}$, thus SDS-PAGE with Coomassie staining can be used to determine labeling efficiency (see Notes 12–14). PAGE will indicate if the reaction was successful, as well as indicate visually the number of reaction products. Successful labeling of $Au_{144}(pMBA)_{60}$ to IgG is indicated by a slight decrease (up-shift) in the electrophoretic mobility of the antibody relative to the antibody alone (this indicates one MPC/IgG). If additional discrete bands corresponding to Coomassie staining are observed above, this is indicative of a higher MPC/IgG biolabel (see Fig. 3 for examples of successful labeling by PAGE).

1. On a Tris–Gly–SDS PAGE gel (see Note 13), the following should be analyzed (see Note 12): the reaction products, IgG control, $Au_{144}(pMBA)_{60}$ control as well as any protein markers if desired. Allow the gel to run at 150 V for approximately 30–45 min.

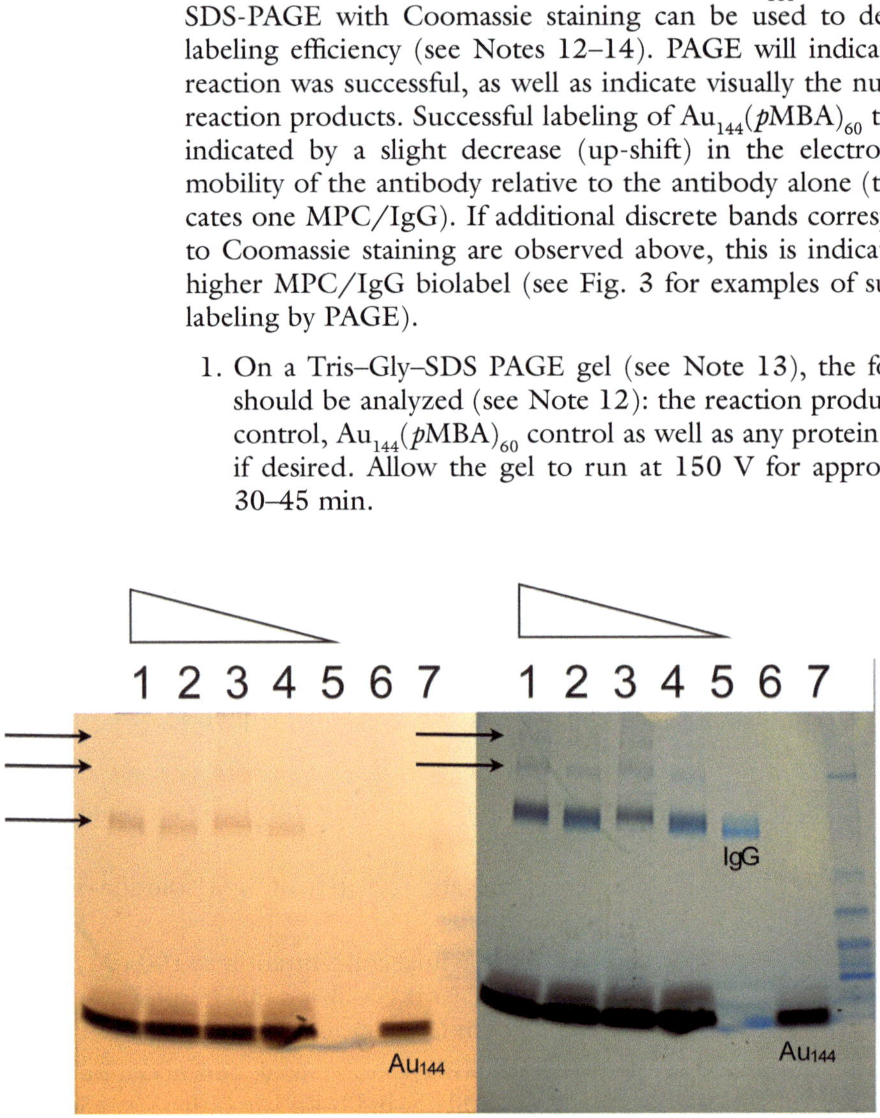


Fig. 3. PAGE analysis of reaction progress. This experiment is a result of varying reactant concentrations. Lanes 1–4 indicate decreasing amounts of reactants. Lane 5 is free IgG. Lane 6 is Au144. Lane 7 is a marker. Shown in the *left panel* is the gel immediately after running. Note how a gel shift is observed, and the particle-IgG conjugate is visible without staining. Shown in the *right panel* is the same gel that has been stained with Coomassie. The IgG band is now visible and correlates with the Au144–IgG conjugate. Also note the presence of more than one higher molecular weight band, indicating different formulations of MPC conjugate.

2. Place the gel on a light box and photograph it prior to staining in order to observe free MPCs and labeled MPCs.
3. Stain with Coomassie Staining Solution for 30 min at room temperature on a rocker or shaking rotator.
4. Decant Coomassie Staining Solution and wash the gel 3–4 times with water to remove residual stain.
5. Destain the gel by adding ~10 mL Destain Solution and rock for 30 min. Decant the liquid and repeat this procedure every 30 min for about 2 h or overnight if desired.
6. Place the gel on a light box and photograph it to visualize the free antibody and antibody conjugates.

### 3.7. Isolation, Characterization and Utilization of Desired IgG: $Au_{144}(pMBA)_{60}$ Conjugate

Generally, if only one stoichiometric product is identified by PAGE, unreacted starting materials can be removed using an appropriately size MWCO (molecular weight cut off filter) or dialysis membrane. If the labeling reaction provides more than one stoichiometric product and they must be separated for the desired application, this can be accomplished by gel purification or chromatography. Since PAGE is utilized for most of the rapid characterization of these conjugates, a gel purification protocol (crush and soak) will be described that allows isolation of various conjugates with minimal optimization. Additionally, this technique will remove any unreacted small molecule starting materials.

1. On a native gel (e.g., no SDS) of the same percentage used for reaction monitoring, run all reaction samples desired to be purified. Run at the same voltage and run times used previously (see Note 15).
2. Using a scalpel, cut out the desired bands from the gel and place in a 1.5 mL Eppendorf® tube (see Note 15).
3. Crush the gel into small pieces using a pipet tip.
4. Add nanopure $H_2O$ to completely cover the gel pieces (see Note 16). Allow elution to proceed overnight at room temperature or 4°C if desired.
5. Centrifuge the sample at 13.2 rcf for 10 min to concentrate the gel matrix at the vertex of the tube.
6. Remove the supernatant and transfer to a new 1.5 mL tube (see Note 17).
7. Freeze and lyophilize the sample to dryness.
8. Reconstitute the IgG–$Au_{144}(pMBA)_{60}$ conjugate in a buffer of choice for labeling experiment. Dialysis can also be performed at this stage if there is concern that any remaining Tris from the gel matrix will disrupt downstream applications.
9. The identity of this conjugate can then be confirmed using MALDI-MS.

Fig. 4. Cryoelectron micrograph of 2 nm Au MPCs conjugated to scFv antibody fragments mixed with the corresponding antigen N9 neuraminidase. The tetrameric structure of N9 neuraminidase results in the binding of four scFv fragments. The larger particles observed are 10 nm AuNP, included as markers in this cryo-EM preparation (reprinted from ref. 48 with permission from the American Chemical Society).

10. Once the bioconjugate has been characterized it can be imaged in an electron microscope. As an example, Fig. 4 shows the N9 neuraminidase tetramer in complex with four $Au_{144}$ conjugated scFv antibody fragments.

## 4. Notes

1. It is necessary to allow dissolution of *p*MBA overnight, using this solution immediately after dissolution will NOT produce $Au_{144}(pMBA)_{60}$. We suspect it is important to allow an equilibrium to be reached between the reduced and oxidized forms of *p*MBA for the successful synthesis of $Au_{144}(pMBA)_{60}$. Also note that *p*MBA will not dissolve in water without the addition of base.
2. In order to prevent nonspecific adsorption of the biolabeled nanoparticle conjugate, it is sometimes necessary to introduce polyethylene glycol (PEG) functionalities onto the conjugate as well. This can be accomplished by performing a place exchange reaction with a thiolated PEG molecule and the nanoparticle

platform of interest. These can be purchased from commercial sources; however, they are usually much cheaper to synthesize in house. We have had success synthesizing PEG-thiols from literature precedents (47). Generally, low feed ratios of incoming PEG-thiol to gold nanoparticle allow for facile product purification with minimal product loss (we have had success using 1:1 and 2:1 feed ratios of incoming thiol to outgoing thiol, higher feed ratios of 4:1 and greater have provided products that were not isolable using standard alcohol precipitation methods).

3. Order of addition is extremely important for the synthesis of these MPCs. Deviating from the order listed is highly likely to yield unintended products.
4. Like many syntheses, MPC syntheses are notoriously fickle in terms of reaction scaling. We have found this reaction scale to reliably produce $Au_{144}(pMBA)_{60}$, while deviations (especially scale-ups) have generated insoluble material.
5. Generally $NaBH_4$ is supplied as a "chunky" solid, and we have found that grinding it into a powder (and storing it as a powder in a desiccator) speeds up dissolution times, which in turn aids in reproducibility. This is because $NaBH_4$ reacts with water. If $NaBH_4$ takes a longer time to dissolve (frequent when using chunks) or if it is not added immediately after preparation (even when using pulverized $NaBH_4$), its reducing ability is diminished. This is important because the molar equivalent of $NaBH_4$ that is used is crucial in obtaining the desired products.
6. As the reduction progresses, we have found that the desired product distribution diverges to two closely sized, different MPCs of undetermined molecular formulae. The reaction should only be allowed to proceed for a maximum of 2 h to avoid the synthesis of materials that are difficult to separate with standard purification methods.
7. When assessing the efficacy of the purification procedure, it is recommended to save an aliquot of crude sample to compare against each pellet from the precipitation. The final pellet should contain pure $Au_{144}(pMBA)_{60}$. Once this is confirmed by MALDI and TEM, a sample should be saved to compare all future preparations against for quick determination that future reaction products are remotely close to $Au_{144}(pMBA)_{60}$ in electrophoretic mobility (and thus size).
8. An excess of NHS is used relative to EDC in order to outcompete NHS hydrolysis, which occurs readily in water especially at neutral pH and higher. The pH should be ~6 prior to the addition of the amine.
9. Both the NHS–EDC and Thiol Place Exchange strategies presented here will require reaction parameter optimization for

each different biomolecule that will be labeled. Depending on the identity of the molecule (DNA, protein, antibody, small molecule) as well as how many free amines or free thiols (respective to the strategies *vide supra*) exist for conjugation, various relative ratios of MPC and biomolecule should be tested for the desired stoichiometry of the biolabeled product (e.g., How many MPCs per biomolecule are desired for the specific labeling application?). It is likely that a range of stoichiometric products will be generated if multiple labeling sites exist on the biomolecule. Additionally, depending on the stability of the biomolecule, both temperature and time parameters should also be investigated for labeling optimization.

10. When performing ligand place exchange reactions on proteins or antibodies it is recommended, if possible, to label adventitious cysteine residues. However, this is not always a possibility; cysteine residues may be in a solvent inaccessible site due to protein folding or they could be participating in a disulfide bond, which if reduced can disrupt the native folding and thus function of the protein. It is possible to engineer terminal cysteine residues into a protein; however, we recommend adding a terminal glycine adjacent to the (now N2 or C2) cysteine residue. Terminal cysteines easily form disulfides over time, which can hinder labeling efficiency as well as require the use of reducing agents (for successful labeling) as time passes. This can lead to the need for constant reoptimization of a labeling strategy, which is not recommended as it wastes valuable time.
11. TCEP is used because it is a mild phosphine-based reductant that does not interfere with the chemistry of the gold cluster. Classic reducing agents like DTT and β-mercaptoethanol are thiolates themselves, which can participate in the ligand place exchange reaction as well as etch the particles; neither of which is desired in a gold conjugation reaction. Thiol based reductants cannot be used for this purpose without the need for an extra purification step to remove them prior to labeling the MPC.
12. Unlike loading dyes for most SDS-PAGE gels on proteins, when using thiolate MPCs the loading dye MUST be NONREDUCING, DO NOT include DTT or β-mercaptoethanol because they will etch the MPC before sample loading and the experimental results will be inconclusive.
13. Gel electrophoresis analysis of reaction completion can be accomplished using commercially available 4–20% gradient Tris–HCl gels (BioRad®) or pouring manually while considering an appropriate gel percentage for the biomolecule of interest (single percentage gels between 8 and 15% acrylamide are usually fine for most proteins depending on their size; larger proteins require smaller percentage gels and vice versa). MPCs will most likely run at the bottom of the gel using these

percentages. However, since they are directly visible by eye without staining, running them off the bottom of the gel is not a large concern. Additionally, if pouring gels manually, we have found that the addition of 5% glycerol (final concentration) in the gel matrix aids MPC migration and resolution.

14. If the biomolecule to be labeled is nucleic acid based (DNA or RNA), 0.5–1× TBE (Tris–Boro–EDTA) is the recommended buffer to use in the gel matrix, running buffer and loading dyes.

15. Ensure that sample loading is high enough based on previous electrophoretic experiments that the IgG–$Au_{144}(pMBA)_{60}$ conjugate is visible by eye. Thus, one or many black/brown bands should be visible in the upper half of the gel after run completion. Two gels of the exact same loading identities (and with IgG and MPC controls) can be run simultaneously, where one is stained with Coomassie Staining Solution (can use less sample on this gel because the detection limit is higher) and the other is not in order to be certain of the location of the conjugates of interest on the gel.

16. Water is recommended here because the product will be isolated by lyophilization. The addition of buffer at this elution step will yield a super concentrated solution of salts and conjugate, which the biomolecule will most likely be unstable in.

17. Using a P-1000 pipet tip for this frequently allows gel pieces to come with the supernatant upon transfer. By placing a P-1000 pipet tip *inside* a P-200 tip, gel contamination in the supernatant can be reduced due to the smaller diameter of the P-200 tip than the P-1000 tip. This also allows for a one step removal of supernatant, as any volume up to 1 mL can be removed.

## References


1. Wang Q, Mercogliano CP, Löwe J (2011) A ferritin-based label for cellular electron cryotomography. Structure 19:147–154
2. Mercogliano CP, DeRosier DJ (2007) Concatenated metallothionein as a clonable gold label for electron microscopy. J Struct Biol 160:70–82
3. Poglitsch C, Meredith G et al (1999) Electron crystal structure of an RNA polymerase II transcription elongation complex. Cell 98:791–798
4. Bowman MC, Ballard TE, Ackerson CJ, Feldheim DL, Margolis DM, Melander C (2008) Inhibition of HIV fusion with multivalent gold nanoparticles. J Am Chem Soc 130:6896–6897
5. Liu J, Taylor DW, Taylor KA (2004) A 3-D reconstruction of smooth muscle alpha-actinin by CryoEm reveals two different conformations at the actin-binding region. J Mol Biol 16: 115–125
6. Chompoosor A, Saha K, Ghosh PS et al (2010) The role of surface functionality on acute cytotoxicity, ROS generation and DNA damage by cationic gold nanoparticles. Small 6: 2246–2249
7. Pan Y, Neuss S, Leifert A et al (2007) Size-dependent cytotoxicity of gold nanoparticles. Small 3:1941–1949
8. Pan Y, Leifert A, Ruau D et al (2009) Gold nanoparticles of diameter 1.4 nm trigger necrosis by oxidative stress and mitochondrial damage. Small 5:2067–2076
9. Alkilany AM, Murphy CJ (2010) Toxicity and cellular uptake of gold nanoparticles: what we have learned so far? J Nanopart Res 12: 2313–2333

10. Fischer H, Chan W (2007) Nanotoxicity: the growing need for in vivo study. Curr Opin Biotechnol 18:565–571
11. Cho WS, Cho M, Jeong J et al (2009) Acute toxicity and pharmacokinetics of 13 nm-sized PEG-coated gold nanoparticles. Toxicol Appl Pharmacol 236:16–24
12. Templeton A, Hostetler M et al (1998) Gateway reactions to diverse, polyfunctional monolayer-protected gold clusters. J Am Chem Soc 120:4845–4849
13. Hostetler MJ, Templeton AC, Murray RW (1999) Dynamics of place-exchange reactions on monolayer-protected gold cluster molecules. Langmuir 15:3782–3789
14. Song Y, Murray RW (2002) Dynamics and extent of ligand exchange depend on electronic charge of metal nanoparticles. J Am Chem Soc 124:7096–7102
15. Templeton AC, Wuelfing WP, Murray RW (2000) Monolayer-protected cluster molecules. Acc Chem Res 33:27–36
16. Daniel M, Astruc D (2004) Gold nanoparticles: assembly, supramolecular chemistry, quantum-size-related properties, and applications toward biology, catalysis, and nanotechnology. Chem Rev 104:293–346
17. Faulk WP, Taylor GM (1971) An immunocolloid method for the electron microscope. Immunochemistry 8:1081–1083
18. Hughes D (2005) Immunogold probes in light microscopy. Methods Mol Biol 295:173–192
19. Hainfeld JF, Powell RD (2000) New frontiers in gold labeling. J Histochem Cytochem 48:471–480
20. Hermann R, Walther P (1996) Immunogold labeling in scanning electron microscopy. Histochem Cell Biol 106:31–39
21. Verkleij A, Koster A, Muller W, Humbel B (1999) Immuno-Gold Labeling in Transmission Electron Microscopy. In: Thiery JP (ed) Molecular mechanisms of transcellular signaling: from membrane receptors to transcription factors. IOP Press, Amsterdam, pp 339–356
22. El-Sayed IH, Huang X, El-Sayed MA (2005) Surface plasmon resonance scattering and absorption of anti-EGFR antibody conjugated gold nanoparticles in cancer diagnostics: applications in oral cancer. Nano Lett 5:829–834
23. Hainfeld J (1990) STEM analysis of Janssen Auroprobe One. In: GW Bailey (ed) Proc. Internat Cong Electron Microsc. San Francisco: 954
24. Kramarcy NR, Sealock R (1991) Commercial preparations of colloidal gold-antibody complexes frequently contain free active antibody. J Histochem Cytochem 39:37–39
25. Bange J, Zwick E (2001) Molecular targets for breast cancer therapy and prevention. Nat Med 7:548–552
26. Copland J, Eghtedari M, Popov V et al (2004) Bioconjugated gold nanoparticles as a molecular based contrast agent: implications for imaging of deep tumors using optoacoustic tomography. Mol Imaging Biol 6:341–349
27. Hainfeld JF, Furuya FR (1992) A 1.4-nm gold cluster covalently attached to antibodies improves immunolabeling. J Histochem Cytochem 40:177–184
28. Sun XJ, Tolbert LP, Hildebrand JG (1995) Using laser scanning confocal microscopy as a guide for electron microscopic study: a simple method for correlation of light and electron microscopy. J Histochem Cytochem 43: 329–335
29. Hainfeld J, Furuya F (1995) Immunogold-silver staining: principles, methods, and applications, 1st edn. CRC, Boca Raton
30. Scopsi L, Larsson LI, Bastholm L, Nielsen MH (1986) Silver-enhanced colloidal gold probes as markers for scanning electron microscopy. Histochemistry 86:35–41
31. Skripkin E, Yusupova G, Yusupov M et al (1993) Synthesis and ribosome binding properties of model mRNAs modified with undecagold cluster. Bioconjug Chem 4:549–553
32. Segond von Banchet G, Heppelmann B (1995) Non-radioactive localization of substance P binding sites in rat brain and spinal cord using peptides labeled with 1.4-nm gold particles. J Histochem Cytochem 43:821–827
33. Lipka JJ, Hainfeld JF, Wall JS (1983) Undecagold labeling of a glycoprotein: STEM visualization of an undecagoldphosphine cluster labeling the carbohydrate sites of human haptoglobin-hemoglobin complex. J Ultrastruct Res 84:120–129
34. Kogot JM, England HJ, Strouse GF, Logan TM (2008) Single peptide assembly onto a 1.5 nm Au surface via a histidine tag. J Am Chem Soc 130:16156–16157
35. Walter M, Akola J, Lopez-Acevedo O et al (2008) A unified view of ligand-protected gold clusters as superatom complexes. Proc Natl Acad Sci USA 105:9157–9162
36. Zhu M, Aikens CM, Hollander FJ et al (2008) Correlating the crystal structure of a thiol-protected Au25 cluster and optical properties. J Am Chem Soc 130:5883–5885
37. Qian H, Eckenhoff WT, Zhu Y et al (2010) Total structure determination of thiolate-protected Au38 nanoparticles. J Am Chem Soc 132:8280–8281

38. Jadzinsky PD, Calero G, Ackerson CJ et al (2007) Structure of a thiol monolayer-protected gold nanoparticle at 1.1 A resolution. Science 318:430–433
39. Ackerson CJ, Jadzinsky PD, Jensen GJ, Kornberg RD (2006) Rigid, specific, and discrete gold nanoparticle/antibody conjugates. J Am Chem Soc 128:2635–2640
40. Aubin-Tam ME, Hamad-Schifferli K (2005) Gold nanoparticle—cytochrome c complexes: the effect of nanoparticle ligand charge on protein structure. Langmuir 21: 12080–12084
41. Aubin-Tam ME, Hamad-Schifferli K (2008) Structure and function of nanoparticle-protein conjugates. Biomed Mater 3:034001
42. Aubin-Tam ME, Hwang W, Hamad-Schifferli K (2009) Site-directed nanoparticle labeling of cytochrome c. Proc Natl Acad Sci USA 106:4095–4100
43. Ackerson CJ, Jadzinsky PD, Sexton JZ et al (2010) Synthesis and bioconjugation of 2 and 3 nm-diameter gold nanoparticles. Bioconjug Chem 21:214–218
44. Lévy R, Thanh NTK, Doty RC et al (2004) Rational and combinatorial design of peptide capping ligands for gold nanoparticles. J Am Chem Soc 126:10076–10084
45. Krpeti Z, Nativo P, Porta F, Brust M (2009) A multidentate peptide for stabilization and facile bioconjugation of gold nanoparticles. Bioconjug Chem 20:619–624
46. Ackerson CJ, Sykes MT, Kornberg RD (2005) Defined DNA/nanoparticle conjugates. Proc Natl Acad Sci USA 102:13383–13385
47. Zhu ZJ, Ghosh PS, Miranda OR et al (2008) Multiplexed screening of cellular uptake of gold nanoparticles using laser desorption/ionization mass spectrometry. J Am Chem Soc 130:14139–14143
48. Ackerson CJ, Jadzinsky PD et al (2010) Synthesis and bioconjugation of 2 and 3 nm-diameter gold nanoparticles. Bioconjug Chem 21(2):214–218

# Part III

## Scanning Probe Microscopy

# Chapter 18

# Atomic Force Microscopy Imaging of Macromolecular Complexes

**Sergio Santos, Daniel Billingsley, and Neil Thomson**


## Abstract

This chapter reviews amplitude modulation (AM) AFM in air and its applications to high-resolution imaging and interpretation of macromolecular complexes. We discuss single DNA molecular imaging and DNA–protein interactions, such as those with topoisomerases and RNA polymerase. We show how relative humidity can have a major influence on resolution and contrast and how it can also affect conformational switching of supercoiled DNA. Four regimes of AFM tip–sample interaction in air are defined and described, and relate to water perturbation and/or intermittent mechanical contact of the tip with either the molecular sample or the surface. Precise control and understanding of the AFM operational parameters is shown to allow the user to switch between these different regimes: an interpretation of the origins of topographical contrast is given for each regime. Perpetual water contact is shown to lead to a high-resolution mode of operation, which we term SASS (small amplitude small set-point) imaging, and which maximizes resolution while greatly decreasing tip and sample wear and any noise due to perturbation of the surface water. Thus, this chapter provides sufficient information to reliably control the AFM in the AM AFM mode of operation in order to image both heterogeneous samples and single macromolecules including complexes, with high resolution and with reproducibility. A brief introduction to AFM, its versatility and applications to biology is also given while providing references to key work and general reviews in the field.


**Key words:** AFM, Biomolecules, Macromolecular complexes, Double-stranded DNA, RNA polymerase, Topoisomerase, Water, Height

## 1. Introduction

Atomic force microscopy (AFM) is part of the scanning probe microscopy (SPM) family (1–7) and since the time of its invention in 1986, there has been a vast amount of literature covering the different modes in which it can be operated and how to obtain high-resolution images (8–12). The basic principle of the AFM is to probe a surface with a sharp tip mounted on a micro-cantilever which is brought into close proximity with the sample surface to

Alioscka A. Sousa and Michael J. Kruhlak (eds.), *Nanoimaging: Methods and Protocols*, Methods in Molecular Biology, vol. 950, DOI 10.1007/978-1-62703-137-0_18, © Springer Science+Business Media, LLC 2013

characterize it (13, 14). The tip has an end curvature in the range of nanometers, and its size ultimately limits the resolution. There are several modes of operation, which can be classified as quasi-static, i.e., DC and dynamic, i.e., AC modes (dAFM), while the instrument can also be operated in the whole range of environmental conditions such as ultra high vacuum, ambient and liquid. In the DC modes, the deflection of the cantilever is static and several types of information such as topography and friction contrast can be obtained. In the dynamic modes, the cantilever is vibrated in the proximity of the surface as opposed to contact modes where the tip is pressed onto the surface and repulsive forces control the interaction throughout. As the tip senses the surface forces, these begin to control the behavior of the cantilever. Dynamic modes were developed almost immediately after the contact mode of operation and these have several advantages, for example, (1) the stored cantilever energy inhibits trapping of the tip on the surface (9, 15, 16) and (2) the surface is typically only intermittently contacted (and only in the normal direction) thus reducing damaging lateral forces (17). The excitation of higher modes and harmonics is quite a recent development in the dAFM repertoire that might allow determination and simultaneous acquisition of mechanical and chemical maps at nanometre length scales (12, 18–24). Nevertheless, in this chapter we focus on dAFM imaging using the fundamental frequency, since optimization of the instrument using this vibrational frequency will assist all future dynamic modes, including multifrequency methods. We show how to optimize the operational parameters and use variations in relative humidity RH to probe macromolecular complexes. There are several reasons why detailed understanding of the dynamics of the cantilever are advantageous when studying biomolecular interactions and complexes. In particular, in the dynamic modes, which are best suited for biomolecular investigations, the nonlinear character of the surface forces can make the operation and the understanding of contrast challenging. Nevertheless, arguably, the complexity of the interaction is also responsible for the versatility of the instrument where different key modes of operation can be used.

### 1.1. Brief Introduction to Dynamic AFM, Versatility and Its Use in Biology

The usefulness of the AFM in biology was recognized almost as soon as it was invented (3, 4, 25, 26). The dynamic AFM (9, 27, 28) modes have been essential for the progression of biological imaging since these methods eliminate damaging shear forces that arise in contact mode (4, 9, 10, 17). The precise control of the xyz-position in AFM is possible thanks to the high sensitivity of the dimensions of piezoelectric materials to applied voltages (1, 5). These are responsible for controlled $x$, $y$ and $z$ motion and are capable of sub-angstrom precision (25). In the most common setup, a laser beam is focused on the end of the cantilever and the deflection is detected by a split photodiode. The signal is then processed by the control system and used as feedback to act on the $z$-piezo actuator. An important difference between the AFM and the Scanning Tunnelling Microscope (STM) is that the AFM can image both electrical insulators and conductors. It should be noticed that this is a key advantage in biological applications where conductivity is rare (1, 4, 29). Furthermore, the AFM can be operated in all environments, including air and liquid (10, 30), allowing biomolecules and biological samples to remain hydrated and often active (3, 4, 25, 31). The operating temperatures also offer a large range of possibilities since these can vary from room temperature or above down to liquid-nitrogen and even lower temperatures (4, 32). Two major dAFM methods exist in the AFM community and these two branches are a historical result of the development of AFM where ultra high vacuum and ambient or liquid imaging developed as two separate disciplines (5, 16, 29). Typically, vacuum imaging was performed with a Frequency Modulation (FM) feedback system (27) whereas an Amplitude Modulation (AM) is preferred feedback for liquid (10) and ambient imaging (9, 28). Good reviews on the emergence of the two branches can be found in the literature (5, 27). In short, AM is typically too slow in vacuum whereas FM loses sensitivity with small values of $Q$ in air and liquid environments. In general, it is likely that these two techniques or modes of operation will be used as complementary tools in research (5, 6, 13, 14, 33). Finally, several extensive reviews can be found in the literature, where the imaging of biomolecular (4, 14, 34), cell (13) and other systems (3, 5, 14, 35, 36) are discussed in detail. The number of applications of the AFM in biology is increasing every year (14).

## 2. Materials

### 2.1. Mica for Sample Immobilization

Muscovite mica substrates are typically used for sample immobilization of macromolecular complexes such as DNA, antibodies, RNA, and enzymes. In this chapter, we specifically discuss DNA, DNA–RNA Polymerase, and DNA–Topoisomerase interactions: preparations are described below.

1. Muscovite mica in 3″ × 1′ sheets (Agar Scientific Ltd., UK).
2. Five minute epoxy glue (ITW Devcon, UK).
3. Steel SPM specimen discs measuring 12 or 15 mm in diameter (Agar Scientific Ltd.).
4. Scotch Magic™ tape.

5. "Punch and die" set for making circular mica discs. A set can be purchased from Precision Brand Products Inc (Downers Grove, USA).

### 2.2. Bacterial Linear and Supercoiled DNA, Topoisomerase VI, and Other Sample Preparation Materials

1. DNA stock solutions should be diluted into a deposition buffer, for example, 4 mM tris(hydroxymethyl)aminomethane (TRIS–HCl) (pH 7.5) and 4 mM $MgCl_2$. The stock solution can be pipetted using small aliquots (1–5 μl) into eppendorf tubes and flash frozen in liquid nitrogen prior to storage in a −80°C freezer. The DNA solution can also be stored at 4°C and be expected to be preserved for several months.
2. *E. coli* RNA polymerase holoenzyme (Epicentre, USA). The holoenzyme should be 100% saturated with sigma subunit ($\sigma^{70}$), enabling specific binding to promoter sites situated on the DNA template. Store at −20°C. This storage leads to no loss of activity being observed over several months.
3. Topoisomerase VI stock solution (gift of JA Taylor, John Innes Centre, Norwich) should be diluted into a deposition buffer, for example, 4 mM TRIS–HCl (pH 7.5) and 4 mM $MgCl_2$. The stock solution can be pipetted using small aliquots (1–5 μl) into eppendorf tubes and flash frozen in liquid nitrogen prior to storage in a −80°C. Small aliquots could be stored at 4°C but it is preferable to make a fresh solution frequently.
4. Milli Q water (Millipore (UK) Ltd., UK).
5. Transcription buffer: 20 mM Tris–HCl at pH 7.9, 50 mM KCl, 5 mM $MgCl_2$, 1 mM DTT.
6. Gilson pipettes (Gilson Inc.,USA).
7. Oxygen-free Nitrogen gas—cylinder exit pressure: 1 bar.

### 2.3. AFM and Other Instrumentation

1. Commercial Multimode AFM equipped with a 14 μm scanner, E-scanner (Digital Instruments, Veeco Metrology Inc., USA). Nanoscope IIIa controller or later version required.
2. Diving board shaped silicon micro-cantilevers of 160 μm in length, nominal spring constant of $k = 42$ N/m and resonant frequency 300 kHz (Olympus, Tokyo, Japan).
3. Low vacuum plastic hood (Nalgene, Rochester, USA). The relative humidity can be controlled by placing the AFM inside the plastic hood. This allows isolation from the surrounding laboratory environment.
4. A digital thermo hygrometer (Omega Engineering Inc., USA) inserted into the plastic chamber for relative humidity and temperature measurements. The quoted error in relative humidity is ±3% RH.

## 3. Methods

### 3.1. Sample Preparation

#### 3.1.1. Preparation of Mica Supports for Sample Immobilization

Samples such as individual biomolecules or macromolecular complexes must usually be supported on a surface or substrate for AFM imaging. Some typical substrates are Muscovite mica, glass, and silicon oxide (4, 29). While glass can be used to image large biological samples such as cells it is too rough for single molecules and mica is the support surface that is typically chosen for biomolecular imaging. The surface of mica is atomically flat over large distances (microns to millimeters) (37) allowing easy discrimination of the molecules by the AFM tip (3, 4, 29, 38, 39). The high signal-to-noise of AFM allows good contrast without the need for additional agents. Mica can be cleaved readily along the basal plane, enabling a contamination-free, fresh surface on which to deposit molecules. It has taken much effort and research to find the appropriate conditions for optimum binding of DNA (39, 40) and this pursuit is still ongoing. The chemical and surface structure of mica is well known and usually divalent counterions, either Group II or transition metals such as Mg(II) or Ni(II), are used to add binding of DNA or proteins. This is important in AFM experiments since, as molecules are deposited onto the surface, these interact with it through physisorption and other surface intermolecular forces which are sometimes required as a known parameter in order to interpret data (34); the net forces form a complex interplay between physisorption and chemisorption processes depending on the chemical structure of the deposited molecule and the chemical structure and thermodynamic state of the surface. It is, however, sometimes not a welcome phenomenon that the perfect atomic flatness of mica might be disturbed by the presence of potassium carbonate crystallites and other salts on its surface (37, 41). Moreover, water layers might form sometimes perturbing the tip–sample interaction (42) shielding it enough as to lose a relatively large amount of the topographical information. We discuss methods later in the chapter to interpret data and choose modes of operation where water perturbations are either minimized or used as an advantage.

1. Punch out circular mica sheets to fit on the steel SPM specimen discs using the 10–15 mm holes on the "punch and die" set and a hammer or mallet.
2. Mix the two component epoxy glue and place 5–10 μl of the mixture in the center of a specimen disc, using a toothpick or a plastic pipette tip.
3. Place the mica disc in the center of the specimen disc and press gently with tweezers so a continuous and thin layer of glue forms between the mica and the specimen (the glue should be

evenly spread on the disc and form a thin and homogeneous layer with no bubbles).

4. Let the glue dry for 24–48 h.

*3.1.2. Preparation of DNA and DNA–Topoisomerase and DNA–RNA Polymerase Complexes*

1. *DNA:* Dilute the DNA solution down to concentrations of 1–4 nM DNA. We have noticed that these concentrations are preferred for optimal binding to the mica surface.
2. *DNA topoisomerase VI:* Pipette DNA and topoisomerase VI solutions into eppendorf tubes at a 1–2 or 1–3 ratio. It is preferable to first prepare the solution at higher concentrations, i.e., 20 mM for DNA and 40–60 mM for topoisomerase VI, to maximize specific binding in the solution. Small aliquots could be stored at 4°C but it is preferable to make a fresh solution frequently.
3. *DNA-RNA polymerase (RNAP):* Initiation complexes can be formed by taking 200 fmol of a linear DNA fragment containing two $\lambda_{PR}$ promoter sites and incubating it at 37°C for 15 min with 400 fmol of *E. coli* $\sigma^{70}$ holoenzyme and 10 µl of transcription buffer. This initial step enables the formation of open promoter complexes (OPC). Transcription elongation can be initiated through the addition of NTPs to the reaction mix. Stalled elongation complexes (SEC) are formed by adding ATP, GTP, and UTP to a final concentration of 200 µM and incubating at room temperature for 15 min. Due to the absence of CTP, the RNAPs will stall at the first downstream occurrence of guanine on the template strand, from each promoter site. If all four NTPs are added to the reaction mix then it is expected that transcription elongation can proceed unhindered until RNAPs traveling in opposite directions begin to interact with each other. Collided complexes (CCs) are formed in one of two ways. In the two-step method, 200 µM of CTP is added to the mix containing SECs, before being left to incubate at room temperature for 15 min. This allows stalling and reinitiation to be studied. The one-step method involves adding all four NTPs to the reaction mix containing OPCs and incubating at RT for 15 min (see Note 1).

*3.1.3. Physisorption of Biomacromolecules onto Mica*

1. Dilute the stock solution of the macromolecules until you obtain 1–4 nM of DNA in the solution, into the deposition buffer.
2. Cleave the mica with the scotch tape.
3. Pipette 10–20 µl of the diluted solution onto the freshly cleaved mica support.
4. Incubate the sample for approximately 20 s before rinsing in excess Milli Q water and subsequently drying in a stream of dry $N_2$ gas (1 bar pressure at a distance of several centimeter).

Rinsing can be achieved by running up to 5 ml of MilliQ water across the mica sample while it is tilted at a 30–45° angle. The degree of rinsing may affect the density of molecules left on the surface after drying.

5. Transport the completed specimen disc onto the piezo tube scanner.

### 3.2. Setting Up the AFM

1. Mount the cantilever chip into the air cantilever holder using fine, square-ended tweezers.
2. Mount the cantilever holder into the AFM head and then the head onto the tube scanner.
3. Align the laser on the back of the cantilever as in standard AFM protocols.
4. Tune the cantilever to find the resonant frequency and set the drive frequency.
5. Bring the AFM tip towards the sample manually using an optical microscope setup. Typical separation remains in the micron scale range before automated approach is used.
6. Ensure the scan size is set to zero on engage, so that the tip–sample interactions can be optimized before imaging is commenced.
7. Use the auto-engage function of the AFM. The tapping engage ratio can be set to over 1 to ensure that the tip is partially withdrawn from the sample immediately after tip engage. A good value is 1.1.
8. Manual approach can then be made by adjusting the set-point accordingly in fine increments. The phase signal can be monitored to assist evaluation of the tip–sample interaction.
9. When scanning commences, increase the scan size in increments to ensure the sample does not contain loose molecules that could contaminate the tip and that the sample is flat with no contamination. E.g., start at 100 nm and then go to 1 μm and then larger scan sizes if necessary.
10. For biomolecular imaging we recommend that the pixel size does not go above 6 nm (e.g., 3 μm scan size at 512 point digitization) since the resolution of the image will be too low and lack of fine feedback control may lead to tip wear and damage.
11. For specific details of how to run the AFM in the noncontact, intermittent contact (tapping) and SASS (small amplitude small set-point) modes, see Subheading 3.4 onwards below.

### 3.3. Humidity Control

#### 3.3.1. Setting Up the Humidity Control Chamber

1. Place the AFM unit inside the chamber's control base (Fig. 1).
2. Place a beaker of desiccant or water on the control chamber base. Adding a little water, i.e., 10 ml, will probably result in a value of relative humidity (RH) lower than 90%. If 50 ml or

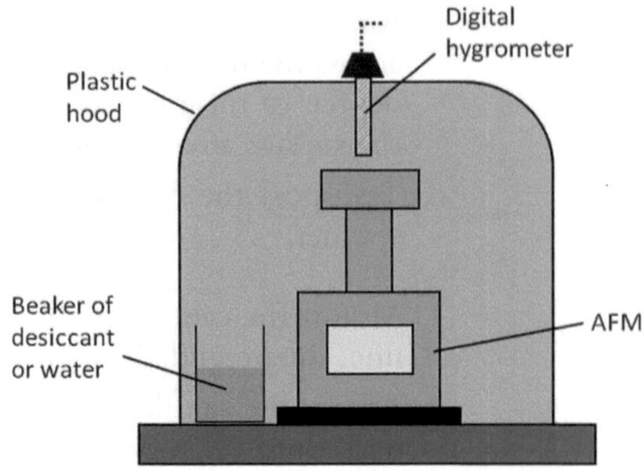


Fig. 1. Experimental setup used for controlled-humidity imaging. The AFM is placed inside a hood to isolate it from the laboratory environment. The humidity is raised and lowered by the addition of a beaker of water or desiccant, respectively. The humidity inside the hood is measured using a digital hygrometer.

more of water at room temperature is added the value of RH will get closer to 95% or more in 0.5–2 h of exposure. The rate at which a given RH is obtained depends on the volume of the control chamber and the amount of water/desiccant added to the beaker. In our experiments we typically use a chamber of approximately 0.01–0.02 $m^3$ (Fig. 1). If a constant value of RH is required, the beaker can be filled with a saturated salt solution. Different soluble salts give rise to different vapor pressures in the chamber. Since conventional AFM imaging is slow, we use pure water, because the RH does not change by more than 1% during acquisition of an image.

3. Set up the AFM unit for imaging as detailed in Subheading 3.2 above up to step 7.
4. Place the lid on top of the chamber's control base.
5. Place the digital hygrometer into the hole at the top of the lid as in Fig. 1.
6. The hole into which the digital hygrometer fits should be insulated preventing vapor to leave or enter the chamber.
7. Use the digital hygrometer to monitor the humidity and also, if required, the temperature. Wait until the relative humidity reaches the appropriate value and stabilizes. With stability here we imply that the relative humidity has reached a steady state where its value does not fluctuate more than approximately 1–2% during the experimental run (see Note 2).

*3.3.2. Biomolecular Studies in Ambient AM AFM with Humidity Control*

1. By following the procedures discussed in the previous section, first set a relatively low value of relative humidity RH (Fig. 2a) by placing a given amount of desiccant in the chamber (see Subheading 3.3.1).

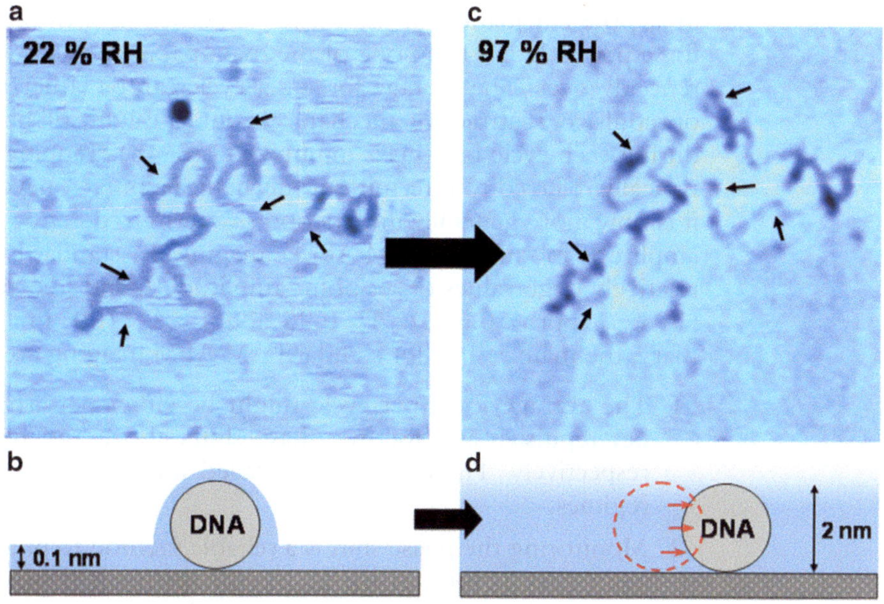


Fig. 2. Summary showing typical conformational changes in closed-circular supercoiled DNA as the humidity is increased, with a corresponding increase in the thickness of the surface water layer. Topological rearrangements in the DNA are observed to induce DNA condensation, kinking of the backbone, and changes in the backbone trajectory. A number of these changes are highlighted with *arrows*. Reproduced from ref. 43 with permission from the PCCP Owner Societies.

2. Take a standard Tapping Mode (TM) AFM image of a single (or multiple) DNA molecule/s (standard imaging procedures are discussed in Subheading 3.4) as shown in Fig. 2a.
3. To observe humidity induced topographic rearrangements of the DNA molecule the relative humidity in the chamber should be increased (Fig. 2b, d). This is achieved by first disengaging the tip from the surface while maintaining the lateral xy-piezo position. Then the desiccant is replaced by a beaker of water or saturated salt solution (Subheading 3.3.1).
4. When the relative humidity has reached a steady state at RH = 95% or above (Fig. 2c, d), take another image of the molecule that has been previously imaged at a low value of RH (see Note 2). Now topologically induced rearrangements of the DNA molecule should be observed to induce DNA condensation, kinking of the backbone, and changes in the backbone trajectory (see Note 3) (43).

### 3.4. Amplitude Modulation Imaging (Tapping Mode)

#### 3.4.1. Introduction to Amplitude Modulation (Tapping Mode) Imaging

Historically, AM AFM has been the main AFM technique applied to biological systems since its invention (5, 25). The first image of single DNA molecules with a Scanning Probe Microscope (SPM) was obtained by Binnig et al. (44) in 1984 with an STM, before the AFM was invented (4). Since that time, the AFM has been used widely to study biomolecular samples and several good reviews can be found in the literature (1, 4, 10, 13, 14, 26, 29, 30). Nevertheless, in air, the elucidation of the existence and control of the average force per cycle has been one of the main parameters allowing current advances. The average force per cycle can be either attractive or repulsive. If attractive, mechanical contact might still be induced during each cycle but this is minimized (45–48). Tapping mode is typically referred to intermittent mechanical contact, nevertheless, it is now more generally understood as a term to describe standard amplitude modulation (AM AFM) imaging. By standard we imply relatively large values of set-point amplitude A and free amplitude $A_0$, i.e., approx $0.3 < A/A_0 < 0.95$ and $1 < A_0 < 100$ nm. In summary, the main results for standard AM AFM imaging are (46):

1. The average force per cycle might be either net attractive or net repulsive and follows phase shifts above and below 90° respectively. These are the so-called attractive and repulsive regimes.
2. Monitoring the phase shift is a reliable method to predict average forces as long as the amplitude set-point is not too low (i.e., $A/A_0 > 0.2$) and/or the energy dissipated in the tip–sample interaction is relatively small compared to the stored energy (49, 50).
3. There might be either one or two physically stable amplitudes for a given cantilever–sample equilibrium position *zc*. When two solutions exist these are termed the L and H state, where the L state corresponds to the cantilever vibrating at a position higher above the sample. Typically, these two states correspond to the attractive and the repulsive regime.
4. We can define a critical value of free amplitude $A_c$ corresponding to the minimum value of $A_0$ for which a transition between force regimes might be observed in amplitude–distance (AD) curves. We describe how to find this value in the next subsections below. The main principle is that below and above $A_c$ the attractive and the repulsive force regimes prevail respectively (42, 46, 51).

#### 3.4.2. The Attractive and Repulsive Force Regimes

1. *Attractive regime:* The attractive force regime is defined as a net attractive average force per cycle where, during one cycle, repulsive forces might or might not be present. It is convention, however, to define the attractive force regime as that for which the phase lag of the cantilever with respect to the drive is larger than 90°. While this convention typically agrees with the above more physically significant definition, there are cases for which it might not (5, 29, 47).
2. *Repulsive regime:* The repulsive force regime is defined as a net repulsive average force per cycle where, during one cycle, attractive forces might or might not be present. It is convention, however, to define the repulsive force regime as that for

which the phase lag of the cantilever with respect to the drive is less than 90°. While this convention typically agrees with the above more physically significant definition, there are cases for which it might not (*see* references above).

3. *Amplitude–Phase–Distance (APD) curves:* In APD curves the tip–surface separation is slowly decreased and then increased while monitoring the behavior of the amplitude A (AD curves, Fig. 3a) phase Φ (AP curves, Fig. 3b) or both (APD curves). In simulations, other distance curves such as force and contact time distance curves can be acquired. An example of APD curves is given in Fig. 3.
4. *Critical amplitude $A_c$:* Force regimes can be monitored experimentally and in simulations via the phase lag (see Table 1 and Fig. 3). Then, the value of $A_c$ can be found by acquiring amplitude

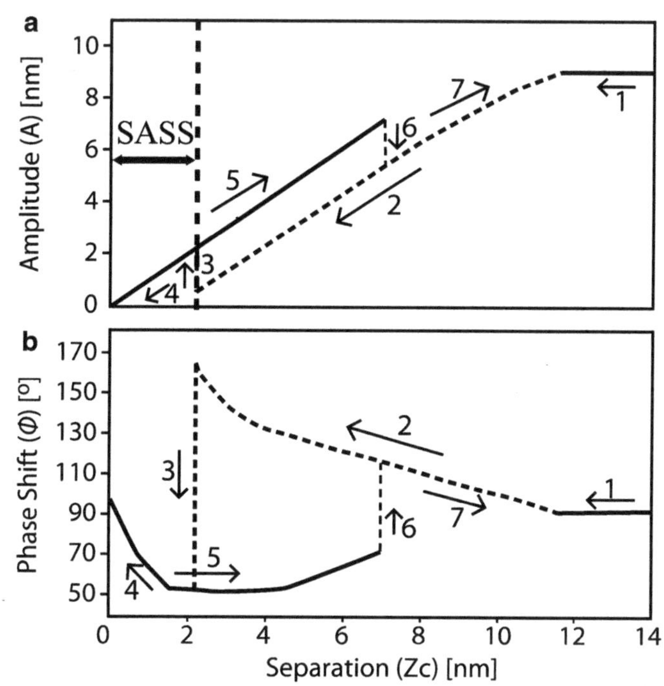


Fig. 3. (**a**) Amplitude–distance curve performed on a mica surface at 40% RH for a silicon rectangular cantilever with nominal spring constant of 40 N/m driven at resonance ($f_0 \approx 300$ kHz) with a free amplitude $A_0 \approx 10$ nm. (**b**) Corresponding phase–distance curve where a step-like discontinuity is observed as the cantilever goes from the L (attractive regime) to the H (repulsive regime) state when approaching the sample and from the H to the L state during retraction. The path followed is marked from 1 to 7, for the cantilever approach and retraction. The region labeled SASS is a region of the parameter space that typically involves perpetual water contact (29, 42, 54) and is discussed in Subheading 3.7. The free amplitude ($A_0$) of the cantilever in this example is nearly 10 nm and corresponds to the value of $z_c$ when the tip begins to be perturbed by the sample. In air, there is always a 1 to 1 correspondence between amplitude and tip–sample separation, i.e., the amplitude decreases in a linear fashion during approach (notwithstanding the perturbations caused by switching force regimes).

**Table 1**
**Outcomes for a hydrophilic tip interacting with a heterogeneous sample, i.e., regions where there is only the mica surface and regions where we have samples such as DNA molecules**

| $A_0$ relative to $A_c$ | Substrate | | Sample (i.e., DNA) | | Are water layers perturbed (CT > 0)? | Mechanical contact |
|---|---|---|---|---|---|---|
| | Case | Phase | Case | Phase | | |
| $A_0 << 1/2A_c$ | $W_{nc}$ | >90° | $W_{nc}$ | >90° | No | No |
| $A_0 \sim 1/2A_c$ | $W_{nc}$ | >90° | $W_c$ | >90° | | |
| | $W_c$ | >90° | $W_{nc}$ | >90° | | |
| | $W_c$ | >90° | $W_c$ | >90° | | |
| $A_0 \sim A_c$ | $W_c$ | >90° | $W_c$ | >90° | | |
| | Rep. | >90° | $W_c$ | >90° | | Yes |
| | $W_c$ | <90° | Rep. | <90° | Yes | |
| | Rep. | <90° | Rep. | <90° | | |
| $A_0 >> A_c$ | $W_c$ | >90° | Rep. | <90° | | |
| | Rep. | <90° | $W_c$ | >90° | | |
| | Rep. | <90° | Rep. | <90° | | |
| | $W_c$ | >90° | $W_c$ | >90° | | No |

Here we use the value of $A_c$ to define four regimes of operation. It is only when $A_0 << 1/2A_c$ that the water layers remain unperturbed throughout. CT stands for contact time. It is only in this case that the tip is in true noncontact on both hydrophobic (or un-wet) and hydrophilic (or wet) areas, i.e., the $W_{nc}/W_{nc}$ case. When $A_0 \sim 1/2A_c$, the water layers are typically perturbed on one and/or the other regions even if no mechanical contact occurs. This involves any combination of $W_c$ and $W_{nc}$, except for $W_{nc}/W_{nc}$, as shown in the table. Increasing the free amplitude to $A_0 \sim A_c$ typically results in a transition to the repulsive regime where intermittent mechanical contact occurs. Here, any combination of Rep. and $W_c$ can result but the Rep./$W_c$ scenario is the most common. When $A_0 >> A_c$, the repulsive regime typically prevails on both samples, that is only the Rep./Rep. case typically follows. The attractive and the repulsive regimes can be monitored or predicted by recording the phase lag where >90 and <90 stand for the attractive and repulsive regimes, respectively. The five common cases that we have emphasized in the article in Figs. 2 ($W_{nc}/W_{nc}$ and $W_{nc}/W_c$) and 4 ($W_c/W_c$, Rep./$W_c$, and Rep./Rep.) are shaded in light gray. Adapted from ref. 42

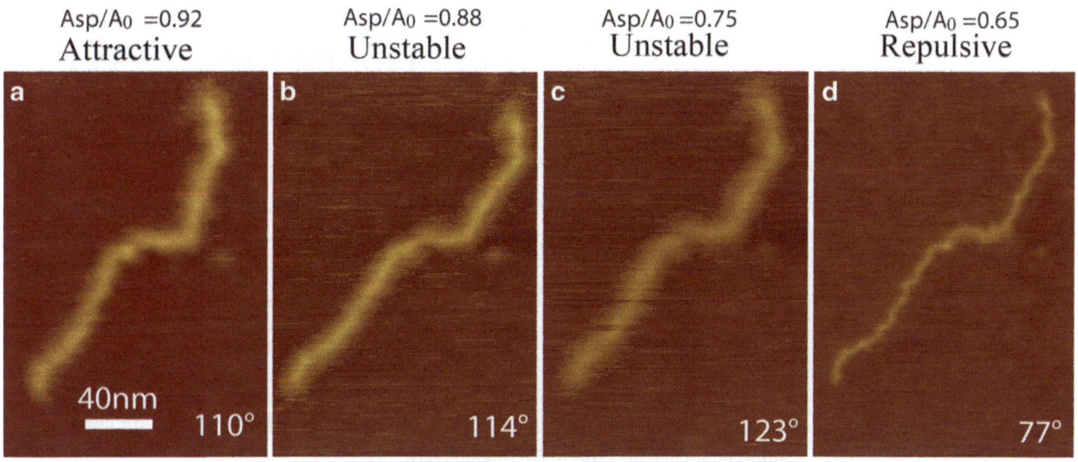


Fig. 4. Topographic images (*Z*-piezo signal) of a single 800 bp dsDNA molecule on mica obtained in a sequence where the value of $A_0 = 10$ nm is larger than the critical value $A_c$ (here $A_c < 10$ nm). The average phase shift Φ for every scan is shown on the *bottom right* in every image and allows differentiating between the L (attractive) and the H (repulsive) state (regimes). Experimental parameters: $A_0 \approx 9$ nm, *k* (spring constant) ≈ 40 N/m, $f = f_0 \approx 300$ kHz (drive and natural frequencies), and *Q* factor ≈ 500.

phase (AP) distance curves (42, 51) and gradually increasing $A_0$ until the phase lag is observed to lie below 90°, i.e., the onset of the repulsive regime is observed. In Fig. 3 the value of $A_c$ has already been reached or surpassed since the repulsive regime is observed to control the interaction on retraction. Thus, in this case, $A_c < 10$ nm. The value of $A_c$ typically lies within a range and this range is a consequence of the stochastic nature of the force transitions (42, 52, 53).

5. *Example of attractive/repulsive interaction regimes and effects on the topography of a single DNA molecule*: Figure 4 shows a sequence of topographic images of a single DNA molecule on mica obtained with the same cantilever-tip used to acquire the APD curve shown in Fig. 3. From a to d, the normalized amplitude set-point ($A/A_0$) has been systematically reduced from 0.92 to 0.65. The same free amplitude of $A_0$ ($A_0 = 10$ nm) as in Fig. 3 for the APD has been used here. Since $A_c < 10$ nm, it is clear that the repulsive regime (H state) should be reached with this value of $A_0$ only by sufficiently decreasing the set-point amplitude, A. In Fig. 4a, the cantilever is vibrating in the L-state with attractive forces on average. In Fig. 4b, c, there is an unstable region where the cantilever is predominantly in the L-state but transiently switches to the H-state. The H state (repulsive regime) has been stably reached at intermediate values of set-point (Fig. 4d) as predicted by the APD curve in Fig. 3. Both apparent molecular height and width display dramatic differences in the L state (attractive regime) as compared to the H state (repulsive regime). This is also shown in Fig. 5.

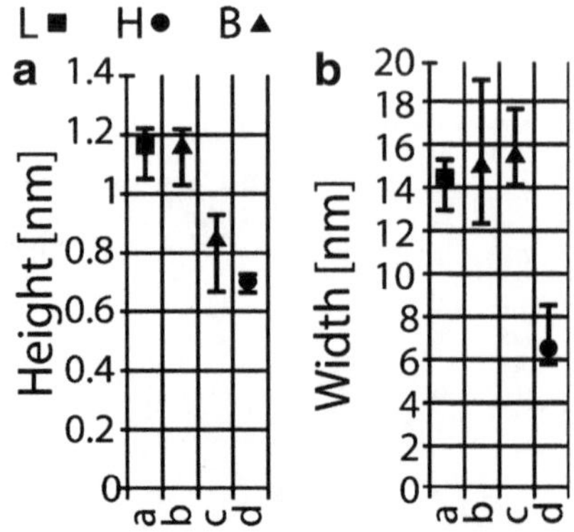


Fig. 5. (**a**) Apparent height and (**b**) width corresponding to Fig. 4. The L (attractive) and H (repulsive) states (force regimes) and severe bi-stable behavior (B) are differentiated by *squares*, *circles*, and *triangles*, respectively.

Note that the noise observed in Fig. 4b, c is not a consequence of an inappropriate choice of integral and proportional gains but a consequence of switching between states (46, 52). That is, even optimization of the feedback gains can be insufficient to induce stability if the cantilever dynamics do not permit one stable oscillation state (55): in these situations no choice of gains makes the noise disappear.

### 3.5. The Effects of the Water Layers on Apparent Height

In the section above we have discussed the effects of varying the free amplitude to reach the attractive and the repulsive force regimes and how these can lead to different degrees of lateral resolution and apparent height of biomolecules. Here, we further distinguish between intermittent water contact and pure noncontact regimes where the water layers on either the tip, the surface (here the mica substrate), or the sample (here the biomolecules) might be disturbed in some cases while in other (pure noncontact mode) might not (42).

Generality can be provided if the concept of critical free amplitude $A_c$ is also used here (42, 45, 46, 56). As stated, $A_c$ corresponds to the minimum value of $A_0$ for which a transition between force regimes might be observed in AD curves (see Fig. 3) (46, 51). Recall from above that the force regimes can be monitored experimentally and in simulations via the phase lag (see Table 1). Thus, the $A_c$ value for a given sample can be found by acquiring amplitude phase (AP) distance curves (51) and gradually increasing $A_0$ until the phase lag is observed to lie below 90°, i.e., the repulsive regime (42). We then proceed to define four types of interaction depending on whether the water layers are disturbed on either the mica surface of the sample (i.e., the DNA molecules in this case) and whether the repulsive regime, where intermittent mechanical contact is made, is reached. These four types of interactions can be described generally in terms of $A_c$. See below.

*3.5.1. Definition of Interactions Where There Is Water on the Tip and/or the Sample*

We write $W_{nc}$, $W_{c,}$ and Rep. for situations where the water layers on the tip are never in contact with the water layers on the sample ($W_{nc}$), the water layers are perturbed but intermittent mechanical contact does not occur ($W_c$), and Rep. when intermittent mechanical contact between the tip and the sample or surface occurs (i.e., a repulsive interaction). Now we write a convention where the first terms refer to the interaction between the tip and the substrate and the second refers to an interaction between the tip and the biomolecular sample. In this way we write:

1. $W_{nc}/W_{nc}$ for an interaction where the water layers are never perturbed on either the sample or the substrate (see Table 1).
2. $W_c/W_c$ for interactions where mechanical contact does not occur but the water layers are being perturbed. This might also include $W_{nc}/W_c$ and other combinations (42).
3. Rep./$W_c$ for interactions where mechanical contact might occur on the sample but not on the substrate. In this case, only the water layers are perturbed on the mica substrate, but this could also occur on the other way around, i.e., $W_c$/Rep. (42).
4. Rep./Rep. for interactions where mechanical contact occurs throughout (42).

*3.5.2. Definition of Interactions Where There Is Water on the Tip and/or the Sample in Terms of $A_c$*

Since the numerical values of $A_0$ and $A$ for interaction forces or regimes taking place might differ from system to system, we write the above points generally in terms of $A_c$. We discuss the possibilities in terms of $A_c$ and with the help of Table 1 and Fig. 6. In Fig. 6, dsDNA molecules are imaged on a mica surface at values of RH above 90%.

1. $A_0 << 1/2A_c$: The water is never perturbed and mechanical contact never occurs (Table 1). An example of this for DNA molecules on mica is shown in Fig. 6a. There is a relative lack of noise in the background (due to the water layers not being perturbed) and the apparent loss in height. Recall that DNA molecules in B-form should be 2 nm in height or diameter. Here we get apparent heights of 0.5 nm or less. While this is an effect of the intrinsic resolution limit in AFM (57), the apparent loss in height is also due to the water layers on the surface shielding the mechanisms of contrast, i.e., the topography, of the DNA molecules (42, 58). This is the noncontact (nc) case where $W_{nc}/W_{nc}$ applies. Apparent loss of height on biomolecules does not have to signify that they are being deformed by the AFM tip (see Note 4).
2. $A_0 \sim 1/2A_c$: With increasing free amplitude (Fig. 6b, c) the water might be perturbed but mechanical contact might not occur (42) (see possibilities in Table 1). In Fig. 6b we show what happens when the set-point is relatively high in these situations. In short, relative stability in the images follows.

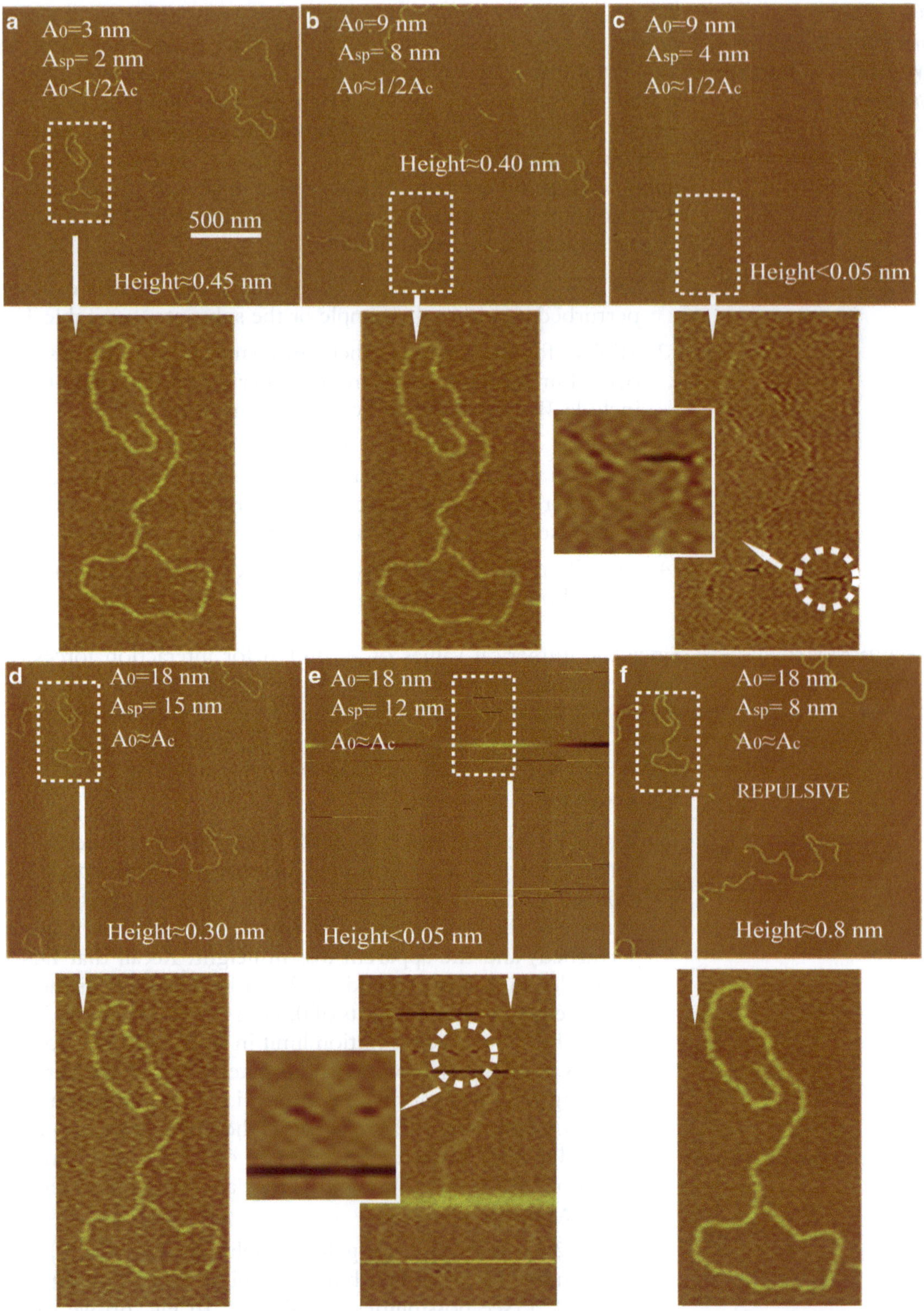


Fig. 6. Topographic images of dsDNA molecules on mica obtained in a sequence where the value of $A_0$ is used **a–c**) below the critical $A_c$ (here $15 < A_c < 19$ nm) and **d–f**) at approximately critical, i.e., $A_0 = 18$ nm. The apparent height and background noise depends on the value of $A_0$ relative to $A_c$ and also on $A/A_0$. Experimental parameters: $k \approx 40$ N/m and $Q$ factor $\approx 500$.

Nevertheless, decreasing the set-point to intermediate values results in noise, loss of height and sometimes height reversal (see inset in Fig. 6c). This could be due to the excitation of sub-harmonics due to water layers interacting and the lack of stability in these situations (see Note 5) (42, 54).

3. $A_0 \sim A_c$: Several situations arise here. (1) The water is always perturbed and the repulsive regime is not reached $W_c/W_c$ (Fig. 6d) or (2) the water is perturbed throughout and the repulsive regime is intermittently reached thus causing noise (Fig. 6e), i.e., any combination of $W_c$ and Rep. (Table 1). Typically, the first situation arises at relatively large values of set-point, $A/A_0$ just below 1 (Fig. 6d). This is due to the fact that here, intermittent contact with the water layers is minimized and transitions to the repulsive regime are inhibited. Nevertheless, both intermittent water contact and transitions to the repulsive regime typically occur when the set-point is lowered (Fig. 6e). This leads to instabilities in terms of the measurements of apparent height and sometimes height reversal (see inset in Fig. 6e). It is sometimes possible however to stably reach the repulsive regime by sufficiently decreasing the set-point (Fig. 6f) at which point the apparent height suddenly increases and the instabilities disappear. This is due to the inhibition of the effects of water shielding on apparent height and the disappearance of instabilities leading to effects such as subharmonics excitation (42, 54). In this particular case (Fig. 6), we found that $13 < A_c < 16$ nm. This implies that it was relatively easy to transition to the repulsive regime stable by sufficiently decreasing the set-point. It is important to understand that values of $A_c$ depend on a number of different parameters for any given AFM experiment (46).
4. $A_0 >> A_c$: The repulsive regime is typically reached throughout, i.e., Rep./Rep. Figure 4b for most values of $A/A_0$. In practice, the situation is similar to that described in Fig. 6f (see Note 4).

### 3.6. The Effect of Peak Forces and Pressures

The values of the peak forces and pressures (stress) depend very strongly on the free amplitude $A_0$ and the tip radius R (Fig. 7) (59). In the figure, average forces have been drawn in continuous lines for all values of A and R because these are relatively independent on R. It can be readily observed that that peak values can be up to ten times higher than average values. Additionally, pressure values, both peak and average, show a much stronger dependence on tip radius than forces (59). Overall, we observe that all magnitudes rapidly decrease with a decrease in free amplitude $A_0$. Thus, it is clear that in order to maintain a sharp tip one needs to work with small values of $A_0$. Nevertheless, from the previous section, this implies that when there is water on the surface, there will be shielding due to the water films on these and lateral resolution and

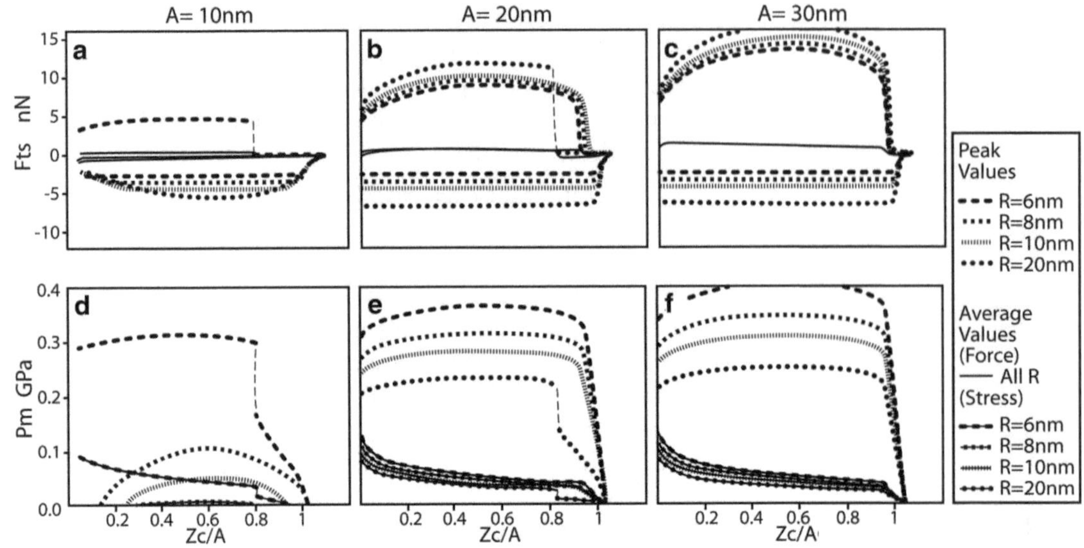


Fig. 7. Relationships between peak forces and average forces per cycle (*top*), and peak and average pressure per cycle as function of normalized tip–sample separations $z_c/A_0$ for a range of $A_0$ and *R*. In the *top row*, average and peak forces are shown for $A_0$ = (**a**) 10 nm, (**b**) 20 nm, and (**c**) 30 nm. Both the maximum (repulsive) and minimum (attractive) peak forces are shown. The average force values are also shown represented by continuous lines for all R for simplicity since these almost completely overlap. The respective average and peak pressure per cycle is shown in the *bottom row* in (**d–f**). In the case of $A_0 = 10$ nm, there is switch from the attractive to the repulsive regime for $R = 6$ nm while the attractive prevails for the rest. For higher values of $A_0$, all the interactions for $A_{sp}/A_0 \leq 0.8$ take place in the repulsive regime. Average forces range from approximately 0.2 nN in (**a**) to 1.5 nN in (**c**). Nevertheless, it is the also the behavior of the force and pressure in the figure more than the actual values that are relevant in this work. The parameters are: $f = f_0 = 300$ kHz, $k = 40$ N/m, $Q = 500$, $g = 40$ mJ, $E = 1.2$ GPa, $E_t = 120$ GPa and the rest as detailed in the figures (59). Adapted from ref. 59.

apparent height might be compromised. In the following, we introduce Small Amplitude Small Set-Point (SASS) imaging which can overcome many of these issues.

*3.6.1. Imaging with Small Set-Points*

When working at very small values of set-point, i.e., $A/A_0 < 0.1$, depending on $A_0$, the tip might be permanently trapped by the water layer. While this has typically been seen as a negative outcome (6, 33), stable oscillations might result in these cases (42, 54). As an example, look at Fig. 8 where (a) noncontact imaging is compared (for a single DNA molecule) with (b–f) SASS in terms of resolution and background noise. In the SASS case, noise is lowered and resolution increases, even though very small values of set point, i.e., A ~ 0.15 nm, are used. Then, we compare in the same figure and in a sequence taken after the first two kinds of image, (g) another noncontact mode image showing that SASS does not perturb the DNA molecule and then with (h) standard tapping mode in the repulsive regime where molecular damage results. In (i) we show a control image in nc mode imaging taken afterwards, demonstrating that the tapping mode imaging has damaged the DNA molecule.

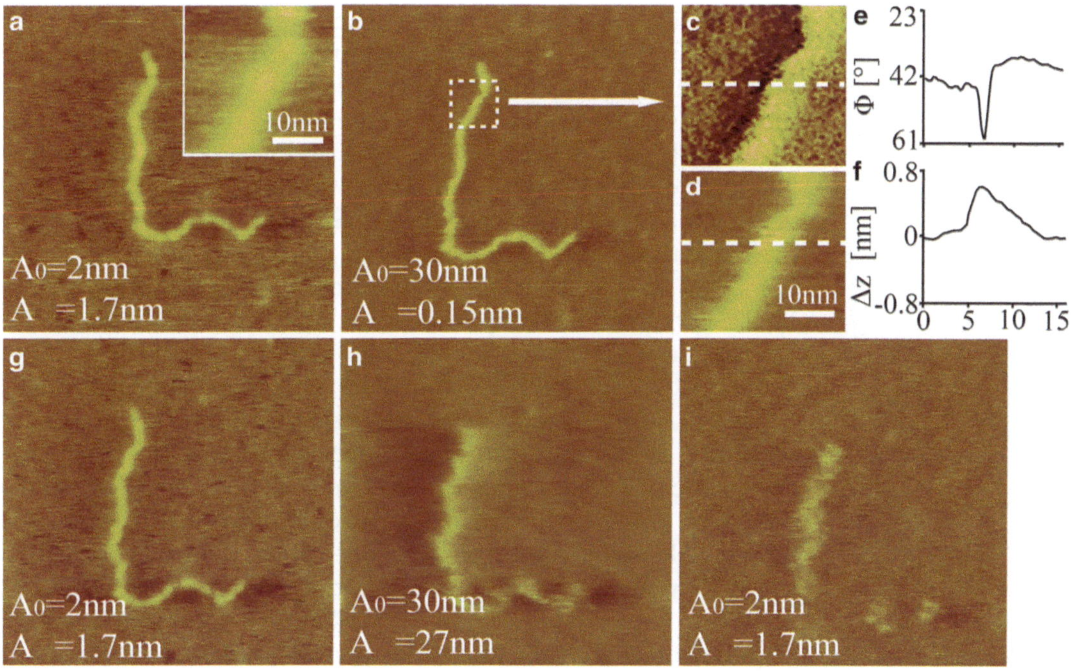


Fig. 8. Sequence of topographic scans that experimentally demonstrate that $A/A_0 << 1$ is light tapping as compared to larger values of $A/A_0$. (**a**) Image obtained in the nc mode by using $A_0 << A_c$ and $A/A_0 > 0.5$ (i.e., $A_0 = 2$ nm $A/A_0 = 0.85$). (**b**) Then the amplitude $A$ has been smoothly decreased to 0.15 nm. Then the free amplitude $A_0$ is set to 30 nm. Thus, this is an example of how a low force interaction regime (SASS) can be smoothly reached by first engaging with a small free amplitude and set-point and then increasing the free amplitude. This also exemplifies the engaging method in SASS. (**c** and **d**) Zoomed views of the phase and topography respectively. (**e** and **f**) Respective cross-sections demonstrating the resolution of this method even for such (relatively) high free amplitudes and the lack of noise compared to nc imagine (Fig. 8a). Recall that in the typical SASS method, the free amplitude maximizing resolution is typically in the range of 2–4 nm whereas here $A_0 = 30$ nm has been used simply to exemplify that the tip and molecule could be preserved in these conditions. (**g**) Another nc mode image taken to ensure that neither tip nor sample had been damaged ($A_0 = 2$ nm and $A/A_0 = 0.85$). (**h**) Standard tapping mode image obtained with a high set-point and in the standard engaging method ($A_0 = 30$ nm and $A_{sp}/A_0 = 0.90$). The standard engaging method consists of engaging with relatively high values of free amplitudes, i.e., $A_0 \geq A_c$. (**i**) Another nc image ($A_0 = 2$ nm and $A/A_0 = 0.85$) taken immediately after (**h**) showing that the sample (and likely also the tip) has undergone severe plastic deformation. We can verify that other surrounding molecules can be imaged with similar (i.e., $A_0 = 2$ nm and $A_{sp}/A_0 = 0.85$) parameters to guarantee that the deformation of the molecule is not an artifact (data not shown here). Adapted from ref. 29.

### 3.7. Small Amplitude Small Set-Point Imaging (SASS) and the Imaging of Macromolecular Complexes

#### 3.7.1. The Standard AM AFM Engaging Method

In the previous section we have shown how to engage the tip in what we term the SASS method and also in the standard AM AFM or Tapping Mode method (Fig. 8). Typically, in standard AM AFM methods, the amplitude $A_0$ (or the driving force $F_0$) is generally first set for the free cantilever and then A is smoothly decreased once the tip is engaged. Unfortunately, in this way, it is not possible to avoid the high interaction regions observed in simulations (Fig. 7) while reaching the repulsive regime. For example, in Fig. 8g the free amplitude was first set to $A_0 = 30$ nm for the free cantilever and then the tip was driven down to the surface to obtain $A = 27$ nm.

*3.7.2. The SASS Engaging Method*

It is possible to reach a region where mechanical intermittent contact might be reached only gently and where the tip is in permanent contact with the water layers (29, 42, 54). This is the method used in SASS imaging. In the SASS engaging method we do similarly as described for Fig. 8b–f. In this way, the tip is always inside the water layer and hard tapping, i.e., $A_0 \approx A_c$ and $A/A_0 \approx 0.5$, is avoided throughout (see Note 4). Here, the explanation of the SASS engaging method follows. It consists in reaching a low-force high-resolution regime without blunting the tip. First a very small free amplitude ($A_0 < 2$–$3$ nm) needs to be set for the free cantilever. Then the user can safely engage by reducing A (i.e., $A = 0.1$ nm). Ensure that the velocity of approach while engaging is minimized to not blunt the tip even with these very small free amplitudes. However, once $A = 0.1$ nm, the tip is generally safe from damage. Finally the driving force should be increased ($F_0$ or $A_0$) sufficiently to enable the tip to track the sample. This is because enough energy needs to enter the cantilever per cycle in order to stabilize the motion. This could be related to the stability criterion that states that enough energy is required to enter the system per cycle in order to stabilize the motion (6). For example $A = 0.1$ nm and $A_0 < 1$ nm might result in severe noise, whereas $A = 0.1$ nm and $A_0 = 1.5$ nm might result in low noise and high resolution (see Note 4).

*3.7.3. The SASS and the Noncontact Mode*

As stated, the SASS method can be used to obtain high-resolution images where background noise is minimized. In particular, the SASS mode operates in a region of perpetual water contact where intermittent mechanical contact might or might not occur (42). The principle behind this mode is based on the possibility to image in a region of parameter space where there is a single amplitude branch. This can be observed in Fig. 3 where it says SASS. In particular, it is not a coincidence during SASS that the L state (or attractive regime) never controls the interaction either on approach or retraction; this allows avoiding uncontrolled changes in cantilever amplitude. This is in fact a single orbital period region (29, 42, 54) where perpetual water contact occurs (see Note 4). A method to reach the SASS region without damaging the tip has been explained in detail in the sections above (the SASS engaging method). In Fig. 9 we show the difference in resolution between the pure nc-mode of operation (Fig. 9a, b) and the SASS mode (Fig. 9c–e) of operation for a system of DNA molecules on mica.

*3.7.4. Reliability and Reproducibility and Low Wear in the SASS Mode*

Typically, standard AM AFM modes involve tip and sample wear. In contrast, the user of SASS will find that many images, up to hundreds, can be taken in the SASS mode with high stability and without noticeable loss of resolution. As an example, Fig. 9 shows a sequence of high-resolution topographic images of DNA topoisomerase

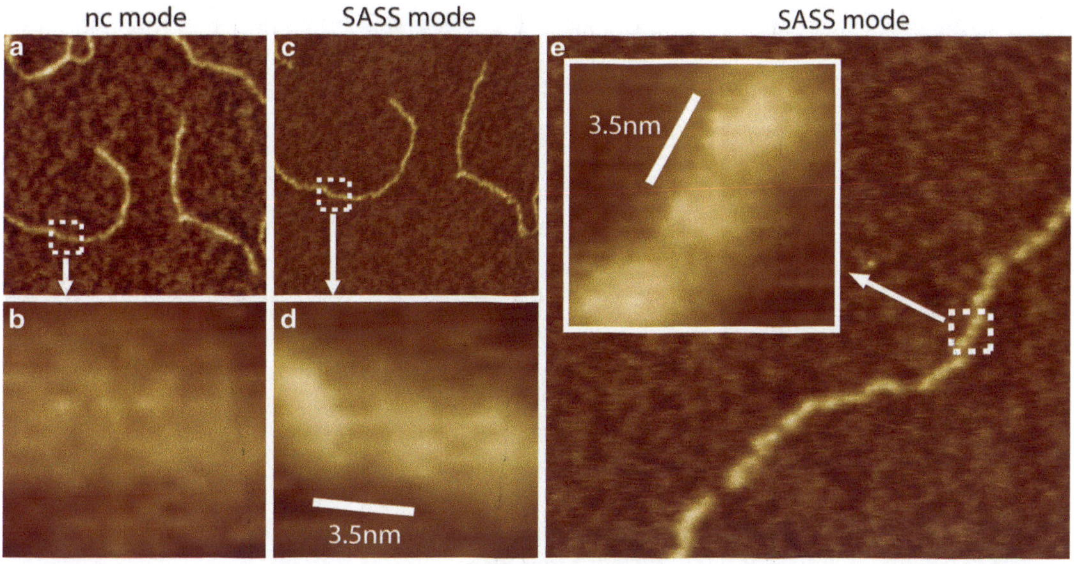


Fig. 9. (**a**) Topographic and (**b**) phase contrast images of 1 kbp DNA fragments using the noncontact mode of imaging and (**c**) and (**d**) the SASS mode respectively. In (**e**), and its inset, we show another example of SASS imaging. Experimental parameters: $f = f_0 \approx 300$ kHz, $k \approx 40$ N/m, $Q$ factor $\approx 550$, $A_0 \approx 4$ nm, and (**a–b**) $A = 3$ nm and (**c–e**) 0.1 nm.

VI complexes imaged in the SASS mode. These are zoomed views (original scans were 1 $\mu m^2$) of a selection of more than one hundred images obtained in a sequence and with the same tip and sample in the SASS mode.

*3.7.5. Imaging Macromolecular Complexes*

The standard AM AFM mode of operation can be used in a complementary way to the SASS mode of operation to obtain information about macromolecular complexes. Figure 10, for example, shows high-resolution images of DNA topoisomerase VI interactions taken using SASS. In Fig. 11, however, standard Tapping Mode images where different classes of *E. coli* RNA polymerase (RNAP) complexes formed for investigations into convergent transcription are shown (see Note 1). By using a time-lapse approach and forming complexes that represented different stages of the process, it was possible to study the outcomes of collisions between two *E. coli* RNA polymerases undergoing transcription elongation from two promoter sites situated on opposite DNA strands (60). AFM was able to provide direct information about the spatial arrangements of the DNA chain and the RNAP protein. By taking contour length measurements of the two RNAPs positions relative to the chain backbone and to each other for each class of complex formed we were able to infer what happened when two converging RNAPs collided. Often, after colliding the RNAPs were observed towards one end of the template in close proximity and in a position closer to the chain terminus than the initial binding site. The interpretation of these data was that one polymerase

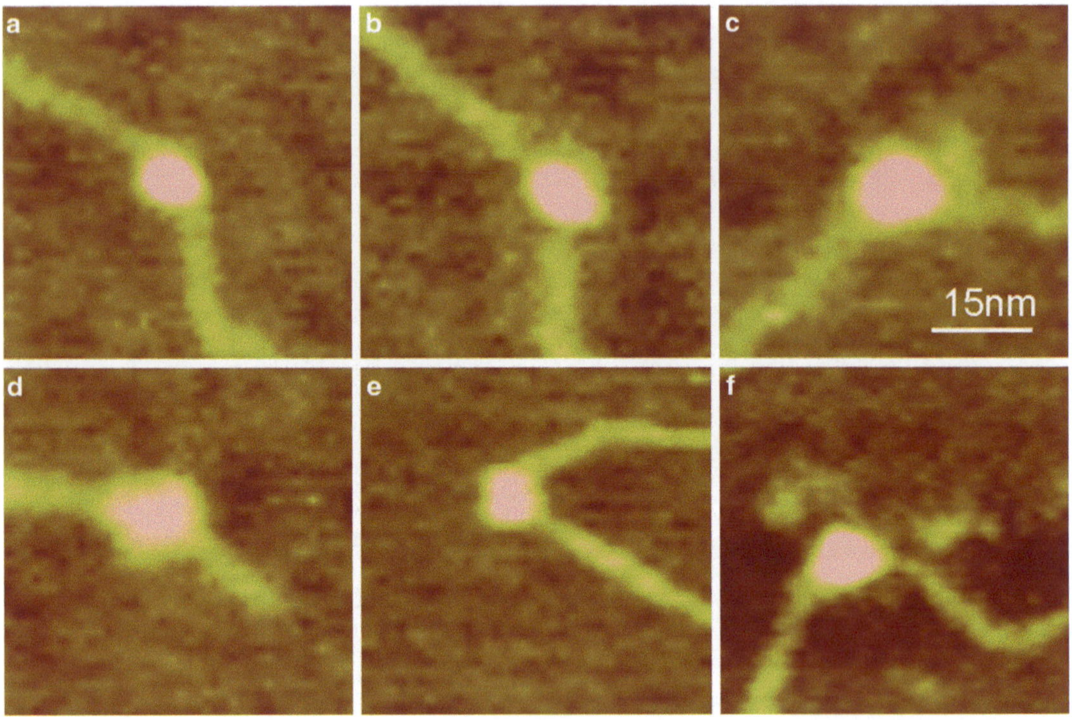


Fig. 10. Topographic scans of linear 1 kbp DNA fragments interacting with *Methanosarcina mazei* topoisomerase VI. All the scans have been taken using the SASS mode with high resolution using the same AFM tip and demonstrate reproducibility. AFM imaging is direct and requires no averaging and indicates that topoisomerase VI has a propensity to bend DNA molecules. Experimental parameters: $f = f_0 \approx 300$ kHz, $k \approx 40$ N/m, $Q$ factor $\approx 550$, $A_0 \approx 4$ nm, and $A = 0.1$ nm.

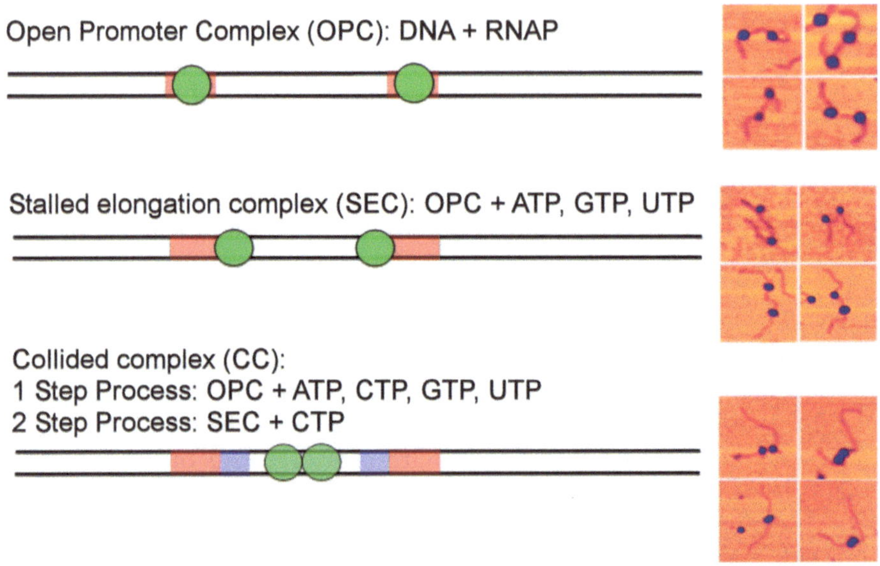


Fig. 11. Schematic diagram showing a summary of the different classes of complex formed for investigations into convergent transcription. RNAPs are shown as *green spheres*, the promoter region as *red rectangles*, and the C-less region as *blue rectangles*. Software zooms of exemplar molecules for each case are shown to the *right*.

transcribes into the other and pushes it backwards along the template, until both RNAPs eventually stall.

The humidity-controlled imaging described earlier may be useful for studying certain aspects of the convergent transcription process. By using this technique it should be possible to distinguish between molecules that are supercoiled and those that are not, by observing an associated conformational change when imaging at high RH (see Note 6). As such, the method could be used to study the importance of DNA supercoiling in transcription regulation. Liu and Wang proposed that if the RNAP had a large transcript, and other ancillary proteins, such as a ribosome, it would experience a significant amount of viscous drag, which would hamper elongation. In response to these observations, they proposed the "twin-supercoiled domain model," in which DNA rotates relative to RNAP then the advancing polymerase would generate positive supercoils in the DNA template directly ahead of it, and negative supercoils behind (61). In the case of transcription originating from two convergently aligned promoters, positive supercoils can be propagated, if the RNAP acts as a kind of "topological barrier" preventing the dissipation of twist. This could lead to the buildup of helical stress in the region between the promoter sites, which could lead to stalling of any translocating RNAPs.

The distance they travel towards each other may cause supercoiling to be propagated in the inter-polymerase region. This could be verified by increasing the humidity to close to 100% RH. On the closed circular plasmids examined here, conformational changes were driven by the release of supercoiling strain. If the imaging of DNA–RNAP complexes at high RH was accompanied by a corresponding conformational change between the proteins then it would confirm the presence of positive supercoiling.

## 4. Notes

1. We have used a DNA template, with convergently aligned promoters on opposite strands to drive RNA polymerases (RNAP) towards each other and observed the outcomes using AFM. This approach can be seen as a simple experimental model for nested genes found in vivo. It allows discrimination between outcomes, and importantly it enables a quantitative comparison of their relative frequencies. Using *E. coli* components for our model, we find that RNA polymerases are unlikely to pass each other but rather collide with a variety of outcomes. The most common occurrence was that one RNAP transcribes into the other and causes the other to be pushed backwards on the DNA template. Afterwards, the two RNAPs stall in close proximity to each other.

2. When increasing or, in any case, varying the relative humidity inside the chamber (see schemes in Fig. 2) the height of the water layer adsorbed onto mica surface might vary, nonlinearly, from fractions of nm to 2 nm or more, at values of relative humidity approaching 100% (62, 63).
3. Conformational switching of supercoiled DNA molecules is not always observed for all molecules on a given surface. Nevertheless, the behavior shown in Fig. 2 is standard behavior. When the relative humidity is increased and the height of the water layers on the surface increases, topological rearrangements can be observed in supercoiled molecules (43).
4. When imaging at relatively high value of relative humidity, i.e., RH > 70–80%, it is best to work in the pure noncontact (nc) mode, i.e., for example $A_0 = 1/4A_c$ (this also implies that the set-point should not be lowered too much, i.e., $A/A_0 > 0.5$) or in a region where the repulsive regime is reached stable, i.e., $A_0 = 3/2A_c$ and $0.3 < A/A_0 < 0.6$–$0.9$. In the nc mode the water layers are not perturbed but lateral resolution and apparent height are compromised (Figs. 6 and 8). In the repulsive regime, resolution can be improved; however, on the other hand the forces might be very large which can lead to tip wear and sample damage. Thus, the other possibility is to work with very small set-points where permanent contact with the water layers might follow, i.e., the SASS mode (Figs. 9 and 10). This last mode has the advantage of increasing resolution and reducing noise because sub-harmonics are not excited and the capillary neck, if formed, is never ruptured. Reproducibility and high resolution are a good combination to study macromolecular complexes and their interactions and the SASS mode is very well suited for it.
5. The typical values of $A_0$ and relative humidity RH for which situations such as those described in Fig. 6 arise might vary. Nevertheless, in general, noise should be expected at $A_0 \sim 1/2A_c$. This might occur at different values of set-point $A/A_0$ depending on the minimum distance of approach. Similar situations could arise when working with high values of $A_0$, i.e., $A_0 >> A_c$ if the set-point is lowered enough. We had previously discussed this phenomenon experimentally (46, 47), and have recently discussed it in terms of water perturbation and the excitation of sub-harmonics (42, 54). Nevertheless, the presence of a sample, i.e., molecules, on the surface, might slightly affect the range for which sub-harmonics are predicted. These might also affect apparent height.
6. By controlling the humidity in which imaging is performed in, we have been able to observe conformational changes in supercoiled DNA molecules at high RH. These are driven by

topological rearrangements of the molecule, as the water layer surrounding it increases in depth. This technique may prove to be useful in studying the role DNA topology plays in convergent transcription.

## Acknowledgments

We thank the group of Tony Maxwell for kindly providing the topoisomerase samples.

### References


1. Binnig G, Quate CF, Gerber C (1986) Atomic force microscope. Phy Rev Lett 56:930–933
2. Quate CF (1994) The AFM as a tool for surface imaging. Surf Sci 299–300:980–995
3. Bustamante C, Keller D (1995) Scanning force microscopy in biology. Physics today 48:33–38
4. Hansma HG, Hoh JH (1994) Biomolecular imaging with the atomic force microscope. Ann Rev Biophy Biomol Struc 23:115–140
5. Garcia R, Perez R (2002) Dynamic atomic force microscopy methods. Surf Sci Rep 47:197–301
6. Giessibl FJ (2003) Advances in atomic force microscopy. Rev Modern Physics 75:949–983
7. Israelachvili J (1991) Intermol. & Surf. Forces. Academic, New York
8. Kienbergera F, Costab TL, Zhua R, Kadad G, Reithmayere M, Chtcheglovaa L, Rankla C, Pachecof BFA, Thalhammerb S et al (2007) Dynamic force microscopy imaging of plasmid DNA and viral RNA. Biomaterials 28:2403–2411
9. Zhong Q, Innlss D, Kjoller K, Elings VB (1993) Fractured polymer/silica fiber surface studied by tapping mode atomic force microscopy. Surf Sci Lett 290:L688–L692
10. Hansma PK, Cleveland JP, Radmacher M et al (1994) Tapping mode atomic force microscopy in liquids. Appl Phy Lett 64:1738–1740
11. Sahin O, Magonov S, Su C, Quate CF, Solgaard O (2007) An atomic force microscope tip designed to measure time-varying nanomechanical forces. Nat Nanotech 2:507–514
12. Proksch R (2006) Multi-frequency, repulsive mode amplitude modulated atomic force microscopy. Appl Phys Lett 89:113121–113123
13. Francis LW, Lewis PD, Wright CJ, Conlan RS (2010) Atomic force microscopy comes of age. Biol Cell 102:133–143
14. Parot P, Dufrene YF, Hinterdorfer P et al (2007) Past, present and future of atomic force microscopy in life sciences and medicine. J Mol Recog 20:418–431
15. Giessibl FJ (1995) Atomic Resolution of the Silicon (111)-(7x7) Surface by Atomic Force Microscopy. Science 267:68–71
16. Weisenhorn AL, Hansma PK, Albrecht TR, Quate CF (1989) Forces in atomic force microscopy in air and water. Appl Phys Lett 54:2651–2653
17. Tamayo J, Garcia R (1996) Deformation, contact time, and phase contrast in tapping mode scanning force microscopy. Langmuir 12:4430–4435
18. Kiracofe D, Raman A (2010) On eigenmodes, stiffness, and sensitivity of atomic force microscope cantilevers in air versus liquids. J Appl Phys 107:033506–033515
19. Turner RD, Kirkham J, Devine D, Thomson NH (2009) Second harmonic atomic force microscopy of living Staphylococcus aureus bacteria *Appl.* Phys Lett 94:043901
20. Stark RW (2004) Spectroscopy of higher harmonics in dynamic atomic force microscopy. Nanotechnology 15:347–351
21. Patil S, Martinez NF, Lozano JR, Garcia R (2007) Force microscopy imaging of individual protein molecules with sub-pico Newton force sensitivity. J Mol Recogn 20:516–523
22. Xu X, Melcher J, Basak S, Reifenberger R, Raman A (2009) Compositional contrast of biological materials in liquids using the momentary excitation of higher eigenmodes in dynamic atomic force microscopy. Phys Rev Lett 102:060801–060804
23. Melcher J, Carrasco C, Xu X et al (2009) Origins of phase contrast in the atomic force microscope in liquids. Proc Natl Acad Sci USA 106:13655–13660

24. Basak S, Raman A, Garimella SV (2006) Hydrodynamic loading of microcantilevers vibrating in viscous fluids. J Appl Phys 99:114906–114915
25. Alessandrini A, Facci P (2005) AFM: a versatile tool in biophysics. Meas Sci Techn 16:R65–R92
26. Gould S, Marti O, Drake B et al (1988) Molecular resolution images of amino acid crystals with the atomic force microscope. Nature 332:332–334
27. Albrecht TR, Grutter P, Horne D, Rugar D (1991) Frequency modulation detection using high Q cantilevers for enhanced force microscope sensitivity. J Appl Phys 69:668–673
28. Martin Y, Williams CC, Wickramasinghe HK (1987) Atomic force microscope-force mapping and profiling on a sub 100-Å scale. J Appl Phys 61:4723–4729
29. Santos S (2011) PhD thesis: Dynamic atomic force microscopy and its applications in biomolecular imaging. University of Leeds
30. Drake B, Prater CB, Weisenhorn AL et al (1989) Imaging crystals, polymers, and processes in water with the atomic force microscope. Science 243:1586
31. Kodera N, Yamamoto D, Ishikawa R, Ando T (2010) Video imaging of walking myosin V by high-speed atomic force microscopy. Nature 468:72–76
32. Gross L, Mohn F, Moll N, Liljeroth P, Meyer G (2009) The Chemical Structure of a Molecule Resolved by Atomic Force Microscopy. Science 325:1110–1114
33. Gan Y (2009) Atomic and subnanometer resolution in ambient conditions by atomic force microscopy. Surf Sci Rep 64:99–121
34. Santos S, Thomson NH (2011) High resolution imaging of Immunoglobulin G (IgG) antibodies and other biomolecules using amplitude modulation atomic force microscopy in air. Humana Press, New York
35. Gerber C, Lang HP (2006) How the doors to the nanoworld were opened. Nat Nanotechn 1:3–5
36. Garcia R, Magerele R, Perez R (2007) Nanoscale compositional mapping with gentle forces. Nat Mater 6:405–411
37. Ostendorf F, Schmitz C, Hirth S et al (2008) How flat is an air-cleaved mica surface? Nanotechnology 19:305705
38. Bezanilla M, Manne S, Laney DE, Lyubchenko YL, Hansma HG (1995) Adsorption of DNA to mica, silylated mica, and minerals: characterization by atomic force microscopy. Langmuir 11:655–659
39. Hansma H, Laney D (1996) DNA binding to mica correlates with cationic radius: assay by atomic force microscopy. Biophysical J 70:1933–1939
40. Vesenka J, Guthold M, Tang CL, Keller D, Delaine E, Bustamante C (1992) Substrate preparation for reliable imaging of DNA molecules with the scanning force microscope. Ultramicroscopy 42–44:1243–1249
41. Ostendorf F, Schmitz C, Hirth S et al (2009) Evidence for potassium carbonate crystallites on Air-cleaved mica surfaces. Langmuir 25:10764–10767
42. Santos S, Verdaguer A, Souier T, Thomson HN, Chiesa M (2011) Measuring the true height of water layers in the nanoscale. Nanotechnology 22:465705–465713
43. Billingsley DJ, Kirkham J, Bonass WA, Thomson NH (2010) Atomic force microscopy of DNA at high humidity: irreversible conformational switching of supercoiled molecules. Phys Chem Chem Phys 12:14727–14734
44. Binnig G, Rohrer H. 1984. Scanning tunneling microscopy. In Trends in Physics, Ed. J Janta, J Pantoflicek, pp. 38-46. The Hague: Eur. Phys. Soc
45. Garcia R, San Paulo A (1999) Attractive and repulsive tip-sample interaction regimes in tapping mode atomic force microscopy. Phys Rev B 60:4961–4967
46. Santos S, Barcons V, Font J, Thomson NH (2010) Bi-stability of amplitude modulation AFM in air: deterministic and stochastic outcomes for imaging biomolecular systems. Nanotechnology 21:225710–225720
47. Santos S, Barcons V, Font J, Thomson NH (2010) Cantilever dynamics in amplitude modulation AFM: continuous and discontinuous transitions. J Phys D: Appl Phys 43:275401–275407
48. Stark RW (2010) Bistability, higher harmonics, and chaos in AFM *Materials Today* (2010) 13(9):24–32
49. Tamayo J, Garcia R (1998) Relationship between phase shift and energy dissipation in tapping-mode scanning force microscopy. Appl Phys Lett 73:2926–2928
50. Santos S, Barcons V, Verdaguer A et al (2011) How localised are energy dissipation processes in the nanoscale? Nanotechnology 22: 345401–345407
51. Zitzler L, Herminghaus S, Mugele F (2002) Capillary forces in tapping mode atomic force microscopy. Phys Rev B 66:155436–155438
52. Garcia R, San Paulo A (2000) Dynamics of a vibrating tip near or in intermittent contact with a surface. Phys Rev B 61:R13381–R13384
53. Marth M, Maier D, Honerkamp J (1999) A unifying view on some experimental effects in tapping-mode atomic force microscopy. J Appl Phys 85:7030–7036
54. Santos S, Barcons V, Verdaguer A, Chiesa M (2011) Sub-harmonic excitation in AM AFM

in the presence of adsorbed water layers. J Appl Phys 110:114902–114911

55. Couturier G, Boisgard R, Nony L, Aimé JP (2003) Noncontact atomic force microscopy: Stability criterion and dynamical responses of the shift of frequency and damping signal. Rev Scientific Instrum 74:2726–2734
56. Garcia R, San PA (2000) Amplitude curves and operating regimes in dynamic atomic force microscopy. Ultramicroscopy 82: 79–83
57. Santos S, Barcons V, Christenson HK, Font J, Thomson NH (2011) The intrinsic resolution limit in the atomic force microscope: implications for heights of nano-scale features. PLoS One 6:e23821
58. Moreno-Herrero F, Colchero J, Baro A (2003) DNA height in atomic force microscopy. Ultramicroscopy 96:167–174
59. Santos S, Thomson NH (2011) Energy dissipation in a dynamic nanoscale contact. Appl Phys Lett 98:013101–013103
60. Crampton N, Bonass WA, Kirkham J, Rivetti C, Thomson NH (2006) Collision events between RNA polymerases in convergent transcription studied by atomic force microscopy. Nucleic Acids Res 34:5416–5425
61. Liu LF, Wang JC (1987) Supercoiling of the DNA-template during transcription. Proc Natl Acad Sci 84:7024–7027
62. Beaglehole D, Christenson HK (1992) Vapor adsorption on mica and silicon: entropy effects, layering, and surface forces. J Phys Chem 96:3395–3403
63. Balmer TE, Christenson HK, Spencer ND, Heuberger M (2008) The effect of surface ions on water adsorption to mica. Langmuir 24:1566–1569

# Chapter 19

# Imaging of Transmembrane Proteins Directly Incorporated Within Supported Lipid Bilayers Using Atomic Force Microscopy

Daniel Levy and Pierre-Emmanuel Milhiet

## Abstract

Structural analysis of transmembrane proteins remains a challenge in biology, mainly due to their difficulty in being overexpressed and the required use of detergents that impair different steps of biochemistry classically used to obtain 3D crystals. In this context, we have developed a new technique for protein incorporation within supported lipid bilayers that only requires a few picomoles of protein per assay. Proteins are directly inserted into a detergent-destabilized bilayer that can be imaged in buffer with atomic force microscopy (AFM) allowing structural analysis down to sub-nanometer lateral resolution. In this chapter, we describe the main guidelines for this technique, from the choice of detergent to the requirements for AFM high-resolution imaging.

**Key words:** Atomic force microscopy, Membrane protein, Lateral segregation, Sub-nanometer resolution, Direct incorporation, supported lipid bilayer, Detergent

## 1. Introduction

Atomic force microscopy (AFM) is a powerful tool in the membrane field. It has been extensively used to analyze the topography of lipid monolayers or bilayers at nanometer resolution, for example, to monitor the interaction of membranes with proteins or drugs (reviewed in ref. 1, 2). Thanks to its ability to work in liquid, the AFM has been especially used to probe the lateral segregation of membrane components in biological membranes and has provided very important clues concerning the mechanism of lipid microdomain formation and in the partition of glycosylphosphatidylinositol-anchored proteins within raft-like microdomains (reviewed in ref. 3). AFM is also a technique of choice in membrane protein research (4). With AFM, topographies of membrane

Alioscka A. Sousa and Michael J. Kruhlak (eds.), *Nanoimaging: Methods and Protocols*, Methods in Molecular Biology, vol. 950,
DOI 10.1007/978-1-62703-137-0_19, © Springer Science+Business Media, LLC 2013

proteins can be acquired at sub-nanometer lateral, and 1 Å vertical, resolutions in near physiological conditions. Furthermore, AFM allows analysis of the flexibility and of the conformational changes of extrinsic domains. By applying loading forces to the AFM tip, biological samples can be nanodissected to provide insights into protein assembly. Finally, force spectroscopy measurements allow delineating the intramolecular interactions that stabilize the proteins into lipid bilayers.

The prerequisites for sub-nanometer resolution imaging are the high density of proteins, limiting protein lateral diffusion that impairs the resolution, and the flatness of the membranes that limits *z*-piezo movements. So far, two types of membrane containing proteins have satisfied these criteria: (1) 2D crystals of membrane proteins obtained by detergent removal and reconstituted at high density, such as bacterial porins, aquaporins, potassium channels, and photosynthetic complexes; and (2) purified native membranes, vesicles or sheets that contain a small number (1–3) of different types of proteins performing highly specialized functions like connexin (5), photosynthetic chromatophore (6), surface layer of bacteria (7) or eye lens membrane (8).

In this chapter, we describe a method for the direct incorporation of transmembrane proteins at high density in planar supported lipid bilayers (SLB) allowing high-resolution AFM imaging. The protocol does not require the formation of 2D crystals of proteins and can be, in principle, applied to any type of protein. Moreover, this method requires less than 1 pmol of purified proteins and, thus, could be applied to eukaryotic membrane proteins known to be difficult to express and purify. Proteins are incorporated in a unique orientation in the membrane allowing physiological protein–protein interactions to be established while symmetrical orientations are often observed in 2D crystals. Since our first report (9), successful incorporation and high-resolution AFM imaging have been obtained with two different proteins of the bacterial photosynthetic apparatus (the light-harvesting complex 2 (LH2) and the light-harvesting complex 1 (LH1)–reaction center (RC), also named core complexes) (10), LacY, a secondary transporter (11), prestin, a motor protein of cochlear outer hair cells (12), and MscS, a mechano-sensitive channel (Personal communication of S. Scheuring, U1006 INSERM, France). Four of these proteins were purified from bacterial membranes (native or recombinant proteins) and one from a stably expressing eukaryotic cell line underlining the versatility of the direct incorporation technique. In addition, incorporation succeeded in different types of SLBs including one containing negatively charged lipids. Here we describe in detail the different steps of incorporation of membrane proteins into SLBs and imaging by AFM, which should provide experimental guidelines for researchers involved not only in AFM imaging of membrane proteins but also in the conception of biomimetic systems containing membrane proteins.

## 2. Materials

### 2.1. Supplies

1. 2 mm glass beads (Sigma-Aldrich, France).
2. 10 mm diameter Teflon pad.
3. Araldite® glue.
4. Muscovite mica sheets (Goodfellow or Electron Microscopy Sciences).
5. Avanti mini extruder (Avanti, USA).

### 2.2. Reagents

1. Lipids: Lipid reagents must be of the highest purity and can be purchased from Avanti Polar Lipids (Alabaster, USA) or other suppliers. Lipid mixtures that we have used include 1,2-dioleoyl-sn-glycero-3-phosphocholine/1,2-dipalmitoyl-sn-glycero-3-phosphocholine (DOPC/DPPC), 1-palmitoyl-2-oleoyl-sn-glycero-3-phosphatidylethanolamine/1-palmitoyl-2-oleoyl-sn-glycero-3-phosphatidylglycerol (POPE/POPG), 1-palmitoyl-2-oleoyl-sn-glycero-3-phosphocholine/sphingomyelin (POPC/SM). Lipid stock solutions are stored at −80°C in $CHCl_3$ solution or in powder and kept under argon 6–12 months according to the supplier's recommendations. Working solutions are prepared at 10 mM in 2:1 $CHCl_3$/MeOH and stored at −20°C.
2. Detergents: Detergent reagents must be of the highest purity and can be purchased from Anatrace. Detergents should be chosen according to biochemical and functional criteria (13) and those successfully used for the incorporation of proteins include *n*-dodecyl-*β*-D-maltopyranoside (DDM), *n*-dodecyl-*β*-D-thiomaltopyranoside (DOTM), and Fos-choline 16 (F16).
3. Proteins: LH1–RC and LH2, described in this chapter, were purified from *Rhodobacter (Rba.) veldkampii* or *Rba. sphaeroides* according to Gubellini et al. (14) and Francia et al. (15), respectively. Protein concentrations were 0.5–1 mg/ml in 0.1% DDM. LacY was purified after expression in *Escherichia coli* according to Merino et al. (16). LacY concentration was 0.75–1 mg/ml in 0.1% DDM. Control of purity and monodispersity of proteins was usually performed by SDS-PAGE gels, native gels, and size exclusion chromatography.
4. Phosphate buffered saline (PBS) at pH 7.4; and 10 mM Tris–HCl, 150 mM KCl, pH 7.4.

### 2.3. Instrumentation

1. Atomic force microscope: For the work described in this chapter we have used a Nano-Bruker Multimode equipped with a Nanoscope IIIA or V controller and a J scanner. OTR4 Olympus tip (0.09 N/m) is convenient for high-resolution imaging. Both tapping and contact mode can be used.

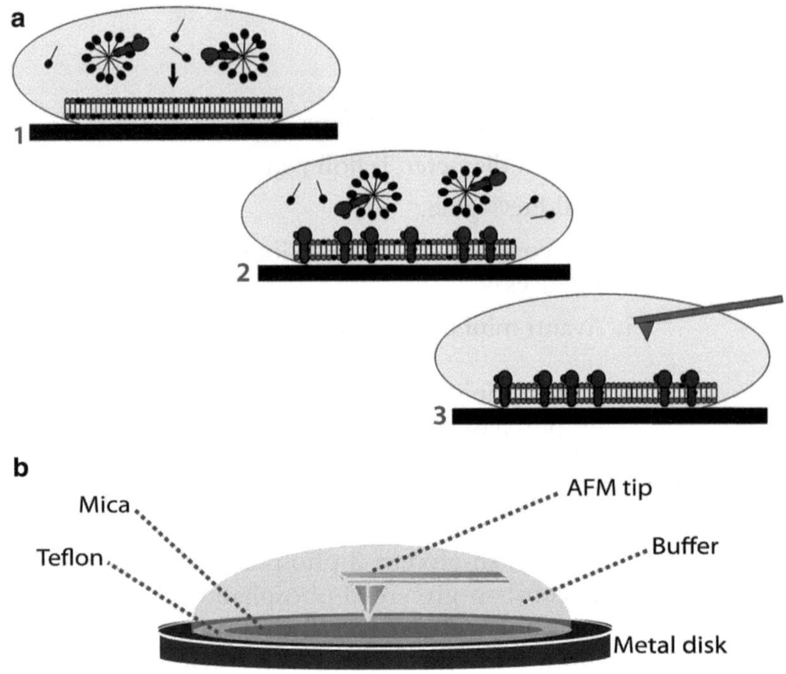


Fig. 1. Principle of the reconstitution of membrane proteins by direct incorporation within supported lipid bilayer. (**a**) Direct incorporation of membrane proteins in SLB. (1) Incubation of solubilized membrane proteins with detergent that partition between the aqueous phase and the SLB, and destabilize the latter. (2) Unidirectional incorporation of membrane proteins in SLB. (3) Detergent removal by extensive rinsing before AFM imaging. (**b**) Sample preparation and imaging. Bilayers are formed on a mica disk glued with Araldite® on a Teflon pad that prevents any buffer leakage during incubation with vesicles and AFM imaging. This Teflon pad is glued at the top of a metal disk that is fixed on the AFM scanner head.

2. Optical reflected light microscope: We have used a Zeiss upright microscope equipped with long working distance objectives, Hg lamp for fluorescence, and an AxioCam MRc Zeiss camera.

## 3. Methods

The principle of the method relies on the direct incorporation of purified membrane proteins into a planar lipid bilayer (Fig. 1a). The concept of direct incorporation is derived from the previous methods of direct incorporation into liposomes for functional studies (reviewed in ref. 17–19). The experimental protocol involves three steps. First, a continuous and flat SLB is formed by fusion of vesicles onto mica. Second, the SLB is incubated with solubilized proteins and detergents at a concentration allowing the destabilization of SLB. Proteins are incorporated at this step in the detergent-destabilized SLB. The time of incorporation is in the range of minutes. Finally, the detergent present in the bulk and in the SLB is removed by extensive rinsing by buffer exchange, and the sample is imaged by AFM (Fig. 1b).

### 3.1. Artificial Supported Lipid Bilayers

#### 3.1.1. Preparation

Different methods have been developed to prepare SLB but the most popular remains the formation of supported membranes by fusion of large unilamellar lipid vesicles (LUVs) on a solid surface (20). LUVs are generally prepared via sonication or extrusion and the vesicle solution is then added on top of the support. Vesicles then adsorb on the substrate before rupturing. Formation of a continuous supported bilayer is a prerequisite to perform efficient incorporation of transmembrane proteins, allowing easier tuning of detergent concentration (see below) and a better reproducibility of experiments. We often use phase-separated lipids bilayer, (e.g., DOPC/DPPC SLBs), in order to assess the formation of a continuous membrane (phase separation is easily observed with AFM (21)). The main steps of the protocol are:

1. Glue a Teflon pad, which prevents any buffer leakage during incubation with vesicles and AFM imaging, at the top of a metal disk that will be later fixed on the AFM scanner head. A mica disk is then glued with Araldite® at the top of this Teflon pad.
2. Dry the lipids that were solubilized in organic solvent under nitrogen flow in a Pyrex® tube at 30°C using a dry block for at least 2 h. If a lipid mixture is used, vortex the tube before drying.
3. Suspend the dried lipids at 0.125 mM in PBS buffer by 1 min strong vortexing under argon in the presence of 2 mm glass beads to form multilamellar vesicles (MLV). Alternatively, longer vortexing without beads can be used.
4. Extrude MLVs using a suitable extruder (e.g., Avanti Mini-Extruder) through a 100 nm diameter polycarbonate membrane to produce LUVs. Their homogeneity can be easily controlled using dynamic light scattering.
5. Cleave the mica disk, deposit LUVs onto it (see Fig. 1b) and allow the LUVs to adsorb and fuse during 2 h incubation. Perform the incubation at a temperature 20°C above the fusion temperature of the lipids allowing them to form a homogeneous fluid phase.
6. Before AFM imaging, wash away non-fused vesicles by successive exchange of buffer.
7. Add 50 μl of PBS buffer to bathe the SLB before detergent destabilization and protein incorporation.

The procedure described above leads to a flat supported lipid bilayer with a root mean square (rms) in the Angstrom range (see an example in Fig. 2a), a big advantage for high-resolution imaging as compared to polymer-supported bilayers that are more corrugated.

It is really important that a water layer with a thickness ranging from 1 to 3 nm separates the bilayer from the substrate (reviewed in ref. 22, 23) and insures SLBs to have the essential feature of free-standing membranes as demonstrated by measuring lipid diffusion using

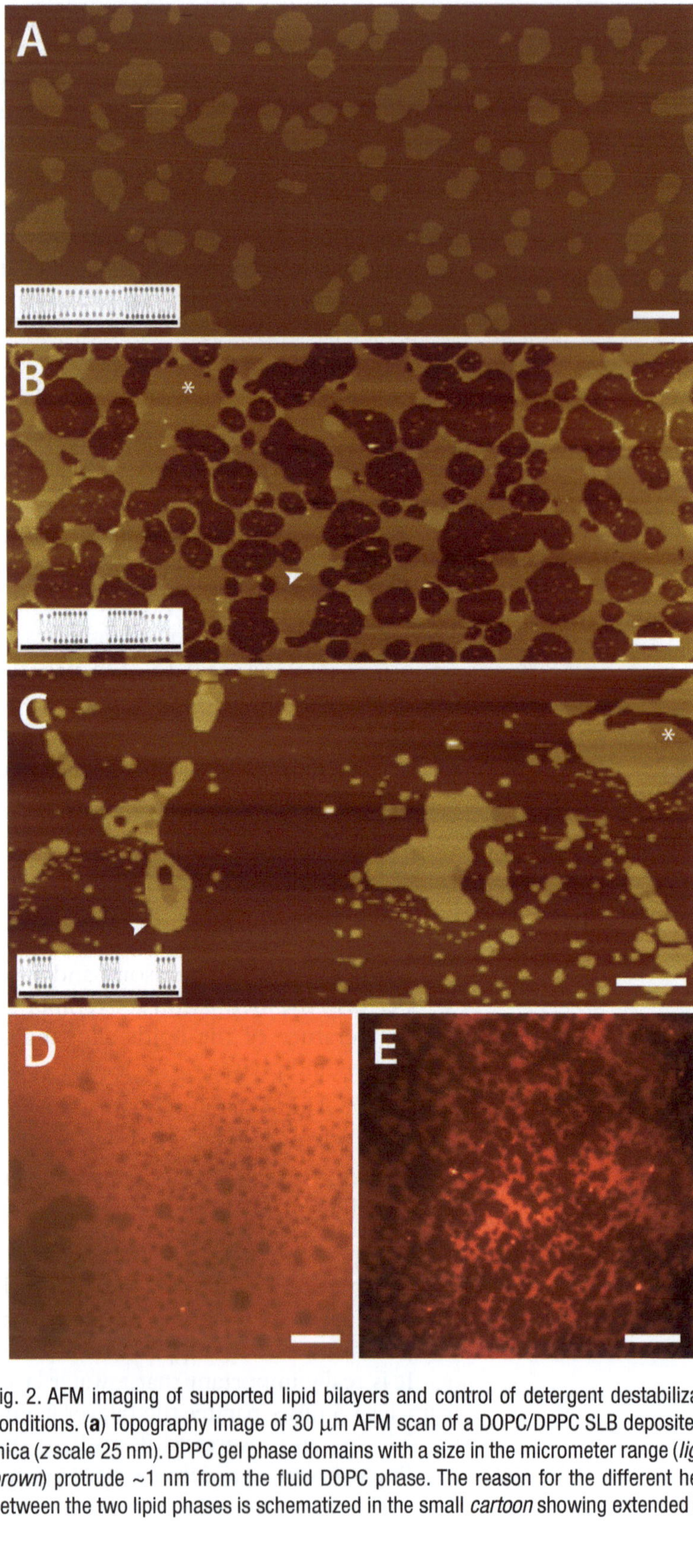


Fig. 2. AFM imaging of supported lipid bilayers and control of detergent destabilization conditions. (**a**) Topography image of 30 μm AFM scan of a DOPC/DPPC SLB deposited on mica (*z* scale 25 nm). DPPC gel phase domains with a size in the micrometer range (*lighter brown*) protrude ~1 nm from the fluid DOPC phase. The reason for the different height between the two lipid phases is schematized in the small *cartoon* showing extended fatty

fluorescence recovery after photobleaching (20) or fluorescence correlation spectroscopy (FCS) (24, 25) (see Note 1).

*3.1.2. Imaging*

1. Use AFM contact mode for SLB imaging.
2. Minimize tip-sample interaction by tuning buffer composition (e.g., 10 mM Tris–HCl, 150 mM KCl, pH 7.4) and the spring constant of the cantilever, which must be small in the 10–200 mN/m range.
3. Make sure that the tip applies a force on the membrane below 100 pN, due to membrane softness.
4. Because AFM scanners are limited to a few tens of micrometers, it can be useful to control SLBs coverage on a large scale using fluorescence microscopy. Lipid phase separation and non-covered areas can be easily observed with an upright microscope by adding a small amount (<0.5%) of lipid fluorescent probe during the preparation of lipid solution (Fig. 2d).

### 3.2. Tuning Membrane Destabilization with Detergent

*3.2.1. Preparation*

A key point of the method is the formation of specific defects in SLB to allow the insertion of solubilized proteins. The rationale is to tune the detergent concentration to form these defects in the SLB but to limit membrane solubilization. We have found that successful insertion of proteins is obtained when SLBs are destabilized by low cmc detergents such as *n*-dodecyl-$\beta$-D-maltopyranoside (DDM, 0.17 mM cmc), *n*-dodecyl-$\beta$-D-thiomaltopyranoside (DOTM, 0.05 mM cmc), or Fos-Choline C16 (F16, 0.013 mM cmc). These detergents are mild detergents and routinely used for the purification of membrane proteins. Other detergents including high cmc detergents like *n*-octyl-β-D-glucopyranoside or LDAO or other non-glycosylated low cmc detergents including Triton X-100 or C12E8 lead to an almost complete solubilization of the SLBs, at least under the experimental conditions described below, and

Fig. 2. (continued) acid chains for DPPC lipids (in *black*) and relaxed chains for DOPC (in *grey*). (**b**) Topography of DOPC/DPPC bilayer (*z* scale, 20 nm) after incubation with the low cmc glycosylated detergent DOTM at 0.075 mM (1.5× cmc). *Brown* areas correspond to solubilized membranes. Both fluid (*white asterisk*) and gel (*white arrowhead*) can still be observed. (**c**) Topography image of DOPC/DPPC bilayer (*z* scale, 20 nm) incubated with the high cmc detergent OG. The membrane is mainly solubilized and gel phase is only preserved (*white arrowhead* and see the inset). Some tiny fluid domains are sometimes observed (*white asterisk*). (**d** and **e**) Fluorescence image of DOPC/DPPC bilayer recorded with an upright microscope, equipped with a 10× long working distance objective, before (**d**) or after (**e**) incubation with 0.2 mM DDM. The darker round-shaped domains in (**d**) correspond to DPPC gel phase domains (most of the lipid fluorescent probe partition in the fluid phase). The dark areas in (**e**) correspond to solubilized membranes. Scale bars are 2 μm (**a**) and 1 μm (**b** and **c**) and 40 μm (**d** and **e**).

should be avoided if possible (the pattern of solubilization is close to that described in Fig. 2c; see also ref. 9, 10). The following protocol has been successfully used:

1. Prepare a SLB as described in Subheading 3.1.1.
2. Prepare 50 μl of a detergent-containing buffer with detergent concentration of 0.6 mM, 0.15 mM, or 0.26 mM for DDM, DOTM, or F16, respectively (see Note 2).
3. Carefully remove 25 μl from the 50 μl of PBS buffer bathing the SLB.
4. Add 25 μl of the detergent solution to the SLB and mix by up-and-down pipeting. The final detergent concentration is 0.3 mM (1.7× cmc), 0.075 mM (1.5× cmc), or 0.13 mM (10× cmc, at 4°C) for DDM, DOTM, or F16, respectively. In the case of a new detergent, incubate SLBs with detergent at final concentrations slightly above the cmc, e.g., 1.05×, 1.2×, and 1.5× cmc.
5. Incubate for 15 min at room temperature to allow the equilibration of the detergent.
6. Remove detergent by gently rinsing with at least 10 drops of detergent-free PBS buffer and repeat this step at least 3 times with 20 min time intervals to favor the exchange between SLB and buffer.

#### 3.2.2. Imaging

Before AFM imaging, it is necessary to wait a few hours, the time required for membrane equilibration. It is also worth noting that topographs of SLB in the presence of detergent cannot be recorded due to the instability of the signal due to membrane remodeling. As shown in Fig. 2b, a low detergent concentration leads to a good preservation of both fluid and gel phase even if the membrane is largely remodeled and partially solubilized (Fig 2b). The percentage of solubilization should be below 50%. Further increase of the detergent concentration leads to, first, almost complete solubilization of the fluid phase (since it is less resistant than the rigid gel phase domains) (Fig. 2c), and, second, to complete lipid solubilization producing free regions of mica devoid of material (data not shown). Thus, we hypothesize that SLBs after detergent removal are representative of SLBs in the presence of detergent. Although this is a simplistic approximation, this allows proposing experimen-

Fig. 3. (continued) preserved after incubation, most of them presenting corrugated shapes that correspond to protein-enriched areas. Some of them are mainly composed of proteins (*white star*) whereas others alternate pure lipid (*black star*) and protein domains. Higher magnification (inset in (**d**) corresponding to a 200 nm scan) shows individual proteins protruding 1.2 nm above the lipid membrane. Scale bars are 2 μm (**a**), 100 nm (**b**), 20 nm (**c**) and 100 nm (**d**). Z scales are 10 nm (**a** and **d**) and 5 nm (**b** and **c**).

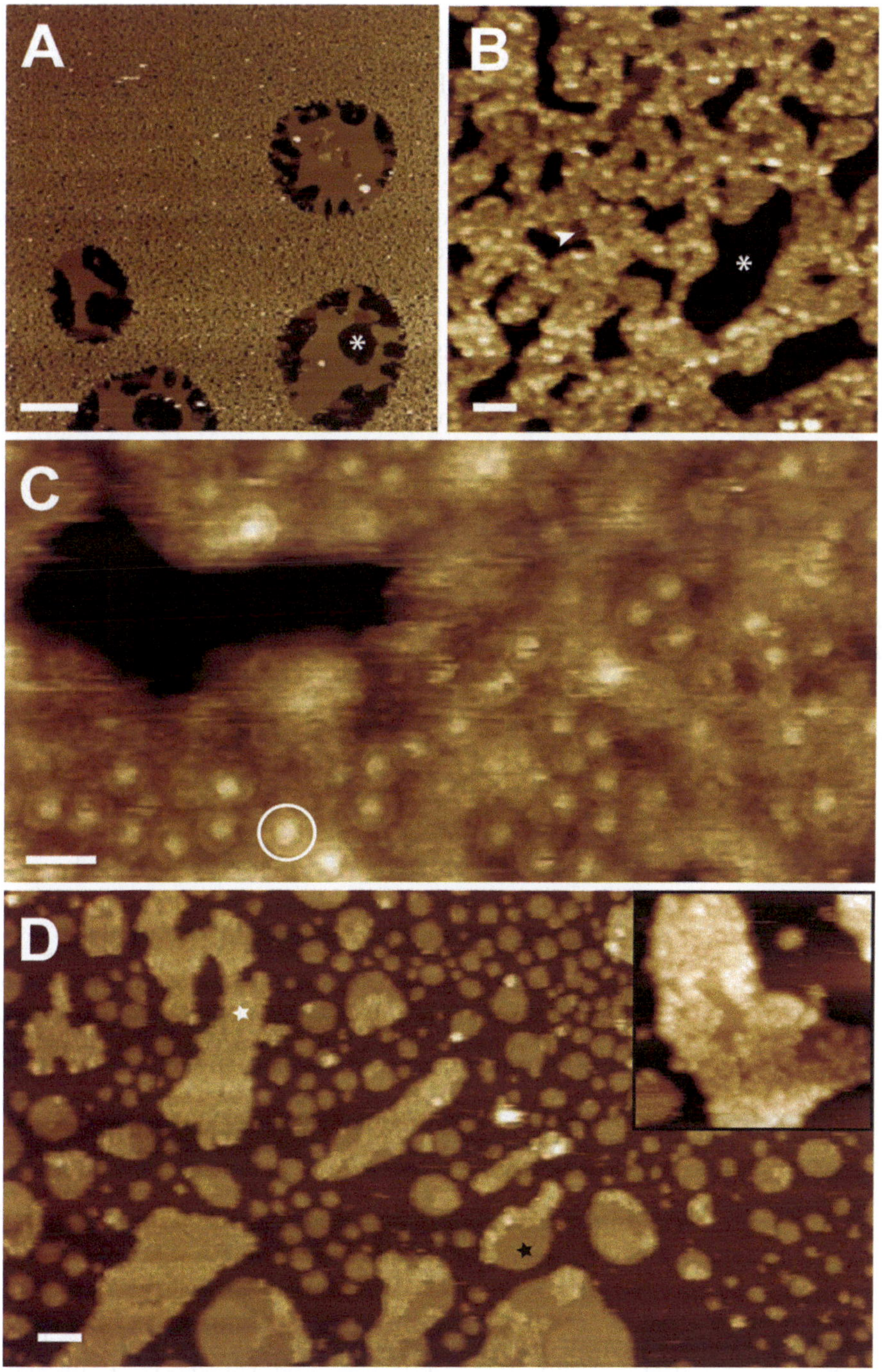


Fig. 3. Incorporation of membrane proteins in supported lipid bilayer. (**a–c**) Incorporation of LH1–RC solubilized within a DOPC/DPPC bilayer. LH1–RC complexes from *Rba. veldkampii* were directly incorporated at 4°C into DOPC/DPPC SLBs using 0.075 mM DOTM. Flat lipid membranes (round-shaped in (**a**) and holes (*white asterisk*)) alternate with corrugated membranes that correspond to lipid domains enriched in proteins. Higher magnification of these domains (**b**) shows corrugated membranes with a few lipid domains (*white arrowhead*) alternating with holes (*white asterisk*). Higher resolution AFM images (**c**) allow delineating core complexes (*white circle*) with the LH1 ring surrounding the RC. (**d**) Incorporation at 4°C of the LacY permease incorporated in 0.34 mM (2× cmc) DDM within POPG/POPE bilayer. Lipid membrane patches were

tal guidelines for the following protein incorporation described in Subheading 3.3.

In addition to AFM imaging, optical microscopy can be used to determine the percentage of remaining membrane. In this case, SLB are previously doped with 0.5% fluorescent lipid (e.g., 1,2-dioleoyl-sn-glycero-3-phosphatidylethanolamine labeled with rhodamine on the polar head) that mixes with fluid phase and is excluded from the gel phase. Representative images show a continuous fluorescence with dark areas corresponding to gel phase domains (Fig. 2d). After detergent incubation and removal, fluorescent fluid phase domains are still observed (Fig. 2e). It is worth noting that most of the fluorescent probes partition in the fluid phase and it is then difficult to discriminate between gel phase domains and solubilized membranes. However, the former is generally round shaped whereas the later is more heterogeneous in size and reveals recognizable features of the detergent/SLB interactions (compare Fig. 2d, e).

### 3.3. Incorporation of Transmembrane Proteins

#### 3.3.1. Incorporation of LH1–RC

Incorporation of membrane proteins is performed at the detergent concentration that has been defined in Subheading 3.2, always higher than the cmc to preserve the solubility of the protein. Proteins are generally purified after membrane-solubilization with low cmc detergents. We describe below the protocol used for the incorporation of LH1–RC, which can be adapted for other membrane proteins as well:

1. Prepare a SLB as described in Subheading 3.1.1.
2. Carefully remove 25 μl from the 50 μl of PBS buffer bathing the SLB.
3. Add 25 μl of buffer containing detergent and proteins at a concentration of 0.150 mM DOTM and 10–100 μg/ml protein (i.e., 0.5–5 μg, 1–10 pmol proteins per assay) (see Notes 3 and 4).
4. Gently mix by up-and-down pipeting of the solution.
5. Incubate for 5–15 min at 20°C for protein incorporation.
6. Remove detergent by extensive rinsing with at least 10 drops of detergent-free buffer and repeat this step at least 3 times with 20 min time intervals.

#### 3.3.2. Incorporation of Fragile Proteins

In the case of fragile proteins, they can be incorporated at 4°C after slight modifications of the protocol described in Subheading 3.3.1. All the steps except detergent removal and AFM imaging must be performed in a cold room.

1. Equilibrate the SLB at 4°C for at least 60 min.
2. Carefully remove 25 μl from the 50 μl PBS drop bathing the SLB.

3. Add 25 μl of buffer with 0.05 mM DOTM (0.5× cmc final concentration).
4. Incubate the bilayer 15 min at 4°C.
5. Remove 25 μl of the solution and add 25 μl of a solution of protein in buffer supplemented with 0.075 mM DOTM concentration (0.05 mM final concentration).
6. Incubate for 15 min at 4°C.
7. Detergent is removed as explained above. Under these conditions, proteins can also be successfully incorporated (10). However, compared to incorporation at 20°C, the resulting SLBs are different, likely due to differences in detergent/SLB interactions during the equilibration time.

#### 3.3.3. Imaging

Extensive and sequential rinsing with detergent-free buffer is necessary to remove detergent and non-incorporated proteins, both precluding high-resolution AFM imaging. It is, however, possible to image the membrane a few hours after incorporation even if membrane dynamics impair image quality (see an example in ref. 9; under these conditions, tapping mode AFM is more suitable). Overnight incubation in detergent-free buffer allows lateral segregation of both lipids and proteins that organize in domains (see below). Such high protein density is a prerequisite for high-resolution imaging. It is noteworthy to mention that the pattern obtained with proteins could be slightly different from that obtained with the detergent control.

Examples of topographs of SLB containing LH1–RC from *Rba. veldkampii* or LacY are depicted in Fig. 3; other examples can be found in our previous reports (9, 10). LH1–RC was incorporated in DOPC/DPPC supported bilayers at 20°C. Large corrugated and interconnected membranes of several microns large are observed, including 100–200 nm areas enriched in proteins that alternate with smooth lipid domains (Fig. 3a, b). As observed with the detergent control, both fluid and gel phase are observed in these areas. Incorporation efficiency is so high that densely packed proteins can be easily observed, even for 1 μm AFM scan and the 10 nm in diameter ring-shaped LH1 clearly identified for 700 nm scan (Fig. 3c).

In the case of incorporation of the permease LacY purified from *E. coli*, we were interested in the incorporation of proteins in a more native environment (11), and we, therefore, used a binary mixture of POPE and POPG that mimics the bacterial membrane. Because of the negative charge of the POPG head, it was necessary to add divalent cations in the buffer in order to favor interaction of vesicles with the negatively charged mica. However, as mentioned in Subheading 3.2, divalent cations increase the stability of the SLB in the presence of detergent by bridging lipid polar heads in

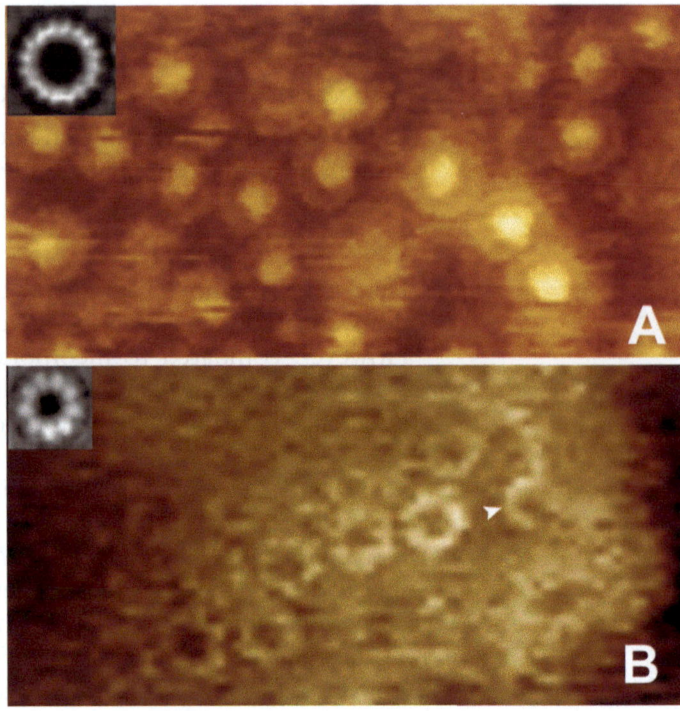


Fig. 4. High-Resolution AFM Imaging of reconstituted proteins. (**a**) 120 nm AFM scan (*z* scale, 10 nm) of LH1–RC proteins from *Rba. veldkampii* incorporated within a DOPC/DPPC bilayer. Details in the LH1 ring can be observed as well as the protruding RC. (**b**) 75 nm AFM scan (*z* scale, 5 nm) of LH2 from *Rba. sphaeroides*. The LH2 ring was sometimes opened by the tip (*white arrowhead*). Insets in (**a**) and (**b**) correspond to image averaging of individual molecules calculated from the trace and retrace topographs using the Xmipp single particle analysis program package (28) (adapted from refs. 9, 10).

contact with the mica. We, therefore, used EDTA to trap $CaCl_2$ and a higher DDM concentration was also required (2× cmc at 4°C). A similar strategy was also used for prestin incorporation and an even higher DDM concentration (3× cmc) was used to compensate $CaCl_2$ effect (12). Using 20 μg/ml LacY, we observed membrane patches with heterogeneous size ranging from a few tens of nanometers to the micron (Fig. 3d). The pattern was very different from those obtained with LH1 and LH2 proteins underlining the fact that it depends on the detergent used for membrane destabilization but also on proteins that are incorporated (it is easy to speculate that the hydrophobicity of the protein can largely influence the incorporation process). In the case of LacY, corrugations covering part or all the membrane patches were observed. Medium magnification revealed the presence of close-packed assemblies protruding 2 nm above the fluid phase, a value in good agreement with the predicted model while higher magnification led us to propose that LacY was monomeric in the lipid bilayer (inset in Fig. 3d).

One of the key properties of this method is the ability of membrane components to laterally segregate after protein incorporation allowing membrane remodeling. This remodeling generally led to lateral segregation of proteins into domains. Proteins are then tightly packed and their diffusion limited, some conditions that are suitable for high-resolution imaging. AFM images depicted in Fig. 4 reveal the substructures of both LH1 from *Rba. veldkampii* and LH2 from *Rba. sphacroides* The LH1 ring is protruding 3 nm above the bilayer and the RC can be easily delineated by the tip. The 16 $\alpha/\beta$-heterodimers of the LH1 ring was resolved by image averaging with Xmipp (http://xmipp.cnb.csic.es/twiki/bin/view/Xmipp/WebHome), a software classically used in electron microscopy image analysis (inset in Fig. 4a). Similarly, the nonameric subunits organization of LH2 was observed, even in raw data (9).

### 3.4. Conclusion and Perspectives

Direct incorporation of transmembrane proteins within SLB appears as a promising technique to obtain structural data in a native environment, i.e., a lipid membrane. Five different types of proteins have been already incorporated but this technique could be applied to most of the proteins from both prokaryotes and eukaryotes, even in a small quantity since only a few picomoles of proteins are required. Since proteins are incorporated in unique orientation, this is a method of choice for the incorporation of membrane protein partners such as LH1 and LH2 proteins and the study of their interaction in the membrane. This could also be used to investigate interactions between the extramembraneous domain of transmembrane proteins and soluble membrane partners, for example, G protein-coupled receptor and G proteins. In this context, such an approach should benefit from the development of super resolution microscopy that allows mapping of proteins in membrane with a lateral resolution of a few tens of nanometers (reviewed in ref. 26).

One of the advantages of the direct incorporation technique is the fact that lipid compositions can be tuned according to the protein as shown in the case of LacY. For eukaryotic membrane proteins, a membrane that preserves the lipid asymmetry is often crucial for the function. This is not yet accessible in the protocol reported here. In this context, interesting results have been obtained in the formation of giant unilamellar vesicles with different composition of the two leaflets, such as the cDICE technique (27). These vesicles can be fused to mica for the formation of asymmetric SLB (unpublished results). Our method could be also improved by filling the holes, formed in the SLB by detergent treatment, with lipids to form continuous bilayers. Formation of continuous membranes is an important parameter for further development in the nanobiotechnology field.

## 4. Notes

1. While formation of the 1–3 nm water layer can be achieved with PBS, it appears that the addition of divalent cations in the buffer favors SLB formation and is sometimes required for vesicle fusion like in the case of negatively charged lipids (see Subheading 3.3). This approach, however, typically precludes protein incorporation (10) (see Subheading 3.3) and we have been able to demonstrate that 5 mM $Ca^{2+}$ in PBS could bridge lipid polar heads with mica, freeze the bilayer, and prevent membrane destabilization by the detergent. However, divalent cations can be eventually trapped by EDTA after SLB formation and before protein incorporation. Alternatively, protein incorporation can be restored by increasing the concentration of detergent.
2. Detergent concentration should be measured with accuracy, especially if proteins have been concentrated after purification, since detergent will interact with the SLB
3. Direct addition in the drop of a concentrated solution of proteins should be avoided because of the common presence of glycerol in the buffer that is often used as cryo-protector in solution. Addition of a small volume of proteins in buffer containing this molecule can induce a local oversolubilization of the membrane since glycerol solution rapidly deposits at the top of the bilayer due to its high density.
4. In the case of proteins purified in another detergent than those reported here, proteins can be directly diluted into the DDM, DOTM or F16 solutions. Final protein concentrations are low and, thus, in most cases, concentrations of detergent present with proteins are negligible. Otherwise, a previous step of detergent exchange should be considered.

## Acknowledgments

The research has been supported by Institut Curie (DL), CNRS (DL and PEM), and INSERM (P.E.M.) and granted by the ANR grant "ANR-06-PCVI-0021-01, AFM -MB-PROT." We are grateful to our collaborators involved in the project: Drs. A. Berquand, J. Hernández-Borrell, C. le Grimellec, F. Gubellini, P. Dosset, L. Picas, and B. Seantier.

## References


1. Goksu EI, Vanegas JM, Blanchette CD et al (2009) AFM for structure and dynamics of biomembranes. Biochim Biophys Acta 1788:254–266
2. El Kirat K, Morandat S, Dufrene YF (2009) Nanoscale analysis of supported lipid bilayers using atomic force microscopy. Biochim Biophys Acta 1798:750–765
3. Giocondi MC, Seantier B, Dosset P et al (2008) Characterizing the interactions between GPI-anchored alkaline phosphatases and membrane domains by AFM. Pflugers Arch 456:179–188
4. Muller DJ, Sapra KT, Scheuring S et al (2006) Single-molecule studies of membrane proteins. Curr Opin Struct Biol 16:489–495
5. Muller DJ, Hand GM, Engel A, Sosinsky GE (2002) Conformational changes in surface structures of isolated connexin 26 gap junctions. EMBO J 21:3598–3607
6. Scheuring S, Lévy D, Rigaud JL (2005) Watching the bacterial photosynthetic apparatus by in situ AFM. Biochim Biophys Acta 1712:109–127
7. Muller DJ, Fotiadis D, Scheuring S et al (1999) Electrostatically balanced subnanometer imaging of biological specimens by atomic force microscope. Biophys J 76:1101–1111
8. Buzhynskyy N, Hite RK, Walz T, Scheuring S (2007) The supramolecular architecture of junctional microdomains in native lens membranes. EMBO Rep 8:51–55
9. Milhiet PE, Gubellini F, Berquand A et al (2006) High resolution AFM of membrane proteins directly incorporated at high density in planar lipid bilayer. Biophys J 91:3268–3275
10. Berquand A, Levy D, Gubellini F et al (2007) Influence of calcium on direct incorporation of membrane proteins into in-plane lipid bilayer. Ultramicroscopy 107:928–933
11. Picas L, Carretero-Genevrier A, Montero MT et al (2010) Preferential insertion of lactose permease in phospholipid domains: AFM observations. Biochim Biophys Acta 1798:1014–1019
12. Kumano S, Murakoshi M, Iida K et al (2010) Atomic force microscopy imaging of the structure of the motor protein prestin reconstituted into an artificial lipid bilayer. FEBS Lett 584:2872–2876
13. Tate CG (2010) Biochemistry. Membrane protein gymnastics. Science 328:1644–1645
14. Gubellini F, Francia F, Busselez J et al (2006) Functional and structural analysis of the photosynthetic apparatus of *Rhodobacter veldkampii*. Biochemistry 45:10512–10520
15. Francia F, Wang J, Venturoli G et al (1999) The reaction center-LH1 antenna complex of *Rhodobacter sphaeroides* contains one PufX molecule which is involved in dimerization of this complex. Biochemistry 38:6834–6845
16. Merino S, Domenech O, Montero MT, Hernandez-Borrell J (2005) Atomic force microscopy study of *Escherichia coli* lactose permease proteolipid sheets. Biosens Bioelectron 20:1843–1846
17. Geertsma ER, Nik Mahmood NA, Schuurman-Wolters GK, Poolman B (2008) Membrane reconstitution of ABC transporters and assays of translocator function. Nat Protoc 3:256–266
18. Rigaud J, Levy D (2003) Reconstitution of membrane proteins into liposomes. Methods Enzymol 372:65–86
19. Rigaud JL, Pitard B, Levy D (1995) Reconstitution of membrane proteins into liposomes: application to energy-transducing membrane proteins. Biochim Biophys Acta 1231:223–246
20. Brian AA, McConnell HM (1984) Allogeneic stimulation of cytotoxic T cells by supported planar membranes. Proc Natl Acad Sci USA 81:6159–6163
21. Seantier B, Giocondi M, Le Grimellec C, Milhiet P (2008) Probing supporting model and native membranes using AFM. Curr Opin Colloid Interface Sci 13:326–337
22. Giocondi MC, Yamamoto D, Lesniewska E et al (2010) Surface topography of membrane domains. Biochim Biophys Acta 1798:703–718
23. Sackmann E (1996) Supported membranes: scientific and practical applications. Science 271:43–48
24. Burns AR, Frankel DJ, Buranda T (2005) Local mobility in lipid domains of supported bilayers characterized by atomic force microscopy and fluorescence correlation spectroscopy. Biophys J 89:1081–1093
25. Chianta S, Ries J, Kahya N, Schwille P (2006) Combined AFM and two-focus SFCS study of raft-exhibiting model membranes. Chemphyschem 7:2409–2418
26. Hell SW (2007) Far-field optical nanoscopy. Science 316:1153–1158
27. Abkarian M, Loiseau E, Massiera G (2011) Continuous droplet interface crossing encapsulation (cDICE) for high throughput monodisperse vesicle design. Soft Matter 7: 4610–4614
28. Marabini R, Masegosa IM, San Martin MC et al (1996) Xmipp an image processing package for electron microscopy. J Struct Biol 116:237–240

# Chapter 20

# Functional AFM Imaging of Cellular Membranes Using Functionalized Tips

**Lilia A. Chtcheglova and Peter Hinterdorfer**

## Abstract

The real-time visualization of specific binding sites on biological samples with high spatial resolution, in order of several nanometers, is an important undertaking in many fields of biology. During the past 5 years, simultaneous topography and recognition imaging (TREC) has become a powerful tool to quickly obtain local receptor nanomaps on complex heterogeneous biosurfaces, such as cells and membranes. In this chapter, we present the TREC technique and explain how to unravel the nano-landscape of cells of the immune system, such as macrophages. We describe the procedures for all steps of the experiment including tip functionalization with Fc fragments via flexible PEG-linker, sample preparation, and localization of Fcγ receptors on macrophages.

**Key words:** Macrophage, Fcγ receptors, Atomic force microscopy, Simultaneous topography and RECognition imaging

## 1. Introduction

The real-time visualization and quantification of receptor binding sites on cell surfaces remains a fundamentally challenging task in molecular cell biology (1). This can be easily achieved by common techniques such as immunostaining. However, the lateral resolution here is limited by the diffraction limit of light (~200 nm). During the last decade advanced optical techniques with improved lateral accuracy (10–30 nm) such as single-molecule optical microscopy (2), near-field scanning optical microscopy (NSOM) (3, 4), stimulated emission depletion microscopy (STED) (5), photoactivated localization microscopy (PALM) (6), or stochastic optical reconstruction microscopy (STORM) (7) have been successfully introduced. Despite fast time resolution in these optical studies no topographical data are attainable. Electron microscopy (EM),

Alioscka A. Sousa and Michael J. Kruhlak (eds.), *Nanoimaging: Methods and Protocols*, Methods in Molecular Biology, vol. 950, DOI 10.1007/978-1-62703-137-0_20, © Springer Science+Business Media, LLC 2013

another technique that is widely used by biologists, provides nm resolution, but often requires heavily processed samples and lacks biochemical specificity. Conversely, atomic force microscopy (AFM) (8) represents a scanning probe microscopy (SPM) technique, which allows nanometer spatial (topographical) resolution investigations of functional biological specimens (e.g., proteins, DNA, membranes, cells, etc.) at physiological conditions and without any rigorous sample preparation protocols.

The recently developed simultaneous topography and recognition imaging technique (TREC) (9–11) provides a topographical image with a separate recognition map from a single scan, with a lateral resolution of 1.5 nm on S-layers (12) and 5–10 nm on different cell surfaces (13–17).

In this chapter, we describe detailed procedures on how the TREC technique is exploited to locally identify Fcγ receptors on gently fixed macrophages (16). Macrophages are referred to as professional phagocytes as they are equipped for rapid and efficient uptake of invading microorganisms at the sites of inflammation (18, 19). One of the major functions of macrophages is pathogen clearance through phagocytosis, a process whereby cells engulf large particles with a diameter larger than 0.5 μm (20). Phagocytosis is a multistep process initiated by the particle recognition. The recognition, in turn, is mediated by a wide variety of cell surface receptors that can bind directly, through cell surface components of microorganisms (e.g., lipopolysaccharides), or indirectly, through opsonins, (e.g., IgG or complement fragments), to the particle. The most studied phagocytic receptors include the Fc receptors (FcR) that bind to the Fc portion of immunoglobulins. The nanomapping of FcγRs (Fc receptors for IgG) is performed on fixed J774.A1 mouse macrophage cell surfaces with magnetically coated AFM tips functionalized with Fc fragments of mouse IgG via long and flexible PEG linkers.

## 2. Materials

### 2.1. AFM System

1. 5500 AFM (Agilent Technologies, USA) equipped with a CCD camera.
2. PicoTREC and MAC Mode III boxes (Agilent Technologies) (see Note 1).
3. Large AFM scanner, MAC bottom sample plate, standard nose cone for contact AFM mode, liquid flow-through cell, and magnetically coated AFM cantilevers known as MAC Levers (Agilent Technologies) (see Note 2).
4. A vibration isolation platform (e.g., an isolation box, where AFM is suspended from the ceiling using bungee cords or a

slab of solid marble, on which the AFM could be placed) to minimize vertical and horizontal mechanical vibrations of the microscope.

#### 2.2. Functionalization of AFM Tips with Fc Fragment via PEG-Linker

1. Fc fragment of mouse IgG (Thermo Fisher Scientific, USA) (see Note 3).
2. A dessicator (5 L).
3. 3-Aminopropyltriethoxysilane (APTES).
4. Triethylamine, ≥99%.
5. Argon gas.
6. 1× Dulbecco's Phosphate Buffered Saline (PBS) without $Ca^{2+}$ and $Mg^{2+}$.
7. 3.3 mg of homemade heterobifunctional PEG cross-linker, Aldehyde–$PEG_{27}$–NHS (21, 22) (see Note 4).
8. Chloroform.
9. Nitrogen gas.
10. 1 M $NaCNBH_3$: dissolve 32 mg solid $NaCNBH_3$ in 450 μL $H_2O$, add 50 μL 100 mM NaOH (see Note 5).
11. 1 M Ethanolamine hydrochloride in water, pH 9.6 adjusted with NaOH.
12. Parafilm.
13. 100 mM NaOH.
14. Petri dish with diameter of 35 mm.

#### 2.3. Components for the Cell Samples

1. J774.A1 cells, mouse macrophage cell line.
2. Growth medium: Dulbecco's modified Eagle's medium (DMEM) high glucose, 1% Penicillin/Streptomycin, and 10% fetal calf serum (FCS).
3. Glass microscope slides with a diameter of 22 mm.
4. 70% isopropanol in water.
5. PFA fixative: 4% paraformaldehyde (PFA) in Hank`s balanced salt solution (HBSS) with $Ca^{2+}$, pH 7.4 (see Note 6).

## 3. Methods

Cautions:

1. Wear gloves and laboratory coat when handling all reagents and chemicals, use gas mask when working with toxic reagents such as paraformaldehyde, $NaCNBH_3$.
2. Additionally, all steps for the tip chemistry protocol (see Subheading 3.1) and cell fixation procedure should be performed in the fume hood.

3. AFM measurements described here are designed to be used with a class II diode laser, which output is of up to 1 mW of visible radiation at 670 nm. DO NOT stare into the laser beam.

Carry out all procedures at room temperature unless otherwise specified.

### 3.1. Functionalization of AFM Tips with Fc Molecules

Attachment of ligand molecules onto AFM tips transforms them into biospecific molecular sensors. The functionalization of AFM tips with ligands is usually performed in three steps: (i) amino ($-NH_2$) groups are produced on the tip surface, (ii) heterobifunctional PEG chains are attached by one end to the amino group on the tip, and (iii) a ligand molecule is coupled to another free functional end of the PEG linker (Fig. 1).

1. Clean AFM cantilevers in chloroform for 10 min and dry with a gentle nitrogen stream. Amino ($-NH_2$) groups are produced on the AFM tips by gas phase silanization with 3-aminopropyltriethoxysilane (APTES) (23). First, APTES is freshly distilled under vacuum. Flood a desiccator (5 L) with argon gas to remove air and moisture. Then place two small plastic trays (e.g., the lids of Eppendorf reaction vials) inside the desiccator, with 30 μL of APTES and 10 μL of triethylamine (see Note 7) separately pipetted into the two trays. Place pre-cleaned AFM cantilevers nearby on a clean inert surface (e.g., Teflon) within the dessicator, and close the dessicator. After 2 h of incubation, remove the trays with APTES and triethylamine. Again, flood the desiccator with argon gas for 5 min. Leave the tips inside for 2 days in order to "cure" the APTES coating.
2. For the attachment of the PEG linker, dissolve 3.3 mg of NHS–$PEG_{27}$–Aldehyde in 0.5 mL of chloroform and transfer with a glass pipette to a small glass reaction chamber.
3. Add 30 μL of triethylamine and carefully immerse the cantilevers in this solution for 2 h at RT. Cover the reaction chamber with an upside-down beaker to avoid chloroform evaporation (see Note 8).
4. Then wash cantilevers extensively (3×) in chloroform, dry with a gentle nitrogen stream and put in a clean dry petri dish, which is then covered with Parafilm. Arrange cantilevers in a circular manner so that the tips are found in the center direction.
5. For ligand coupling, mix a 50 μL aliquot of Fc fragment stock solution (~1 mg/mL in PBS, freshly thawed from −25 °C) with 150 μL of 1× PBS. Pipette the protein solution onto the cantilevers as a small drop to cover all tips with the liquid.
6. Immediately add 2 μL of 1 M $NaCNBH_3$ (freshly prepared each time) and mix carefully with a pipette. Allow the reaction to be carried out for 1–2 h.

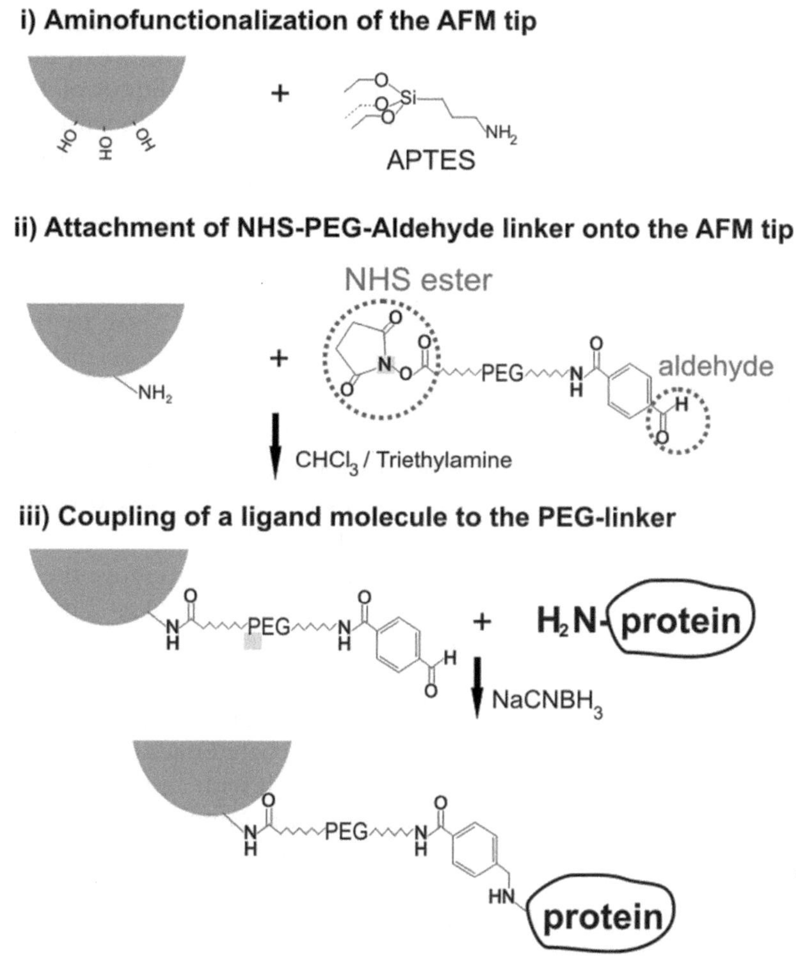


Fig. 1. AFM tip functionalization with Fc fragments via NHS–PEG–Aldehyde linker. (i) Aminofunctionalization of silicon nitride ($Si_3N_4$) AFM tips via silanization with 3-aminopropyltriethoxysilane (APTES) in gas phase. (ii) Use of heterobifunctional NHS–PEG–Aldehyde linker for the flexible attachment of underivatized protein onto the AFM tip. (iii) A ligand molecule is coupled to another free functional end of PEG-linker. The C=N double bond is usually fixed by a reaction with sodium cyanoborohydride ($NaCNBH_3$).

7. To inactivate free aldehyde groups on the tips, add 5 μL of 1 M ethanolamine hydrochloride (pH 9.6 pre-adjusted with NaOH and stored in aliquots at −25 °C) for 10 min.
8. Finally, wash the cantilevers with 1× PBS (AFM working buffer) (3×) and store at 4 °C until used (see Note 9).

### 3.2. Cell Preparation for AFM Measurements

1. Wash glass slides with 70% isopropanol or ethanol and subsequently with distilled water or 1× PBS, put the glass slides in petri dishes. If any liquid drops remain on the slides, pump them carefully away.

2. Cells are grown on the glass slides. When the cells achieve the desired confluence state (~50%), they are "gently" fixed with 4% PFA (16). Warm an aliquot of 4% PFA fixative, pump away growth media, wash the cells with HBSS (1×), incubate in pre-warmed 4% PFA for 30–60 min at 37 °C, and then wash (3×) the fixed cells with HBSS containing $Ca^{2+}$. The fixed cells can either be immediately used for AFM measurements or stored at 4 °C for several days (see Note 10).

### 3.3. Functional AFM Imaging on Cell Membranes

1. Connect the PicoTREC box to the head electronics, MAC box, and AFM controller according to the PicoTREC user manual.
2. Turn on the AFM controller, MAC box, computer, lamps, and other major electronics and allow to warm up for at least 30 min in order to minimize the drift of electrical signals during the experiment.
3. Use the stepper motor to move the sample plate, ensuring that there is enough distance (1.5–2 mm) between the sample plate and AFM scanner. This will prevent a sudden breakage of the AFM cantilever during the further montage.
4. Prior to mounting the cell probe, place some reflective material, such as a gold-coated mica or aluminum sheet, onto the sample plate and cover it with a mica sheet. Mount the glass slide with fixed cells on the sample plate, carefully fix the liquid cell with two clips on the glass slide and fill it immediately with ~600 μL HBSS with $Ca^{2+}$. Take care that the cells do not dry out. Finally, mount the sample plate with the cell probe in the AFM.
5. Carefully mount the functionalized AFM cantilever on the AFM scanner, taking care that the cantilever does not dry out.
6. Orient the scanner, mount it onto the microscope stage, fix, and plug it in. Check with the CCD camera that the cantilever has not broken during the scanner mounting. If any air bubbles are observed, remove the scanner, carefully blot dry with the corner of a piece of filter paper, and then repeat this step again.
7. Set up the instrument for MAC Mode.
8. Ensure that the TREC servo on the front of the PicoTREC box is in the OFF position.
9. Choose an appropriate calibration file for the AFM scanner.
10. Set one image channel to "Topography" (both in trace and retrace direction) and a second channel to "Aux In BNC" (both in trace and retrace direction) (see Note 11).
11. Perform a frequency sweep by tuning the functionalized MAC Lever far away from the sample surface (~60 μm) (Fig. 2), and choose and set the starting drive amplitude (the amplitude meter on the MAC box should read ~6 V (MAC1.2 box) or

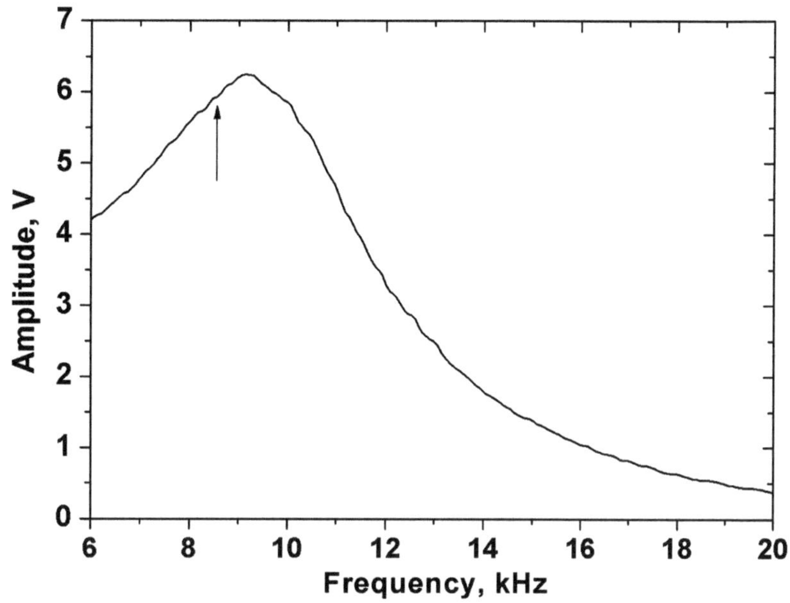


Fig. 2. Typical amplitude–frequency (resonance or tuning) curve acquired in buffer (HBSS) with a MAC Lever. Tuning curves are recorded by varying the excitation frequency from 5 to 20 kHz; tip-sample separation distance is 60 μm. A distinct resonance peak is obtained at ~10 kHz (using a cantilever with a nominal spring constant of 0.15 N/m). A frequency of ~8 kHz is used as excitation frequency for imaging (indicated by an *arrow*).

~1.5–2 V (MACIII box)) (see Note 12). As a rule of thumb we choose the excitation frequency 20% lower than the resonance frequency at large distances (resulting in ~8 kHz excitation frequency, indicated by an arrow in Fig. 2).

12. Perform a rough approach to the surface. Withdraw the tip and place the tip over a cell of interest with the help of the CCD camera. Adjust the drive signal to typically attain either ~6 V on the MAC1.2 box or ~1.5–2 V on the MACIII box.

13. Approach the sample surface and begin imaging a whole cell (typically at a scan size of ~60 × 60 μm$^2$). Consequently reduce the scanning area to ~2 × 2 μm$^2$ by choosing relatively flat cell surface (Fig. 3 left). Use a maximum lateral scan speed of ~3 μm/s; this will result in a total recording time of ~12 min per image (with a resolution of 512 lines per image). Scan using gain settings as high as possible, but without producing excessive noise. Start with a very low imaging force (high amplitude set-point so that the tip slightly touches the surface; one can check this by opening the "real time cross-section" window and follow the topography cross section in trace and retrace) and, consequently, reduce the amplitude set-point to get a fairly good overlay of trace and retrace cross-section lines. Avoid reducing the set-point further; this can result in strong

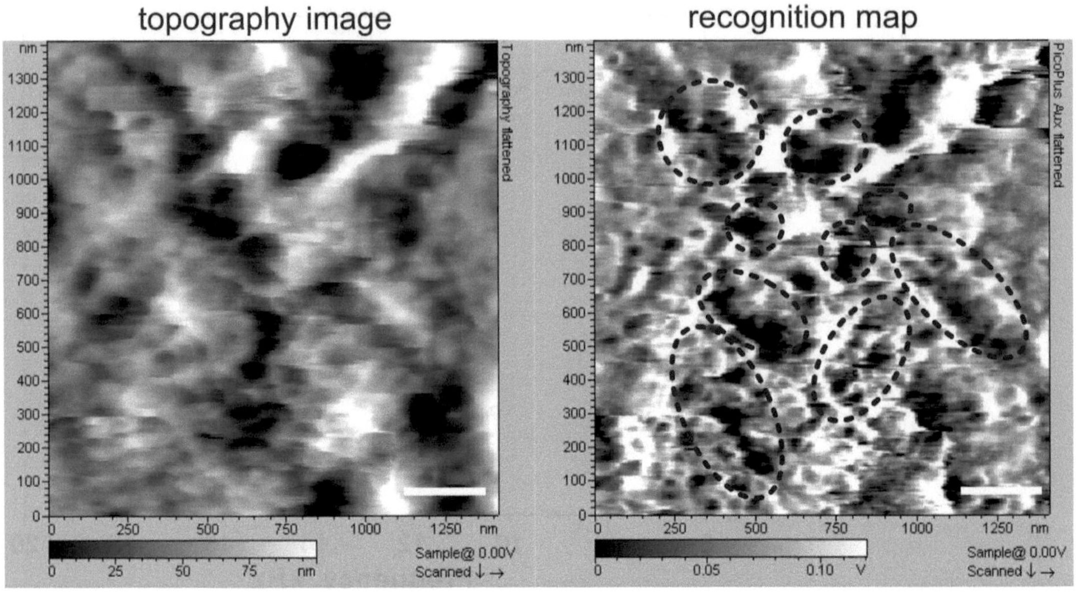


Fig. 3. Simultaneously recorded Topography (*left*) and RECognition (*right*) images on J774.A1 cells with a Fc functionalized tip. Recognition map of Fcγ receptor domains (*black spots* embedded by *black circles*) represents an amplitude reduction due to specific binding between Fc molecules on the AFM tip and FcγRs on the cell surface. Scale bars are 250 nm.

cross talk of topography information into the recognition image (24). Setting the proper imaging amplitude is critical (24): it must be large enough to stretch the PEG linker, slightly deflect the cantilever as it oscillates over the target molecule, and detect the binding interactions. Nevertheless, the oscillation amplitude should remain small enough so that the binding does not rupture on each oscillation. Therefore, an imaging amplitude of ~8–15 nm (this typically corresponds to ~1.5–2.5 V on the amplitude meter of the MAC1.2 box) is recommended for TREC imaging (see Subheading 3.4 below to determine the free oscillation amplitude in nm). Generally, the proper amplitude regime for the observation of recognition events differs from one functionalized cantilever to another, depending upon the length of the linker molecule, on the exact location of the linker molecule on the tip apex, and the size of the attached molecule. Therefore, actual settings will vary, and you might have to change the drive amplitude and approach again.

14. Recognition is observed as dark "hot" spots in the "Aux in BNC" image channel (Fig. 3 right). Normally the recognition maps remain unchanged during 1 h of continuous scanning. If no recognition signal is visible after adjusting the gains and amplitude after several image frames, the functionalized MAC Lever must be replaced. We suggest preparing as many functionalized MAC Levers as possible (a minimum of 5) to ensure at least one "good" MAC Lever is available for TREC imaging.

15. Perform the blocking experiment (see Note 13). Inject Fc fragments at high concentration (~0.8 mg/mL) very slowly and gently into the fluid cell while scanning the cell surface. The first scan after injection might not demonstrate immediate changes in the recognition map. After 2–3 scans, one can observe subsequent disappearance of dark spots with imaging time as the free Fc fragments will bind to Fcγ receptors on the macrophage surface (16) (see Note 14). After blocking experiments, simultaneously recorded topography images should remain unchanged—indicating that the blocking does not affect membrane topography.
16. As an additional check on the specificity of the recognition events, the oscillation amplitude can be varied (24). A decrease of the amplitude results in lower contrast of the recognition spots, since the linker is less stretched. An increase of the amplitude leads to a sudden disappearance of the recognition spots, since the ligand is no longer able to bind continuously to the receptor's binding sites. To perform such an experiment, vary the drive amplitude, but maintain the ratio of set-point amplitude to free amplitude.

### 3.4. Determination of the Free Oscillation Amplitude

Amplitude–distance cycles are used to adjust the free oscillation amplitude and to determine the optimal amplitude reduction value (i.e., the set-point value) for driving the feedback loop during imaging. The amplitude–distance cycles should be collected on solid surfaces such as mica or glass.

1. Open the spectroscopy window. Approach the surface. Perform a complete amplitude–distance cycle (put Limit to OFF; control the *z*-piezo limits so as to not crash the tip). The distance between the first contact and the zero amplitude reflects the peak-to-peak amplitude of free oscillation. This gives the ratio of nm/V conversion.
2. Adjust the drive signal and the set-point in order to get a peak-to-peak set-point amplitude of ~8–15 nm (corresponds to ~1.5–2.5 V on the MAC1.2 box). The determined value of the amplitude is the proper setting, which should be used for TREC (Fig. 4).
3. Withdraw the tip ~60 μm from the surface and determine the actual amplitude above the surface (~6 V on the MAC1.2 box).

### 3.5. Half-Amplitude Imaging (for Picoscan Software)

In contrast to the "normal" MAC mode imaging, where the peak-to-peak value of oscillating amplitude ($A$) is used as a feedback parameter, one can utilize the lower part of the oscillation (i.e., half-amplitude, $A_{1/2}$) to drive a feedback loop to obtain the topography image, whereas the upper part of the oscillation is used to create the recognition map (Fig. 5). Thus, by using the half-amplitude

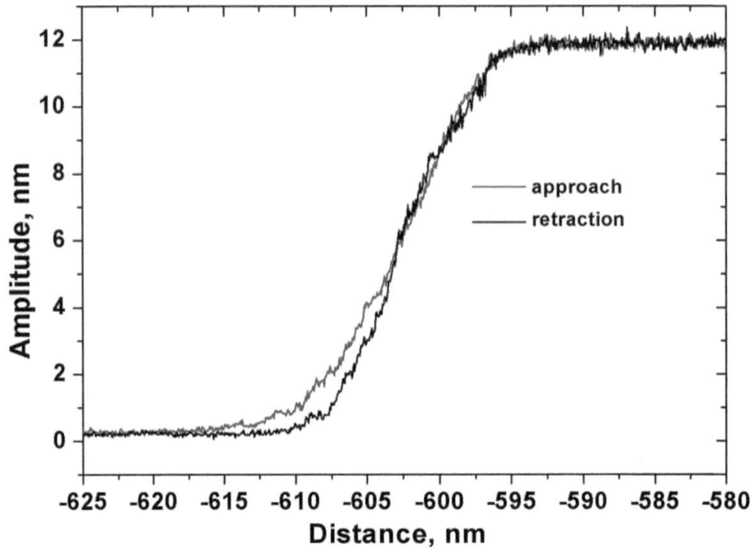


Fig. 4. Amplitude–distance curve acquired in buffer on a glass slide with a MAC Lever. The free oscillation amplitude is set to ~12 nm at an oscillation frequency of ~8 kHz. The oscillation amplitude is shown for both the approach (*light grey*) and the retraction (*black*) of the cantilever as a function of the cantilever-surface distance. When the tip is far away from the surface, the oscillation amplitude (i.e., free amplitude) is not influenced by the sample and remains constant during approach and retraction. As the tip approaches the surface, the amplitude decreases until it is fully dampened to zero at full contact. Upon retraction the amplitude starts to increase until it reaches the free amplitude again.

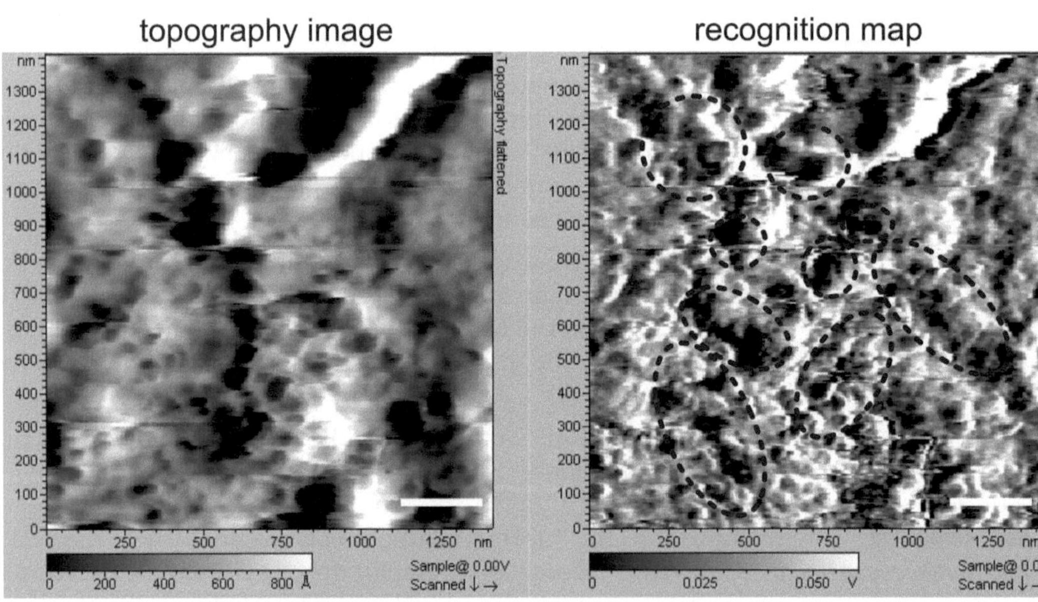


Fig. 5. Example of TREC data collected on J774.A1 cells with a Fc functionalized tip by using half-amplitude feedback. Scale bars are 250 nm. TREC images are collected on the same place as in Fig. 3. Exactly the same "hot" recognition domains (embedded by *black circles*) are observed as in the recognition map recorded in full-amplitude feedback (see Fig. 3 *right*).

feedback, topographical images represent the correct heights as the feedback is not taking into account deviations of the oscillation amplitude due to recognition/adhesion events (24).

1. Turn the servo on the front of the PicoTREC box to the ON position.
2. In the AC mode panel change to Contac mode.
3. Decrease slowly the amplitude set-point until the AFM tip is gently touching the surface again (amplitude set-point will be reduced by about 50% or lower).
4. Set the gains as high as possible and continue to image in this TREC mode as usual.

## 4. Notes

1. PicoTREC from Agilent Technologies is the first AFM option for molecular functional imaging (i.e., detection and mapping molecular binding events while simultaneously generating topographical images in real time). This molecular recognition tool has been designed for use with Agilent's patented MAC Mode on the 5500 AFM, 5500 inverted light microscope (ILM) and very recently on the 6000 ILM systems.
2. MAC Mode is a gentle nondestructive AFM imaging technique, in which a MAC Lever is driven by an oscillating magnetic field, which can be applied directly to the MAC Lever from either above (Top-MAC) or below (MAC Stage) the cantilever. For Top-MAC Mode, use a special nose cone with a standard sample plate, and for MAC Stage, use a standard nose cone for contact mode with a MAC bottom sample plate, respectively.
3. Protein solutions for the tip functionalization protocol should be consistent with the following requirements: proteins should be as pure as possible; no serum samples; the protein solution should not contain other proteins or additives and stabilizers such as BSA; no sodium azide, no glycerol, no detergents; buffer solution should not contain primary amines such Tris.
4. Synthesis of the Aldehyde–$PEG_{27}$–NHS linker consists of several steps (21). Very briefly, the commercial symmetric adduct *O*,*O*′-bis(2-Aminopropyl)-poly(ethylene glycol) 1500, $NH_2$–PEG–$NH_2$ (abbreviated as $NH_2$–PEG–$NH_2$) is reacted with glutaric anhydride and the asymmetric product $NH_2$–PEG–COOH (*N*-glutaryl derivative of $NH_2$–PEG–$NH_2$) is isolated in pure form by three chromatographic steps (22). Subsequently this general linker precursor is reacted with *N*-succinimidyl

4-formylbenzoate in chloroform/triethylamine to obtain Aldehyde–PEG–COOH (21). In the last step, the terminal carboxyl (–COOH) is transformed into NHS-ester by activation with *N,N,N′,N′*-tetramethyl (succinimidyl) uranium tetrafluoroborate (TSTU) in *N,N*-dimethylformamide (DMF)/pyridine (21).

5. *Caution*: $NaCNBH_3$ is toxic, it must be used in a fume hood, and use a gas mask when weighing $NaCNBH_3$.
6. PFA should be handled under a fume hood with required glove protection. Weigh 4 g of PFA and add 100 mL of HBSS with $Ca^{2+}$. Heat stirring solution at 60 °C until PFA dissolves. When solution becomes clear, remove flask from the heating plate, cool to RT and filter through 0.22 μm filter. Adjust pH to 7.4 and store the fresh PFA solution in aliquots at −25 °C.
7. Instead of triethylamine one can use 1,1-diisopropylethylamine (23).
8. When the AFM tips are not used immediately for further ligand binding, they can be stored in a petri dish under argon atmosphere at RT for several days.
9. If the ligand functionalized AFM tips are not used immediately for AFM measurements, they can be stored in buffer solution at 4 °C for several weeks without any loss of binding activity of Fc fragments. Remember that, depending on the ligand stability, this period of time can vary from 1 day to several months.
10. We find that the “gentle” cell fixation procedure does not practically affect the binding activity of Fcγ receptors (16). However, for other cell receptors, an appropriate cell fixation method (concentration of fixative, time of fixation, etc.) should be properly determined.
11. Depending on the PicoTREC box connection, RECognition data will be displayed either in the “Aux In BNC” or “Aux In 5” image window. For users of PicoView software this channel is called “CSAFM/Aux BNC.” Additional channels such as “Amplitude,” “Phase,” and “Deflection” are helpful to control the quality of imaging.
12. Since the spring constant and the dimensions of the cantilevers vary significantly, the optimal excitation frequency is determined for each individual MAC lever. Be aware that by using the bottom MAC plate the generating magnetic field is not uniform. Place the MAC lever first where the magnetic field has a maximum value. Recognition imaging is performed best with a clean strong resonance peak (~6 V on the MAC1.2 box). If the resonance peak is noisy or weak, replace the functional MAC lever.

13. It is recommended that the recognition imaging and block experiments are performed by using half-amplitude amplitude as a feedback parameter to obtain proper recognition signal (see Subheading 3.5).
14. As an alternative to the blocking of cell receptors, one could perform the blocking of the ligand-coated AFM tip. In our case, one can use protein A that binds specifically to Fc fragments.

## References


1. Hinterdorfer P, Dufrêne YF (2006) Detection and localization of single molecular recognition events using atomic force microscopy. Nat Methods 3:347–355
2. Schmidt T, Schütz GJ, Baumgartner W et al (1996) Imaging of single molecule diffusion. Proc Natl Acad Sci USA 93:2926–2929
3. Koopman M, Cambi A, de Bakker BI et al (2004) Near-field scanning optical microscopy in liquid for high resolution single molecule detection on dendritic cells. FEBS Lett 573:6–10
4. Ianoul A, Street M, Grant D, Pezacki J et al (2004) Near-field scanning fluorescence microscopy study of ion channel clusters in cardiac myocyte membranes. Biophys J 87:3525–3535
5. Willing KI, Rizzoli SO, Westphal V et al (2006) STED microscopy reveals that synaptotagmin remains clustered after synaptic vesicle exocytosis. Nature 440:935–939
6. Betzig E, Patterson GH, Sougrat R et al (2006) Imaging intracellular fluorescent proteins at nanometer resolution. Science 313:1642–1645
7. Rust M, Bates M, Zhuang X (2006) Sub-diffraction-limit imaging by stochastic optical reconstruction microscopy (STORM). Nat Methods 3:793–795
8. Binnig G, Quate CF, Gerber C (1986) Atomic force microscope. Phys Rev Lett 56:930–933
9. Stroh CM, Ebner A, Geretschläger M et al (2004) Simultaneous topography and recognition imaging using force microscopy. Biophys J 87:1981–1990
10. Ebner A, Kienberger F, Kada G et al (2005) Localization of single avidin-biotin interactions using simultaneous topography and molecular recognition imaging. Chemphyschem 6: 897–900
11. Stroh C, Wang H, Bash R et al (2004) Single-molecule recognition imaging microscopy. Proc Natl Acad Sci USA 101:12503–12507
12. Tang J, Ebner A, Badelt-Lichtblau H, Völlenkle C et al (2008) Recognition imaging and highly ordered molecular templating of bacterial S-layer nanoarrays containing affinity-tags. Nano Lett 8:4312–4319
13. Chtcheglova LA, Waschke J, Wildling L et al (2007) Nano-scale dynamic recognition imaging on vascular endothelial cells. Biophys J 93:L11–L13
14. Chtcheglova LA, Atalar F, Ozbek U et al (2008) Localization of the ergtoxin-1 receptors on the voltage sensing domain of hERG $K^+$ channel by AFM recognition imaging. Pflugers Arch 456:247–254
15. Chtcheglova LA, Waschke J, Wildling L et al (2010) AFM functional imaging on vascular endothelial cells. J Mol Recognit 23:589–596
16. Ahmad SF, Chtcheglova LA, Mayer B et al (2011) Nanosensing of Fcγ receptors on macrophages. Anal Bioanal Chem 399:2359–2367
17. Chtcheglova LA, Hinterdorfer P (2011) Simultaneous topography and recognition imaging on endothelial cells. J Mol Recognit 24:788–794
18. Aderem A (2002) How to eat something bigger than your head. Cell 110:5–8
19. Underhill DM, Ozinsky A (2002) Phagocytosis of microbes: complexity in action. Annu Rev Immunol 20:825–852
20. Aderem A, Underhill DM (1999) Mechanisms of phagocytosis in macrophages. Annu Rev Immunol 17:593–623
21. Ebner A, Wildling L, Kamruzzahan ACM et al (2007) A new, simple method for linking of antibodies to atomic force microscopy tips. Bioconjug Chem 18:1176–1184
22. Riener CK, Stroh CM, Ebner A et al (2003) Simple test system for single molecule recognition force microscopy. Anal Chim Acta 479:59–75
23. Crampton N, Bonass WA, Kirkham J, Thomson NH (2005) Formation of aminosilane-functionalized mica for atomic force microscopy imaging of DNA. Langmuir 21:7884–7891
24. Preiner J, Ebner A, Chtcheglova L et al (2009) Simultaneous topography and recognition imaging: physical aspects and optimal imaging conditions. Nanotechnology 20:215103–215121

# Chapter 21

# Near-Field Scanning Optical Microscopy for High-Resolution Membrane Studies

**Heath A. Huckabay, Kevin P. Armendariz, William H. Newhart, Sarah M. Wildgen, and Robert C. Dunn**

## Abstract

The desire to directly probe biological structures on the length scales that they exist has driven the steady development of various high-resolution microscopy techniques. Among these, optical microscopy and, in particular, fluorescence-based approaches continue to occupy dominant roles in biological studies given their favorable attributes. Fluorescence microscopy is both sensitive and specific, is generally noninvasive toward biological samples, has excellent temporal resolution for dynamic studies, and is relatively inexpensive. Light-based microscopies can also exploit a myriad of contrast mechanisms based on spectroscopic signatures, energy transfer, polarization, and lifetimes to further enhance the specificity or information content of a measurement. Historically, however, spatial resolution has been limited to approximately half the wavelength due to the diffraction of light. Near-field scanning optical microscopy (NSOM) is one of several optical approaches currently being developed that combines the favorable attributes of fluorescence microscopy with superior spatial resolution. NSOM is particularly well suited for studies of both model and biological membranes and application to these systems is discussed.

**Key words:** NSOM, SNOM, Lipid membranes, High spatial resolution, Sub-diffraction-limited resolution, Fluorescence microscopy

## 1. Introduction

Near-field scanning optical microscopy (NSOM), also known as scanning near-field optical microscopy (SNOM), is a scanning probe technique developed to surpass the spatial resolution constraints that traditionally limit conventional optical microscopy (1–11). As shown in Fig. 1, NSOM uses fiber optic probes to funnel light down to the nanometric dimension. By scanning these probes near a sample surface, fluorescence measurements can be taken with a spatial resolution of tens of nanometers, which represents an order of magnitude improvement over that obtainable

Alioscka A. Sousa and Michael J. Kruhlak (eds.), *Nanoimaging: Methods and Protocols*, Methods in Molecular Biology, vol. 950,
DOI 10.1007/978-1-62703-137-0_21, © Springer Science+Business Media, LLC 2013

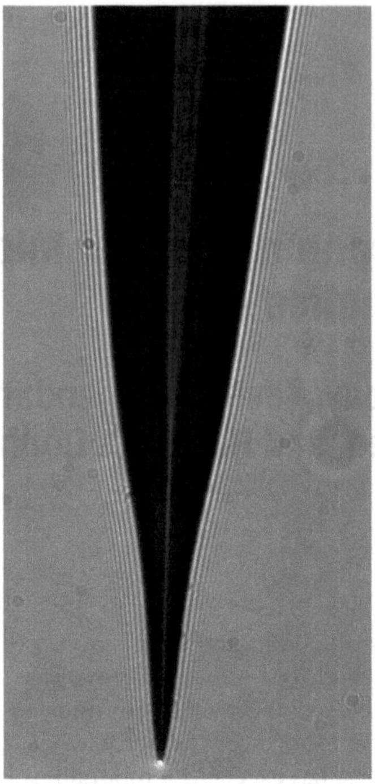

Fig. 1. A magnified view of a typical aluminum-coated NSOM probe fabricated using the heating and pulling method. A small spot of light can be seen emerging from the distal end of the probe. The probe has been fabricated with a double taper geometry which, for general NSOM imaging, strikes a balance between light throughput and small aperture size.

with conventional fluorescence microscopy (1–11). NSOM, therefore, helps bridge the gap in spatial resolution between far-field optical approaches and high-resolution techniques such as electron microscopy.

While electron microscopy has been extraordinarily successful at directly probing biological structures, the increased spatial resolution comes at the expense of working conditions, optical contrast mechanisms, and often very high costs. Fluorescence microscopy, on the other hand, is comparatively inexpensive and easily implemented on fixed or viable biological tissues. It also has an extensive history of fluorescent probe and method development, and exhibits spectroscopic, polarization, and dynamic capabilities. These attributes prove especially powerful for biological studies and it is these applications that have driven many of the technological developments experienced in fluorescence microscopy. A major goal in developing NSOM, therefore, was to create a fluorescence method that retains the favorable attributes associated with conventional fluorescence microscopy but with a spatial resolution approaching that attainable with electron microscopy.

Figure 2 shows a NSOM fluorescence image which demonstrates the sub-diffraction-limited spatial resolution possible with this

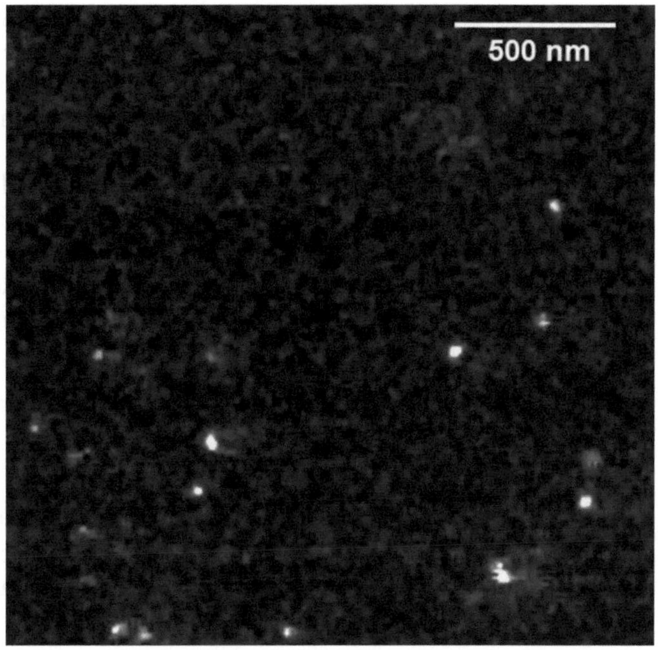


Fig. 2. NSOM fluorescence image of a 2 μm × 2 μm region of a DPPC monolayer doped with the fluorescent lipid marker $DiIC_{18}$. Each emission feature represents fluorescence from a single probe molecule in the membrane and has a FWHM of ~20–30 nm. This illustrates the high spatial resolution and single molecule detection limits of NSOM. Because the reporter molecule is much smaller than the NSOM probe, the FWHM of each emission feature reflects the size of the particular NSOM aperture used in the imaging.

technique. Each bright spot in Fig. 2 represents fluorescence from a single dye molecule dispersed in a lipid monolayer. Since the individual dye molecules are much smaller than the NSOM aperture, the feature size reflects the aperture diameter of the particular NSOM probe used for imaging (see Note 1). Each single molecule emission feature in Fig. 2 has a full-width half-maximum of ~20–30 nm, illustrating the high spatial resolution capabilities of NSOM (7). This example, therefore, demonstrates both the single molecule fluorescence detection limits of NSOM and its high spatial resolution, which is an order of magnitude better than that obtained with conventional optical approaches (1, 7, 10, 11). Moreover, since a feedback mechanism is implemented to maintain the tip–sample gap during scanning, NSOM provides a force mapping of surface topography much like atomic force microscopy (AFM). For biological applications and especially those involving membranes, the simultaneously measured NSOM fluorescence and topography images provides unique capabilities for directly comparing the location of optical markers with sample structure. These attributes offer intriguing possibilities for studying nanometric biological structures such as lipid rafts in membranes.

While commercial instrumentation is marketed for NSOM measurements, it is straightforward to modify existing scanning probe microscopes to implement NSOM. Here we discuss in detail how our particular NSOM instrumentation is constructed and the methods that we have learned for optimizing NSOM performance. Before outlining the details of this approach, however, we discuss how NSOM probes are fabricated. Resolution is largely dictated by the quality of the NSOM probe and, in our view, it is very important that the NSOM tips be fabricated in-house. We, therefore, provide a short primer on the fiber optic probe fabrication and metal coating conditions that we have found consistently lead to NSOM probes with well-defined, small apertures.

## 2. Materials

There are many ways of incorporating NSOM capabilities onto existing AFM platforms. Listed below are the major pieces of instrumentation used in our laboratory for this along with that needed for NSOM tip fabrication.

1. Digital Instruments Nanoscope IIIa AFM controller and work station (Now Bruker, http://www.bruker-axs.com): Bioscope backplane electronics module, signal access module, photon counting module.
2. Zeiss Axiovert 100 inverted fluorescence microscope equipped with high numerical aperture objective (FLUAR, 60×/1.45 oil) or equivalent.
3. *X–Y* closed-loop piezo sample scanner with central void to accommodate microscope objective (Mad City Labs, Nano H100, http://www.madcitylabs.com).
4. Avalanche photodiode (APD) single photon counting module (EG&G SPCM-AQR).
5. Interference filters and dichroic beamsplitter (Chroma, http://www.chroma.com).
6. Long working distance microscope (Infinity K2, http://infinity-usa.com) with CCD camera (Cohu, http://www.cohu.com/) or equivalent to monitor tip position above the sample.
7. Lock-in amplifier for detecting the shear-force feedback signal (Stanford Research Systems, SR830 DSP lock-in amplifier, http://www.thinksrs.com).
8. Sutter Instruments P-2000 micropipette puller or equivalent with arms modified to accept single-mode optical fiber (www.sutter.com).
9. Fiber strippers (www.thorlabs.com).
10. Thermal evaporation chamber with deposition monitor for coating aluminum onto probes.

## 3. Methods

### 3.1. Tip Fabrication

As with all scanning probe techniques, the quality of the measurement is largely dictated by the quality of the probe. This seems to be especially true for NSOM where the formation of a sub-wavelength aperture for the delivery of light can be problematic. There are two main approaches for fabricating tapers in the fiber optics used for NSOM tips: one is through chemical etching while the other utilizes a heating and pulling method (12, 13). Each technique offers distinct advantages and disadvantages. The chemical etching approach uses chemical etchants, such as hydrofluoric acid, to form tapers in the fiber optic probes. The main advantage of this technique is that very large taper angles can be fabricated which increases light throughput efficiency (13–17). This approach is also inexpensive and easily scaled for simultaneous probe production. However, even with the newer tube etching approaches, this technique often produces probes with rough, pitted, or jagged surfaces which degrade the quality of the metal coating used to confine light within the aperture. Since the quality of the metal coating is the key metric for NSOM tip performance, this is a major obstacle for the chemical etching approach. For this reason, we find that the heating and pulling method for tip formation remains the best approach for standard NSOM tip fabrication (18, 19). This is the most prevalent approach for fabricating NSOM probes and, in our experience, the easiest method for reproducibly fabricating probes once the proper procedures are in place. We will, therefore, limit our discussion to fiber optic probes heated and pulled to a taper and then coated with aluminum.

#### 3.1.1. Heating and Pulling Fiber Optic Probes

The heating and pulling method utilizes a heat source to soften the fiber and a pulling mechanism to draw the fiber down to the point of rupture. This approach is simple and once the appropriate parameters are found, it is very repeatable. While the method does require additional instrumentation, it yields very reproducible tapers with exquisitely smooth and clean surfaces that enable deposition of high-quality metal films (see Note 2).

1. To fabricate NSOM probes we use single mode optical fiber with 125 μm cladding. We have used single mode fiber from a number of vendors and find no difference for probe fabrication.
2. Approximately 5 mm of the jacket is removed with fiber strippers and the exposed fiber optic cleaned with ethanol. The fiber is mounted in a Sutter Instruments P-2000 micropipette puller with arms modified to accept single-mode optical fiber. These pullers incorporate a $CO_2$ laser for the heat source and programmable control over all important parameters for taper

formation such as laser power, pull strength, and pull delay. These parameters enable a large range of taper shapes to be fabricated in fiber optic probes. For example, high heat and hard pulls result in long tapers with very small tip apertures while low heat and soft pulls result in short tapers with larger tip apertures (18, 19). Unfortunately, this approach cannot create short tapers with small apertures, a combination that incorporates efficient light throughput with high spatial resolution. Etching methods can produce this geometry, but as mentioned earlier have other issues that make metal coating difficult. We have explored both approaches in our laboratory and believe the heating and pulling method produces quality probes with small apertures more reliably.

3. The exposed fiber is aligned in the path of the heating laser and clamped in place. The fiber is then heated and pulled using parameters that lead to the desired tip shape (see Note 3). It is important that the laser and fiber be aligned properly to obtain a symmetric taper. This should be checked from time to time by positioning a piece of thermal fax paper behind the fiber while momentarily firing the laser on low power. The laser spot burned on the fax paper should have a shadow from the fiber exactly bisecting the spot. If the line is high or low on the spot, asymmetric tapers can result.
4. To ensure tip to tip consistency, we only use pulled fibers from one arm of the puller and discard the other. The pulled fibers are immediately stored in an enclosed container to reduce dust contamination until loaded into the coating chamber.

#### 3.1.2. Aluminum-Coated NSOM Probes

Tapered fiber optic probes alone are not sufficient to act as NSOM probes since light escapes from the sides of the taper region. A reflective coating must be deposited on the taper to confine light within the probe until exiting a well-defined aperture at the distal end of the probe (3, 18). Without the reflective film, escaping light can excite fluorescence in the sample which leads to complicated patterns in the fluorescence image (9).

1. To confine light within the taper until exiting a well-defined aperture at the probe end, a reflective film of 50–100 nm of aluminum is coated around the sides. Aluminum has superior reflectivity properties in the visible region of the spectrum and, thus, requires the least amount of material to confine excitation light (20). This is important as the overall dimensions of the NSOM probe affects how close the probe aperture can approach the sample surface. While aluminum has the best optical properties, it unfortunately has a propensity to form oxides during the coating process (21). The oxides tend to dramatically degrade both the reflective properties of the coated film and the resulting performance metrics of the NSOM tip (22–25).

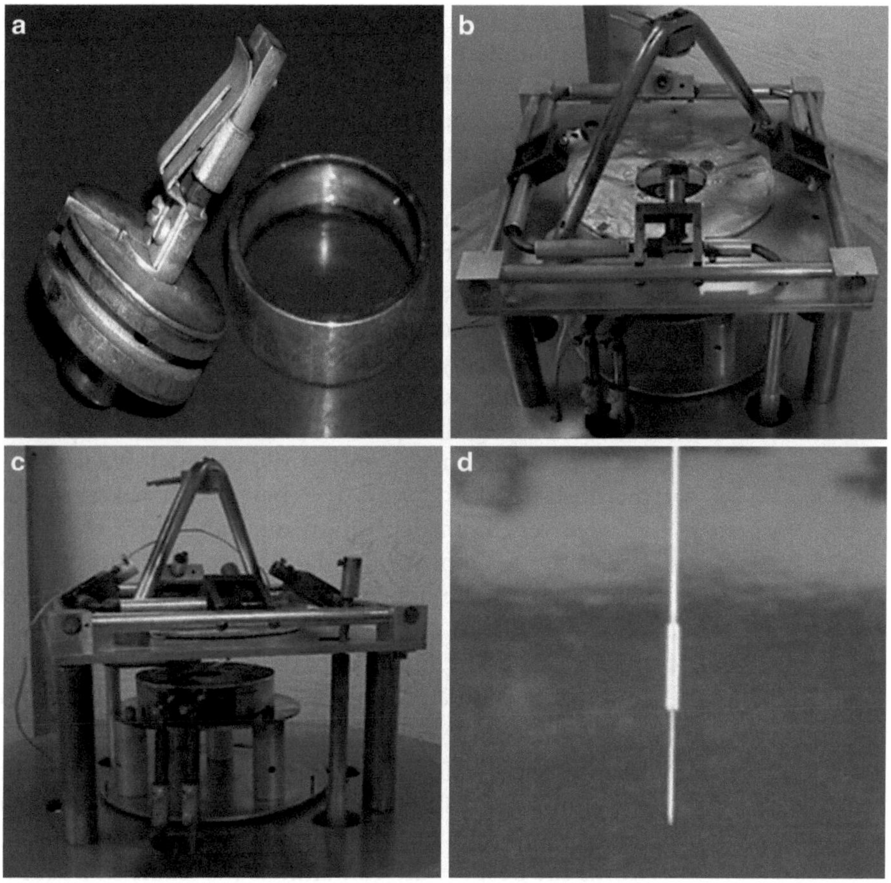


Fig. 3. (**a**) A fiber-chuck is used to hold the NSOM probe in the evaporative coater. The extra length of fiber from the NSOM probe is sealed inside the spin-chuck with a removable metal ring. (**b**) Four rotating mounts position the fiber chucks above a central opening, housing the thermal aluminum evaporation element. Mounted above the rotating tips is a quartz crystal microbalance to measure deposition rate and total film thickness. (**c**) The aluminum evaporation element is surrounded by a metal housing to reduce heating experienced at the tips. A manual shutter is used to block the source when melting and degassing the Al before coating the tips. (**d**) A metal jacket is glued to an aluminum-coated NSOM probe to assist in mounting the tip in the NSOM head.

In our opinion, the metal-coating step in fabricating NSOM probes represents the largest single barrier that must be overcome before carrying out quality NSOM measurements. Fortunately, much is known about coating aluminum and with care, high-quality, thin reflective Al films can be easily deposited onto tapered fiber optics.

2. To reduce oxide formation during film deposition, the coating needs to be carried out under a good vacuum. Achieving vacuum quickly is also important to reduce the batch time required for tip fabrication. We use a custom built thermal evaporation chamber (see Fig. 3) that uses a turbo pump to achieve pressures of $10^{-6}$ Torr in a few hours (see Note 4). Turbo pumps are desirable over diffusion pumps since they are cleaner and capable of achieving vacuum quickly.

3. The pulled fibers are mounted in a custom-built spinning mechanism that rotates the probes while coating to evenly deposit the metal film around the taper. The fiber is mounted into the fiber chuck shown in Fig. 3a. The all metal chuck is vacuum compatible and contains a spring clamp that immobilizes the fiber and a spool where the excess fiber is wound within and covered with the ring shown. Approximately 1 cm of the tapered probe extends from the top of the chuck for coating.
4. As shown in Figs. 3b, c, four fiber chucks can be mounted in the rotating platform. The rotating mounts spin the tips during coating to evenly expose the sides of the taper. The coating must be deposited such that only the sides of the taper are coated, leaving the aperture at the end of the tip free for light to exit. Given these constraints, thermal evaporative coating methods tend to work the best. Thermal evaporation can achieve very high deposition rates and since the deposition is line of sight, simply orienting the tips away from the aluminum source leaves the aperture uncoated. The tips, therefore, are tilted away from the aluminum source at an angle of approximately 35 °. The fiber probes are held 20 cm from the aluminum source and a rotating feedthrough is used to spin the tips at a rate of approximately 5 Hz during deposition.
5. In-house fabricated tantalum boats are used to hold and heat the aluminum. We have tried various materials and geometries for holding and heating the aluminum and found tantalum works the best for us. Tantalum is malleable so boat fabrication is simple and it is able to handle the large currents necessary for fast evaporation rates without boat failure. The boats should be as small as possible to reduce chamber heating, which can also have deleterious effects on the Al coating quality. In our configuration, we use boats with dimensions ~1 × 5 cm cut from a 0.015 in.-thick tantalum sheet. Fresh boats are used for each coating and loaded with approximately 5 in. of 1.0 mm-diameter aluminum wire (99.999%, Aldrich).
6. It is essential that the chamber incorporate a shutter above the aluminum source. For quality films, the aluminum needs to be melted, degassed, and impurities burned-off before opening the shutter and exposing the tips to the aluminum. The aluminum should be heated sufficiently so that once the shutter is opened, a high deposition rate is immediately achieved.
7. We have found that enclosing the heating apparatus (electrodes, tantalum boat, and aluminum) in a metal housing helps reduce ambient heating and leads to smoother aluminum films on the probes.
8. Pinholes in the aluminum coating often indicate the presence of contamination on the pulled fiber (20). It is important to

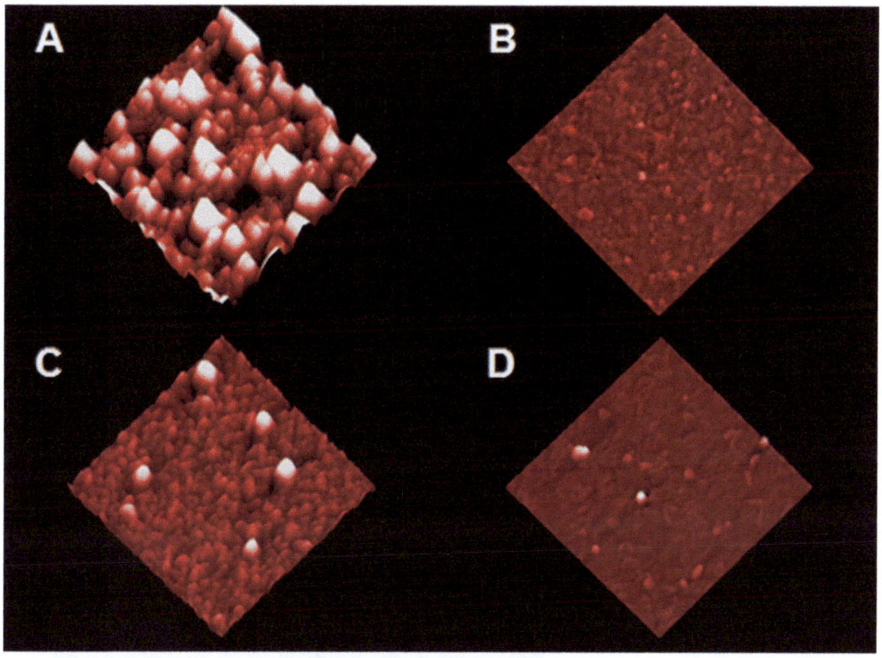


Fig. 4. 2 μm × 2 μm AFM images of the aluminum coatings applied to NSOM probes in vacuums of (**a**) $1 \times 10^{-3}$ Torr and (**b**) $3 \times 10^{-6}$ Torr. As seen in these images, a decrease in the chamber pressure results in a considerable decrease in roughness of the Al coating. Figures (**c**) and (**d**) show similar AFM measurements on NSOM tip coatings deposited at rates of 30 Å/s and 2,000 Å/s, respectively. These results show that the quality of the Al coating also depends on the rate of deposition, with faster rates preferable over slow depositions. Adapted from ref. 20 with permission.

decrease the time between pulling the fiber and getting it into vacuum to reduce the chances of dust attaching to the surface.

9. As mentioned, several parameters are important for obtaining smooth, highly reflective aluminum coatings on NSOM probes. Obviously, obtaining a good vacuum is important since this will reduce the presence of oxygen which is known to degrade coating quality by forming aluminum oxide grains. As an example, Figs. 4a, b compare AFM topography images of NSOM probes coated at $1 \times 10^{-3}$ Torr and $3 \times 10^{-6}$ Torr, both coated at the same deposition rate. The 2 μm × 2 μm images reveal a dramatic reduction in grain size as the ambient pressure is reduced. Similarly, deposition rate is found to significantly affect the final grain structure in the aluminum coating. Figure 4c, d compares AFM measurements from NSOM tips coated at low and high deposition rates. Clearly, higher rates correspond to smaller grains and smoother coatings which translate into higher film reflectivity (20, 26). These results are summarized in Table 1 which tabulates the mean surface roughness of the aluminum coating under various conditions.
10. Once coated with 50–100 nm of aluminum, a metal collar is glued onto the end of the NSOM probe as shown in Fig. 3d.

**Table 1**
**Mean surface roughness of Al coatings of NSOM probes deposited under various conditions as measured by AFM**

| Coating condition | Surface roughness mean (Std. Dev.) (nm) |
|---|---|
| Pressure = $1 \times 10^{-3}$ Torr | 68.1(24.9) |
| Pressure = $1 \times 10^{-4}$ Torr | 34.6(2.7) |
| Pressure = $1 \times 10^{-5}$ Torr | 20.7(2.6) |
| Pressure = $3 \times 10^{-6}$ Torr | 16.1(2.2) |
| Rate = 30 Å/s | 36.3(7.6) |
| Rate = 700 Å/s | 18.3(2.0) |
| Rate = 2,000 Å/s | 10.5(2.0) |

Smoother, high-quality Al coatings are achieved at lower pressures and faster deposition rates

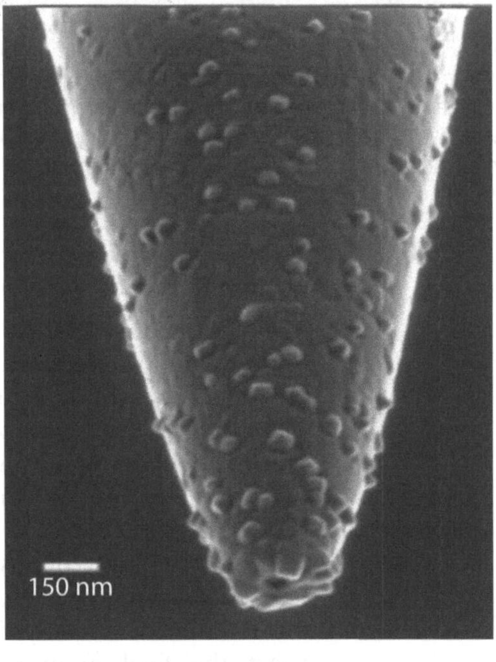


Fig. 5. A scanning electron microscopy image of a NSOM probe fabricated using optimal conditions outlined in the text.

The collar enables the probe to be firmly held in the NSOM head without damage. Since the tip will be oscillated for feedback, it is important that the collar be glued firmly on the probe so the dithering amplitude gets transferred efficiently to the probe.

Figure 5 shows an electron micrograph of a NSOM probe coated with aluminum under optimal conditions given in

Table 1. The reduced grain size in the aluminum coating increases film reflectivity, thus decreasing the amount needed to confine light within the taper region. This reduces the overall dimensions of the probe end which helps in positioning the probe near the sample surface, especially for samples with complicated morphology as often found in biological tissues.

### 3.2. Near-Field Scanning Optical Microscope

The major components needed for implementing NSOM measurements are already available and just need to be adapted. Most homebuilt and commercial NSOMs are built on optical microscopy platforms and use scan electronics developed for AFM. Commercial optical microscopes equipped with high numerical aperture optics aid in both sample visualization and signal collection. For biological applications, inverted fluorescence microscopes provide a very flexible and stable platform for NSOM measurements (27).

Figure 6 shows one of our NSOM instruments built on a Zeiss Axiovert 100 inverted fluorescence microscope equipped with a high numerical aperture objective (FLUAR, 60×/1.45 oil). The design shown in Fig. 6 shares many common features with the Digital Instruments Bioscope AFM with slight modifications made to accommodate NSOM measurements (27). The NSOM head

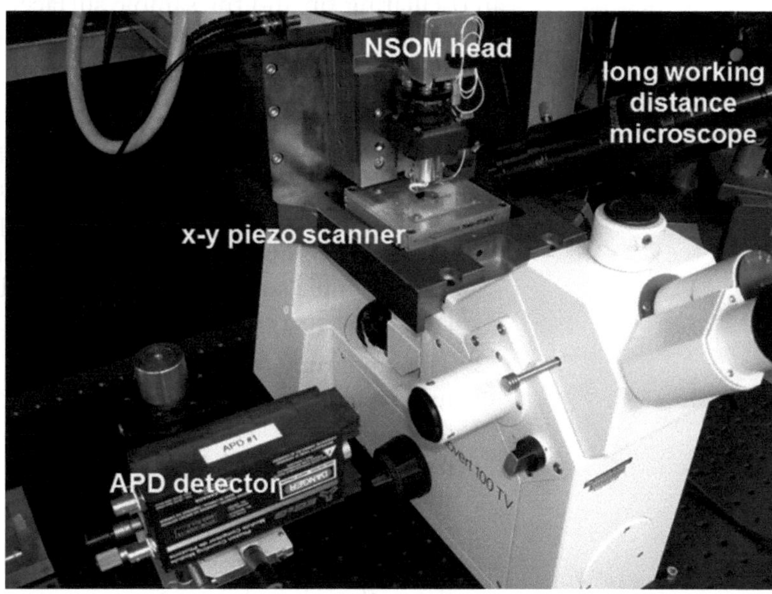


Fig. 6. A NSOM instrument constructed on an inverted fluorescence microscope. The NSOM head contains the *z*-piezo to control the tip–sample gap and the shear-force feedback detection system. The head is mounted in a motorized dovetail for coarse adjustment of the head toward the sample. The sample is mounted below on a closed-loop *x*–*y* piezo stage that scans the sample under the probe. Fluorescence from the sample is collected through the microscope, passed through emission filters, and imaged onto the active area of an avalanche photodiode (APD) detector.

holds the fiber optic probe and the necessary laser, detection, and piezo actuators to implement the tip–sample feedback control. It is held in a dovetail mount connected to a motorized *z*-motion stage, which is used to control the large-scale adjustment of the tip toward the sample. The coarse *z*-motion stage is used to bring the NSOM tip within the 5 μm travel of the *z*-piezo tube which holds the NSOM probe. The *z*-piezo tube controls the tip–sample gap while the sample itself is held on a separate *x*–*y* piezo stage that scans the sample under the NSOM tip. This arrangement keeps the NSOM probe centered on the microscope objective which ensures that the fluorescence signal remains aligned on the small active area of the APD detector. Shown in Fig. 6 is a long working distance microscope (Infinity K2) with CCD camera to monitor the approach of the NSOM probe toward the sample surface (see Note 5). The microscope is controlled using a Digital Instruments Nanoscope IIIa controller equipped with a Bioscope backplane electronics module, signal access module, and photon counting module. A lock-in amplifier is also used to detect the oscillating shear-force tip feedback signal (not shown in Fig. 6).

#### 3.2.1. Shear-Force Feedback for Tip–Sample Gap Regulation

To maintain the tip close to the sample surface while scanning, a feedback mechanism must be implemented. Most schemes introduced for NSOM revolve around force measurements between an oscillating tip and the sample surface. A popular method is the shear-force approach where the tip is dithered laterally at its resonant frequency while the amplitude of the oscillation is monitored (28–30).

1. Figure 7 shows a NSOM head built for shear-force feedback. The tip is mounted on a piezo bimorph that is affixed to the end of a *z*-piezo tube and encased in a protective metal housing. The *z*-piezo tube is used to move the tip vertically with respect to the sample surface while the piezo bimorph oscillates the tip laterally at its resonant frequency.
2. The amplitude of the oscillating tip is monitored from the side using a fiber-coupled laser to shadow the tip onto a split photodiode detector. The lateral dithering of the NSOM tip in the path of the feedback laser produces an oscillating signal on each side of the split photodiode detector. To reduce the length of the NSOM tip extending from the bimorph, the housing of the split photodiode has been removed and the substrate ground flat to the edge of the active areas as seen in Fig. 7. Minimizing the length of the NSOM probe emerging from the bimorph is desirable since it increases the resonance frequency of the tip, thus producing more stable feedback conditions (see Note 6).
3. Each side of the detector is amplified and sent to the differential input of a lock-in amplifier which is referenced to the signal used to drive the bimorph. The DC output of the lock-in

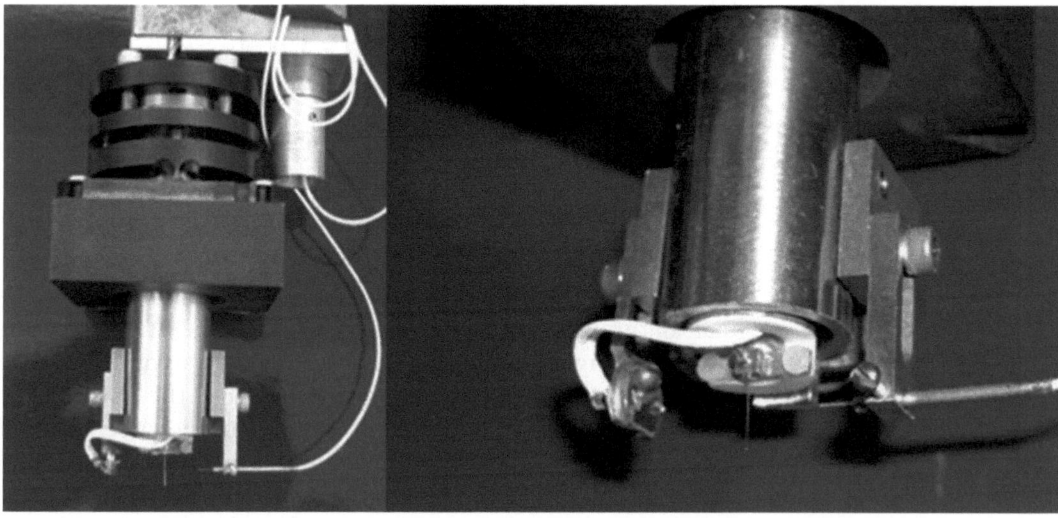

Fig. 7. The NSOM head is used to control the z-position of the near-field probe using a shear-force feedback mechanism. The probe is mounted onto a piezo bimorph held at the end of a z-piezo tube. The bimorph is used to dither the NSOM probe laterally at its resonance frequency. The probe oscillation amplitude is monitored from the side by using a fiber-coupled laser source that shadows the tip onto a split photodiode detector. Each region of the detector is amplified and sent to the differential input of a lock-in amplifier referenced to the tip oscillation frequency.

amplifier, therefore, is proportional to the oscillation amplitude of the NSOM probe which will decrease when the tip interacts with the sample. This approach is easily integrated with commercial AFM electronics that support tapping mode AFM measurements.

4. To generate the feedback signal, most commercial AFM instruments use an error signal. Therefore, the DC output of the lock-in amplifier must be subtracted from a software-controlled "set point" voltage that is set by the user to define where on the force curve the tip should be positioned. Most AFM manufacturers sell a signal access module that enables access and control of all the relevant signals necessary to implement a shear-force feedback scheme.

### 3.3. NSOM Fluorescence Imaging

Fluorescence continues to be the most prevalent and useful contrast mode utilized in NSOM measurements. For the arrangement shown in Fig. 6, excitation light exiting the NSOM probe excites fluorescence in the sample which is collected from below using a high NA objective. The collected light is passed through filters to remove the excitation light and then imaged onto the active area of a high quantum efficiency APD detector. The output of the photon counting module is sent to the AFM controller where a photon counting chip bins the counts in the appropriate pixel.

1. The NSOM head containing the NSOM probe is slowly lowered toward the sample using the coarse adjust-motorized z-stage.

It is important to have light exiting the NSOM probe which can be seen from the CCD camera coupled with the long working distance microscope. As the tip nears the sample surface, a reflected spot from the surface will also be visible. Carefully adjusting the motorized $z$-stage to bring the spot exiting the NSOM probe and the reflected spot close together on the monitor enables quick positioning of the NSOM probe within the small travel of the $z$-piezo tube.

2. Once the NSOM probe is near the surface, the detector needs final alignment. The small active area of the APD presents some challenges in aligning the fluorescence signal. We use a flipper mirror to direct a red light source into the optical fiber and through the NSOM probe to mimic fluorescence. The red light exiting the NSOM probe enables the APD to be aligned with the optical filters. With stable feedback and careful movements, this also enables the focus to be adjusted while the tip is engaged on the sample surface which is critical for maximizing the fluorescence signal when using a high NA objective.
3. We use a closed-loop $x$–$y$ piezo stage for sample scanning. Like all fluorescence measurements, sample photobleaching often proves limiting in NSOM studies. Open loop piezo scanners suffer from piezo creep and hysteresis which requires several "training" scans before collection of a reliable image. For AFM measurements this is usually not a problem, but for NSOM measurements repeatedly scanning the same area increases photobleaching and, therefore, degrades the signal. Closed-loop scanners that actively adjust for creep and hysteresis enable efficient collection of NSOM fluorescence signals without the repeated prior scans required with open loop stages. Moreover, these capabilities are especially important for single point spectroscopic or dynamic measurements where the NSOM probe needs to be positioned and held above a precise location of the sample.

#### 3.3.1. NSOM Fluorescence Imaging of Membranes

NSOM is well suited for structural studies on model membranes and biological membranes where the two-dimensional nature of the sample enables the close approach of the NSOM aperture (31–36). The high spatial resolution of NSOM combined with the simultaneous fluorescence and force signals proves especially informative in membrane systems. These signals can be directly correlated to understand membrane organization as a function of constituents and conditions.

For model membrane studies, the Langmuir-Blodgett (LB) technique has been used extensively to transfer lipid films onto solid supports where they are amenable for study with surface techniques such as NSOM. The LB technique enables precise control over the lipid film surface pressure, which is related to the first derivative of free energy with respect to surface area. As such, these

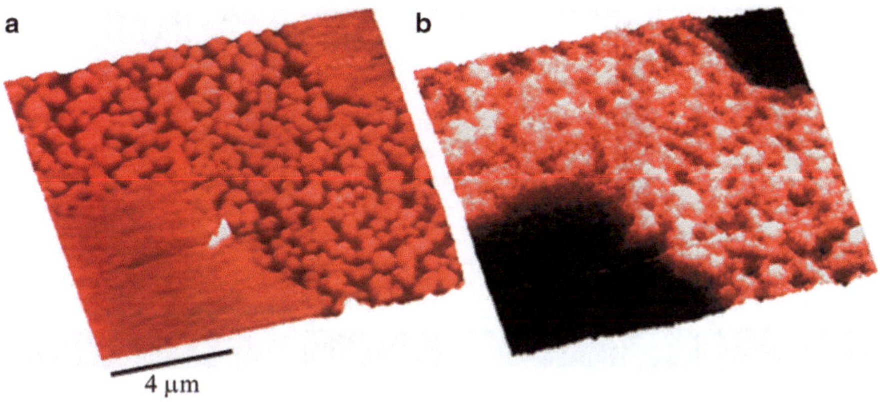


Fig. 8. Simultaneously collected NSOM topography (**a**) and fluorescence (**b**) images of a DPPC monolayer doped with the fluorescent membrane probe $DiIC_{18}$. Small ~8 Å level height differences observed in the topography image reflect the coexisting liquid-expanded (LE) and liquid-condensed (LC) phases. The bright regions in the NSOM fluorescence image mark the LE domains which incorporate the fluorescent lipid probe. Adapted from ref. 31 with permission.

studies provide a direct link between film properties and thermodynamics. Moreover, the LB technique enables films to be fabricated over a wide range of temperatures, pressures and film compositions which further enhances the properties that can be explored.

Figure 8 shows representative NSOM measurements on a DPPC monolayer transferred onto mica using the LB technique. This film was transferred in the phase coexistence region of the pressure isotherm where less ordered liquid-expanded (LE) phase coexists with the more ordered liquid-condensed (LC) domains. As studied extensively with AFM, these phase differences result in ~5–8 Å height changes which can be seen in the NSOM shear force image shown in Fig. 8a (31). The NSOM fluorescence image shown in Fig. 8b maps the location of the fluorescent lipid analog $DiIC_{18}$, which has been doped into the DPPC film at 0.25 mol% to stain LE regions of the membrane.

These measurements help illustrate the utility, and some of the advantages, of NSOM for studying the structure of lipid monolayers. First, force capabilities enable resolution of Angstrom level height changes associated with phase coexistence, while high lateral resolution in the fluorescence image can detect small domains beyond the resolution of conventional optical techniques. Second, simultaneous force and fluorescence mapping provides a direct, pixel by pixel comparison between surface topography and fluorescence properties. This enables the unambiguous assignment of domains to specific phases which avoids artifacts due to defects in the membrane. Third, the sub-diffraction-limited measurements afforded by using NSOM fluorescence can spatially map the locations of fluorescent probes or fluorescently tagged species with resolution reaching tens of nanometers. These capabilities can help

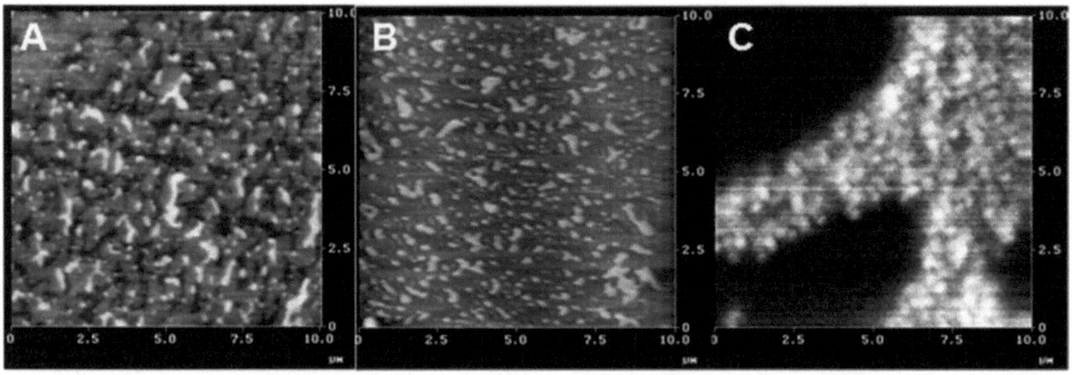


Fig. 9. (**a**) AFM topography image of a DPPC bilayer membrane on a mica substrate formed using the Langmuir-Blodgett/Langmuir-Schäefer (LB/LS) technique. The 10 μm × 10 μm region shows three quantized height levels due to the stacking of liquid-expanded (LE) and liquid-condensed (LC) from each leaflet. The convolution of height information from each leaflet forming the bilayer makes it impossible to assign the phase structure present in each leaflet. NSOM topography (**b**) and fluorescence (**c**) images on a similar bilayer, however, can be used to measure the phase distribution in each side of the lipid bilayer by controlling which leaflet incorporates the fluorescent probe. While the NSOM topography (**b**) suffers the same convolution problem that AFM has, the NSOM fluorescence image (**c**) shows distinct structures reflective of coexisting lipid phases similar to the monolayer shown in Fig. 8. (All images are 10 μm × 10 μm.) Adapted from ref. 31 with permission.

in studies of biologically relevant membrane microdomains such as those encountered in lipid raft studies.

Given the correspondence between the topography and fluorescence images seen in Fig. 8, one might argue that AFM measurements of topography are sufficient to measure phase distributions in these simple systems with high resolution. However, assigning phase based on height changes alone can be ambiguous in monolayer systems and becomes difficult to impossible in more complicated systems.

For example, Fig. 9 shows an AFM image of a DPPC bilayer formed through the sequential transfer of monolayers onto a mica substrate. Each DPPC monolayer was transferred at the same phase coexistence region of the pressure isotherm as that shown in Fig. 8. Three quantized height level changes are observed in the AFM image (see Fig. 9a) which corresponds to the stacking of LE and LC domains from each monolayer that forms the bilayer. This convolution of height information from each monolayer forming the bilayer makes assigning phase distributions in each leaflet impossible. However, high-resolution NSOM fluorescence measurements can resolve the submicron domains present in the leaflets by simply controlling which leaflet incorporates the fluorescent dye. Figure 9b, c show the NSOM force and fluorescence from a DPPC bilayer in which only the top layer has been doped with the fluorescent probe. While the NSOM force image suffers the same limitation as the AFM measurements, the NSOM fluorescence image clearly reveals similar domain structure as that seen in monolayers. Similar measurements with the dye loaded in the bottom leaflet shows comparable behavior.

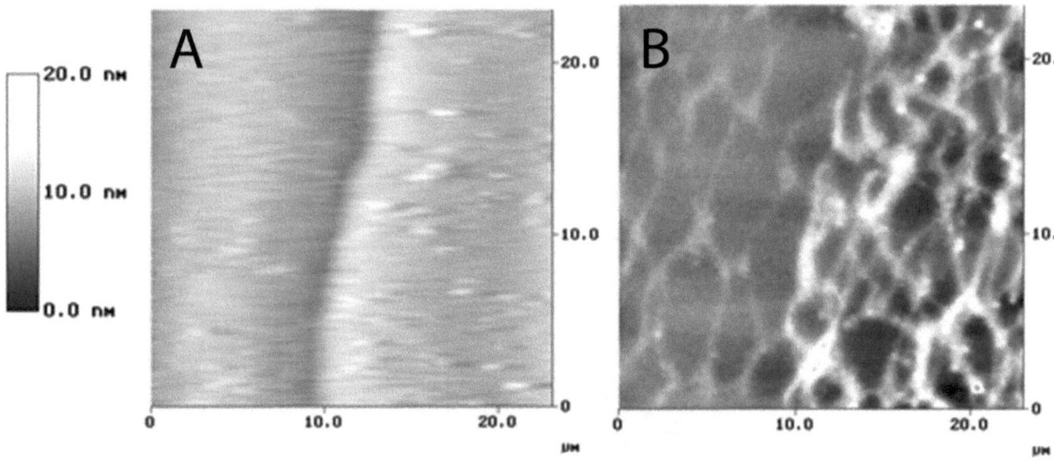


Fig. 10. 23 μm × 23 μm NSOM topography (**a**) and fluorescence (**b**) images of a monolayer of the replacement lung surfactant Survanta. The surfactant has been compressed beyond the collapse pressure which has resulted in buckling of the monolayer. This is evident from the change in topography in the force image and corresponding transition in phase structure observed in the NSOM fluorescence image. (Both images are 23 μm × 23 μm.) Adapted from ref. 37 with permission.

For model membrane studies, the simultaneous collection of both force and fluorescence information with NSOM can also provide valuable insights into membrane collapse and buckling. For example, Fig. 10 shows a monolayer of the replacement lung surfactant, Survanta, used in the treatment of respiratory distress syndrome. The monolayer has been compressed to the point of collapse where the monolayer has buckled into a multilayer structure. This is evident in the NSOM force and fluorescence images shown in Fig. 10 which reveal a transition in the film structure. Examination of the NSOM force image (see Fig. 10a) reveals a ~4–5 nm height change in the sample which coincides with a large change in the phase structure seen in the NSOM fluorescence image (see Fig. 10b) (37). These and similar measurements show that the collapse of these films results in the continuous folding of the monolayer into a multilayer structure. The complexity seen in the NSOM fluorescence image arises from the stacked monolayers in the folded regions. By comparing the structure seen in the NSOM fluorescence images at folded edges correlated with the force images, it was determined that the folds were formed from a continuous monolayer film and did not arise from the breaking and stacking of individual monolayer film units (37).

As illustrated in the previous examples, the high spatial resolution and simultaneous collection of both topographic and fluorescence signals with NSOM becomes increasingly informative and useful as the complexity of the membrane system increases. For natural biological membranes, these capabilities can help unravel complex structural details that would otherwise be hidden or ambiguous in studies relying solely on either fluorescence or

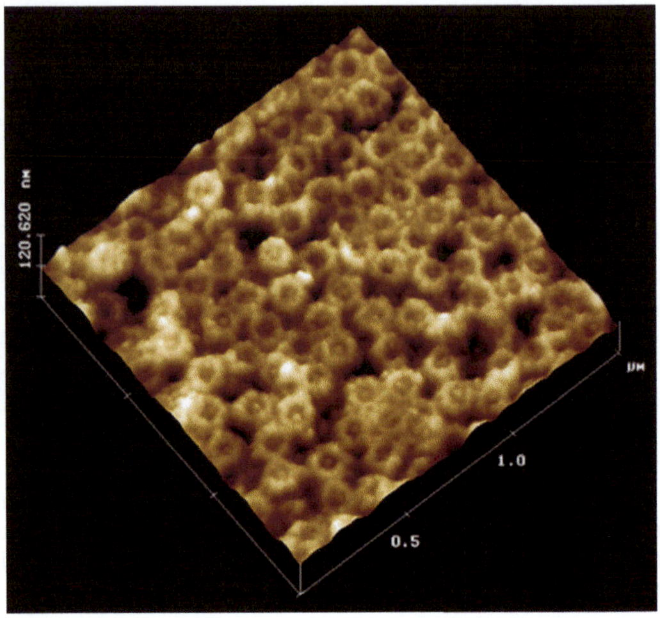


Fig. 11. 2 µm × 2 µm AFM image of the nuclear envelope from oocytes of *Xenopus laevis*. The donut-like structures observed in the membrane represent nuclear pore complexes (NPC) which form the only direct pathway across the membrane. A mass is observed in the pore region of many of the NPCs. The identity of this mass is the subject of much speculation, but some believe it represents vault complexes localized to the pore. Adapted from ref. 38 with permission.

topography information alone. To help illustrate this point, NSOM studies of the nuclear envelope isolated from the nuclei of *Xenopus laevis* oocytes are briefly discussed (38).

The nuclear envelope contains large protein complexes known as nuclear pore complexes (NPC) that form the only direct route of passage across the nuclear envelope. These NPCs are easily seen by AFM images such as that shown in Fig. 11. Each toroidal structure in the image is an individual NPC in the membrane. Close examination reveals that some NPCs have a mass located in the central pore region, the identity of which has been the subject of much speculation. Some studies suggested they represent vault complexes, which are large ribonucleoproteins whose specific functional roles remain elusive.

To directly probe the relationship between NPCs and vault complexes, NSOM studies were carried out with immunofluorescently labeled vault complexes. The NSOM images in Fig. 12 show the NSOM topography, fluorescence and overlay images from this study. The NSOM force image in Fig. 12a reveals the locations of NPCs in the nuclear membrane, albeit with lower resolution than comparable AFM measurements. This can be compared with the small fluorescent features seen in the NSOM fluorescence image (see Fig. 12b) which maps the locations of the individually labeled vault complexes. NSOM, therefore, provides a

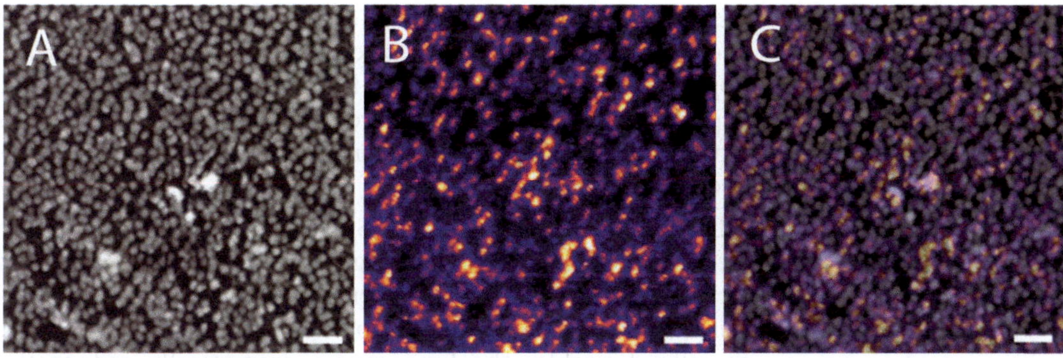


Fig. 12. NSOM topography (**a**) fluorescence (**b**) and overlay (**c**) images of the nuclear envelope in which vault complexes have been immunofluorescently labeled. Individual NPCs are observed in the NSOM topography image, albeit with lower resolution than the AFM measurement shown in Fig. 11. Features in the NSOM fluorescence image reflect labeled vault complexes which can be compared directly with the location of NPCs observed in the force image. The overlay image shows the colocalization of NPCs and vault complexes with less than 100 nm resolution. Scale bar in each image is 1 μm. Adapted from ref. 5 with permission.

novel method for colocalization where the force image maps the locations of NPCs and the NSOM fluorescence image localizes the vault complexes, all with ~50 nm spatial resolution (5, 38). Figure 12c displays an overlay of the NSOM force and fluorescence images which clearly shows that vault proteins are localized with NPCs.

The above studies help illustrate the utility of employing NSOM measurements for the analysis of biological membranes. While fluorescence microscopy has found widespread use in understanding membrane organization and structure, its limited spatial resolution restricts its ability to explore membrane properties at a more fundamental level. The small length scale and complexity of structures encountered in lipid monolayers and bilayers necessitate the use of high-resolution techniques such as NSOM. NSOM combines the favorable attributes of fluorescence microscopy and AFM which seems particularly well suited for deciphering the structure and organization of biological membranes at the nanometric level.

## 4. Notes

1. Transmission measurements should never be used or trusted to quantify resolution of a NSOM tip. There is extensive literature illustrating the artifacts inherent in such measurements. The gold standard for characterizing resolution of the tip aperture remains single molecule NSOM fluorescence measurements such as that shown in Fig. 2.
2. It is very important to develop in-house fiber optic tip fabrication capabilities. Tailoring the tip shape and characteristics to the

performance metrics desired is often required. Moreover, for biological samples, which often have complicated surface topography and a propensity to foul NSOM tips, several probe replacements are usually required before optimal imaging is obtained. This can quickly become cost prohibitive with commercial probes.

3. The particular shape of the NSOM probe is dictated by the performance characteristics sought. As mentioned, using high heat and hard pulling parameters yields the smallest tip apertures but very long taper regions. While small apertures are usually desirable, the long tapers dramatically reduce the efficiency of light delivered to the tip aperture. For NSOM measurements with resolution in the ~50–100 nm range, we find the two-stage taper shape shown in Fig. 1 to be the best option for obtaining quality results (7). The shallow taper angle near the distal end of the tip produces a small aperture while the larger taper angle before that enables efficient coupling of light toward the aperture.
4. The major components of the chamber include: (1) Leybold Turbovac 151 backed with an Alcatel mechanical roughing pump; (2) Inficon XTM/2 Deposition monitor positioned above the tips is essential for controlling both the rate of deposition and total film thickness.
5. The long working distance microscope is an important component of any NSOM microscope. By monitoring the light from the NSOM probe and its reflection off the sample surface, the NSOM probe can be quickly and conveniently positioned within the travel of the *z*-piezo tube. It is also useful for monitoring the condition of the NSOM probe during imaging.
6. Stable tip feedback is critical for obtaining high-quality NSOM fluorescence images. In addition to the usual approaches of adjusting feedback gains and scan angles to explore the best imaging conditions, adjusting power through the NSOM probe can also have a dramatic effect on tip feedback performance. Power-induced heating of the metal coating leads to NSOM tip expansion, which can sometimes be used to improve tracking of the sample topography (7, 39).

## Acknowledgments

We gratefully acknowledge support from NSF (CBET 1133814). KPA gratefully acknowledges the support from the NIH Dynamic Aspects of Chemical Biology Training Grant (T32 GM08545). SMW gratefully acknowledges support provided by the Madison and Lila Self Foundation.

## References


1. Betzig E, Chichester RJ (1993) Single molecules observed by near-field scanning optical microscopy. Science 262:1422–1425
2. Betzig E, Isaacson M, Barshatzky H, Lewis A, Lin K (1988) Near-field scanning optical microscopy (NSOM). Proc SPIE-Int Soc Opt Eng 897:91–99
3. Betzig E, Trautman JK (1992) Near-field optics: microscopy, spectroscopy and surface modification beyond the diffraction limit. Science 257:189–195
4. Buratto SK (1996) Near-field scanning optical microscopy. Curr Opin Solid State Mater Sci 1:485–492
5. Dickenson NE, Armendariz KP, Huckabay HA, Livanec PW, Dunn RC (2010) Near-field scanning optical microscopy: a tool for nanometric exploration of biological membranes. Anal Bioanal Chem 396:31–43
6. Dunn RC (1999) Near-field scanning optical microscopy. Chem Rev 99:2891–2927
7. Erickson ES, Dunn RC (2005) Sample heating in near-field scanning optical microscopy. Appl Phys Lett 87:1–3
8. Flanders BN, Dunn RC (2007) Near-field scanning optical microscopy of lipid membranes. Interface Sci Technol 14:117–140
9. Shiku H, Dunn RC (1999) Near-field scanning optical microscopy. Anal Chem 71:23A–29A
10. Trautman JK, Macklin JJ, Brus LE, Betzig E (1994) Near-field spectroscopy of single molecules at room temperature. Nature 369:40–42
11. Xie XS, Dunn RC (1994) Probing single molecule dynamics. Science 265:361–364
12. Valaskovic GA, Holton M, Morrison GH (1995) Parameter control, characterization, and optimization in the fabrication of optical fiber near-field probes. Appl Opt 34:1215–1228
13. Hoffmann P, Dutoit B, Salathe R-P (1995) Comparison of mechanically drawn and protection layer chemically etched optical fiber tips. Ultramicroscopy 61:165–170
14. Mononobe S, Naya M, Saiki T, Ohtsu M (1997) Reproducible fabrication of a fiber probe with a nanometric protrusion for near-field optics. Appl Opt 36:1496–1500
15. Lambelet P, Sayah A, Pfeffer M, Philipona C, Marquis-Weible F (1998) Chemically etched fiber tips for near-field optical microscopy: a process for smoother tips. Appl Opt 37: 7289–7292
16. Chuang Y-H, Sun K-G, Wang C-J, Huang JY, Pan C-L (1998) A simple chemical etching technique for reproducible fabrication of robust scanning near-field fiber probes. Rev Sci Instrum 69:437–439
17. Mononobe S, Ohtsu M (1997) A model based on geometrical construction in designing a pencil-shaped fiber probe for near-field optics. J Lightwave Technol 15:1051–1055
18. Betzig E, Trautman JK, Harris TD, Weiner JS, Kostelak RL (1991) Breaking the diffraction barrier: optical microscopy on a nanometric scale. Science 251:1468–1470
19. Williamson RL, Miles MJ (1996) Melt-drawn scanning near-field optical microscopy probe profiles. J Appl Phys 80:4804–4812
20. Hollars CW, Dunn RC (1998) Evaluation of thermal evaporation conditions used in coating aluminum on near-field fiber-optic probes. Rev Sci Instrum 69:1747–1752
21. Hass G, Waylonis JE (1961) Optical constants and reflectance and transmittance of evaporated aluminum in the visible and ultraviolet. J Opt Soc Am 51:719–722
22. Barna A, Barna PB, Radnoczi G, Reicha FM, Toth L (1979) Formation of aluminum thin films in the presence of oxygen and nickel. Phys Status Solidi A 55:427–435
23. Krueger WH, Pollack SR (1972) Initial oxidation of aluminum thin films at room temperature. Surface Sci 30:263–279
24. Krueger WH, Pollack SR (1972) Room temperature adsorption of water by aluminum thin films. Surface Sci 30:280–298
25. Hass G, Hunter WR, Tousey R (1956) Reflectance of evaporated aluminum in the vacuum ultraviolet. J Opt Soc Am 46:1009–1012
26. Dhere NG, Arsenio TP, Patnaik BK (1975) Purity and morphology of aluminum films. Thin Solid Films 30:267–279
27. Hollars CW, Dunn RC (2000) Probing single molecule orientations in model lipid membranes with near-field scanning optical microscopy. J Chem Phys 112:7822–7830
28. Betzig E, Finn PL, Weiner JS (1992) Combined shear force and near-field scanning optical microscopy. Appl Phys Lett 60:2484–2486
29. Toledo-Crow R, Yang PC, Chen Y, Vaez-Iravani M (1992) Near-field differential scanning optical microscope with atomic force regulation. Appl Phys Lett 60:2957–2959
30. Valaskovic GA, Holton M, Morrison GH (1995) Biological near-field scanning optical microscopy: instrumentation and sample issues for shear-force feedback. Ultramicroscopy 57:212–218
31. Hollars CW, Dunn RC (1998) Submicron structure in L-α-Dipalmitoylphosphatidylcholine

monolayers and bilayers probed with confocal, atomic force, and near-field microscopy. Biophys J 75:342–353

32. Shiku H, Dunn RC (1998) Direct observation of DPPC phase domain motion on mica surfaces under conditions of high relative humidity. J Phys Chem B 102:3791–3797
33. Johnston LJ (2007) Nanoscale imaging of domains in supported lipid membranes. Langmuir 23:5886–5895
34. Ianoul A, Johnston LJ (2007) Near-field scanning optical microscopy to identify membrane microdomains. Methods Mol Biol 400:469–480
35. Hwang J, Gheber LA, Margolis L, Edidin M (1998) Domains in cell plasma membranes investigated by near-field scanning optical microscopy. Biophys J 74:2184–2190
36. Dunn RC, Holtom GR, Mets L, Xie XS (1994) Near-field fluorescence imaging and fluorescence lifetime measurement of light harvesting complexes in intact photosynthetic membranes. J Phys Chem 98:3094–3098
37. Sibug-Aga R, Dunn RC (2004) High-resolution studies of lung surfactant collapse. Photochem Photobiol 80:471–476
38. Dickenson NE, Moore D, Suprenant KA, Dunn RC (2007) Vault ribonucleoprotein particles and the central mass of the nuclear pore complex. Photochem Photobiol 83:686–691
39. Dickenson NE, Erickson ES, Mooren OL, Dunn RC (2007) Characterization of power induced heating and damage in fiber optic probes for near-field scanning optical microscopy. Rev Sci Instrum 78:686–691

# Part IV

## Complementary Techniques

# Chapter 22

# Correlative Fluorescence and EFTEM Imaging of the Organized Components of the Mammalian Nucleus

Michael J. Kruhlak

## Abstract

The cell nucleus contains many distinct subnuclear compartments, domains, and bodies that vary in their composition, structure, and function. While the cellular constituents that occupy the subnuclear regions may be well known, defining the structural details of the molecular assembly of the constituents has been more difficult. A correlative fluorescence and energy-filtering transmission electron microscopy (EFTEM) imaging method has the ability to provide these details. The correlative microscopy method described here allows the tracking of subnuclear structures from specific cells by fluorescence microscopy and then, using electron energy loss imaging in the transmission electron microscope, reveals the ultrastructural features of the nuclear components along with endogenous elemental information that relates directly to the biochemical composition of the structure. The ultrastructural features and composition of well-characterized PML bodies and interchromatin granule clusters are compared to those of ligand-activated glucocorticoid receptor (GR) foci, with GR foci containing fibrogranular nucleic acid-containing features and PML bodies being devoid of nucleic acid.

**Key words:** Correlative microscopy, Energy-filtering transmission electron microscopy, Nucleus, PML bodies, Glucocorticoid receptor

## 1. Introduction

Correlative microscopy is a powerful research tool that combines the advantages of different light and electron imaging modalities to obtain information on the composition and structure of the specimen, across resolution scales, present within a biological sample. A variety of imaging techniques are available to address biologically relevant questions. The recent development of nanoimaging methods circumventing the diffraction limit of light microscopy have expanded this repertoire. Indeed, the super-resolution details obtained from these exciting imaging methods are often correlated with the same structures viewed by electron

Alioscka A. Sousa and Michael J. Kruhlak (eds.), *Nanoimaging: Methods and Protocols*, Methods in Molecular Biology, vol. 950, DOI 10.1007/978-1-62703-137-0_22, © Springer Science+Business Media, LLC 2013

microscopy (1, 2). While each imaging method often has unique advantages in providing information concerning a biological sample, individually the methods are unable to satisfy the demanding need for specific detailed information of a cellular constituent within a contextual ultrastructure. Fluorescence microscopy and transmission electron microscopy (TEM) are complementary and allow relevant macromolecular assemblies to be examined from the micron to the nanometer scale while preserving spatial orientation and allowing detection of rare structural and functional complexes. Numerous combinations of correlative fluorescence and electron microscopic techniques have been described (3, 4), with recent approaches aimed at improving the fidelity of immunogold labeling (5), increasing the sensitivity and spatial precision (6), using quantum dot probes that have intrinsic properties that make them visible for both fluorescence and EM imaging (7–9) and designing of proteins with genetically encoded epitope tags suitable for EM imaging (10, 11), all with the aim of pinpointing specific components in a nonspecifically stained ultrastructural background. Correlative microscopy also improves the efficiency of quantitative EM imaging while aiding researchers not familiar with the more demanding requirements of electron microscopy by optimizing sample labeling by light microscopic analysis. Here the combination of fluorescence microscopy with energy-filtering transmission electron microscopy (EFTEM) (12) will be described, with the aim to examine the ultrastructural details of subnuclear compartments and domains within the mammalian nucleus thought to be involved in regulating gene expression.

The mammalian cell nucleus is a highly organized organelle containing a diversity of compartments, bodies, and domains (13–15). These intricate structures, of which the nucleolus is the primary example, are dynamic and adjust their resident constituents depending on the gene expression profile or cell cycle stage of growth. A valuable composite list of these subnuclear structures has been compiled from morphological information obtained using light microscopy (13). A few structures have definitive links to cancer, such as promyelocytic leukemia (PML) bodies (16). The functional significance of the majority of these subnuclear structures, including PML bodies, has not been fully elucidated.

Obtaining ultrastructural information has been fruitful for a handful of these subnuclear structures, with the brightfield imaging mode of the transmission electron microscope being the imaging method mainly employed. To generate contrast in biological samples for TEM, heavy-atom contrasting agents are needed. However, these labeling agents are non-specific and do not readily allow biochemical information from the sample to be obtained. In addition, the contrasting agents used may mask the underlying details of the structure. Interpreting the functional significance of these structures based on morphology alone is difficult. An imaging method that provides biochemical information about the constituents along with structural detail is highly desirable. Capturing the inelastically scattered beam electrons transmitted through the specimen using the electron energy loss imaging mode of an EFTEM microscope is able to provide such biochemical information. Importantly, the contrast in the resultant image is generated from lighter biologically relevant elements contained within the sample itself, such as phosphorus and nitrogen (17). Thereby, the energy loss information obtained is specific to the endogenous elements in the sample, effectively eliminating the concern of nonspecificity from exogenous labeling reagents. The electron energy loss imaging mode of EFTEM is particularly well suited for imaging the cell nucleus (18) and has been used successfully to examine the ultrastructure of nuclear components such as viral replication compartments (19), chromatin containing DNA breaks (20), transcription factories (21, 22), interchromatin granule clusters (23), and PML bodies (24).

The promyelocytic leukemia (PML) tumor suppressor protein is the main constituent of nuclear bodies known as ND10 or PML bodies, which are 0.1–1.0 μm in size. The PML containing subnuclear domains do not contain RNA or DNA, are involved in various biological responses, such as DNA damage (25) and cellular stress (26) responses, and house a wide range of transcriptional regulatory factors (16). There are 10–30 PML bodies in the typical mammalian nucleus and they have distinct ultrastructural features when examined by EFTEM (24).

The glucocorticoid receptor (GR) is a nuclear receptor that functions as a ligand-dependent transcription factor in mediating the transcription of target genes. The unliganded GR resides in the cytoplasm, but upon hormone binding the ligand-bound GR translocates to the nucleus. Once in the nucleus, the activated GR binds to glucocorticoid target genes and modulates their expression. The synthetic hormone dexamethasone is an efficient agonist of GR that strongly transactivates the nuclear receptor and leads to accumulation of activated GR in the nucleus including concentration within numerous intranuclear foci (27). It is unclear what the GR foci represent but they may be related to the organization of the activated GR in modulating expression of particular target genes.

The nucleus contains other subnuclear compartments that have been studied both by light and electron microscopy. Interchromatin granule clusters (IGCs), also known as nuclear speckles (28), are easily recognizable subnuclear compartments that contain clusters of small nuclear ribonucleoprotein (snRNPs) granules surrounded by blocks of chromatin (23). IGCs do not contain DNA (29) and are not involved in transcribing genes, however, active genes may associate with IGCs for efficient processing of transcripts (30, 31). Although the functional significance of IGCs requires further investigation, the structural features of IGCs are well characterized and serve as a subnuclear landmark when examining the ultrastructure of the cell nucleus.

Here the details of a correlative fluorescence and EFTEM microscopy technique, which has been used with a high rate of success, will be outlined. The technique is used to identify particular cells by epifluorescence microscopy, embed and thin-section the same cells, image the distribution of desired proteins in the thin sections by immunofluorescence, and then examine the same sections by EFTEM. The technique also allows thin serial sections to be obtained for potential 3D analysis, and may be complementary to imaging biological samples by EFTEM tomography (32, 33). In particular, the correlative fluorescence and EFTEM microscopy method is used to examine subnuclear structures within intact cells exhibiting defined metabolic states, such as the accumulation of the activated ligand-bound glucocorticoid receptor in distinct foci and how these subnuclear structures compare to PML bodies and IGCs, both in terms of ultrastructure and biochemical composition. Net elemental maps from the sample will be collected, while maintaining spatial information. Comparing the phosphorus to nitrogen ratio of molecular components allows discrimination of protein-based from nucleic acid-based structures.

## 2. Materials

### 2.1. Cell Culture

1. Human cervical carcinoma (HeLa) cells.
2. Dulbecco's-modified Eagle's medium (DMEM) supplemented with 10% fetal calf serum, 100 U/ml penicillin, 100 μg/ml streptomycin, and 2 mM L-glutamine.
3. Gridded chamber coverglass (MatTek) (see Note 1).
4. Dexamethasone 20 μg/ml stock solution, 100 nM final concentration (see Note 2).

### 2.2. Immunolabeling

1. 1×PBS.
2. 16% paraformaldehyde solution, EM grade: the day of the experiment prepare a fresh 4% paraformaldehyde solution by diluting in 1×PBS.
3. 0.5% Triton X-100 in 1×PBS.
4. Blocking buffer—1% bovine serum albumin in 1×PBS; 0.05% goat serum in 1×PBS.
5. Humidified chamber—Nalgene utility sample box with moist Kimwipes.
6. Primary antibodies: anti-PML mouse monoclonal antibody and anti-glucocorticoid receptor rabbit polyclonal antibody (see Note 3).

7. Secondary antibodies: goat anti-mouse IgG (H+L) highly cross-absorbed Alexa Fluor 546, stock 2 mg/ml solution, goat anti-rabbit IgG $F(ab')_2$ Alexa Fluor 488, stock 2 mg/ml solution (see Note 4).

### 2.3. Wide-Field Epifluorescence Microscopy

A research grade wide-field epifluorescence microscope, equipped with high quality objective lenses, fluorescence filter cubes, and a scientific digital camera for acquiring images from the labeled sample, is needed for the fluorescence microscopy part of the method. The specific equipment used in the method described here is listed below.

1. Inverted wide-field epifluorescence microscope (Zeiss Axiovert 200 M).
2. Upright wide-field epifluorescence microscope (Zeiss Axioplan2).
3. Fluorescence filter cubes designed for the fluorophores used to label the sample. Filter cubes specific for Alexa 488 and Alexa 546 were used for the samples described here (see Note 5).
4. 10× Plan-Neofluar (N.A. 0.30) objective lens.
5. 63× C-apochromat (N.A. 1.2) water immersion objective lens.
6. 63× Plan-apochromat (N.A. 1.4) oil immersion objective lens.
7. MetaMorph image acquisition and processing software.
8. Adobe Photoshop image processing software.

### 2.4. Embedding and Mounting the Sample

1. 8% glutaraldehyde solution, EM grade: the day of the experiment prepare a fresh 2% glutaraldehyde solution by diluting in 1×PBS.
2. 95% ethanol.
3. Quetol 651/NSA resin embedding kit (Electron Microscopy Sciences).
4. Flat embedding capsules.
5. Nalgene utility sample box.
6. Orbital shaker.
7. Digital incubator.
8. Heat plate.
9. Multi-angle vise and jeweler's saw.
10. Stereo microscope (Leica MZ9.5).
11. General supplies: tube rack, safety razor blade, transparent tape, cotton tip applicators, permanent black marker (fine point), ProBond 5 min epoxy resin (Elmers Products), Dumont #3 C Tweezers (Electron Microscopy Sciences), glass microscope slides, and #1.5 coverslips (18×18 mm).

#### 2.5. Ultramicrotomy

1. Maxtaform copper London finder grids, 400 mesh (Electron Microscopy Sciences).
2. Parallel/Divider Bar 100 PB Copper grids for serial sections (Electron Microscopy Sciences).
3. Leica ultracut E ultramicrotome.
4. Diamond knife.
5. Grid storage box.

#### 2.6. Energy-Filtering Transmission Electron Microscopy

1. CM120 transmission electron microscope (FEI/Philips) equipped with a GIF100 post-column spectrometer (Gatan) or FEI Tecnai 30 (FEI Company) transmission electron microscope equipped with a Gatan Tridiem electron energy loss spectrometer.
2. Dumont #3C Tweezers (Electron Microscopy Sciences).
3. Digital Micrograph image acquisition and processing software (Gatan).

## 3. Methods

The method described below includes the main steps for correlative fluorescence and EFTEM imaging of subnuclear structures. The steps include the immunolabeling of cells prior to embedding the cells in resin for EFTEM imaging. However, the method can be extended to include live cell imaging to monitor and, if desired, functionally alter cells prior to fixation and immunolabeling (20). The ultimate goal of the method is to image the ultrastructure of interesting subnuclear structures, but an important secondary reason for the multiple detailed steps in performing the correlative microscopy method is to track cells from the fluorescence imaging through the EM preparation steps, imaging of the thin sections by fluorescence microscopy, and onto the EFTEM imaging. That way, both the cells with structures of interest can be tracked and the fidelity of the sample preparation can be monitored.

#### 3.1. Cell Culture

1. Grow cells to 70% confluency in gridded chamber coverglass dishes in 2–3 mL of DMEM growth medium.
2. Incubate with fresh medium containing 200 nM dexamethasone for 60 min at 37°C prior to fixing and immunolabeling the cells (see Note 6).

#### 3.2. Immunolabeling of Cells (See Note 7)

1. After 5 min of fixation in freshly prepared 4% paraformaldehyde in 1×PBS, rinse the cells with three successive changes of 1×PBS.
2. Permeabilize the cells by incubating with 0.5% triton X-100 in 1×PBS for 5 min at room temperature.

3. Rinse the cells three times with 1×PBS.
4. Incubate with blocking buffer for 60 min at room temperature.
5. Remove blocking buffer, and rinse three times with 1×PBS.
6. Add primary antibodies and incubate in a humidified chamber for 60 min at room temperature.
7. Rinse three times with 1×PBS.
8. Add secondary antibodies and incubate in a humidified chamber in the dark for 60 min at room temperature (see Note 4).
9. Rinse three times with 1×PBS.
10. Leave cells in 1×PBS.

#### 3.3. Imaging the Pre-embedded Immunolabeled Cells

1. Using the 63× C-apochromat (N.A. 1.2) water immersion objective lens, view the quality of the immunolabeling using the appropriate fluorescence filter sets.
2. Choose an optical configuration suitable for exciting and collecting fluorescence for two channels, the green fluorescence from the Alexa 488 fluorophore and the red fluorescence from the Alexa 546 fluorophore (see Note 8).
3. Adjust the exposure times for the individual fluorescence channels and acquire images of the desired cells.
4. Record the location of the cells of interest from the gridded coverglass using brightfield illumination (see Note 9).

#### 3.4. Embedding the Cells

1. Remove 1×PBS from the chamber dish.
2. Post-fix with freshly prepared 2% glutaraldehyde in 1×PBS for 5 min at room temperature.
3. Rinse three times with 1×PBS.
4. Replace 1×PBS with 10% ethanol in $ddH_2O$, and incubate for 60 min at room temperature.
5. Replace 10% ethanol with 25% ethanol in $ddH_2O$, and incubate for 60 min at room temperature.
6. Replace 25% ethanol with 50% ethanol in $ddH_2O$, and incubate for 60 min at room temperature.
7. Replace 50% ethanol with 75% ethanol in $ddH_2O$, and incubate for 60 min at room temperature.
8. Replace 75% ethanol with 95% ethanol, and incubate for 60 min at room temperature.
9. Replace with 95% ethanol and incubate overnight at room temperature.
10. Replace with 95% ethanol and incubate for 60 min at room temperature.
11. Place the chamber dishes in a non-humidified Nalgene utility box (see Note 10).

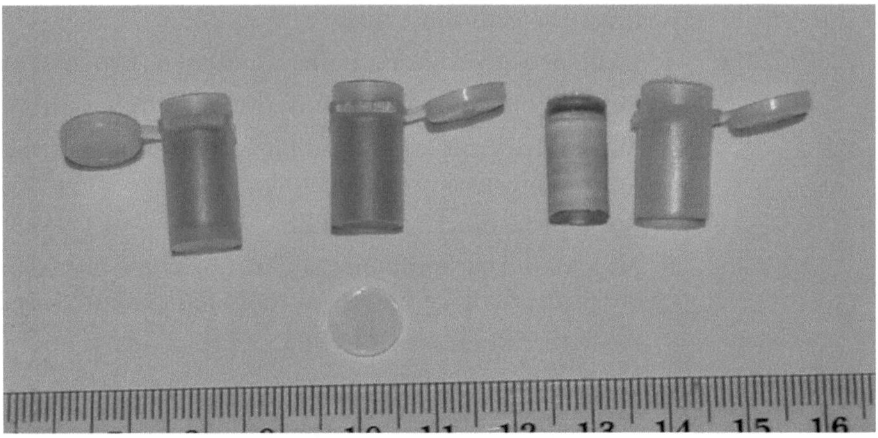

Fig. 1. Preparing Quetol mounting "bullets" in flat bottom embedding capsules.

12. Replace 95% ethanol with Quetol 651 solution.
13. Place the lid on the utility box to cover the chamber dishes and incubate for 2 h at room temperature with slow shaking on an orbital shaker.
14. Replace with Quetol 651 mixture, including 2% DMP-30, for 2 h at room temperature in the covered utility box with shaking.
15. Repeat step 14 twice.
16. Replace Quetol 651 mixture, including 2% DMP-30 and place enough mixture in the chamber coverglass dishes so that approximately 3 mm thick layer of resin fills the dish above the coverslip.
17. Place the chamber dishes (in the utility box but without the lid) in a digital incubator set at 65°C for 48 h to allow the Quetol resin to polymerize and harden.
18. Remove the chamber dishes from the oven and allow to cool to room temperature.
19. The chamber dishes can be stored in the dark at room temperature until ready to mount the cells for sectioning.

#### 3.5. Preparing Quetol Mounting Bullets

1. Place multiple flat bottom embedding capsules in a tube rack.
2. Use the left over, or replaced, Quetol 651 mixture (including 2% DMP-30), to fill the flat bottom embedding capsules.
3. Place the tube rack with capsules in the digital incubator set at 65°C for 48 h to allow the Quetol resin to polymerize and harden.
4. Remove the tube rack and allow to cool to room temperature.
5. Use a safety razor blade to cut off the bottom of the individual embedding capsules and push the hardened Quetol "bullet" out of the capsule from the bottom (see Fig. 1).

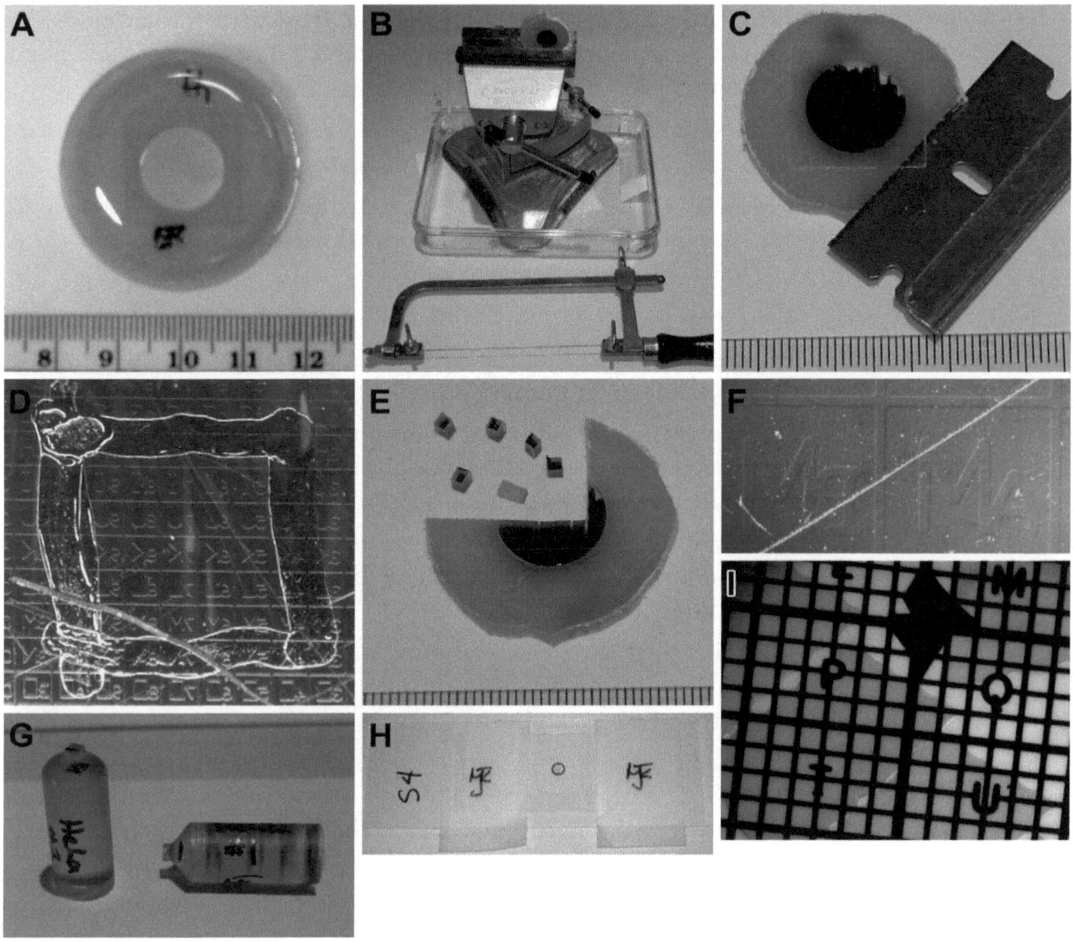


Fig. 2. Sample handling and preparation steps, from embedding cells to imaging thin sections.

### 3.6. Mounting the Cells for Sectioning

1. Write the name of the sample on the hardened Quetol resin inside the chamber dish (Fig. 2a) (see Note 11).
2. Mark the backside (the side opposite the cells attached to the coverglass) of the embedding plastic with a black permanent marker.
3. Place the chamber dish in the vise and use the jeweler's saw to cut away the perimeter walls (Fig. 2b) (see Note 12).
4. Remove the walless chamber dish from the vise and use a safety razor blade to slide under the coverslip and remove it from the embedding resin (Fig. 2c) (see Note 13).
5. Once the coverglass is removed, examine the sample using a stereo microscope to check whether any remnants of the coverglass remain, and remove them with the razor blade.
6. With a permanent marker, trace the boundaries of the specific gridded area of the plastic containing the cells of interest (Fig. 2d).

7. Clamp the plastic disc in the vise, and with the jeweler's saw cut out small ~1.5 × 1.5 mm blocks from the area of the embedding plastic that contains the cells of interest (Fig. 2e).
8. Using forceps, place the individual blocks on a microscope slide with the cells side up, black marker side down.
9. Place the slide on the stage of the upright epifluorescence microscope and examine using the 10× Plan-Neofluar (N.A. 0.30) objective lens.
10. Focus to the surface of the block using brightfield illumination.
11. Double-check that the area of the coverslip grid containing the coordinates of interest is present in the block.
12. Switch to epifluorescence illumination to check that the cells of interest survived the embedding and mounting within the individual blocks (see Note 14).
13. Mark the name of the sample on the side of a pre-made Quetol "bullet" where the individual blocks will be mounted using 5 min epoxy glue (Fig. 2g).
14. Mix equal parts of the 5 min epoxy glue and put a small drop on the top of the Quetol bullet using the wooden end of a cotton-tipped applicator.
15. Using forceps, pick up the sides of the block to be mounted and place it in the drop of epoxy on top of the bullet, black side down. Avoid getting epoxy on the forceps.
16. Use the wooden end of the cotton-tipped applicator to spread the epoxy onto the four side walls of the block, without spreading the epoxy on the top surface, to be sure it will be securely attached to the Quetol bullet once the epoxy hardens.

### 3.7. Sectioning the Mounted Blocks with an Ultramicrotome

1. Trim the block into a keystone shape using a safety razor blade.
2. The region of interest can be observed in the block when mounted in the ultramicrotome and can be rotated to provide optimal orientation for serial sectioning.
3. Obtain a ribbon of 5–15 serial ultrathin (50–70 nm) sections on the surface of the water in the diamond knife.
4. Manipulate the position of the ribbon of sections so that they are perpendicular to the parallel grid bars. As the ribbon is picked up, the region of interest should be carefully positioned so that it is not located over the grid bars (see Note 15).
5. Allow the grids to air dry and place in a grid holder box.
6. Although not strictly necessary, a thin (5 nm) carbon film can be applied to the specimen section using a Auto 306 evaporator. The thin carbon coating will help stabilize the sections under the electron beam in the EFTEM microscope.

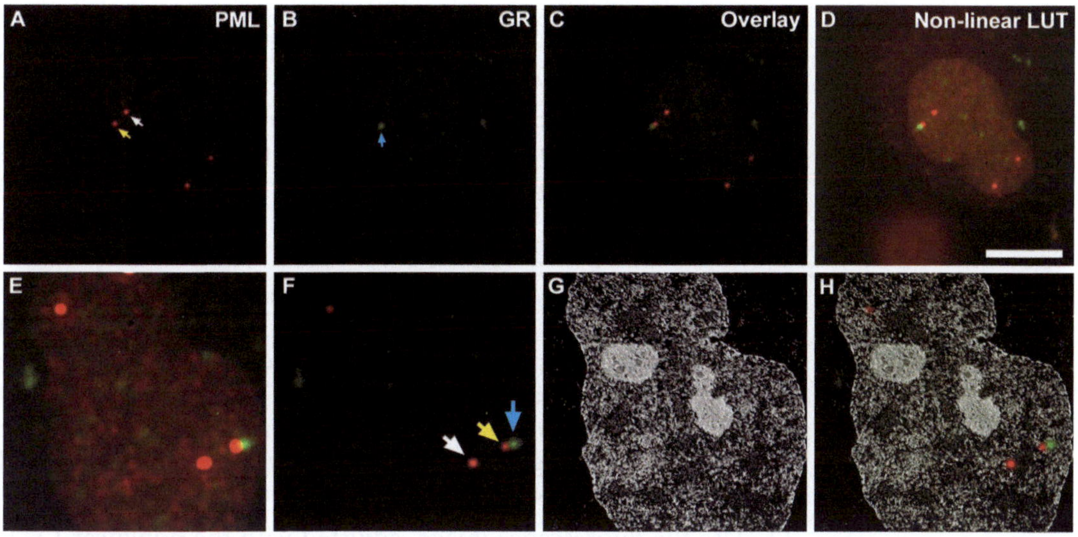


Fig. 3. Aligning the fluorescence image map to the low magnification EFTEM image. Immunofluorescence images of PML ((**a**), *red*) and GR ((**b**), *green*) from ultrathin sections of double immunolabeled HeLa cells. The *white arrow* in (**a**) denotes one of the four PML bodies present in this ultrathin section, whereas, the *yellow arrow* denotes a PML body adjacent to the GR focus, denoted by the *blue arrow* in (**b**), as seen in the overlay image in (**c**). A nonlinear look-up table (Nonlinear LUT), with the brightness and contrast excessively adjusted for both the PML and GR intensities, to help display the boundary of the nucleus is shown in (**d**). A resized, rotationally transformed, and cropped version of the nonlinear LUT overlay image to match the magnification of the low magnification EFTEM image is shown in (**e**). The corresponding resized, rotationally transformed, and cropped but linearly adjusted (histogram stretched) overlay image is shown in (**f**). The *colored arrows* denote the same PML bodies as in (**a**) and GR focus as in (**b**). In (**g**), a low magnification (×7,000) net phosphorus EFTEM image of the same nucleus, with the fluorescence image overlaid on the net phosphorus image in (**h**). Scale bar, 10 μm.

### *3.8. Imaging the Thin Sections by Fluorescence Microscopy*

1. Using clean forceps, place the grid on an ethanol-cleaned microscope slide with the sections on the top side away from the microscope slide.
2. Place a clean #1.5 coverslip over the grid.
3. Tape the left and right sides of the coverslip to the microscope slide to hold the coverslip in place (Fig. 2h).
4. Place a small drop of immersion oil on the coverslip and examine with the 63× Plan-apochromat (N.A. 1.4) objective lens on either the inverted or upright epifluorescence microscope.
5. Using brightfield illumination, focus on the grid and find the location of the sections (Fig. 2I) (see Note 16).
6. Adjust for optimal focus using epifluorescence excitation and acquire fluorescence images from the cells of interest (Fig. 3a, b) (see Note 17) (see Note 18).
7. Record the grid coordinates in the EM finder grids where the cells of interest are located, or the location of the serial sections on the grids.
8. Remove the grid from the microscope slide and place into a grid holder box.

9. With fluorescence images collected from the embedded and sectioned cells, using image processing software, such as MetaMorph or Adobe Photoshop, to adjust the brightness, contrast, and linearity (gamma) of the image intensity to exaggerate the low intensity pixels within the image and save as separate files (Fig. 3d). This nonlinear look-up table (LUT) will be useful as a guide to resize and align with images collected during EFTEM acquisition.
10. The pixel size within both the fluorescence image and the EFTEM images, collected at different magnifications, should be known. The respective pixels can be used to calculate the difference in magnification between the fluorescence and EFTEM image(s) and either digitally zoom or resample the fluorescence image to use as an overlay mask on the acquired EFTEM image (see Note 19).
11. Use the fluorescence image, adjusted with the nonlinear LUT, to view the boundaries of the nucleus and any structural clues to account for any differences in rotation between fluorescence image and the EFTEM image. The processed fluorescence image is used as a map to locate the position of the structures of interest within the EFTEM image (Fig. 3e) (see Note 20).

#### 3.9. EFTEM Imaging of the Thin Sections

1. Stabilize the sections with a low flux of beam electrons at low magnification (see Note 21).
2. Check the alignment of the EFTEM microscope, including beam and energy selecting filter.
3. Acquire energy loss image series (one pre-edge and one post-edge image), with the energy selecting slit set at a width of 20 eV, for both phosphorus ($L_{II,III}$ edge; pre-edge 120 eV and post-edge 155 eV) and nitrogen ($N_K$ edge; pre-edge 385 eV and post-edge 415 eV) at different magnifications (7,000×, 12,000×, 30,000×, 50,000×, etc.). Frame the different regions of the nucleus containing the structures of interest, in this case, the PML bodies and GR focus (see Note 22).
4. The processed fluorescence image can be used as a map to help locate the subnuclear structures of interest within the energy loss images.
5. Steps 2 and 3 are repeated for each serial section containing the cell and subnuclear structures of interest.

#### 3.10. Image Processing and Alignment of Fluorescence and EFTEM Images

1. Net elemental maps are generated from the EFTEM energy loss image series by dividing the respective element-enhanced post-edge image by the pre-edge image, using the Digital Micrograph EFTEM image acquisition and processing software (Fig. 4) (see Note 23).
2. Linearly adjust the brightness and contrast (histogram stretch) for the EFTEM net elemental images and export them as

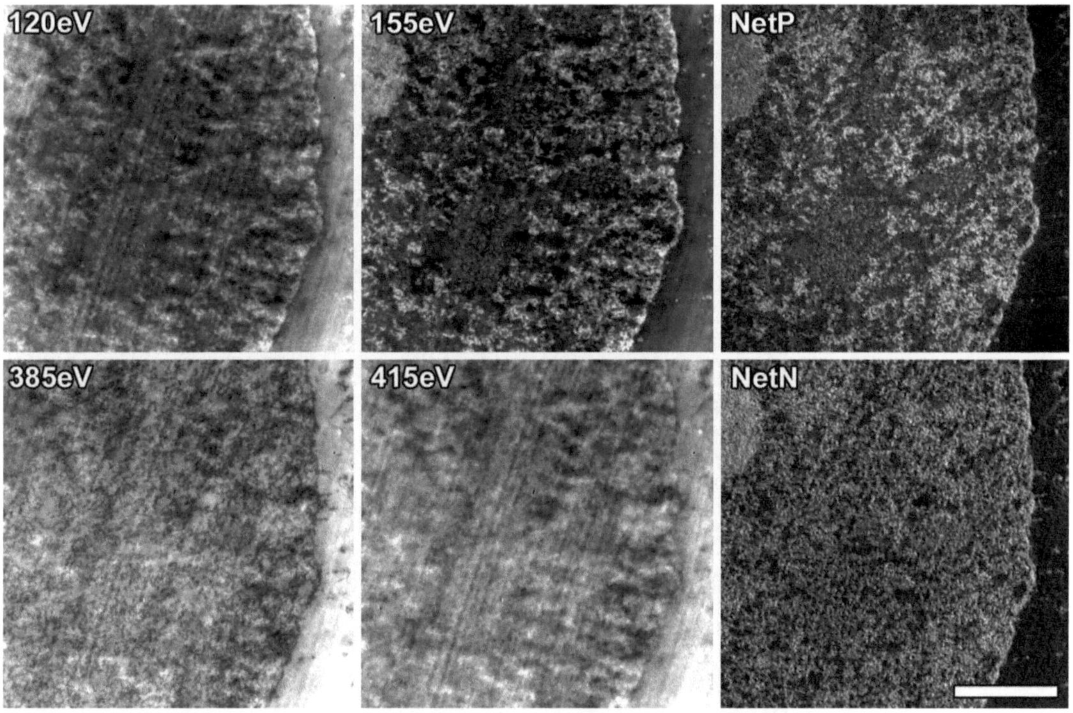


Fig. 4. EFTEM energy loss image series and corresponding net element image maps. Pre-edge and post-edge images are collected for each specific element (120 eV and 155 eV, respectively, for phosphorus and 385 eV and 415 eV, respectively, for nitrogen) and the net element image map can be obtained by dividing the post-edge image by the pre-edge image. The raw images, without any contrast adjustment, collected at ×12,000 magnification are shown. Scale bar, 1.0 μm.

TIFF files so that they can be arranged into figures using Adobe Photoshop.

3. The transformed (resized and rotated) and contrast-adjusted (histogram stretched) fluorescence image can be overlaid onto the low magnification (7,000×) EFTEM net phosphorus image (Fig. 5b).
4. The corresponding net phosphorus and net nitrogen images acquired at higher magnifications, 12,000× (Fig. 5c, d, respectively), and different regions of interest at 30,000× (Fig. 5e–h, respectively), can then be matched to specific regions in the lower magnification image map.
5. The net phosphorus and net nitrogen images can be pseudo-colored and an overlay image generated to compare the phosphorus to nitrogen content of the ultrastructural features within the subnuclear structures of interest (Fig. 6). The phosphorus to nitrogen ratio represented in the nuclear structures can be used to distinguish nucleic acid-based from protein-based components (17). By comparing the phosphorus and nitrogen distribution in the acquired images, it is apparent that the PML body contains very little phosphorus but considerable

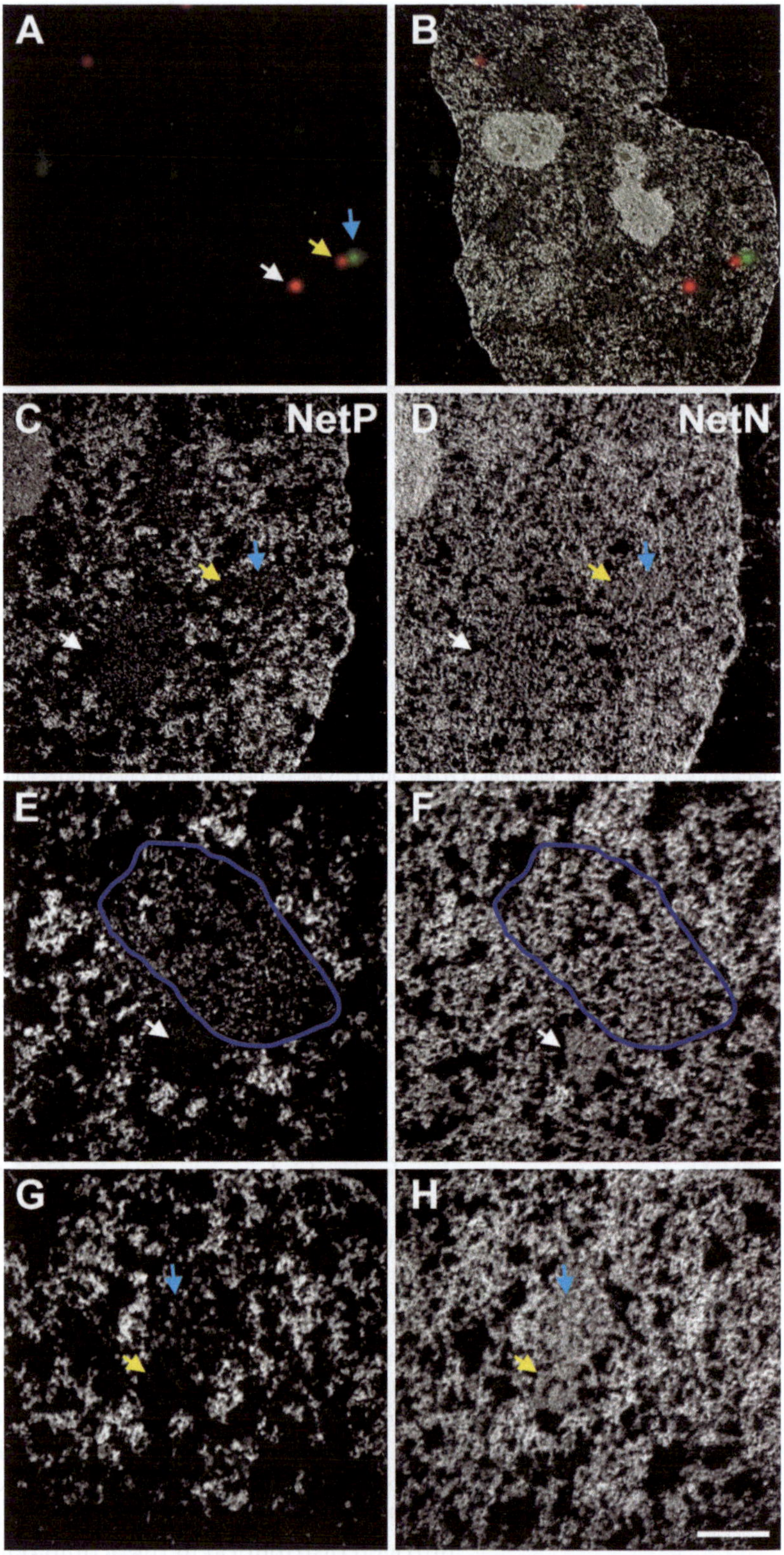
A
B
C
NetP
D
NetN
E
F
G
H

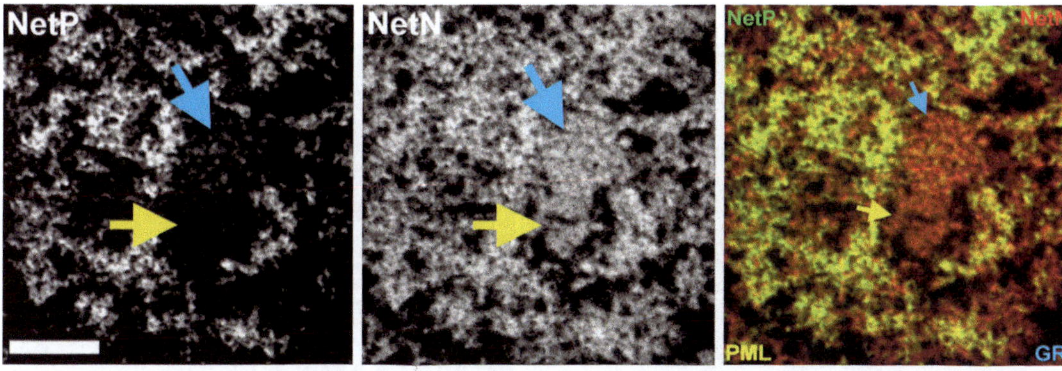


Fig. 6. Comparison of net phosphorus and net nitrogen signals in PML bodies and GR focus. High magnification (×30,000) EFTEM net phosphorus (NetP) and net nitrogen (NetN) images of the same PML body (*yellow arrow*) and GR focus (*blue arrow*) as shown in Fig. 5 but from a subsequent serial section. As highlighted in the NetP (*pseudo-colored green*) and NetN (*pseudo-colored red*) overlay image (*right panel*), the PML body contains very little phosphorus but considerable nitrogen indicating that it does not contain nucleic acid but is protein-based, whereas the GR focus contains phosphorus-rich fibrogranular structures present within a nitrogen-rich surrounding, indicating that nucleic acid is present within the GR focus that is surrounded by a significant amount of protein. Scale bar, 500 nm.

nitrogen, indicating that it does not contain nucleic acid but is protein based, whereas the GR focus contains phosphorus-rich fibrogranular structures present within a nitrogen-rich surrounding, indicating that nucleic acid is present within the GR focus that is surrounded by a significant amount of protein.

6. Lastly, if serial sections were collected and imaged, then generate a composite figure of the images from the different sections to compare the net phosphorus and net nitrogen image maps, and the ultrastructural features, from the subnuclear structures of interest within the respective sections (Fig. 7).

### 3.11. Discussion

Correlative fluorescence and EFTEM imaging has the unique advantage of providing biochemical information, along with nanometer scale ultrastructural detail, of subnuclear structures in specific

Fig. 5. Net element EFTEM images of the ultrastructure of PML bodies, a glucocorticoid receptor focus and an interchromatin granule cluster. (**a**) The fluorescence overlay image acquired from the thin section, the *white* and *yellow arrows* denoting PML bodies and the *blue arrow* denoting the GR focus. An overlay of the fluorescence image with the low magnification (×7,000) net phosphorus EFTEM image is shown in (**b**). The higher magnification (×12,000) net phosphorus and net nitrogen image maps are shown in (**c**, **d**), respectively, with the *arrows* denoting the location of the PML bodies (*white* and *yellow arrows*) and GR focus (*blue arrow*). Even higher magnification (×30,000) net phosphorus (**e**, **g**) and net nitrogen (**f**, **h**) image maps of the two regions of the nucleus containing PML bodies are shown. In (**e**, **f**), the PML body (*white arrow*) is next to an interchromatin granule cluster (*denoted by the encircling purple region of interest*). The PML body contains very little phosphorus but considerable nitrogen ((compare (**e**) to (**f**)), whereas the IGC contains multiple RNP granules that provide signal in the phosphorus map (**e**). The PML body in (**g**, **h**) (*yellow arrow*) is beside the GR focus (*blue arrow*). Again, the PML body has very little phosphorus, but significant nitrogen signal, whereas the GR focus contains both phosphorus-rich fibrogranular structures as well as surrounding nitrogen signal. Scale bar, 500 nm.

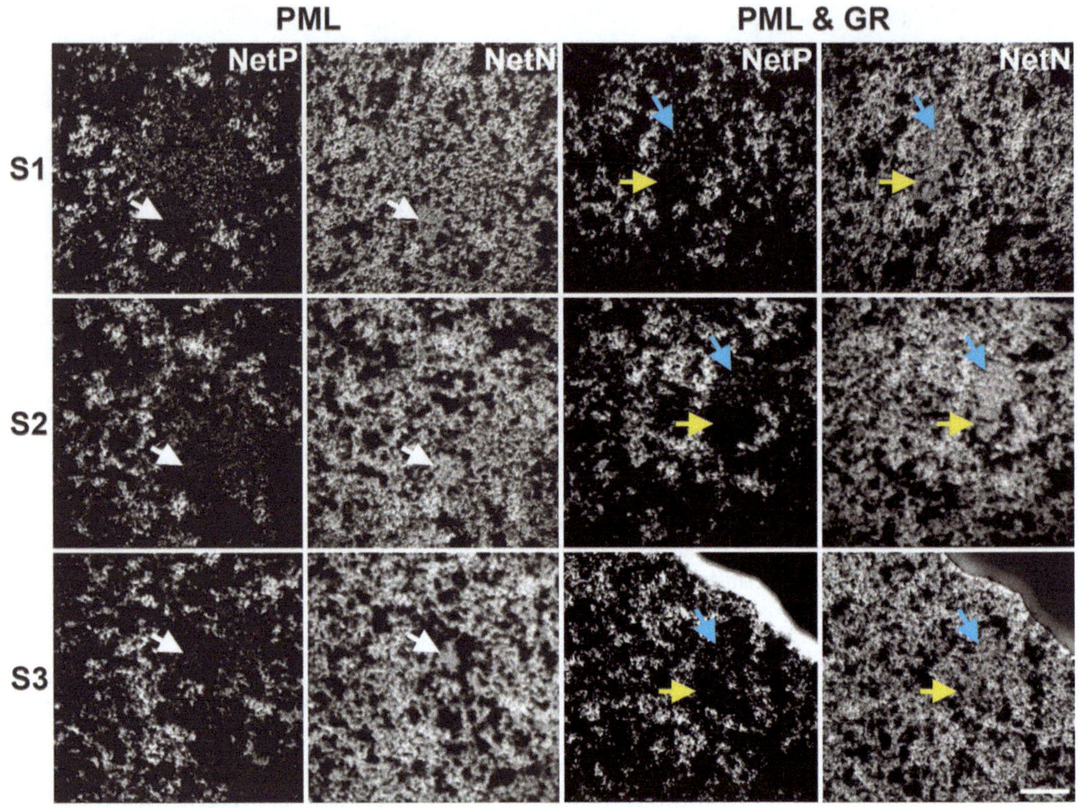


Fig. 7. Serial sections of EFTEM images of PML bodies, GR focus, and IGC. High magnification (×30,000) net phosphorus (NetP) and net nitrogen (NetN) images from serial sections (rows S1, S2, and S3) of the regions containing the PML body (*white arrow*) and IGC (*left two columns*) and PML body (*yellow arrow*) with GR focus (*blue arrow*) (*right two columns*). Scale bar, 500 nm.

cells under defined metabolic states. The contrast provided in the EFTEM images, which are used to generate net element image maps, comes directly from the endogenous elements comprising the molecular structures of the nucleus. This specificity prevents any masking from exogenous contrasting agents that are needed in conventional TEM. Although the comparison of net phosphorus and net nitrogen ratios within EFTEM images allows one to distinguish nucleic acid-based from protein-based components, it does not allow the discrimination between DNA and RNA as the phosphorus to nitrogen ratios are equivalent for the two molecules. However, the DNA and RNA are organized differently in the nucleus, with DNA packaged in well-conserved nucleosomal chromatin structure, whereas RNA is bound by numerous RNP binding proteins. Therefore, the morphological features of the two nucleic acids help aid in sorting out the composition of subnuclear structures.

## 4. Notes

1. Gridded coverglass chambers are only needed if relocating specific cells on the glass coverslip is required.
2. Dexamethasone is needed to activate the gluococorticoid receptor so that it translocates to the nucleus allowing the GR foci to become visible, otherwise it is not needed for the general correlative microscopy method.
3. The mouse anti-PML monoclonal 5E10 and rabbit anti-GR polyclonal antibodies were gifts from Dr. R. van Driel and Dr. T. Archer, respectively.
4. Fluoronanogold secondary antibodies can be substituted at this step if gold labeling is desired.
5. High quality filter cubes are preferable, especially when imaging the thin embedded sections, since there will be limited fluorescence signal and it is desirable for most of the signal to be collected by the camera.
6. This step is necessary for the glucocorticoid receptor to bind agonist and translocate to the nucleus. Dexamethasone is a strong agonist for GR and the 200 nM final concentration will saturate the available population of GR insuring that most of the receptor will translocate to the nucleus.
7. Pre-embedment immunolabeling is performed in this method rather than post-embedment immunolabeling of the thin sections. The immunolabeling steps do not alter the nuclear ultrastructure (21).
8. If desired, confocal imaging instead of wide-field fluorescence microscopy can be performed at this stage of the method.
9. To help provide greater contrast when viewing the coverslip grid, the condenser aperture can be restricted (closed slightly) and focus adjusted away from the cells.
10. Placing the chamber dishes in a non-humidified Nalgene utility box minimizes the spread of vapors and will help contain any potential spills from the Quetol reagents. The Quetol 651 mixture is made following the reagent ratios suggested by the manufacturer's instructions. The mixture should be made in advance of starting the embedding procedure and should be handled in a chemical fume hood, including during the embedding steps.
11. At this point, the cells can be viewed using the epifluorescence microscope to check whether the immunofluorescence labeling has survived the embedding procedure.
12. Removing the side walls of the chamber dish helps by removing the circular ridge at the bottom perimeter of the dish and,

thereby, allowing the razor blade to lay flat against the bottom of the chamber dish to prevent unnecessary leverage when trying to remove the delicate coverslip.

13. A heat plate can be used to slightly warm the chamber dish to help remove the coverglass.
14. Brightfield differential interference contrast illumination helps when visualizing the distribution of the cells in the embedded block, providing information as to where the block should be trimmed before sectioning with the ultramicrotome, but is not necessary (see Fig. 2f).
15. Parallel/Divider Bar 100 PB Copper grids are well suited for imaging serial sections as the spacing of the grid bars minimizes the likelihood of the cell of interest being placed on a grid bar which prevents the cell from being imaged using the energy-filtering transmission electron microscope.
16. The grid shown in the figure is generally used for single sections, whereas, the Parallel/Divider Bar 100 PB Copper grids are well suited for imaging serial sections (see Note 15).
17. Be efficient when exposing the sections to excitation light, and avoid using excess excitation light intensity, as the fluorescence signal in the sections will fade quickly relative to intact cells.
18. A sample of multi-fluorescent beads can be used to check the alignment of the different fluorescence filter cubes within the microscope, and can be done in advance.
19. Record all of the image processing steps performed on the fluorescence images, particularly the image resizing and rotation transformation, so that they can be applied to the linearly adjusted (histogram stretched) fluorescence images used in the overlay with the final EFTEM net element images, (Fig. 3f).
20. In this sample, immunogold labeling was not used to pinpoint the location of the glucocorticoid focus, but rather the known structure of the PML body was used as an internal fiduciary mark to locate the GR focus. The PML body and GR focus happen to be adjacent to one another in the sample presented.
21. This step can be skipped if a thin film of carbon has been deposited on the section.
22. The energy loss images presented in Figs. 4, 5, 6, and 7 were acquired using a Zeiss 902 EFTEM microscope with a prism-mirror-prism in-column configuration for energy loss imaging. More modern EFTEM microscopes generally employ a Gatan GIF electron imaging spectrometer.
23. The energy loss images series presented in Fig. 4 exhibit some nonuniformity in section thickness, such as from the diamond knife ridges and, therefore, are not strictly quantitative. If the number of atoms from a particular element, such as phosphorus,

were to be measured then the nonuniformity of the section thickness would need to be accounted for when generating the net element image map (33, 34). Nonetheless, the images are still useful to highlight regions in the nucleus containing an abundance of the specific element relative to background.

## Acknowledgments

The EFTEM images were collected with generous support from David Bazett-Jones. Ying Ren helped prepare the samples, and Michael Hendzel and Alioscka Sousa provided helpful critical reading of the manuscript. Maria Aronova and Richard Leapman generously provided advice, and access to multiple EFTEM microscopes. This research was supported in part by the Intramural Research Program of the NIH, National Cancer Institute, Center for Cancer Research.

### References

1. Betzig E, Patterson GH, Sougrat R, Lindwasser OW, Olenych S, Bonifacino JS, Davidson MW, Lippincott-Schwartz J, Hess HF (2006) Imaging intracellular fluorescent proteins at nanometer resolution. Science 313:1642–1645
2. Watanabe S, Punge A, Hollopeter G, Willig KI, Hobson RJ, Davis MW, Hell SW, Jorgensen EM (2011) Protein localization in electron micrographs using fluorescence nanoscopy. Nat Methods 8:80–84
3. Sosinsky GE, Giepmans BN, Deerinck TJ, Gaietta GM, Ellisman MH (2007) Markers for correlated light and electron microscopy. Methods Cell Biol 79:575–591
4. Giepmans BN (2008) Bridging fluorescence microscopy and electron microscopy. Histochem Cell Biol 130:211–217
5. Kireev I, Lakonishok M, Liu W, Joshi VN, Powell R, Belmont AS (2008) In vivo immunogold labeling confirms large-scale chromatin folding motifs. Nat Methods 5:311–313
6. Kukulski W, Schorb M, Welsch S, Picco A, Kaksonen M, Briggs JA (2011) Correlated fluorescence and 3D electron microscopy with high sensitivity and spatial precision. J Cell Biol 192:111–119
7. Deerinck TJ, Giepmans BN, Smarr BL, Martone ME, Ellisman MH (2007) Light and electron microscopic localization of multiple proteins using quantum dots. Methods Mol Biol 374:43–53
8. Giepmans BN, Deerinck TJ, Smarr BL, Jones YZ, Ellisman MH (2005) Correlated light and electron microscopic imaging of multiple endogenous proteins using Quantum dots. Nat Methods 2:743–749
9. Nisman R, Dellaire G, Ren Y, Li R, Bazett-Jones DP (2004) Application of quantum dots as probes for correlative fluorescence, conventional, and energy-filtered transmission electron microscopy. J Histochem Cytochem 52:13–18
10. Grabenbauer M, Geerts WJ, Fernandez-Rodriguez J, Hoenger A, Koster AJ, Nilsson T (2005) Correlative microscopy and electron tomography of GFP through photooxidation. Nat Methods 2:857–862
11. Shu X, Lev-Ram V, Deerinck TJ, Qi Y, Ramko EB, Davidson MW, Jin Y, Ellisman MH, Tsien RY (2011) A genetically encoded tag for correlated light and electron microscopy of intact cells, tissues, and organisms. PLoS Biol 9:e1001041
12. Ren Y, Kruhlak MJ, Bazett-Jones DP (2003) Same serial section correlative light and energy-filtered transmission electron microscopy. J Histochem Cytochem 51:605–612
13. Spector DL (2001) Nuclear domains. J Cell Sci 114:2891–2893
14. Spector DL (2006) SnapShot: cellular bodies. Cell 127:1071
15. Zhao R, Bodnar MS, Spector DL (2009) Nuclear neighborhoods and gene expression. Curr Opin Genet Dev 19:172–179

16. Lallemand-Breitenbach V, de The H (2010) PML nuclear bodies. Cold Spring Harb Perspect Biol 2:a000661
17. Bazett-Jones DP, Hendzel MJ (1999) Electron spectroscopic imaging of chromatin. Methods 17:188–200
18. Bazett-Jones DP, Li R, Fussner E, Nisman R, Dehghani H (2008) Elucidating chromatin and nuclear domain architecture with electron spectroscopic imaging. Chromosome Res 16:397–412
19. Spencer CA, Kruhlak MJ, Jenkins HL, Sun X, Bazett-Jones DP (2000) Mitotic transcription repression in vivo in the absence of nucleosomal chromatin condensation. J Cell Biol 150: 13–26
20. Kruhlak MJ, Celeste A, Dellaire G, Fernandez-Capetillo O, Muller WG, McNally JG, Bazett-Jones DP, Nussenzweig A (2006) Changes in chromatin structure and mobility in living cells at sites of DNA double-strand breaks. J Cell Biol 172:823–834
21. Eskiw CH, Rapp A, Carter DR, Cook PR (2008) RNA polymerase II activity is located on the surface of protein-rich transcription factories. J Cell Sci 121:1999–2007
22. Eskiw CH, Fraser P (2011) Ultrastructural study of transcription factories in mouse erythroblasts. J Cell Sci 124:3676–3683
23. Hendzel MJ, Kruhlak MJ, Bazett-Jones DP (1998) Organization of highly acetylated chromatin around sites of heterogeneous nuclear RNA accumulation. Mol Biol Cell 9: 2491–2507
24. Boisvert FM, Hendzel MJ, Bazett-Jones DP (2000) Promyelocytic leukemia (PML) nuclear bodies are protein structures that do not accumulate RNA. J Cell Biol 148:283–292
25. Dellaire G, Bazett-Jones DP (2004) PML nuclear bodies: dynamic sensors of DNA damage and cellular stress. Bioessays 26:963–977
26. Eskiw CH, Dellaire G, Mymryk JS, Bazett-Jones DP (2003) Size, position and dynamic behavior of PML nuclear bodies following cell stress as a paradigm for supramolecular trafficking and assembly. J Cell Sci 116:4455–4466
27. Htun H, Barsony J, Renyi I, Gould DL, Hager GL (1996) Visualization of glucocorticoid receptor translocation and intranuclear organization in living cells with a green fluorescent protein chimera. Proc Natl Acad Sci USA 93:4845–4850
28. Lamond AI, Spector DL (2003) Nuclear speckles: a model for nuclear organelles. Nat Rev Mol Cell Biol 4:605–612
29. Thiry M (1993) Differential location of nucleic acids within interchromatin granule clusters. Eur J Cell Biol 62:259–269
30. Pandit S, Wang D, Fu XD (2008) Functional integration of transcriptional and RNA processing machineries. Curr Opin Cell Biol 20:260–265
31. Spector, D. L., Lamond, A. I. (2011) Nuclear speckles. *Cold Spring Harb Perspect Biol. 3.*
32. Leapman RD, Kocsis E, Zhang G, Talbot TL, Laquerriere P (2004) Three-dimensional distributions of elements in biological samples by energy-filtered electron tomography. Ultramicroscopy 100:115–125
33. Aronova MA, Kim YC, Harmon R, Sousa AA, Zhang G, Leapman RD (2007) Three-dimensional elemental mapping of phosphorus by quantitative electron spectroscopic tomography (QuEST). J Struct Biol 160:35–48
34. Leapman RD, Aronova MA (2007) Localizing specific elements bound to macromolecules by EFTEM. Methods Cell Biol 79:593–613

# Chapter 23

# High Data Output Method for 3-D Correlative Light-Electron Microscopy Using Ultrathin Cryosections

**Katia Cortese, Giuseppe Vicidomini, Maria Cristina Gagliani, Patrizia Boccacci, Alberto Diaspro, and Carlo Tacchetti**

## Abstract

Investigation of intracellular dynamics requires a detailed description of the molecular topography and ultrastructural morphology of the cell, for example, the position of a protein in relation to a given compartment of the cell and the morphology of the compartment. Standard fluorescence light microscopy (FLM) localizes proteins in living or fixed cells with a resolution of few hundreds of nanometers, but the unlabeled cellular context is partially missing. Electron microscopy (EM) techniques, such as immuno-EM, reveal protein topology with a few tens of nanometer resolution and retain the cellular context. However, EM analysis shows shortcomings compared to FLM, such as, lower statistical output, applicability only to fixed cells, and higher technical difficulties. To bridge the gap between fluorescent cell imaging and EM, several laboratories have developed methods for correlative light-electron microscopy (CLEM). In CLEM, a limited number of fluorescently labeled cell compartments are first imaged by light microscopy and then visualized and analyzed by EM. Recently, two different CLEM approaches using the EM cryo-immunogold method have been developed to extend the analysis to a high number of regions of interest and to correlate the topology of specific antigens. In this chapter, we describe one of these methods, the High Data Output CLEM (HDO-CLEM) approach. The major benefits of HDO-CLEM are the possibility to (1) correlate several hundreds of events at the same time, (2) perform three-dimensional (3D) correlation, (3) immunolabel both endogenous and recombinantly tagged proteins at the same time, and (4) combine the high data analysis capability of FLM with the high precision-accuracy of transmission electron microscopy in a CLEM hybrid morphometric analysis. We have identified and optimized critical steps in sample preparation, defined routines for sample analysis and retracing of regions of interest, developed software for semi/fully automatic 3D FLM reconstruction and defined preliminary conditions for a hybrid light/electron microscopy morphometry approach.

**Key words:** Confocal microscopy, Correlative microscopy, Cryosection, Electron microscopy, Immunofluorescence, Immunogold, Morphometry

Alioscka A. Sousa and Michael J. Kruhlak (eds.), *Nanoimaging: Methods and Protocols*, Methods in Molecular Biology, vol. 950, DOI 10.1007/978-1-62703-137-0_23, © Springer Science+Business Media, LLC 2013

## 1. Introduction

In modern cell biology, imaging techniques such as fluorescent light microscopy (FLM), electron microscopy (EM) and electron tomography (ET) provide the most powerful tools to study cellular structure and function. All approaches are endowed with positive and negative aspects. FLM has the advantage to be performed on both living and fixed cells, and can provide results on a large set of data. However, the relatively poor resolution of standard far-field fluorescence microscopes reduces the possibility to precisely identify the nature of a given fluorescent signal. For example, a fluorescent spot might represent an entire organelle, a subdomain of an organelle, or even aggregates of proteins. In addition, the relationship between the fluorescent spot and the surroundings cellular unlabeled structures are missing. On the other hand, immuno-electron microscopy (immuno-EM) allows the localization of proteins with a resolution of tens of nanometers and an exquisite depiction of subcellular structures of all stained compartments, labeled and unlabeled. However, EM methods are static in nature and cannot track highly dynamic processes such as membrane trafficking events. Moreover, the localization of certain rare cellular structures within the enormous volume of a cell, or group of cells, may be difficult by EM and may limit the analysis to a small set of data. Although comparatively new FLM techniques, usually referred as nanoscopy approaches (1, 2), have almost closed the resolution gap between FLM and EM, the correlation of the information obtained by EM and FLM can answer many important open questions in cell biology, mostly linked to the definition of the relationships between labeled and unlabeled compartments.

Thus, ideally it would be very useful to combine the strengths of both FLM and EM, while observing the very same sample by both techniques (3, 4). This combined approach is referred to as correlative light-electron microscopy (CLEM). Many EM laboratories have spent considerable efforts in the development of easy, robust and quantitative methods to perform CLEM. Some of these approaches focus on the retracing aspect of the regions of interest (ROIs), identified first by FLM and successively retraced by EM (5, 6). Others have focused on the development of new correlative probes, which can be visualized both by FLM and EM (5, 7–10). While still others have focused on improving the live-cell imaging aspect of CLEM, or optimal fixation conditions (5, 11–13). The choice of a specific CLEM method ultimately depends on the given scientific question to answer.

The High Data Output CLEM (HDO-CLEM) method (3, 4) has been designed with the aims of: (1) maximizing labeling efficiency, (2) performing analysis on large data sets, (3) obtaining good preservation of morphology, and (4) allowing 2D and 3D analysis.

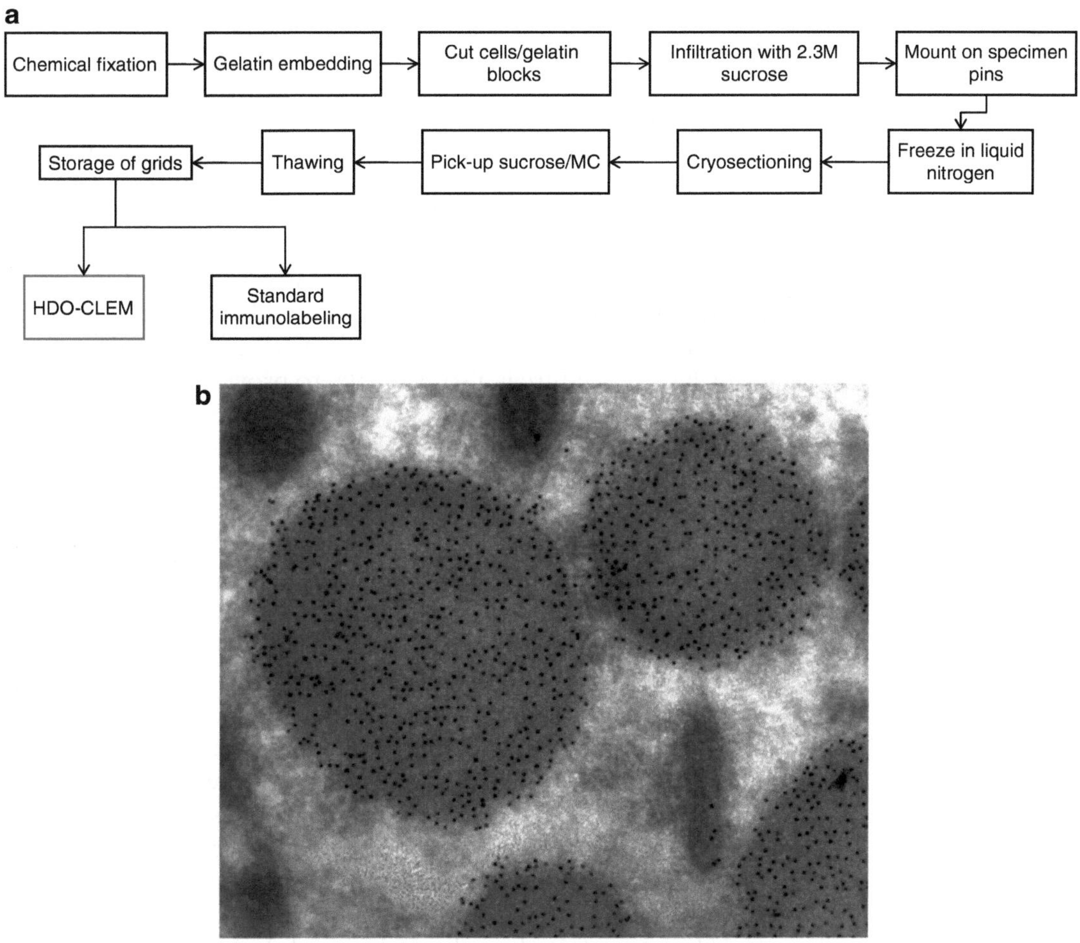


Fig. 1. (**a**) Flow diagram of a typical immuno-EM protocol using the Tokuyasu technique. (**b**) Example of Tokuyasu immunolabeling. Scale bar 0.5 μm. Reproduced with permission from ref. 4.

The approach is based on the use of ultra/semithin Tokuyasu cryosections (14), and integrates immunogold and immunofluorescence labeling. The Tokuyasu method is based on sucrose embedding of aldehyde fixed samples and it allows a reasonable balance between morphology and antigenicity retention. The use of sucrose prevents ice crystal formation during the freezing procedure of aldehyde fixed samples (14). In non-crystalline ice (vitreous ice) the freezing procedure has minimal impact on the ultrastructure and allows ultrathin sectioning of the sample. With the introduction of the Tokuyasu cryosectioning method and the subsequent improvements (15, 16) (Fig. 1), immuno-EM has become widely accessible to many EM laboratories and has set the standard for antigen localization at the nanometer resolution. The labeling procedures adopted in the HDO-CLEM method follow a stepwise approach. The antigens of interest are first labeled with a specific primary antibody and then, simultaneously, by a specific secondary fluorescent antibody and Protein A-gold. The procedure is suitable for both single and double labeling approaches.

The HDO-CLEM protocol has been optimized to correlate the highest amount of different cells, structures, or events in a single experiment. The possibility to apply the method to semithin cryosections, or ribbons of serial cryosections, makes the method suitable for 3D studies, both by EM tomography and FLM, respectively. FLM 3D relies on free software, MicroscoBioJ (3, 4), and a collection of Image-J (17) based plug-ins, able to create 3D multicolor volume rendering from serial 2D images. A high-content data analysis procedure is also available, to provide large-scale FLM morphometric data sets, with an accuracy close to the EM. The method involves the correction of the FLM data by a systematic error obtained from the FLM/EM correlation of a significant cohort of samples.

Here we report a step-by-step description of the HDO-CLEM method we have developed, as we have applied it to correlate FLM and EM analysis of Russell bodies (18, 19). Russell bodies are dilated ER cisternae, generally separated from the normal ER network, containing condensed aberrant immunoglobulins (Ig) originally identified in plasmocitoma cells. We used HeLa cells stably transfected with a mouse immunoglobulin μ chain, lacking the first constant domain (μ-ΔCH1), together with a wild-type immunoglobulin λ light chain, displaying spherical dilated RER cisternae resembling the Russell bodies identified in plasmocitoma cells (rough Russell bodies, RRBs). RRBs are about 1 μm in diameter and contain protein aggregates of mutant immunoglobulin M. Conversely, when μ-ΔCH1 is transfected in the absence of λ, it aggregates in the smooth ERGIC tubular compartments (smooth Russell bodies, SRBs), as shown by the association with the ERGIC marker ERGIC-53. SRBs are composed of narrow (100 nm wide, n-SRB) or dilated (200 nm wide, d-SRB), curled, tubular cisternae.

## 2. Materials

### *2.1. Reagents*

1. 0.1 M phosphate-buffered saline (PBS) at pH 7.4.
2. 4% Paraformaldehyde.
3. 4% Paraformaldehyde/0.4% glutaraldehyde.
4. 0.15% Glycine.
5. 2.3 M sucrose.
6. 2% Methylcellulose.
7. 12% Gelatin.

8. 10% Sodium azide solution.
9. 10% Bovine serum albumin (BSA) stock solution.
10. 4% Uranyl acetate.
11. 0.3 M oxalic acid.
12. 2% Uranyl acetate/0.15 M oxalic acid.
13. 25% Ammonium hydroxide
14. 0.4% Uranyl acetate/1.8% 25 ctp methylcellulose.
15. 50% Glycerol in water.
16. Formvar powder.
17. 40,6-Diamidino-2-phenylindole (DAPI) (1 mg/ml stock solution, use 100 ng/ml).

### 2.2. Antibodies

1. Rabbit anti-IgG (μ chain) (Zymed) primary antibodies (1:200).
2. Rabbit anti-calreticulin (Stressgen) primary antibodies (1:200).
3. Cy3-conjugated donkey anti-rabbit secondary antibody (Zymed).
4. Cy2-conjugated donkey anti-rabbit secondary antibody (Zymed).
5. 10 and 15 nm Protein A-gold (PAG).

### 2.3. Sectioning and Imaging Equipment

1. T5293 DUMONT Biology tweezers # 5 stainless.
2. Aluminum specimen pins for ultra-cryo-microtomy (local manufacturer).
3. Diamond knife, cryo, dry, 3 mm—45 °, 1 mm/s cutting speed (Drukker or Diatome).
4. Leica EM FCS Ultracut ultramicrotome (Leica Microsystem).
5. Finder gold EM grids (Agar).
6. Microscope coverslips (Nr. 0/1, 20 × 20 mm).
7. Microscope slides (1.0–1.2 mm thick, 25.4 × 76.2 mm size).
8. A spectral laser scanning microscope equipped with an HeNe laser (1 mW, 543 nm), a multiline Ar laser (100 mW, 457–476–488–514 nm) and a near ultra-violet laser diode (405 nm), such as a Olympus FV1000, with a transmitted light detector for recording brightfield images. A PlanApo 60×/1.42 oil objective (Olympus).
9. Transmission electron microscope capable of recording tomographic tilt series, such as a Tecnai 12-G2 Bio-Twin EM (FEI Company). This microscope is equipped with a LaB6 emitter for high brightness, a CompuStage for accurate movement of the specimen on five axes ($X$, $Y$, $Z$, $\alpha$, $\beta$), a SIS Mega View III CCD camera (Olympus-SIS) for digital image acquisition, and Kodak film 4489 for recording some EM micrographs.

***2.4. Software***

1. ImageJ (http://rsbweb.nih.gov/ij/) free software package (17) (see Note 1).
2. MicroSCoBioJ (http://imagejdocu.tudor.lu/doku.php?id=plugin:stacks:microscobioj:start) free software package (3, 4) (see Note 2).
3. TurboReg (http://bigwww.epfl.ch/thevenaz/turboreg/) free software package (see Note 3).
4. MosaicJ (http://bigwww.epfl.ch/thevenaz/mosaicj/) free software package (see Note 4).
5. IMOD (http://bio3d.colorado.edu/imod/) free software package (20) (see Note 5).

## 3. Methods

Outline of the methods: (1) preparation of reagents, (2) preparation of semi/ultrathin cryosections of cell samples by Tokuyasu technique (Fig. 1), (3) immunofluorescence and immunogold labeling of semi/ultrathin cryosections (Fig. 2), (4) image acquisition and processing (ImageJ and MicroScoBioJ software) (Fig. 3), and (5) hybrid morphometry (Fig. 3).

***3.1. Preparing Reagents***

1. 0.1 M PBS: Weigh 80 g NaCl, 2 g KCl, 14.4 g $Na_2HPO_4{\cdot}2H_2O$, and 2.3 g $NaH_2PO_4{\cdot}H_2O$, and add distilled water (Milli Q) up to 1,000 ml. Dilute 1:10 with distilled water prior to use.
2. 4% Paraformaldehyde: Dissolve 4 g of granular parafolmaldehyde in 100 ml of PBS. This solution must be prepared under the chemical fume hood. Pre-heat the PBS to 60°C to allow the parafolmaldehyde to completely dissolve, then add 50 μl of 1N NaOH while stirring.
3. 4% Paraformaldehyde/0.4% glutaraldehyde: Mix 70% glutaraldehyde and 4% paraformaldehyde in a 1:175 v:v ratio.
4. 0.15% Glycine: To 30 mg of glycine, add PBS up to a 20 ml final volume.
5. 2.3 M sucrose: Dissolve 78.73 g of sucrose in 50 ml of PBS and stir until the sucrose is completely dissolved. Add PBS to a 100 ml final volume and store 50 ml aliquots at 4°C.
6. 2% Methylcellulose: Add 4 g of methylcellulose to 196 ml of distilled water while stirring at 90°C. Rapidly cool to 10°C on ice while stirring, and then add water to a final volume of 200 ml. Seal with parafilm and stir overnight at 4°C. Store methylcellulose at 4°C in the dark, for up to 3 months.
7. 12% Gelatin: Dissolve 12 g of gelatin powder (local food brand is satisfactory) in 75 ml of PBS by first stirring the solution for 10 min at room temperature and then for 4–6 h at 60°C. When

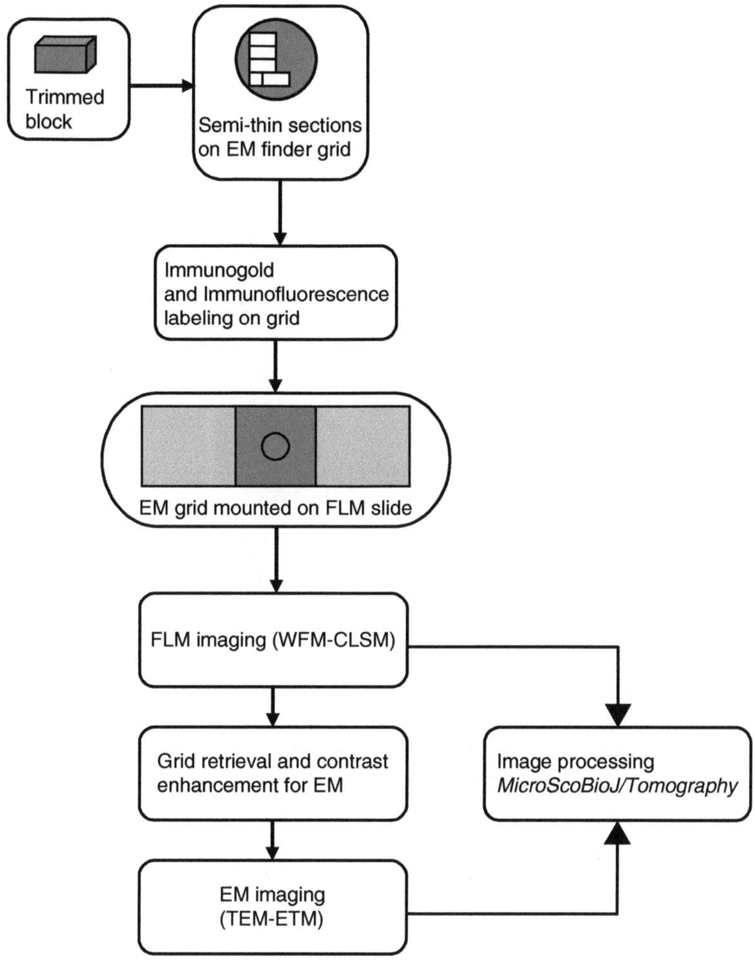


Fig. 2. Flow diagram of a typical HDO-CLEM experiment.

the gelatin is completely dissolved, cool the solution to 37°C, add 200 μl of a 10% sodium azide solution, and finally add PBS to a 100 ml final volume. Aliquot and store at 4°C, up to several months.

8. 10% BSA stock solution: Add 10 g of serum albumin in 100 ml of distilled water (Milli Q) and gently stir overnight, to prevent foaming, at 4°C. Set the pH to 7.4 with 1N NaOH, and then add sodium azide to a 0.02% final concentration. Centrifuge the solution for 1 h at 100,000 × *g*, retrieve the supernatant, and store in small aliquots at 4°C.
9. 4% Uranyl acetate: Dissolve 4 g of uranyl acetate in 100 ml of distilled water by stirring at room temperature.
10. 0.3 M oxalic acid: Dissolve 3.78 g of oxalic acid in 100 ml of distilled water by stirring at room temperature.
11. 2% Uranyl acetate/0.15 M oxalic acid: Mix 4% uranyl acetate and 0.3 M oxalic acid in a 1:1 v:v ratio. Adjust the pH to 7–8 by adding drops of 25% ammonium hydroxide while stirring.

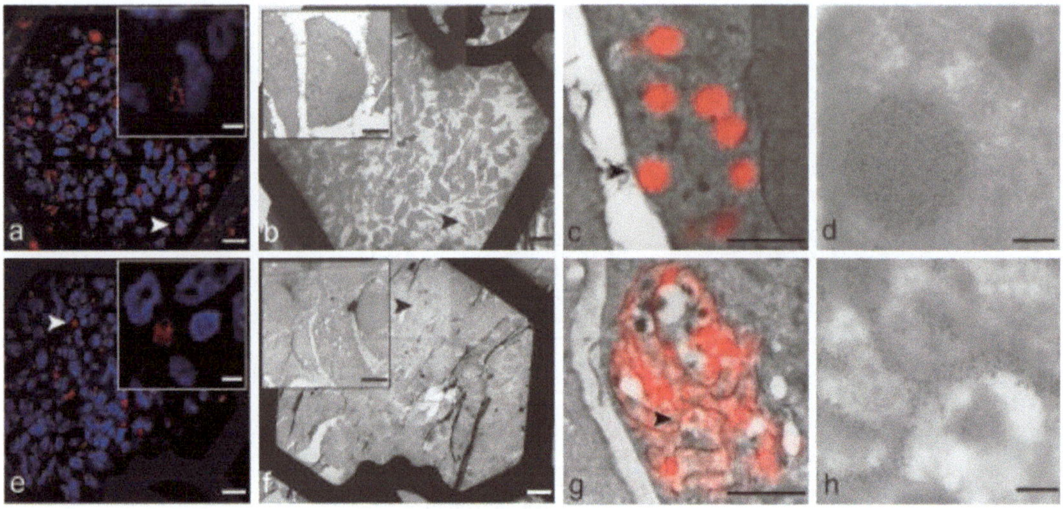


Fig. 3. Qualitative HDO-CLEM analysis on rough Russell Bodies (RRBs, **a**–**d**) and smooth Russell Bodies (SRBs, **e**–**h**). Smooth Russell Bodies do not have ribosomes on the limiting membrane. Panel represents CLSM imaging (2,048 × 2,048 pixels, pixel size 100 nm) of a single mesh grid. Imaging is able to reveal DAPI staining of nuclei (*blue*) and Cy3-labeled RRBs (**a**) and SRBs (**e**) subcellular structures (*red*). Insets represent high-sampling images (512 × 512 pixel, pixel size 50 nm) of the ROIs indicated by the *white arrows*. Grid mesh bars and letters are depicted in the brightfield channel (*gray*). (**b**, **f**) TEM mosaics of the same mesh previously observed at CLSM, (**a**) and (**e**) respectively. Mosaics are obtained using 3 × 3 low-magnification images (×31,000). The same ROIs identified in the CLSM images are easily retraced within TEM mesh images (*black arrows*). Insets represent medium magnification TEM images (×33,500) of the ROIs indicated by the *black arrow*. (**c**, **g**) High-sampling CLSM images (512 × 512 pixels, pixel size 25 nm) and high-magnification TEM images (×311,500) are used to build CLEM overlay images of the analyzed ROIs. (**d**, **h**). Higher magnification of the area pointed by the *black arrows* in (**c**) and (**g**) clearly shows immunolabeling for Igμ in RRB and SRB, respectively. Scale bars: 20 μm (**a**, **b**, **e**, and **f**), 4 μm (insets **a**, **b**, **e**, and **f**), 2 μm (**c** and **g**), 0.2 μm (**d**, **h**). Reproduced with permission from ref. 4.

12. 0.4% Uranyl acetate/1.8% 25 ctp methylcellulose: Gently mix 1 ml of 4% uranyl acetate to 9 ml of 2% methylcellulose solution. Store the solution at 4°C in the dark, for up to 3 months.
13. 50% Glycerol in water: Add 100 ml of glycerol to 100 ml of deionized water while stirring. Aliquot to 100 ml bottles and autoclave.
14. Formvar: Weigh 1.1 g of formvar and place in a 100 ml volumetric flask. Add chloroform to a 75 ml volume, and stir until dissolved. Subsequently, add more chloroform to a 100 ml final volume and let rest overnight at room temperature (see Note 6). Store formvar solution in a clean bottle for up to 3 months.

### 3.2. Preparation of Ultrathin Cryosections for HDO-CLEM

Hela cells, treated to induce Russell bodies formation (18), are plated on tissue culture dishes until they reach 80% confluence. Unless otherwise indicated, all steps are performed at room temperature.

1. Fix cells by adding a mixture of 4% paraformaldehyde/0.4% glutaraldehyde to an equal volume of culture medium for 2 h at room temperature (see Note 7).

2. Quench free aldehyde groups by incubating the cells for 10 min in 0.15% glycine at room temperature.
3. Add a small volume of 1% gelatin in PBS. Scrape off the cells from the tissue culture dishes with a cell scraper.
4. Transfer the cells to a 1.5 ml microcentrifuge tube, pellet the cells for 1 min at 13,000 rpm (16,100×*g*) and remove the supernatant.
5. Resuspend the cells with 1 ml of 12% gelatin for 10 min at 37°C.
6. Pellet the cells for 1 min at 13,000 rpm and carefully remove the supernatant (gelatin 12%).
7. Resuspend the pellet in 1 ml of 12% gelatin.
8. Repeat the centrifugation step and remove the supernatant.
9. Transfer the pellet to 4°C for 15 min.
10. Carefully cut the microcentrifuge tube, at the level of the pellet, using a razor blade and remove the solidified gelatin/cell pellet from the microcentrifuge tube.
11. Using a sharp razor blade cut small (~ 1 $mm^3$) gelatin blocks with the aid of a stereo-microscope (see Note 8) while taking good care not to dry the sample during the cutting procedure. Rapidly proceed to the next step.
12. Submerge the small blocks in 2.3 M sucrose solution, and slowly rotate tubes overnight at 4°C (see Note 9).
13. Carefully retrieve the specimen blocks from the 2.3 M sucrose solution, using stainless biology tweezers # 5, and transfer to the top of the aluminum pins.
14. Remove the excess sucrose by gently touching with a piece of filter paper while leaving enough sucrose to glue the block to the pin.
15. Position the specimen in the best orientation for sectioning (see Note 10).
16. Rapidly freeze pins by dipping in liquid nitrogen and store in liquid nitrogen (see Note 11).
17. The formvar film is made on a microscope slide. The surface is cleaned by wiping the slide with a chamois and next with lens paper. The Formvar solution is poured gently without forming air bubbles in an adapted separating funnel provided with two stop-cock valves in its outlet. The upper one is used to regulate the flow, the second one to switch on or off. The level of the Formvar solution in the funnel is about 2/3 of the length of the slide. The slide is carefully placed in the funnel with a pair of tweezers and the top of the funnel is covered to reduce evaporation of the chloroform. Then the lower valve is opened with one turn. The upper valve is set in such a way that the

Formvar/chloroform solution flows out of the funnel in 10s. Remove the slide from the funnel after 30 s and loosen the film at the edges of the slide by cutting the film 0.5 mm from the edges with a razor blade. Fill a beaker with clean distilled water and slowly bring the slide in vertical position into the water. The film will float on the water surface. The thickness of the film should be about 50 nm, which will give it a silver-white appearance on the water surface. The grids can be placed on the film, avoiding irregularities in the film. Once the film is covered with grids, a microscope slide provided with a sticker is gently pressed on the film into the water. The film with grids will attach to the sticker. Be sure to submerge the slide far enough into the water so that the film is totally submerged before the slide can be removed from the water and dried.

18. Cut semithin (200–300 nm) or ultrathin cryosections (60–65 nm) with a diamond knife using the cryo-ultramicrotome. The temperature of sample, knife and chamber are set at −100°C for semithin sections, and −120°C for ultrathin cryosections, respectively. The cutting speed is set at 1–2.5 mm/s.
19. Collect ribbons of flat cryosections on a droplet of a 1:1 mixture of 2.5 M sucrose and 2% methylcellulose and transfer to formvar film-coated, carbon stabilized, gold finder EM grids.
20. Store the cryosections at 4°C, for up to 1 month.

### 3.3. Immunofluorescence and Immunogold Labeling for HDO-CLEM

Every step of the labeling procedure is performed by floating the section-side grid on drops of the different solutions placed on a strip of parafilm. Unless otherwise indicated, all steps are performed at room temperature.

#### 3.3.1. Immunolabeling (See Note 12)

1. Sequentially incubate grids with 2% gelatin in PBS for 20 min at 37°C, to remove the methylcellulose coat, with 1% BSA in PBS for 10 min at 37°C, to quench nonspecific protein binding sites, and then with rabbit anti-calreticulin antibody diluted in 1% BSA in PBS for 30 min.
2. Rinse samples in 1% BSA in PBS.
3. Incubate for 45 min with the Cy3-conjugated donkey anti-rabbit antibody, 40,6-diamidino-2- phenylindole (DAPI), and Protein A-gold (PAG) 10 nm, in 1% BSA in PBS for 20 min.

#### 3.3.2. Double Labeling

1. For double calreticulin/Igμ chain labeling, after the calreticulin labeling in the previous section is complete, fix the sample for 5 min with 1% glutaraldehyde (GA) in PBS in order to prevent any cross-reactivity of antibodies. In a double-labelling procedure GA fixation is required in between the two labeling sequences. GA inactivates free IgG binding sites for protein A that are introduced during the first round of labelling. GA fixation

is less successful in denaturizing IgG molecules when a bridging IgG or IgG-gold is used.

2. Incubate for 30 min with rabbit anti-Igμ chain.
3. Rinse the sample with 1% BSA in PBS
4. Incubate with the Cy2-conjugated donkey anti-rabbit secondary antibody in the presence of DAPI for 45 min.
5. Incubate with PAG 15 nm for 20 min.
6. Fix with 1% glutaraldehyde in PBS for 2 min.
7. Lastly, wash in water.

*3.3.3. Preparing Grids for FLM Observations*

1. Coat formvar film-coated, carbon stabilized, gold finder EM grids with a 200 nm thin layer of methylcellulose (see Note 13).
2. Place the gold finder grid on a 200 μl drop of 50% glycerol in water present on a microscope coverslip (see Note 14).
3. Sandwich the grid between the coverslip and a microscope slide.
4. Sonicate the assembled microscope slide for 10 min in distilled water to remove potential air bubbles in the glycerol medium (see Note 15).

*3.3.4. Retrieval of Grids and EM Preparation*

Once FLM imaging is completed, samples are processed for EM. The described sample preparation is fully compatible with ET.

1. Carefully remove the grids from the glass slide.
2. Extensively wash grids in water to remove glycerol and methylcellulose.
3. To increase contrast for EM analysis, incubate cryosections with 2% uranyl acetate/0.15 M oxalic acid (200 nm semithin cryosections for 5 min, or 60 nm ultrathin cryosections for 10 min), followed by incubation of all cryosections with 0.4% uranyl acetate/1.8% 25 ctp methylcellulose for 5 min.

***3.4. High Data Output (HDO) CLEM***

Cryosections of pelletted cells permit high sensitivity labeling and allow multi-modal (FLM/EM) imaging of an elevated number of investigated structures, cells, or events. Moreover, different consecutive physical sections of the same structure or cell can be retraced in the long ribbon and used for three-dimensional analysis (Fig. 4).

1. *FLM imaging: multichannel image.* Set the CLSM spectral imaging parameters as follows: Excite DAPI using the 405 nm laser diode, and collect fluorescence in the 425–475 nm spectral region; Excite Cy3 using the 543 nm HeNe laser, and collect fluorescence in the 555–655 nm spectral region; Excite the Cy2 dye using the 488 nm Ar laser line, and collect fluorescence in the 500–540 nm spectral region. Use 543 nm HeNe laser to acquire the brightfield image. Use sequential individual-channel scans to prevent spectral crosstalk.

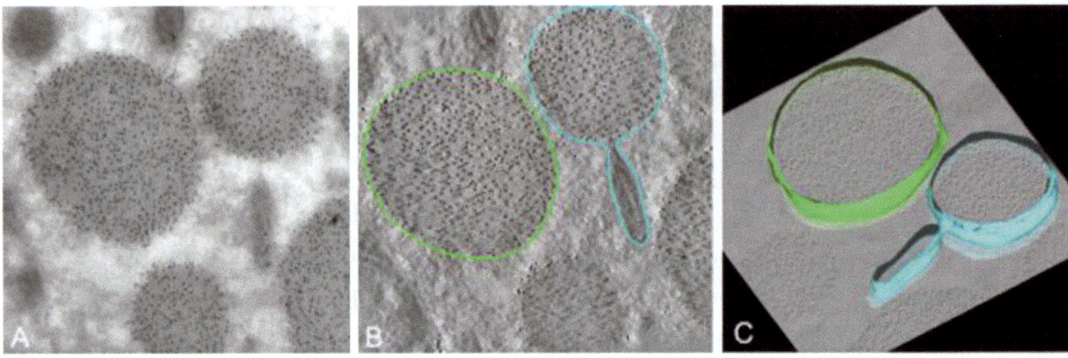


Fig. 4. Electron tomography. (**a**) Tilt series (0 ° image) of RRBs immunogold and immunofluorescence labeled for Igμ protein (single-axis tomogram −65 °, +65 ° with 1 ° increments, image size 1,024 × 1,024, pixel size 2.11 nm). (**b**) Tomograms calculated from the tilt series (**a**); computer-assisted tracing of the membrane contours on the tomogram was used to generate the 3D model of the RRBs. (**c**) 3D modeling and rendering of the RRBs depicted in (**a**). Note the connection of the RRB (*blue*) with rough ER visible in (**a**) and (**b**). Scale bar: 0.5 μm. Reproduced with permission from ref. 4.

2. *FLM imaging: regions of interest identification.* Collect multichannel (DAPI/Cy3/brightfield or DAPI/Cy2/Cy3/brightfield) partially overlapping images of the whole ribbon of serial cryosections at the maximum field of view while still satisfying the Nyquist sampling criterion (i.e., 211 μm × 211 μm, 2,048 × 2,048 pixels, in our CLSM) (see Note 16). The set of images can be used to produce a mosaic of the entire ribbon. A high-quality mosaic can be obtained using the MosaicJ plugin (21). The plugin requires a set of overlapping images of the entire ribbon, along with their user-provided coarse mosaic. The solution then refines this initial mosaic in a fully automatic fashion. The mosaic can be used to identify regions of interest (ROIs) to successively retrace for the EM imaging. The ROI can be a single cell, structure, or event, or can be a consecutive series of physical sections of the same cell or structure (serial ROI). The pattern of the nuclei helps to identify the physical section of the same cell, and thereby the same ROI in different cryosections.
3. *FLM imaging: high-quality.* The ROIs previously identified in the mosaic may be further imaged using different settings. If the pixel sampling used during the imaging for the ROI identification does not satisfy the Nyquist criterion (see Note 16) the image obtained does not preserve the finer details of the structures. As a consequence, any process (e.g., registration and segmentation) performed on this image will be compromised. With the help of the ribbon mosaic all the ROIs are retraced and acquired using a pixel size of 50–25 nm (see Note 16).
4. *FLM image processing: registering serial ROIs.* The TurboReg plugin is used to align and register the series of multichannel (DAPI/Cy3 or DAPI/Cy2/Cy3) images representing the

consecutive physical sections of a cell or structure. Registration is performed using a consecutive pair of images (a reference image and a source image): the second image (source) is registered to the first image (reference) and the result is used as the new reference image for the third image, and so on. Nuclei patterns are used to specify the landmarks for the registration of the respective channel. Successively, the same transformation that is applied to the nuclei channel will be applied to the remaining channels. The "rigid body rotation" or "affine" registration type (22) and the automatic mode of Turboreg are used most of the time. Occasional dramatic differences in the nuclei pattern occur between consecutive cryosections, and thus require the manual mode of registration, disengaging the automatic refinement procedure.

5. *FLM image processing: 3D surface rendering.* Surface rendering is a method used to display a 2D projection of an iso-surface representing a 3D data set (Fig. 3). The first step for surface rendering is the extraction of the surface (iso-surface) from the 3D data set at the boundary between the background and the structure to display. The iso-surface is usually approximated by a set of polygonal meshes (in our case triangles). The Mesh Maker plugin implements the Marching Cube algorithm, one of the most common algorithms used to extract iso-surfaces from a 3D data set (23). The algorithm requires a value (iso-value) that specifies if a pixel in the image belongs to the background or to the structure to display. A user-friendly interface is developed to facilitate the choice of the iso-value and to introduce the sampling interval (voxel size) of the 3D data set (see Note 17). A separate iso-surface has to be extracted for each channel (nuclei and Russell bodies, for example) of the 3D data set. Given that the iso-surfaces can be easily displayed using the Mesh Viewer plugin, up to four different iso-surfaces can be viewed simultaneously. The plugin provides different navigation functions, like zooming, translation and rotation.
6. *EM imaging: single mesh mosaic.* EM high-resolution mosaic single mesh maps of serial cryosections containing Russell Bodies are obtained using the "multiple image alignment" (MIA) function of the Olympus SYS iTEM software embedded in the Megaview camera. A more time-consuming manual acquisition can also be performed.
7. *EM imaging: ROI retracing.* The ROIs within cryosections are manually tracked with the help of the EM and FLM mosaics. Number and letter landmarks on the grid mesh, and the nuclei pattern, allow an easy recognition of the ROIs. The number and letter landmarks present within the grid mesh are visualized in the brightfield channel of the FLM mosaic. EM digital image or EM micrographs of the ROIs are recorded. For nanoscale three-dimensional analysis ET imaging is also performed.

8. *CLEM image processing: overlay.* A CLEM overlay image is obtained by rescaling and registering the FLM image to the EM images. Rescaling of the FLM image is obtained using basic function of ImageJ (Image→Scale). The scale factor is given by the ratio of EM image pixel size and FLM image pixel size. Registration is obtained using manual mode and "rigid body rotation" as a function of the Turboreg plug-in. FLM and EM images are set as source and reference, respectively. Superimposition of the grayscale EM image and color-coded FLM image is obtained using the basic function of ImageJ (Plugins→Color Functions→Color Composite Merge).
9. *Electron tomography.* If required, tomograms of selected ROIs can be generated using the IMOD software package (20). PAGs introduced during immunolabeling (see Subheadings 3.3.1 and 3.3.2) are used as fiducial markers (Fig. 5).

### 3.5. Hybrid Morphometry: A Systematic Error-Based Approach

Electron microscopy provides detailed information about cellular ultrastructure, membrane shapes and organelle architecture. However, certain cellular structures are rare and, therefore, difficult to locate by EM within the enormous volume of a cell or group of cells. Even if such structures can be located occasionally, a meaningful quantitative analysis requires observation of multiple copies. FLM is ideal for the localization of rare and dispersed structures within the cell, but it suffers from poor spatial resolution and, therefore, it lacks fine structural information. These complementary advantages and disadvantages set the motivation for hybrid morphometry CLEM approaches combining large data sets (FLM) and detailed morphometry information (EM). In summary, statistical morphometric features are inferred from a large FLM data set and, subsequently, a small subset of the data is correlated with EM data to evaluate the quality of the FLM analysis. Because of the higher spatial resolution of EM with respect to FLM, when a morphometric property of the very same object is measured by the two modalities, the EM value can be considered as the "real" one, as it is essentially unaffected by error introduced by poor spatial resolution. Broadly speaking, comparing the FLM and EM morphometric values measured on a cohort of correlated structures allows an estimate of the "systematic error" introduced by the FLM. Such a systematic error can be used to better evaluate and weight the results obtained by the FLM morphometric analysis performed on a statistical relevant (large) data set. Here, we describe step-by-step the hybrid morphometric analysis to study the surface area of smooth and rough Russell bodies.

1. *FLM image analysis: surface area estimation.* FLM imaging misses the unlabeled surrounding cellular structures. In most cases this is an undesirable property, but becomes an important characteristic used for automatic surface-area estimation.

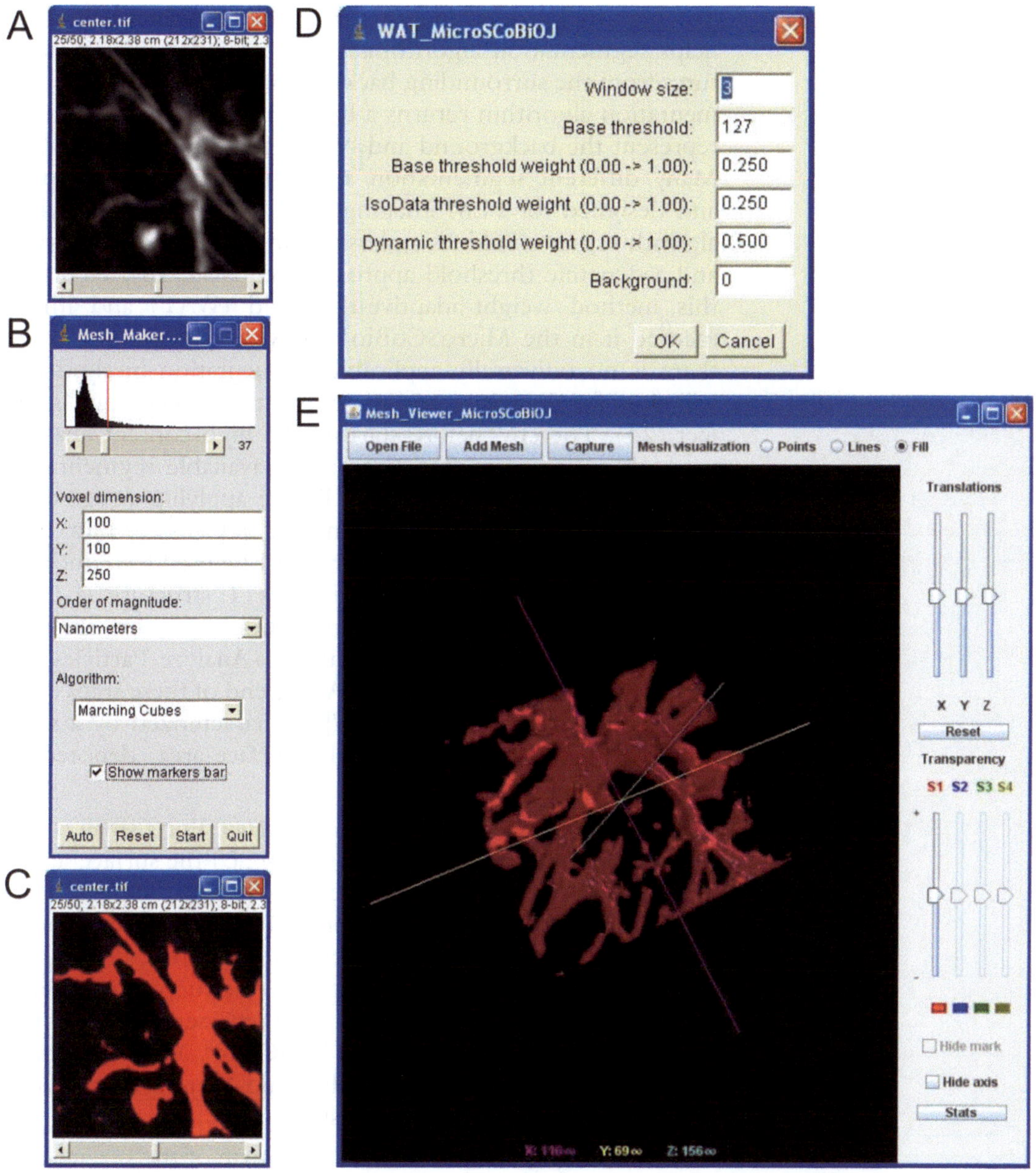


Fig. 5. MicroSCoBioJ interface. (**a**) Grayscale FLM 3D stack containing the cellular structure to display. (**b**) Mesh Maker MicroSCoBioJ user-friendly interface. Using the following menu one can choose: segmentation threshold, voxel dimensions, and algorithm (marching cubes or marching tetratrahedra). (**c**) Binary 3D stack (*black*: background, *red*: structure) obtained by the segmentation algorithm included in the Mesh Maker plugin and successfully used to generate the iso-surface mesh. Note that the iso-surface mesh calculation might take several minutes depending on the size of the 3D stack (up to 1 h for very large 3D stack); a progress bar will display the percentage of the performed calculation. (**d**) Mesh Viewer MicroSCoBioJ user-friendly interface. It is possible to visualize the iso-surface mesh as points, lines or filled surface. (ES) WAT MicroSCoBioJ user-friendly interface.

The high background to structure contrast of the FLM image helps segmentation algorithms to separate the labeled structures from the surrounding background. In summary, the segmentation algorithm returns a binary image, in which "zero" represent the background and "one," the object or structure. Many different segmentation algorithms have been applied and designed for FLM imaging (24). Here we use a hybrid algorithm that combines a classical single threshold approach and a dynamic threshold approach (see Note 18). We called this method weight-adaptive-threshold (WAT) and implemented it in the MicroSCoBioJ software package. However, there is no universally applicable segmentation method that will work for all images, and no segmentation technique is perfect (24). Thereby, before starting the high-content analysis, we highly recommend to test all the available segmentation algorithms provided by ImageJ. Before applying a segmentation algorithm to raw FLM images, a Gaussian-blur filter (Process→Filter→Gaussian Blur) must be used to reduce the noise (see Note 19). The binary images (1: structure, 0: background) obtained from the segmentation algorithm are used to measure the surface-area (Analyze→Analyze Particles) for each structure of interest (SOI). At the end of these steps every structure of interest $\{SOI_i: i \in U\}$ is characterized by a value representing an estimation of the surface-area, denoted by $SOIi^{FLM}$.

2. *EM image analysis: surface-area estimation.* For a small subset of the structures of interest $\{SOIj: j \in S \subset U\}$ the surface-area is also estimated using the correlated EM images. Manual edge-detection (membrane contour) on EM images is used to measure the surface-area of the structures of interest, denoted by $SOIj^{EM}$. At the end of these steps every structure of interest $\{SOIj: j \in S\}$ is characterized by a value representing an estimation of the surface-area, denoted by $SOIj^{EM}$. Manual segmentation is a time-consuming procedure and, therefore, unrealistic when dealing with thousands of structures.

3. *Calculation of systematic error.* An estimate of the surface-area of the investigated structure is obtained by the average of the FLM measurements, $SA = \sum_{i \in U} SOI_i^{FLM}/N$, where $N = \#U$ denotes the number of structures analyzed. Even if the estimation of the surface-area system is obtained on a large data set one has to keep in mind that FLM has poor spatial resolution compared to EM. To understand the accuracy of the surface-area estimate *SA* we compute the average of the relative errors assuming the EM measures the true values: $E = \sum_{j \in S} (SOI_j^{FLM} - SOI_j^{EM})/(SOI_j^{EM} \times M)$, where $M = \#S$ denotes the number of structures correlated. Generally, the value *E* can be

considered as an index of the systematic error introduced using FLM for the morphometric analysis.

## 4. Notes

1. ImageJ is a public domain Java image processing program. It can display, edit, analyze, process, save and print many different types of microscopy images. ImageJ was designed with an open architecture that provides extensibility via Java plug-ins.
2. MicroscoBioJ is a collection of three plug-ins designed to work in the ImageJ software package: (1) Mesh Maker computes triangle meshes from a three-dimensional (3D) fluorescence microscope data set; (2) Mesh Viewer is a 3D viewer which is able to visualize the triangle meshes created by MeshMaker; (3) weight-adaptive-threshold (WTA), performs a segmentation algorithm on a fluorescence microscope image. The Mesh Viewer plugin requires the installation of the Java3D library, release 1.4.0_01 (http://java3d.java.net/binary-builds-old.html).
3. TurboReg is written as a plugin for ImageJ. The purpose of this plugin is to register—i.e., to align or to match—two images, one of them being called the source image and the other the target image. Three alignment modes are available: manual, automatic, and batch. In all three cases, the user is given the opportunity to interactively specify some landmarks, which establishes the initial correspondence between the images. In the automatic and batch modes, the landmarks of the source image are then automatically refined to better match those of the target image. In the manual mode, this automatic refinement procedure is disengaged.
4. MosaicJ is a plugin designed to work in the ImageJ software package. The purpose of this plugin is to facilitate the assembly of a mosaic of overlapping individual images. It provides a semi-automated solution where the initial rough positioning of the images must be performed by the user, and where the final delicate adjustments are performed by the plugin. The MosaicJ plugin requires that the plugin TurboReg is installed.
5. IMOD is a set of image processing, modeling and display programs used for tomographic reconstruction and for 3D reconstruction of EM serial sections and optical sections. The package contains tools for assembling and aligning data within multiple types and sizes of image stacks, viewing 3-D data from any orientation, and modeling and display of the image files.
6. Chloroform is highly volatile, and therefore the formvar solution must always be kept within sealed containers.

7. Aldehyde-based fixative can modify the antigen configuration and reduce the antibody binding capabilities. Therefore, labeling efficiency in immuno-EM can vary depending on the fixation protocol used relative to the specific antigen studied and antibody used. Thus, optimization of fixation conditions for a given antibody in a non-CLEM setup is highly recommended before starting CLEM studies. In our experience, for CLEM studies it is advisable to use low concentrations of glutaraldehyde in the fixative (from 0.025% to 0.2%). These concentrations often result in optimal ultrastructural preservation and limited autofluorescence, allowing easier confocal imaging of flat ribbons of semithin cryosections, and good antigen preservation.
8. Each block should be small enough to be accommodated on the tip of an aluminum pin for cryo-ultramicrotomy (about 1 $mm^3$).
9. Ice crystal formation during freezing of the specimens will damage the ultrastructure of the cell and make sectioning difficult, since good cryosections can be cut only from vitreous frozen blocks.
10. The mounting procedure should be as rapid as possible. Water evaporating from the specimen during air exposure results in an increase in sucrose concentration so that sucrose crystals may form. It is advisable to perform this transfer step with a dissection microscope in a cold room. If a suitable cold room is unavailable, use the dissection microscope at room temperature and mount the blocks rapidly using a box filled with ice.
11. Use cryovials for storage, with the caps punched to make a hole to prevent high internal pressure. Cryovials are then stored in cryoboxes in liquid nitrogen.
12. Use of the rabbit anti-calreticulin antibody labels the endoplasmic reticulum, and use of the rabbit IgG anti-Igμ chain labels the two forms of Russell Bodies (round shaped and convoluted).
13. The methylcellulose coat plays different roles. It protects the section during FLM imaging, ensuring a high rate of success for the subsequent EM imaging session, and protects the formvar film-coating and cryosections during the disassembling of the grid from the temporary FLM slide.
14. Gold grids are preferred to other materials to avoid changes in the spectral properties of DAPI.
15. During FLM imaging, the section side of the EM grid is facing toward the objective. The glycerol/water layer between the cryosections and the coverslip ensures optimal imaging conditions, avoiding potential spherical aberration induced by a mismatch of refractive index.

16. Pixel size (or sampling distance) for FLM imaging has to be chosen according to the Nyquist sampling criterion. This criterion determines the minimal sampling distance needed to capture all the spatial information from the microscope into the image. When the sampling distance is larger than the Nyquist distance, spatial information about the image is lost. As a rule of thumb, the Nyquist distance equals half of the spatial resolution of the system. For our confocal system configuration a spatial resolution of ~200 nm is expected, thereby a pixel size of 100–80 nm is recommended in the normal case. Using a smaller pixel size, no additional spatial information is afforded, and the image is said to have been oversampled. The extra pixels do not theoretically contribute to the spatial resolution, but can often help improve the accuracy of feature measurements taken from the image. To ensure adequate sampling for successive image processing, a pixel size equal to a third of the expected spatial resolution is suggested.
17. The voxel size of the FLM 3D data set is given laterally by the pixel size of the FLM image and axially by the physical section of the cryosection.
18. The algorithm we used is based on the concept of threshold. A threshold value is computed for each pixel of an object image. If the pixel intensity is higher than the threshold, the pixel belongs to the object. The IsoData (25) algorithm identifies a global threshold for all pixels of an image, using statistical properties of the image histogram. Unfortunately, this algorithm is not able to accommodate changes in the distribution of different fluorescence concentrations in a structure. Chow and Kaneko (26) solved this problem but dynamically changing the threshold over the image: each local pixel threshold is obtained by statistical investigation of the intensity values of the local neighbourhoods of such pixel. The statistic that is most appropriate depends largely on the input image: the more complex is the statistic, the better the results. For this work we combined the IsoData threshold ($\mathrm{Th}_{\mathrm{IsoD}}$), with the dynamic threshold ($\mathrm{Th}_{\mathrm{d}}$), to find a weight -adaptive threshold $\mathrm{Th}_{\mathrm{wa}}(x, y) = w_{\mathrm{IsoD}}\mathrm{Th}_{\mathrm{IsoD}} + w_{\mathrm{IsoD}}\mathrm{Th}_{\mathrm{d}}(x, y) + w_{\mathrm{base}}\mathrm{Th}_{\mathrm{base}}$ with the sum of the weights $w$ equals to one. $\mathrm{Th}_{\mathrm{base}}$ can be used to increase or decrease the final threshold on the base of prior information. Moreover, further information, such as background, can be directly added to the algorithm.
19. The sigma value for the Gaussian filter has to be chosen according to the imaging conditions. In particular, the sigma value has to be in the order of the resolution of the FLM system. In such a regime, only the frequencies dominated by noise are eliminated in the image, while frequencies containing information about the structure are preserved. Assuming a pixel size of

50 nm and a resolution of ~200 nm, a sigma of four pixels is suggested. Alternatively, it is possible to use a deconvolution algorithm (27) instead of the Gaussian filter. These algorithms simultaneously reduce noise, blurring and enhance the contrast of the images. It has been demonstrated that segmentation and edge-detection algorithms applied on deconvolved images leads to superior results (28).

## Acknowledgments

The Authors thank all members of Centro di Ricerca MicroSCoBio for support and discussions.

### References

1. Hell SW (2007) Far-field optical nanoscopy. Science 316:1153–1158
2. Watanabe S, Punge A, Hollopeter G et al (2011) Protein localization in electron micrographs using fluorescence nanoscopy. Nat Methods 8:80–84
3. Vicidomini G, Gagliani MC, Canfora M et al (2008) High data output and automated 3D correlative light-electron microscopy method. Traffic 9:1828–1838
4. Vicidomini G, Gagliani MC, Cortese K et al (2010) A novel approach for correlative light electron microscopy analysis. Microsc Res Tech 73:215–224
5. Kukulski W, Schorb M, Welsch S et al (2011) Correlated fluorescence and 3D electron microscopy with high sensitivity and spatial precision. J Cell Biol 192:111–119
6. Sartori A, Gatz R, Beck F et al (2007) Correlative microscopy: bridging the gap between fluorescence light microscopy and cryo-electron tomography. J Struct Biol 160:135–145
7. Ellinger A, Meisslitzer-Ruppitsch C, Rohrl C, Neumuller J, Pavelka M (2009) Photooxidation technology for correlated light and electron microscopy. J Microsc 235:322–335
8. Gaietta G, Deerinck TJ, Adams SR et al (2002) Multicolor and electron microscopic imaging of connexin trafficking. Science 296:503–507
9. Grabenbauer M, Geerts WJC, Fernadez-Rodriguez J et al (2005) Correlative microscopy and electron tomography of GFP through photooxidation. Nat Methods 2:857–862
10. Takizawa T, Suzuki K, Robinson JM (1998) Correlative microscopy using FluoroNanogold on ultrathin cryosections. Proof of principle. J Histochem Cytochem 46:1097–1102
11. Brown E, Verkade P (2010) The use of markers for correlative light electron microscopy. Protoplasma 244:91–97
12. van Rijnsoever C, Oorschot V, Klumperman J (2008) Correlative light-electron microscopy (CLEM) combining live-cell imaging and immunolabeling of ultrathin cryosections. Nat Methods 5:973–980
13. Shu X, Lev-Ram V, Deerinck TJ et al (2011) A genetically encoded tag for correlated light and electron microscopy of intact cells, tissues, and organisms. PLoS Biol 9:e1001041
14. Tokuyasu KT (1980) Immunochemistry on ultrathin frozen sections. Histochem J 12:381–403
15. Slot JW, Geuze HJ (2007) Cryosectioning and immunolabeling. Nat Protoc 2:2480–2491
16. van Donselaar E, Posthuma G, Zeuschner D, Humbel BM, Slot JW (2007) Immunogold labeling of cryosections from high-pressure frozen cells. Traffic 8:471–485
17. Abramoff MD, Magalhaes PJ, Ram SJ (2004) Image processing with image. J Biophotonics Int 11:36–42
18. Mattioli L, Anelli T, Fagioli C et al (2006) ER storage diseases: a role for ERGIC-53 in controlling the formation and shape of Russell bodies. J Cell Sci 119:2532–2541
19. Russell W (1890) An address on a characteristic organism of cancer. Br Med J 2:1356–1360

20. Kremer JR, Mastronarde DN, McIntosh JR (1996) Computer visualization of three-dimensional image data using IMOD. J Struct Biol 116:71–76
21. Thevenaz P, Unser M (2007) User-friendly semi-automated assembly of accurate image mosaics in microscopy. Microsc Res Tech 70:135–146
22. Thevenaz P, Ruttimann UE, Unser M (1998) A pyramid approach to subpixel registration based on intensity. IEEE Trans Image Process 7:27–41
23. William EL, Harvey EC (1987) Marching cubes: a high resolution 3D surface construction algorithm. Comput Graphics 21:163–168
24. Dima AA, Elliott JT, Filliben JJ et al (2011) Comparison of segmentation algorithms for fluorescence microscopy images of cells. Cytometry A 79:545–559
25. Ridler T, Calvard S (1978) Picture thresholding using an iterative selection method. IEEE Trans Syst Man Cybernet 8:629–632
26. Chow CK, Kaneko T (1972) Automatic boundary detection of the left ventricle from cineangiograms. Comput Biomed Res 5: 388–410
27. Bertero M, Boccacci P, Desidera G, Vicidomini G (2009) Image deblurring with Poisson data: from cells to galaxies. Inverse Probl 25(12)
28. Vicidomini G, Boccacci P, Diaspro A, Bertero M (2009) Application of the split-gradient method to 3D image deconvolution in fluorescence microscopy. J Microsc 234:47–61

# Chapter 24

# Correlative Optical and Scanning Probe Microscopies for Mapping Interactions at Membranes

## Christopher M. Yip

## Abstract

Innovative approaches for real-time imaging on molecular-length scales are providing researchers with powerful strategies for characterizing molecular and cellular structures and dynamics. Combinatorial techniques that integrate two or more distinct imaging modalities are particularly compelling as they provide a means for overcoming the limitations of the individual modalities and, when applied simultaneously, enable the collection of rich multi-modal datasets. Almost since its inception, scanning probe microscopy has closely associated with optical microscopy. This is particularly evident in the fields of cellular and molecular biophysics where researchers are taking full advantage of these real-time, in situ, tools to acquire three-dimensional molecular-scale topographical images with nanometer resolution, while simultaneously characterizing their structure and interactions though conventional optical microscopy. The ability to apply mechanical or optical stimuli provides an additional experimental dimension that has shown tremendous promise for examining dynamic events on sub-cellular length scales. In this chapter, we describe recent efforts in developing these integrated platforms, the methodology for, and inherent challenges in, performing coupled imaging experiments, and the potential and future opportunities of these research tools for the fields of molecular and cellular biophysics with a specific emphasis on the application of these coupled approaches for the characterization of interactions occurring at membrane interfaces.

**Key words:** Scanning probe microscopy, Super-resolution microscopy, Fluorescence microscopy, Supported lipid bilayers, Combinatorial microscopy, Total internal reflection microscopy

## 1. Introduction

Recent advances in both scanning probe and optical microscopies have yielded correlative imaging platforms capable of characterizing molecular-scale structures and dynamics. Taking full advantage of the high spatial resolution (nm) and in situ imaging capabilities of scanning probe microscopy and the ever-evolving field of single-molecule optical microscopy, these strategies have contributed significantly to our understanding of intermolecular and interfacial

Alioscka A. Sousa and Michael J. Kruhlak (eds.), *Nanoimaging: Methods and Protocols*, Methods in Molecular Biology, vol. 950, DOI 10.1007/978-1-62703-137-0_24, © Springer Science+Business Media, LLC 2013

dynamics and processes, especially at membrane interfaces. The rationale for combining these platforms lies largely in addressing the key limitations of the individual platforms, especially in the context of biological systems:

1. Scanning probe microscopy (SPM) imaging is predicated on local differences in tip-sample interactions. As such, features that lack topographical contrast are buried within a membrane, or that are mobile on the time-scale of the SPM imaging process (seconds-minutes) can be challenging to resolve.
2. The SPM cannot directly measure molecular conformation or orientation. It simply provides image contrast into which must be fit an appropriate structural model, often based on crystallographic data that may not be available. This statement, of course, has to be couched in the context of SPM imaging of single molecules compared to ordered two-dimensional arrays for which ensemble imaging approaches can be applied to facilitate high-resolution mapping. Postacquisition image processing akin to approaches used in electron microscopy for determining single molecule structures and conformations have been applied with great success by a number of workers (1–4).
3. Correlating topographical features in an SPM image with specific molecular species or complexes can be challenging, relying often solely on relative changes in feature size, shape, or volume.
4. Optical microscopy remains conventionally diffraction-limited although recent advances in approaches such as photo-activation localization microscopy (PALM), stochastic optical reconstruction microscopy, stimulated emission-depletion microscopy (STED), and structured light approaches (5–7), among others (8–10), are now enabling sub-diffraction limited imaging in biological samples, including live cells (11–17).
5. Fluorescence techniques necessarily require the presence of a naturally occurring fluorophore, or more typically, integration of an extrinsic fluorescent moiety, either via genetic engineering in the case of fluorescent proteins, or covalent modification. Such approaches, while powerful, do require careful consideration of the potential that this modification may alter the behavior and characteristics (structure/conformation/interactions) of the host molecule (18).

The functional integration of SPM with optical microscopy and spectroscopy (19–36) affords a particularly powerful variant of "combinatorial microscopy" (37). Typically the SPM is mounted onto a research-grade inverted microscope through an adapter plate that allows one to independently position the SPM scanner and tip at specific locations on the sample of interest. Such platforms afford access to all conventional optical imaging modes, including transmission, reflectance, epi-, and confocal illumination, while enabling the broad suite of scanning probe imaging modalities. We note the obvious requirement of an optically transparent sample.

A wide-field optical-SPM system effectively decouples SPM topography and fluorescence detection. This allows one to use the wider field of view of the optical microscope to optically resolve events that may be occurring outside the SPM's imaging frame and that occur on time scales (seconds to milliseconds) not normally accessible to conventional SPM. This also provides an opportunity to see whether the scanning motion of the SPM tip itself is affecting the sample. It is well known that the scanning motion of the tip can cause localized sample erosion or increased mass transfer into the SPM imaging frame. Accordingly, one typically captures several SPM images of the same region of interest, using different image sizes in order to ensure that there are no tip-induced scan artifacts.

The time required to capture an SPM image can be calculated from the tip scan rate and the image size. For example, ~8.5 min would be needed to acquire a typical 512 pixel × 512 pixel SPM image at a scan rate of 1 Hz. Scanning faster would reduce the time required to acquire an image; however, the range of allowable scan rates for a given experiment and sample reflects a complex interplay of a number of factors, including sample topography, tip characteristics (quality factor, spring constant), feedback control loop (integral and proportional gains), cantilever drive frequency and amplitude if operating in intermittent contact mode, and the size of the area being imaged. This can be compared to the near video-rate or faster capture speeds associated with wide-field fluorescence imaging systems. This temporal consideration has been addressed through the pioneering efforts of the Ando group and others, which have led to the development of innovative high-speed, video-rate SPMs for characterizing cellular and biomolecular structures and dynamics (38–45). Commercial fast-scanning SPM systems have begun to emerge in the market.

The integration of an SPM with an inverted microscope involves adapting the optical microscope stage to accommodate the SPM scan head. With newer SPM platforms incorporating flat plate rather than conventional tube-scanners, this is a straightforward solution allowing full access to the optical light path of the microscope. Such a platform enables facile integration of confocal and total internal reflection fluorescence imaging modes. In our lab, we modified a Olympus Fluoview 500 confocal/TIRF microscope by removing the existing fiber-coupled TIRF illuminator and replacing it with an open-beam configuration comprising multiple laser lines, shutters, and polarization optics (46, 47) (Fig. 1). This flexible architecture is ideal for other emerging high-resolution optical techniques, such as PALM, STED, and STORM, all of which require either precise control over the excitation scheme (PALM, STORM: pulse width, timing, wavelength) or specific shaping of the excitation beam (STED). All of these approaches

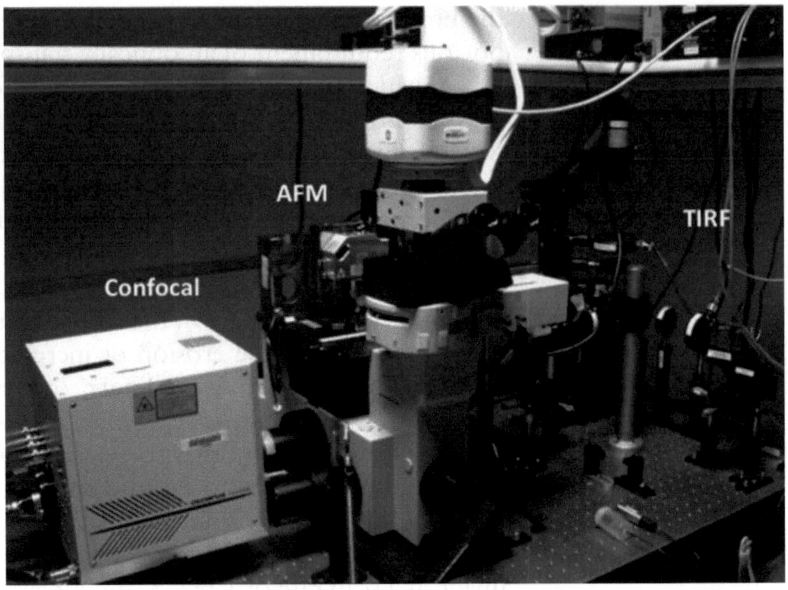


Fig. 1. Our combined confocal-pTIRFM-SPM microscope comprising an Olympus IX-70 inverted microscope, an Olympus Fluoview 500 confocal microscope, a home-built 3-line TIRF a Digital Instruments Bioscope scanning probe microscope and a Photometrics Evolve EMCCD (note that water-cooling is not shown in this image) (47).

can be implemented without adversely affecting operation of the SPM. Indeed, a number of groups have recently reported the use of correlated super-resolution optical-SPM imaging, including studies of lipid bilayers (48), and protein aggregates (49, 50), clearly heralding a new era in truly correlated single-molecule super-resolution imaging.

As will be described later, it is important that the optical and SPM imaging axes be collinear to facilitate image registration between the two imaging modalities. While direct pixel-to-pixel mapping of the optical and SPM fields of view could be accomplished using a stage scanning strategy (19, 23, 51, 52), this is not a particularly convenient solution due to issues with imaging time, SPM feedback control, and the potential for photobleaching.

From an image and feature correlation perspective, features seen in the fluorescence images, especially if sequential imaging is performed, may not be seen in the SPM datasets. This is an important consideration especially if one is contemplating mapping structural rearrangements, such as membrane domain dynamics or aggregation by both topography (SPM), and fluorescence. Conversely, the SPM provides an ideal tool to resolve what structures give rise to a fluorescence signal within a diffraction-limited optical imaging frame, details which are not accessible using conventional optical means.

### 1.1. Total Internal Reflection Fluorescence Microscopy

Total internal reflection fluorescence microscopy (TIRFM) has been shown to be particularly useful for examining surface-specific phenomena, including cell adhesion, endo- and exo-cytosis (53–68). This technique is predicated on the following: When light passes from a material with a high refractive index, $n_2$, into a medium with a lower refractive index, $n_1$, the amount of light transmitted through and reflected by the interface is dependent on $n_1$, $n_2$, and the incident angle at the interface, $\theta_i$. At the critical angle, $\theta_c$, as defined by Snell's law, total internal reflection occurs, resulting in the generation of an evanescent field that extends into the lower index medium.

$$\text{Snell's Law}: \sin(\theta_t) = \frac{n_1}{n_2}\sin(\theta_i)$$
$$\text{at the critical angle}, \theta_c \rightarrow \theta_t = 90^\circ \qquad (1)$$
$$\therefore \theta_c = \sin^{-1}\left(\frac{n_2}{n_1}\right)$$

Due to the exponentially decaying intensity, $I(z)$, of the evanescent field, TIRFM is a powerful approach for characterizing phenomena and structures present within ~200 nm of a surface.

$$I(z) = I_0 e^{-z/2d_p} \qquad (2)$$

where $I_0$ : intensity at the interface;

$z$: distance from the interface

$d_p$ : wavelength dependent penetration depth

$$d_p = \frac{\lambda}{4\pi\sqrt{n_2^2\sin^2\theta - n_1^2}}, \theta \geq \theta_c$$

$\lambda$ : wavelength

$\theta, \theta_c$ : incident and critical angle

Of the three conventional TIRFM platforms (prism-, objective-, and coupled waveguide), all of which use camera-based imaging (46), only the objective- and coupled waveguide approaches can be easily integrated with the SPM, with objective-based TIRF using high-numerical aperture objectives (NA > 1.35) being the most popular configuration (25, 69, 70). Exploiting the polarization angle of the incident light in TIR illumination is an innovative approach for examining the orientation of the fluorophore (and by extension the molecules) relative to the surface normal (71–74). This strategy has been used successfully to examine live cell membrane dynamics (74, 75), cell–cell interactions (76), molecular dynamics at surfaces (77, 78) and in supported lipid bilayers (79, 80), molecular motor dynamics in live cells (81–87), and changes in the local order of the membrane (71, 74, 88).

### 1.2. Coupling SPM and Optical Microscopy

The potential inherent in combining SPM with optical microscopy can been clearly seen in recent studies of various biological structures, interactions, and processes (19–23, 26, 28–31, 33, 36, 69, 70, 89–104). Confocal-SPM has been used to study the effect of local mechanical stimuli by the SPM tip on calcium wave initiation and propagation in live bone cells (105, 106), and the local mechanical properties of live cells (107, 108), while confocal-TIRFM-SPM has been used to study changes in cytoskeletal dynamics and local adhesion of live cells (109–113). Combined TIRFM-SPM is also ideal for single molecule mechanobiology studies, including investigations of forced unfolding (114) and photo-induced conformational changes (91). These latter examples clearly illustrate the potential of these correlative approaches for the field of mechanobiology (115, 116).

Correlated confocal-TIRFM-SPM imaging is well suited for characterizing molecular dynamics in membranes, protein–membrane interactions (26, 27, 35, 117) and structural changes in model phase-segregated membranes induced by ligand association (19, 20, 118–122). We have shown the power of coupled pTIRFM-SPM in our studies of the interactions of the model antimicrobial peptide indolicidin with supported lipid bilayers wherein we were able to clearly resolve peptide concentration- and lipid-dependent rearrangements (27), processes that appeared to be linked with peptide hydrophobicity. We have recently begun to explore the possibility of using pTIRFM in combination with SPM to examine differences or temporal changes in fluorescent membrane probe orientational order and their correlation with SPM topography (123, 124).

### 1.3. Correlated Confocal-SPM Imaging of Molecular Self-Assembly on Phase-Separated Membrane Bilayers (125)

To illustrate the potential of correlated confocal-SPM imaging, we examined the formation of J-aggregates of the small molecule chromophore, pseudoisocyanine (PIC), on supported planar lipid bilayers prepared on mica from either pure 1,2-distearoyl-*sn*-glycero-3-phospho-ethanolamine (DSPE), or a mixed gel-fluid-phase composition comprising 1,2-distearoyl-*sn*-glycero-3-phosphocholine (DSPC) and 1,2-dioleoyl-*sn*-glycero-3-phosphocholine (DOPC). PIC is known to self-assemble into a specific supramolecular architecture known as a J-aggregate, a motif that can be uniquely identified by a bathochromic shift in PIC emission relative to the monomer (126). However, questions remain regarding the critical dimension associated with the onset of J-aggregate behavior, which is thought to be due to exciton propagation and how the structure of the J-aggregate itself impacts this process.

As can be seen in Fig. 2, fusion of SUV suspensions of DSPC/DOPC and DSPE lipids to mica led to the formation of molecularly smooth islands, as determined by in situ SPM. To identify the gel- and fluid-phase domains in the mixed DSPC/DOPC bilayers, Rho-DHPE was used. This Rhodamine-B headgroup-labeled lipid preferentially segregates to the fluid-phase domains in the DSPC-DOPC domains.

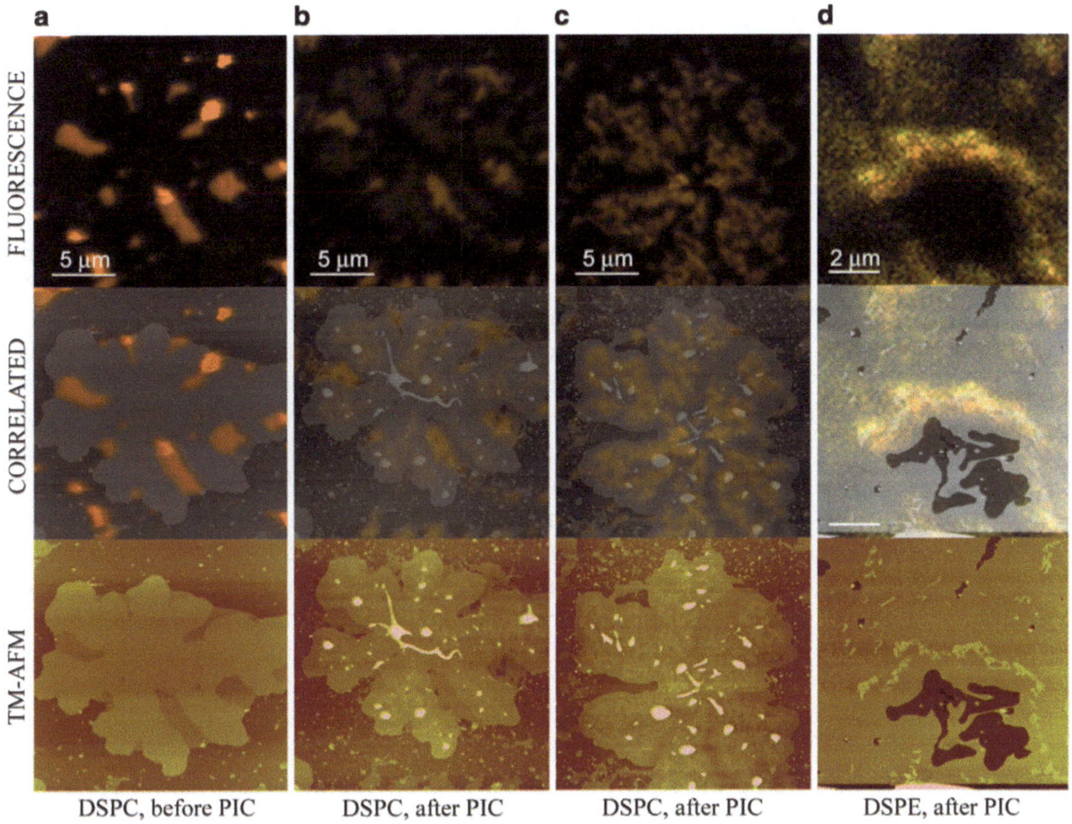


Fig. 2. Combinatorial confocal fluorescence-SPM imaging of pseudoisocyanine (PIC) J-aggregate formation on supported lipid bilayers. (*Top*) Confocal; (*Middle*) Correlated confocal-SPM; (*Bottom*) in situ SPM images showing preferential growth of PIC J-aggregates on (**a**–**c**) DSPC gel-domain of a DSPC-DOPC-Rho-DHPE bilayer, before/after PIC addition. Scale bars: 5 μm; (**d**) DSPE bilayer after PIC addition. Scale bar: 2 μm.

This was confirmed by correlated confocal-SPM imaging, which revealed that the Rho-DHPE labeled fluid-phase DOPC domains were ~5 nm tall while the gel-phase DSPC domains extended ~6.5 nm above the mica surface. The addition of a 100 μl aliquot of a 364 μM aqueous solution of PIC resulted in the formation of small aggregates on the surface of the lipid bilayers with preferential formation of the aggregates on the gel-state DSPC domains for the phase-separated DPSC/DOPC bilayers. This suggested a specific interaction between the PIC molecules and the charge and/or physical state of the gel-phase lipids. To assess whether the PIC molecules had indeed assembled into a specific motif, we then examined the simultaneously acquired confocal microscopy images. These images clearly illustrated that the aggregates were indeed fluorescent and, using hyperspectral imaging, also confirmed that they had in fact self-assembled into the proposed J-aggregate motif on the gel phase domains. Interestingly the same effect was also seen on the single component DSPE bilayer, again suggestive of a templating effect due to the charge and headgroup packing of the DSPE lipids. Close inspection and registration of the confocal and SPM data sets revealed the presence of irregularly shaped nonfluorescent PIC aggregates ~20–100 nm tall and ~50 nm in diameter on the lipid bilayers. These data present an interesting insight into J-aggregate self-assembly. Specifically, these data argue that these small nonfluorescent aggregates are not large enough to facilitate generation of the excitonic signal characteristic of a J-aggregate. Indeed, this suggests that correlated confocal-SPM imaging represents a particularly powerful means of addressing the critical stages of nucleation and growth of supramolecular structures and that mapping fluorescence against topography provides a unique strategy for tracking functional behavior, a characteristic that would be difficult to assess purely from topographical data provided by the SPM.

Combinatorial microscopies based on the integration of scanning probe microscopy with the breadth of optical microscopy tools represent exceptionally powerful approaches for examining molecular phenomena in real-time and with nanometer-scale resolution. This integration is straightforward from a mechanical point of view and the potential of such systems is limited by the imagination of the user. While early adopters of these combinatorial platforms have been restricted in some respects by differences in the spatial and temporal resolution of the two techniques, recent advances in fast-scanning SPM and super-resolution optical imaging are helping to enhance the capabilities and potential of these tools.

## 2. Materials

For our specific confocal-TIRF-SPM platform, we use the following:

1. HEPES buffer, pH 7.4.
2. Ultra-pure Milli-Q water (Millipore).
3. Mica Grade V-1 0.275–0.325 mm thick, 25 mm diameter discs (Structure Probe Inc, www.2spi.com).
4. Inverted optical microscope (IX-70) equipped with 4×, 10×, 20×, and a 60× 1.45 NA PlanApo objectives (Olympus Canada).
5. Olympus Fluoview 500 confocal microscope (Olympus Canada).
6. Optical train for TIRF imaging, as described previously (47), including free-space launched 532, 473 and 405 nm diode-pumped solid-state (DPSS) lasers (Extreme Lasers, USA; Power Technology, USA), vibration isolation optical table (Newport Corporation, USA), optical stages, dichroic mirrors, visible wavelength Glan-Taylor linear polarizers prism, achromatic half-wave plates (Thorlabs, USA), Uniblitz mechanical shutters

(Vincent Associates, USA), and motorized rotation mount and ESP motion controller (Newport Corporation, USA).

7. Nanoscope IIIA scanning probe microscope system equipped with a Dimension scan head, phase extender and signal access modules, and a glass cantilever holder for fluid imaging (Digital Instruments Bioscope (now Bruker Nano), California).
8. SPM tips: model DNP-S (for fluid imaging) (Bruker Nano, California).
9. ImageJ (http://rsbweb.nih.gov/ij/).
10. μ-Manager microscope automation software (http://www.micro-manager.org; http://valelab.ucsf.edu/~MM/MMwiki/).
11. Evolve 512 EMCCD camera equipped with a water-cooling system (Photometrics http://www.photometrics.com) (see Note 1).
12. Willco glass bottom dishes: GWSt-5030 (Willco Wells, The Netherlands, http://www.willcowells.com).
13. UV-curable optical adhesive: Norland #63 (NOA63, Cranbury, NJ).
14. UV curing lamp: Bondwand (Edmund Scientific, New Jersey).
15. All lipids were purchased from Avanti Polar Lipids (Alabaster, AL, USA) in either chloroform solution or in powder form: 1,2-distearoyl-*sn*-glycero-3-phosphocholine (DSPC), 1,2-dioleoyl-*sn*-glycero-3-phosphocholine (DOPC), 1,2-distearoyl-*sn*-glycero-3-phospho-ethanolamine (DSPE), and stored in chloroform at a stock concentration of 10 mg/ml. Lissamine-Rhodamine-B headgroup-labeled dihexadecyl phosphatidylethanolamine (Rho-DHPE) was purchased from Molecular Probes (Invitrogen) and stored in chloroform at a stock concentration of 1–10 mg/ml.

## 3. Methods

### 3.1. Mica-Glass Substrate Preparation Method for Confocal-TIRF-SPM Imaging

1. Cleave a 1″ round piece of mica to expose a fresh surface, either with a razor blade or with adhesive tape. There should be no visible step edges on the surface of the mica.
2. Place ~1.5 μl of a UV-curable optical adhesive in the middle of a clean round 1″ glass coverslip or a 50 mm Willco-dish. Ensure there are no bubbles in the adhesive droplet.
3. Carefully place the thin mica sheet on the glass surface with the adhesive drop. Press down lightly on the middle of the mica sheet with a gloved finger to help spread the adhesive. The adhesive drop should spread to a very thin layer between these two materials.

4. Place the mica-glass substrate under a UV lamp ($\lambda = 354$ nm) for ~15 min to permanently cure the adhesive.
5. The substrate is now ready for liposome vesicle fusion to form a supported lipid bilayer (see Note 2).

### 3.2. Lipid Bilayer Preparation by Vesicle Fusion (See Note 3)

1. Dissolve the desired molar ratio of lipids and fluorescent probe in chloroform in a clean glass test tube.
2. Remove the chloroform by vacuum rotary evaporation to afford a dry film.
3. Rehydrate the dry lipid film in 10 mM aqueous pH 7.4 HEPES buffer with ultra-pure Milli-Q water to a final liposome solution concentration of ~1 mM.
4. Small unilamellar vesicles (SUVs) are formed from the MLV suspension by sonicating the hydrated mixture for 20 min. The SUV suspensions can be stored at 4 °C for up to 2 weeks prior to use.
5. To form a bilayer, 100 μL of the SUV suspension and 400 μL HEPES buffer are sequentially deposited onto a freshly cleaved mica surface in the confocal-TIRF-SPM sample chamber and left to stand for 3 min. Excess vesicles are removed from the sample by flushing the chamber ten times with 1 mL aliquots of HEPES buffer.
6. External agents and their perturbing effects on membrane morphology and dynamics can be studied by direct injection of the agent dissolved in aqueous buffer during confocal-TIRF-SPM imaging.

### 3.3. Confocal-TIRF-SPM Imaging Protocol

The following instructions are specific to an older Digital Instruments Bioscope system equipped with a Dimension scan head mounted on an Olympus IX-70 base. The Olympus base is equipped with a custom-built multi-line laser-based TIRF illumination system and an Olympus Fluoview 500 confocal microscope. From the SPM perspective, we would note that each vendor has specific instructions for setting up their SPM systems. These instructions are also specific for performing SPM imaging in fluid.

1. Mount a new cantilever into the SPM fluid cell and fit the fluid cell to the SPM scan head.
2. Place the SPM scan head into its support frame on the inverted microscope. Ensure that there is sufficient clearance between the scan head and the microscope objective to prevent accidental contact between the two. At this point, it is usually easier to align the SPM- and optical axis of the microscope without the sample being present.
3. With a low magnification (4× or 10×) air objective, focus on the tip and roughly align the SPM laser in air on the SPM tip.

Using the SPM stage positioners, position the SPM so that the SPM tip is aligned with the center of the optical microscope's field of view. If the optical microscope is equipped with a photoframe reticule, try to center the SPM tip in the reticule. If your microscope is so equipped, re-center the SPM tip using higher magnification objectives. If necessary, you may have to lower the SPM scan head so that it is in the focal plane of the objective.

4. For TIRF imaging, rotate the TIRF objective into place on the optical microscope nosepiece (see Note 4). Re-center the SPM tip in TIRF objective's field of view. This centering should be performed in air and in the absence of immersion oil. For epifluorescence or confocal imaging, the same procedure can be used (see Note 5).
5. Align the SPM laser on the SPM tip and center the reflected laser spot on the SPM's photodetector. Note that this is done in air and it will be necessary to realign the laser once the SPM tip is immersed in fluid (see Note 6).
6. Raise the SPM scan head so that there is enough space between the scan head and the objective to insert the sample.
7. Place a drop of immersion oil on the top lens of the imaging objective.
8. Secure the sample chamber to the microscope stage and focus on the substrate–solution interface using brightfield illumination. Surface defects or calibration spheres adsorbed onto the sample surface may facilitate this step. Do not focus on the SPM tip.
9. Configure the optical system for TIRF illumination, including the angle of incidence of the laser beam. This process will depend on the specifics of the TIRF platform.
10. Translate the sample to the region of interest using the optical microscope stage controls. Since the SPM stage is centered on the microscope sample stage, via step 3, the SPM tip should remain centered on the optical field of view.
11. Gradually lower the SPM tip towards the sample surface using the SPM scan head approach motor drive until the SPM tip is almost in focus. Since the optical microscope is focused on the sample surface, this procedure effectively brings the tip into the focal plane of the microscope. Realign the SPM laser as necessary.
12. The sample is now ready to be imaged by simultaneous or sequential confocal-TIRF-SPM microscopy. Periodic realignment of the SPM laser and/or refocusing the objective may be necessary to compensate for thermal and/or mechanical drift during extended-duration imaging. Care must be exercised to avoid crashing the sample into the SPM tip during refocusing

and withdrawing the SPM tip prior to manual refocusing may be helpful. Commercial autofocus systems may be employed for extended-duration imaging although care must be exercised to ensure physical compatibility with the specific microscope frame and objectives.

### 3.4. Image Collection and Analysis

Image collection can be performed using the manufacturers' SPM and optical imaging software. Since the field of the view of the SPM is significantly smaller than that of the optical microscope, it is important to identify specific fiduciary markers that can be clearly identified in both imaging modes. In the context of membrane bilayers, these may be open membrane-free areas that provide both topographical contrast (a "hole") and lack fluorescent contrast. Fluorescent calibration spheres are also useful.

#### 3.4.1. Scanning Probe Microscopy

Typical imaging parameters for the SPM Digital Instruments Bioscope scanning probe microscope using a Nanoscope IIIA controller, a dual-range Dimension scanner having a maximum lateral scan area of ~90 μm × 90 μm, using 120 μm long oxide-sharpened silicon nitride V-shaped cantilevers (type DNP-S, Veeco) using the Nanoscope 5.31 software:

1. 512 × 512 pixel images at scan rates of between ~1 and 2 Hz using a tip oscillation frequency of ~8.5 kHz.
2. The tip oscillation frequency is cantilever-dependent and should be selected based on the manufacturer's protocol, which typically involves obtaining a power spectrum as a function of frequency and voltage, and identifying an appropriate drive frequency and amplitude. This should be determined once the tip is close to the sample surface and immersed in fluid.
3. All the raw SPM data files should be separated, and then subjected to appropriate flatten and plane-fit filters, as necessary.

#### 3.4.2. Epi-Fluorescence and TIRF Microscopy

On our platform, all epi-fluorescence and TIRF-images are acquired using the open-source microscope automation software μManager (127) using a water-cooled Evolve 512 EMCCD camera (512 × 512 pixels, pixel size: 16 μm × 16 μm) (see Note 7).

#### 3.4.3. Image Analysis

1. Postacquisition image analysis is performed using the Nanoscope software (SPM) or the open-source package public domain software ImageJ (128).
2. Determining the fluorescence and SPM image overlap can be performed using registration plug-ins in ImageJ or other commercial image processing packages. The quality of the registration is highly dependent on the presence of fiducial points present in both the optical and SPM images. This can be facilitated by the use of calibration microspheres.

## 4. Notes

1. TIRFM images are usually acquired with electron-multiplying CCD (EMCCD) digital cameras. These cameras are usually cooled with a heatsink fitted with a small fan. Vibrations from the fan can cause mechanical noise in the SPM images and we would recommend the use of either a water-cooling or a remote fan and ducting system.
2. Depending on the lipids being used for the experiments, it may be necessary to pretreat the mica with a $CaCl_2$ wash to create a positive charge on the mica surface. This will help negatively charged lipids to spread on the mica surface. We have also seen that for some lipids with particularly high phase transition temperatures, it may be necessary to heat the substrate and/or lipid mixtures prior to fusion in order to facilitate spreading and stable bilayer formation.
3. Small unilamellar vesicles can also be formed by extrusion through a polycarbonate membrane. Commercial tools, such as Avanti's Mini-Extruder (Avanti Polar Lipids, Part # 610000) are available.
4. Objective-based TIRFM imaging uses oil immersion high-numerical aperture objectives (NA ≥ 1.35) designed to be aberration-free when used with 170 μm thick glass cover slips.
5. Ideally, the SPM field of view should also be in the center of the optical field of view. This is particularly important for TIRF imaging due to the fact that the nonuniform nature of the evanescent field generated by the TIR condition. The intensity of the evanescent field at the substrate decays following a Gaussian profile in the $x$–$y$ plane of the microscope. Accordingly $I_0$ is a function of $(x, y)$ in the imaging plane and $I(z)$ is actually $I(x, y, z)$.
6. An important consideration in the coupled imaging platform is the possibility of interference due to the laser used in the SPM optical lever detection system. As many of the older SPM platforms employ a ~670 nm laser, which could interfere with the emitted fluorescence from the sample or inadvertently excite fluorophores in the sample, it is important to consider installing band-pass filters at the appropriate locations along the optical emission path. To address this concern, newer SPM platforms are typically equipped with near-IR lasers for the optical detection system although these could complicate experiments involving near-IR excitable fluorophores.
7. For multicolor TIRF microscopy, either a single filter cube equipped with a multi-bandpass dichroic mirror and a multi-bandpass emission filter can be used or several separate filter cubes mounted in the microscope filter turret for each excitation

wavelength may be used. A single filter cube is particularly useful as it eliminates the time required to switch between cubes and prevents pixel-shifting artifacts that can sometimes occur when using multiple filter cubes.

## References


1. Engel A, Schoenenberger CA, Muller DJ (1997) High resolution imaging of native biological sample surfaces using scanning probe microscopy. Curr Opin Struct Biol 7:279–284
2. Pogoryelov D, Reichen C, Klyszejko AL et al (2007) The oligomeric state of c rings from cyanobacterial F-ATP synthases varies from 13 to 15. J Bacteriol 189:5895–5902
3. Muller DJ, Engel A (2007) Atomic force microscopy and spectroscopy of native membrane proteins. Nat Protoc 2:2191–2197
4. Muller DJ, Sapra KT, Scheuring S et al (2006) Single-molecule studies of membrane proteins. Curr Opin Struct Biol 16:489–495
5. Gustafasson MGL (2000) Surpassing the lateral resolution limit by a factor of two using structured illumination microscopy. J Microsc 198:82–87
6. Gustafsson MGL (2005) Nonlinear structured-illumination microscopy: wide-field fluorescence imaging with theoretically unlimited resolution. Proc Natl Acad Sci USA 102:13081–13086
7. Fedosseev R, Belyaev Y, Frohn J, Stemmer A (2005) Structured light illumination for extended resolution in fluorescence microscopy. Optics Lasers Engin 43:403–414
8. Yildiz A, Selvin PR (2005) Fluorescence imaging with one nanometer accuracy: application to molecular motors. Acc Chem Res 38:574–582
9. Toprak E, Enderlein J, Syed S et al (2006) Defocused orientation and position imaging (DOPI) of myosin V. Proc Natl Acad Sci USA 103:6495–6499
10. Park H, Toprak E, Selvin PR (2007) Single-molecule fluorescence to study molecular motors. Q Rev Biophys 40:87–111
11. Schwartz MA (2011) Super-resolution microscopy: a new dimension in focal adhesions. Curr Biol 21:R115–R116
12. Leung BO, Chou KC (2011) Review of super-resolution fluorescence microscopy for biology. Appl Spectrosc 65:967–980
13. Henriques R, Griffiths C, Hesper Rego E, Mhlanga MM (2011) PALM and STORM: unlocking live-cell super-resolution. Biopolymers 95:322–331
14. Toomre D, Bewersdorf J (2010) A new wave of cellular imaging. Annu Rev Cell Dev Biol 26:285–314
15. Thompson MA, Biteen JS, Lord SJ, Conley NR, Moerner WE (2010) Molecules and methods for super-resolution imaging. Methods Enzymol 475:27–59
16. Betzig E, Patterson GH, Sougrat R et al (2006) Imaging intracellular fluorescent proteins at nanometer resolution. Science 313:1642–1645
17. Gould TJ, Verkhusha VV, Hess ST (2009) Imaging biological structures with fluorescence photoactivation localization microscopy. Nat Protoc 4:291–308
18. Shaw JE, Epand RF, Epand RM, Li Z, Bittman R, Yip CM (2006) Correlated fluorescence-atomic force microscopy of membrane domains: structure of fluorescence probes determines lipid localization. Biophys J 90:2170–2178
19. Burns AR (2003) Domain structure in model membrane bilayers investigated by simultaneous atomic force microscopy and fluorescence imaging. Langmuir 19:8358–8363
20. Shaw JE, Slade A, Yip CM (2003) Simultaneous in situ total internal reflectance fluorescence/atomic force microscopy studies of DPPC/dPOPC microdomains in supported planar lipid bilayers. J Am Chem Soc 125:11838–11839
21. Kassies R, Van der Werf KO, Lenferink A et al (2005) Combined AFM and confocal fluorescence microscope for applications in bio-nanotechnology. J Microsc 217:109–116
22. Burns AR, Frankel DJ, Buranda T (2005) Local mobility in lipid domains of supported bilayers characterized by atomic force microscopy and fluorescence correlation spectroscopy. Biophys J 89:1081–1093
23. Kellermayer MSZ, Karsai A, Kengyel A et al (2006) Spatially and temporally synchronized atomic force and total internal reflection fluorescence microscopy for imaging and manipulating cells and biomolecules. Biophys J 91:2665–2677
24. Neugebauer U, Rosch P, Schmitt M et al (2006) On the way to nanometer-sized information of the bacterial surface by tip-enhanced Raman spectroscopy. Chemphyschem 7:1428–1430

25. Shaw JE, Epand RF, Sinnathamby K et al (2006) Tracking peptide-membrane interactions: Insights from in situ coupled confocal-atomic force microscopy imaging of NAP-22 peptide insertion and assembly. J Struc Biol 155:458–469
26. Slade AL, Schoeniger JS, Sasaki DY, Yip CM (2006) In situ scanning probe microscopy studies of tetanus toxin-membrane interactions. Biophys J 91:4565–4574
27. Shaw JE, Alattia JR, Verity JE, Prive GG, Yip CM (2006) Mechanisms of antimicrobial peptide action: Studies of indolicidin assembly at model membrane interfaces by in situ atomic force microscopy. J Struct Biol 154:42–58
28. Lin WC, Blanchette CD, Ratto TV, Longo ML (2006) Lipid asymmetry in DLPC/DSPC-supported lipid bilayers: A combined AFM and fluorescence microscopy study. Biophys J 90:228–237
29. Frankel DJ, Pfeiffer JR, Surviladze Z et al (2006) Revealing the topography of cellular membrane domains by combined atomic force microscopy/fluorescence imaging. Biophys J 90:2404–2413
30. Chiantia S, Kahya N, Ries J, Schwille P (2006) Effects of ceramide on liquid-ordered domains investigated by simultaneous AFM and FCS. Biophys J 90:4500–4508
31. Haupt BJ, Pelling AE, Horton MA (2007) Integrated confocal and scanning probe microscopy for biomedical research. Sci World J 6:1609–1618
32. Johnston LJ (2007) Nanoscale imaging of domains in supported lipid membranes. Langmuir 23:5886–5895
33. Coban O, Burger M, Laliberte M, Ianoul A, Johnston LJ (2007) Ganglioside partitioning and aggregation in phase-separated monolayers characterized by bodipy GM1 monomer/dimer emission. Langmuir 23:6704–6711
34. Coban O, Popov J, Burger M, Vobornik D, Johnston LJ (2007) Transition from nanodomains to microdomains induced by exposure of lipid monolayers to air. Biophys J 92:2842–2853
35. Shaw JE, Epand RF, Hsu JCY et al (2008) Cationic peptide-induced remodelling of model membranes: direct visualization by in situ atomic force microscopy. J Struc Biol 180:563–578
36. Trache A, Meininger GA (2005) Atomic force-multi-optical imaging integrated microscope for monitoring molecular dynamics in live cells. J Biomed Opt 10:064023–064021–064017.
37. Axelrod D, Omann GM (2006) Combinatorial microscopy. Nat Rev Mol Cell Biol 7:944–952
38. Casuso I, Kodera N, Le Grimellec C, Ando T, Scheuring S (2009) Contact-mode high-resolution high-speed atomic force microscopy movies of the purple membrane. Biophys J 97:1354–1361
39. Miyagi A, Tsunaka Y, Uchihashi T et al (2008) Visualization of intrinsically disordered regions of proteins by high-speed atomic force microscopy. Chemphyschem 9:1859–1866
40. Ando T, Uchihashi T, Kodera N et al (2008) High-speed AFM and nano-visualization of biomolecular processes. Pflugers Arch 456: 211–225
41. Ando T, Uchihashi T, Kodera N et al (2007) High-speed atomic force microscopy for observing dynamic biomolecular processes. J Mol Recognit 20:448–458
42. Yokokawa M, Yoshimura SH, Naito Y et al (2006) Fast-scanning atomic force microscopy reveals the molecular mechanism of DNA cleavage by ApaI endonuclease. IEE Proc Nanobiotechnol 153:60–66
43. Kodera N, Kinoshita T, Ito T, Ando T (2003) High-resolution imaging of myosin motor in action by a high-speed atomic force microscope. Adv Exp Med Biol 538:119–127
44. Ando T, Kodera N, Naito Y et al (2003) A high-speed atomic force microscope for studying biological macromolecules in action. Chemphyschem 4:1196–1202
45. Ando T, Kodera N, Takai E et al (2001) A high-speed atomic force microscope for studying biological macromolecules. Proc Natl Acad Sci USA 98:12468–12472
46. Shaw JE, Oreopoulos J, Wong D, Hsu JCY, Yip CM (2006) Coupling evanescent-wave fluorescence imaging and spectroscopy with scanning probe microscopy: challenges and insights from TIRF-AFM. Surf Interface Anal 38:1459–1471
47. Oreopoulos J, Yip CM (2008) Combined scanning probe and total internal reflection fluorescence microscopy. Methods 46:2–10
48. Kuo C, Hochstrasser RM (2011) Super-resolution microscopy of lipid bilayer phases. J Am Chem Soc 133:4664–4667
49. Duim WC, Chen B, Frydman J, Moerner WE (2011) Sub-diffraction imaging of huntingtin protein aggregates by fluorescence blink-microscopy and atomic force microscopy. Chemphyschem 12:2387–2390
50. Cordes T, Strackharn M, Stahl SW et al (2010) Resolving single-molecule assembled patterns with superresolution blink-microscopy. Nano Lett 10:645–651
51. Gradinaru CC, Martinsson P, Aartsma TJ, Schmidt T (2004) Simultaneous atomic-force and two-photon fluorescence imaging of

biological specimens in vivo. Ultramicroscopy 99:235–245

52. Chon JWM, Gu M (2004) Scanning total internal reflection fluorescence microscopy under one-photon and two-photon excitation: image formation. Appl Opt 43:1063–1071
53. Axelrod D, Burghardt TP, Thompson NL (1984) Total internal reflection fluorescence. Ann Rev Biophys Bioeng 13:247–268
54. Reichert WM, Truskey GA (1990) Total internal-reflection fluorescence (tirf) microscopy.1. Modeling cell contact region fluorescence. J Cell Sci 96:219–230
55. Burmeister JS, Truskey GA, Reichert WM (1994) Quantitative-analysis of variable-angle total internal-reflection fluorescence microscopy (Va-tirfm) of cell substrate contacts. J Microsc-Oxf 173:39–51
56. Thompson NL, Lagerholm BC (1997) Total internal reflection fluorescence: applications in cellularbiophysics. Curr Opin Biotech 8:58–64
57. Oheim M, Loerke D, Stuhmer W, Chow RH (1998) The last few milliseconds in the life of a secretory granule—Docking, dynamics and fusion visualized by total internal reflection fluorescence microscopy (TIRFM). Eur Biophys J Biophy Lett 27:83–98
58. Oheim M, Stuhmer W (2000) Tracking chromaffin granules on their way through the actin cortex. Eur Biophys J Biophys Lett 29:67–89
59. Loerke D, Preitz B, Stuhmer W, Oheim M (2000) Super-resolution measurements with evanescent-wave fluorescence excitation using variable beam incidence. J Biom Optics 5:23–30
60. Oheim M, Stuhmer W (2002) Multiparameter evanescent-wave Imaging in biological fluorescence microscopy. IEEE J Quantum Electr 38:142–148
61. Toomre D, Manstein DJ (2001) Lighting up the cell surface with evanescent wave microscopy. Trends Cell Biol 11:298–303
62. Yasushi S, Uyemura T (2002) Total internal reflection fluorescence microscopy for single-molecule imaging in living cells. Cell Struct Funct 27:357–365
63. Axelrod D (2001) Total internal reflection fluorescence microscopy in cell biology. Traffic 2:764–774
64. Moerner WE, Fromm DP (2003) Methods of single-molecule fluorescence spectroscopy and microscopy. Rev Sci Instrum 74:3597–3619
65. Mashanov GI, Tacon D, Knight BAE, Peckham M, Molloy JE (2003) Visualizing single molecules inside living cells using total internal reflection fluorescence microscopy. Methods 29:142–152
66. Axelrod D (2003) Total internal reflection fluorescence microscopy in cell biology Methods Enzymol 361:1–33
67. Wazawa T, Ueda M (2005) Total internal reflection fluorescence microscopy in single molecule nanobioscience. Adv Biochem Eng Biotechnol 95:77–106
68. Schneckenburger H (2005) Total internal reflection fluorescence microscopy: technical innovations and novel applications. Curr Opin Biotech 16:13–18
69. Mathur AB, Truskey GA, Reichert WM (2000) Total internal reflection microscopy and atomic force microscopy (TIRFM-AFM) to study stress transduction mechanisms in endothelial cells. Crit Rev Biom Eng 28:197–202
70. Mathur AB, Truskey GA, Reichert WM (2000) Atomic force and total internal reflection fluorescence microscopy for the study of force transmission in endothelial cells. Biophysi J 78:1725–1735
71. Thompson NL, McConnell HM, Burghardt TP (1984) Order in supported phospholipid monolayers detected by the dichroism of fluorescence excited with polarized evanescent illumination. Biophysi J 46:739–747
72. Axelrod D (1989) Fluorescence polarization microscopy. Methods Cell Biol 30:333–352
73. Timbs MM, Thompson NL (1990) Slow rotational mobilities of antibodies and lipids associated with substrate-supported phospholipid monolayers as measured by polarized fluorescence photobleaching recovery. Biophys J 58:413–428
74. Sund SE, Swanson JA, Axelrod D (1999) Cell membrane orientation visualized by polarized total internal reflection fluorescence. Biophys J 77:2266–2283
75. Letinic K, Sebastian R, Toomre D, Rakic P (2009) Exocyst is involved in polarized cell migration and cerebral cortical development. Proc Natl Acad Sci USA 106:11342–11347
76. Soderholm N, Vielfort K, Hultenby K, Aro H (2011) Pathogenic Neisseria hitchhike on the uropod of human neutrophils. PLoS One 6:e24353
77. Ha T, Glass J, Enderle T, Chemla DS, Weiss S (1998) Hindered rotational diffusion and rotational jumps of single molecules. Phys Rev Lett 80:2093–2096
78. Ha T, Laurence TA, Chemla DS, Weiss S (1999) Polarization spectroscopy of single fluorescent molecules. J Phys Chem B 103: 6839–6850
79. Osborne MA (2005) Real-time dipole orientational imaging as a probe of ligand-protein interactions. J Phys Chem B 109: 18153–18161

80. Tokimoto T, Bethea TRC, Zhou M, Ghosh I, Wirth MJ (2007) Probing orientations of single fluorescent labels on a peptide reversibly binding to the human delta-opioid receptor. Appl Spectrosc 61:130–137
81. Forkey JN, Quinlan ME, Goldman YE (2000) Protein structural dynamics by single-molecule fluorescence polarization. Prog Biophys Mol Biol 74:1–35
82. Forkey JN, Quinlan ME, Shaw MA, Corrie JET, Goldman YE (2003) Three-dimensional structural dynamics of myosin V by single-molecule fluorescence polarization. Nature 422:399–404
83. Yildiz A, Forkey JN, McKinney SA et al (2003) Myosin V walks hand-over-hand: Single fluorophore imaging with 1.5-nm localization. Science 300:2061–2065
84. Forkey JN, Quinlan ME, Goldman YE (2005) Measurement of single macromolecule orientation by total internal reflection fluorescence polarization microscopy. Biophys J 89: 1261–1271
85. Quinlan ME, Forkey JN, Goldman YE (2005) Orientation of the myosin light chain region by single molecule total internal reflection fluorescence polarization microscopy. Biophys J 89:1132–1142
86. Rosenberg SA, Quinlan ME, Forkey JN, Goldman YE (2005) Rotational motions of macromolecules by single-molecule fluorescence microscopy. Acc Chem Res 38: 583–593
87. Jacobs DT, Weigert R, Grode KD, Donaldson JG, Cheney RE (2009) Myosin Vc is a molecular motor that functions in secretory granule trafficking. Mol Biol Cell 20:4471–4488
88. Burghardt TP (1984) Model-independent fluorescence polarization for measuring order in a biological assembly. Biopolymers 23: 2383–2406
89. Schabert F, Knapp H, Karrasch S, Haring R, Engel A (1994) Confocal scanning laser—scanning probe hybrid microscope for biological applications. Ultramicroscopy 53: 147–157
90. Lal R, Proksch R (1997) Multimodal atomic force microscopy: biological imaging using atomic force microscopy combined with light fluorescence and confocal microscopies and electrophysiologic recording. Int J Imaging Syst Tech 8:293–300
91. Hugel T, Holland NB, Cattani A et al (2002) Single-molecule optomechanical cycle. Science 296:1103–1106
92. Nishida S, Funabashi Y, Ikai A (2002) Combination of AFM with an objective-type total internal reflection fluorescence microscope (TIRFM) for nanomanipulation of single cells. Ultramicroscopy 91:269–274
93. Yip CM, Darabie AA, McLaurin J (2002) A beta 42-peptide assembly on lipid Bilayers. J Mol Biol 318:97–107
94. Kodama T, Ohtani H, Arakawa H, Ikai A (2004) Atomic force microscope equipped with confocal laser scanning microscope for the spectroscopic measurement of the contact area in liquid. Chem Phys Lett 385:507–511
95. Yamada T, Afrin R, Arakawa H, Ikai A (2004) High sensitivity detection of protein molecules picked up on a probe of atomic force microscope based on the fluorescence detection by a total internal reflection fluorescence microscope. FEBS Lett 569:59–64
96. Franz CM, Muller DJ (2005) Analyzing focal adhesion structure by atomic force microscopy. J Cell Sci 118:5315–5323
97. Gaiduk A, Kuhnemuth R, Antonik M, Seidel CAM (2005) Optical characteristics of atomic force microscopy tips for single-molecule fluorescence applications. Chemphyschem 6: 976–983
98. Shaw JE, Epand RF, Epand RM et al (2006) Correlated fluorescence-atomic force microscopy of membrane domains: structure of fluorescence probes determines lipid localization. Biophys J 90:2170–2178
99. Meller K, Theiss C (2006) Atomic force microscopy and confocal laser scanning microscopy on the cytoskeleton of permeabilised and embedded cells. Ultramicroscopy 106:320–325
100. Madl J, Rhode S, Stangl H et al (2006) A combined optical and atomic force microscope for live cell investigations. Ultramicroscopy 106:645–651
101. Chiantia S, Ries J, Kahya N, Schwille P (2006) Combined AFM and two-focus SFCS study of raft-exhibiting model membranes. Chemphyschem 7:2409–2418
102. Blanchette CD, Lin WC, Ratto TV, Longo ML (2006) Galactosylceramide domain microstructure: Impact of cholesterol and nucleation/growth conditions. Biophys J 90:4466–4478
103. Garcia-Saez AJ, Chiantia S, Schwille P (2007) Effect of line tension on the lateral organization of lipid membranes. J Biol Chem 282(46):33537–33544
104. Gaiduk A, Kuhnemuth R, Felekyan S et al (2007) Fluorescence detection with high time resolution: From optical microscopy to simultaneous force and fluorescence spectroscopy. Microsc Res Techn 70:433–441
105. Mathur AB, Truskey GA, Reichert WM (2000) Atomic force and total internal reflection

fluorescence microscopy for the study of force transmission in endothelial cells. Biophys J 78:1725–1735

106. Charras GT, Horton MA (2002) Single cell mechanotransduction and its modulation analyzed by atomic force microscope indentation. Biophys J 82:2970–2981
107. Terebiznik MR, Vieira OV, Marcus SL et al (2002) Elimination of host cell PtdIns(4,5) P(2) by bacterial SigD promotes membrane fission during invasion by Salmonella. Nat Cell Biol 4:766–773
108. Flannagan RS, Harrison RE, Yip CM, Jaqaman K, Grinstein S (2010) Dynamic macrophage "probing" is required for the efficient capture of phagocytic targets. J Cell Biol 191:1205–1218
109. Trache A, Meininger GA (2005) Atomic force-multi-optical imaging integrated microscope for monitoring molecular dynamics in live cells. J Biomed Opt 10(6): 064023
110. Trache A, Trzeciakowski JP, Meininger GA (2010) Mg2+ modulates integrin-extracellular matrix interaction in vascular smooth muscle cells studied by atomic force microscopy. J Mol Recognit 23:316–321
111. Trache A, Lim SM (2010) Live cell response to mechanical stimulation studied by integrated optical and atomic force microscopy. J Vis Exp 4;(44). pii: 2072
112. Lim SM, Kreipe BA, Trzeciakowski J, Dangott L, Trache A (2010) Extracellular matrix effect on RhoA signaling modulation in vascular smooth muscle cells. Exp Cell Res 316: 2833–2848
113. Trache A, Lim SM (2009) Integrated microscopy for real-time imaging of mechanotransduction studies in live cells. J Biomed Opt 14:034024
114. Sarkar A, Robertson RB, Fernandez JM (2004) Simultaneous atomic force microscope and fluorescence measurements of protein unfolding using a calibrated evanescent wave. Proc Natl Acad Sci USA 101: 12882–12886
115. Delcea M, Schmidt S, Palankar R et al (2010) Mechanobiology: correlation between mechanical stability of microcapsules studied by AFM and impact of cell-induced stresses. Small 6:2858–2862
116. Stadler B, Blattler TM, Franco-Obregon A (2010) Time-lapse imaging of in vitro myogenesis using atomic force microscopy. J Microsc 237:63–69
117. Epand RM, Vuong P, Yip CM, Maekawa S, Epand RF (2004) Cholesterol-dependent partitioning of PtdIns(4,5)P2 into membrane domains by the N-terminal fragment of NAP-22 (neuronal axonal myristoylated membrane protein of 22 kDa). Biochem J 379:527–532
118. Hollars CW, Dunn RC (1998) Submicron structure in L-alpha-dipalmitoylphosphatidylcholine monolayers and bilayers probed with confocal, atomic force, and near-field microscopy. Biophys J 75:342–353
119. Yuan C, Johnston LJ (2002) Phase evolution in cholesterol/DPPC monolayers: atomic force microscopy and near field scanning optical microscopy studies. J Microsc 205:136–146
120. Burgos P, Lu Z, Ianoul A, Hnatovsky C et al (2003) Near-field scanning optical microscopy probes: a comparison of pulled and double-etched bent NSOM probes for fluorescence imaging of biological samples. J Microsc 211:37–47
121. Tokumasu F, Hwang J, Dvorak JA (2004) Heterogeneous molecular distribution in supported multicomponent lipid bilayers. Langmuir 20:614–618
122. Ianoul A, Street M, Grant D et al (2004) Near-field scanning fluorescence microscopy study of ion channel clusters in cardiac myocyte membranes. Biophys J 87:3525–3535
123. Oreopoulos J, Yip CM (2009) Combinatorial microscopy for the study of protein-membrane interactions in supported lipid bilayers: Order parameter measurements by combined polarized TIRFM/AFM. J Struct Biol 168:21–36
124. Oreopoulos J, Epand RF, Epand RM, Yip CM (2010) Peptide-induced domain formation in supported lipid bilayers: direct evidence by combined atomic force and polarized total internal reflection fluorescence microscopy. Biophys J 98:815–823
125. Mo GC, Yip CM (2009) Supported lipid bilayer templated J-aggregate growth: role of stabilizing cation-pi interactions and headgroup packing. Langmuir 25:10719–10729
126. Fidder H, Terpstra J, Wiersma DA (1991) Dynamics of Frenkel excitons in disordered molecular aggregates. J Chem Phys 94:6895–6907
127. Stuurman N, Amodaj N, Vale RD (2007) Micro-manager: open source software for light microscope imaging. Microsc Today 15:42–43
128. Collins TJ (2007) ImageJ for microscopy. Biotechniques 43:25–30

# Chapter 25

# Nanoimaging Cells Using Soft X-Ray Tomography

**Dilworth Y. Parkinson, Lindsay R. Epperly, Gerry McDermott, Mark A. Le Gros, Rosanne M. Boudreau, and Carolyn A. Larabell**

## Abstract

Soft X-ray microscopy is ideally suited to visualizing and quantifying biological cells. Specimens, including eukaryotic cells, are imaged intact, unstained and fully hydrated, and therefore visualized in a near-native state. The contrast in soft X-ray microscopy is generated by the differential attenuation of X-rays by the molecules in the specimen—water is relatively transmissive to this type of illumination compared to carbon and nitrogen. The attenuation of X-rays by the specimen follows the Beer–Lambert law, and therefore both linear and a quantitative measure of thickness and chemical species present at each point in the cell. In this chapter, we will describe the procedures and computational methods that lead to 50 nm (or better) tomographic reconstructions of cells using soft X-ray microscope data, and the subsequent segmentation and analysis of these volumetric reconstructions. In addition to being a high-fidelity imaging modality, soft X-ray tomography is relatively high-throughput; a complete tomographic data set can be collected in a matter of minutes. This new modality is being applied to imaging cells that range from small prokaryotes to stem cells obtained from mammalian tissues.

**Key words:** Cell structure, Microscopy, Nanoimaging, Reconstruction, Soft X-ray tomography, Segmentation, Visualization

## 1. Introduction

Soft X-ray tomography (SXT) is a relatively recent addition to the spectacular array of imaging techniques available to biologists (1). As with all modalities, SXT fills a niche as being the optimal imaging technique for answering a particular set of questions (2). Specifically, SXT is uniquely suited to high-resolution, quantitative imaging of cells that are intact, unstained and fully hydrated—in other words, SXT excels at visualizing cells, including eukaryotic cells, that are held in a near-native state, and are, therefore, highly representative of the in vivo fully functional cell (1, 2).

The interior of a cell is highly organized and finely structured (1). This is especially true in the case of eukaryotic cells that are

Alioscka A. Sousa and Michael J. Kruhlak (eds.), *Nanoimaging: Methods and Protocols*, Methods in Molecular Biology, vol. 950, DOI 10.1007/978-1-62703-137-0_25, © Springer Science+Business Media, LLC 2013

partitioned into functionally discrete subcellular spaces termed organelles (3–5). In a 2-dimensional projection image of a cell these structural features are superimposed, making interpretation difficult, if not impossible. As a consequence, the subcellular architecture must be imaged in 3-dimensions. Since microscopes can only produce 2-dimensional images this may appear at first sight to be a major hurdle. However, in practice it is a relatively straightforward process to calculate 3-dimensional views from 2-dimensional data by the application of methods such as deconvolution or tomography (6). In the case of tomography, this simply means taking 2-dimensional images of the specimen from a number of different perspectives, usually at angular intervals around a rotation axis. Provided a sufficient number of images are taken, a fully isotropic, 3-dimensional representation of the specimen can be calculated from 2-dimensional imaging data (7). For SXT, a series (or "stack") of images of the specimen are taken in this manner using a soft X-ray microscope (8). This contains all of the information necessary to calculate a volumetric reconstruction of the entire field of view over which imaging data was collected (9). In the case of small prokaryotic cells, the field of view may contain many tens or even hundreds of cells. For larger cells, such as eukaryotic cells, this number is obviously lower. For yeast, each field of view typically contains a maximum of six cells, and for higher order cells this will typically be a single cell. However, since SXT data collection is relatively rapid—standard exposure times for each image are on the order of 100 ms; a data set contains 90 or 180 images—the specimen throughput is still exceptionally high compared to other high-resolution cellular imaging modalities, even for large eukaryotic cells (9). Consequently, a large number of cells can be characterized in a relatively short space of time. For example, in two recently published studies quantifying yeast morphology, data from hundreds of high-resolution tomographic reconstructions were used in the analyses (5, 10). Neither of these types of nanoimaging studies could have been carried out using alternative techniques, such as electron tomography.

The specimen in SXT is illuminated using photons with energies that lie in the so-called water window region of the electromagnetic spectrum (11). The attenuation of these photons follows the Beer–Lambert Law, and is therefore linear, quantitative and dependent on chemical species and thickness. Carbon and nitrogen—major constituents of most biomolecules—attenuate the transmission of X-rays an order of magnitude more strongly than water (12). This characteristic produces exquisite contrast in SXT images of cells (5, 8, 13, 14). As a result, subcellular structures in an SXT reconstruction can be readily identified and "segmented" (i.e., isolated from the other contents in the reconstruction) based on their shape and the measured linear absorption coefficient (LAC) at each point in the cell (8, 12, 13, 15, 16)). Each organelle type has a characteristic LAC. This reflects the organelle's biochemical composition. For example, yeast vacuoles have a low concentration of biomolecules and relatively high water content. Their LAC is, therefore, small compared with organelles that have low water content and are densely packed with biomolecules, for example the nucleus. The LAC values allow individual organelles to be identified and isolated, and provide key information on the internal structure of the organelle. Segmentation of the subcellular contents allows valuable quantitative analyses to be carried out. For example, it is easy to determine key parameters such as the volume and density of the organelles at key stages in the cell cycle, or how these parameters are affected by exposure to external influences, such as the presence or absence of specific molecules, temperature, confluence, and so on. This information is enormously valuable in studies ranging from basic cell biology to applied research aimed at generating new pharmaceuticals or in optimizing the economics of biofuel production.

We will now describe the methods and steps needed to image a sample of cells using SXT, beginning with specimen preparation.

## 2. Materials

Currently, soft X-ray microscopes are primarily located at synchrotron light sources. However, "table top" soft X-ray sources with suitable characteristics are being actively developed (17–19). We fully expect that in the near future soft X-ray microscopes using illumination from these types of small-scale sources will become commonplace in most research institutions.

Cell types suitable for SXT imaging span the entire spectrum of cell biology; from small bacteria cultured in the laboratory to large eukaryotic cells purified directly from primary mammalian tissue (1). The materials needed for the preparation of cells for soft X-ray microscopy are wide-ranging and highly specific to each cell type, and will, therefore, not be listed here. However, the following materials and instruments are in common use on the soft X-ray microscope XM-2 at the Advanced Light Source, Berkeley, CA (the optical layout of this microscope is shown in Fig. 1), and are generally useful for carrying out the technique irrespective of the photon source, or soft X-ray microscope design.

1. Brightfield Light Microscope: Zeiss Axiovert 40 CFL with Condenser 4.0 ($a = 53$ mm) and Zeiss Objective A-Plan 10×/0.25 Ph 1.
2. Glass capillaries (Hilgenberg GmbH 1409190).
3. Quartz capillaries (Hilgenberg GmbH 1406785).

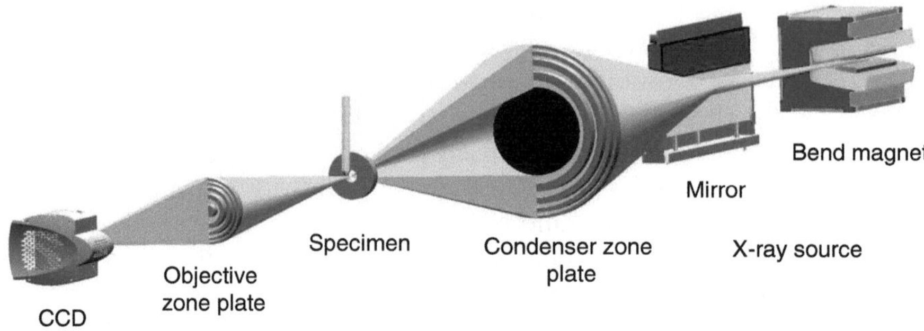


Fig. 1. Schematic representation of XM-2, a soft X-ray microscope for nanoimaging biological cells (located at the Advanced Light Source, Lawrence Berkeley National Laboratory, Berkeley, California). This is the first such microscope in the world to be built specifically for biological and biomedical research (for more information visit http://ncxt.lbl.gov).

4. Capillary tube puller (Sutter Instrument Co. Models P-97 & P-2000).
5. 100 nm colloidal gold nanoparticles (BB International, UK).
6. Filters: glue (Dow Corning 732 multipurpose sealant) and 20 and 15 μm sieves (BioDesign Cellmicrosieve).
7. Microloader tips (Eppendorf 930001007).
8. A multi-processor computer: Apple cluster with 20 nodes, each with two dual-core processors (see Note 1 for more details).
9. Image processing software: Matlab (Mathworks), ImageJ (National Institutes of Health, USA), XMIPP (Centro Nacional de Biotecnología, Madrid, Spain), ASPIRE (U. Michigan), and SPARX (Baylor College of Medicine) (see Note 2).
10. 3-D visualization and analysis software: Amira™ (Visage Imaging, Germany) and Avizo (VSG, France) (see Note 3).

## 3. Methods

### 3.1. Data Collection

#### 3.1.1. Specimen Holders

When collecting imaging data for tomographic reconstruction the ideal specimen holder allows arbitrary access to a minimum of 180° of rotation (360° is even better). Use of grids and other mounting systems developed for electron microscopy hinders full rotation over angular ranges as large as these. Therefore, when possible it is preferable to use a mounting system such as thin-walled glass capillary tubes. In this chapter, we will only describe methods for imaging cells in glass or quartz capillary tubes. This approach allows full-rotation tomography, and gives optimum results. Capillary geometry allows data to be collected around a complete circle, and hence eliminates the issue of having to deal with the "missing wedge" of data inherent in other types of specimen holders.

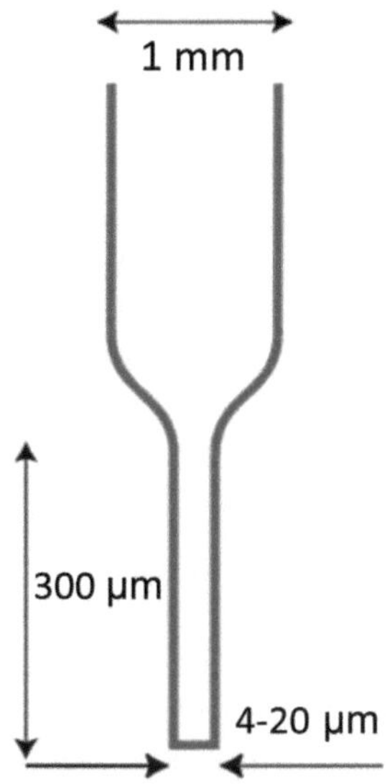


Fig. 2. Dimensions and geometry of a capillary specimen holder.

The "missing wedge" is at best a frustration that causes an unavoidable artifact in the tomographic reconstruction. At worst, this incompleteness in data makes segmentation difficult and interpretations ambiguous. Capillary specimen holders are made from glass or quartz tubes that have had one end pulled to a diameter of 2–10 μm, and have a length of more than 300 μm suitable for being imaged in the soft X-ray microscope, see Fig. 2.

#### 3.1.2. Adding Fiducial Markers to Capillary Specimen Holders

In order to align the images to a common frame of reference (see section below), it is beneficial to coat the capillary specimen holders with 100 nm gold fiducial markers, prior to loading them with cells. This can be done easily using commercially available gold nanoparticles.

1. To coat the capillary specimen holder, simply dip it in the solution of 100 nm gold nanoparticles and slowly withdraw.
2. While the outside of tube is being coated, gently blow air through the capillary to prevent the solution from getting inside.

#### 3.1.3. Specimen Preparation

Prior to being imaged using SXT, lower order cells do not require any special preparation. Indeed, cells such as prokaryotes or yeast can simply be taken from their growth chamber and pipetted directly into a specimen holder and then mounted in the soft X-ray microscope. In some cases the growth media is very high in carbon and nitrogen, and it may be advantageous to transfer the cells into phosphate-buffered saline solution; however, this is not a prerequisite.

For higher order mammalian cells, particularly adherent cells, more care must be taken to ensure the good health of the cells when they are imaged. In general, we have found the following protocol to produce the best results when imaging adherent cells.

*3.1.4. Thawing Adherent Cells to Plating*

1. Remove a mammalian cell stock vial from liquid nitrogen temperature storage. If the vial has been submerged in nitrogen, assume nitrogen has seeped into the vial and do not allow it to warm in your hands. Touch the vial as little as possible and always point the cap away from you and others. Place the vial in a preheated 37°C water bath and allow it to thaw for 2–2.5 min. Do not allow the cells to stay at 37°C once the vial is thawed.
2. The outside of the thawed vial is sterilized by dunking, spraying, or dripping with 70% ethanol. Open the vial under sterile conditions and transfer the contents to preheated 37°C media. Freshly frozen cells can be fragile so transfer cells slowly into and out of serological pipette and use care to avoid introducing bubbles.
3. Cell number/cm$^2$ differs depending on the size of the cell and the particular requirement of cell line but, generally, plate cells at ~15% density in a plastic cell culture plate, dish or flask. Some cell lines may require a surface coating such as fibronectin, laminin, etc.; however, the actual plating density and surface of growth is cell line specific, and must be determined on a case-by-case basis.

*3.1.5. Suspending Adherent Mammalian Cells*

Prior to being imaged, adherent mammalian cells are suspended in solution. Once suspended, adherent mammalian cells are treated identically to prokaryotes or lower order eukaryotes, such as yeasts.

1. Filter trypsin, media, and soybean trypsin inhibitor using a 0.2 μm filter to remove debris or other particulate matter.
2. Use trypsin to detach adherent cells for loading into a sample holder (see Note 4). In our hands, we see the best results when cells are detached from plates that are 75–80% confluent.
3. Generally, for a 75 cm$^2$ flask, aspirate media, then quickly wash serum from the cell layer with 3 mL of 0.25% trypsin and aspirate wash. Add back 1 mL 0.25% trypsin, distribute, and incubate at 37°C for 2 min or until cells move when the plate is tilted.
4. Stop trypsin by adding 10 mL of media containing 10% serum.
5. Pellet cells by centrifugation (0.3 rcf for 4–10 min) and resuspend the pellet in 400 μL to 1 mL of media.

*3.1.6. Specimen Loading in Capillary Sample Holders*

Cells must be loaded into the narrow, drawn region of the specimen holder and debris must be kept to a minimum. To help achieve selection of single cells and to minimize debris, size selection via fluorescence-activated cell sorting (FACS) or filtration is recommended. Filtration by size is described below; discussing FACS of cells is out of the scope of this chapter and may be a protocol that needs to be carried out at a specialist resource, depending on availability.

1. Using a clean air source, blow air into filters and Eppendorf tubes to remove debris, and filter all liquids using a 0.2 μm filter.
2. Construct filter set-ups by gluing (Dow Corning RTV Sealant 732) 1.5 $cm^2$ squares of BioDesign Cellmicrosieve (20 and 15 μm) onto halved p1000 pipetman tips that have been cut in half to remove the pointed tips. Fit the filter set-ups into 1.5 mL Eppendorf tubes and transfer suspended cells onto filters. Selection for size is done by centrifugation (0.2 rcf for 10 s) using first 20 μm, then 15 μm filters.
3. After final filtration, remove filter set-ups from Eppendorf tubes and pellet cells (0.3 rcf for 4 min). For efficient loading, a high cell titer is recommended. Resuspend the pellet into a volume of loading media (usually growth media) that is 1× to 2× volume of the pellet.
4. To load cells into a glass capillary tube, add 0.98 μL of high titer cell solution into a microloader tip. Thread the microloader tip into the back end of a capillary tube, place the tip of the microloader in the region of the capillary tube where the glass begins to taper into the portion that has been pulled to a diameter of between 2 and 10 μm and dispense fluid. About 0.1 μL of fluid will fill the pulled tip.
5. Under a brightfield light microscope (10× or higher magnification air lens), check that cells are in the region of the glass capillary tube where the sample holder has narrowed to 15 μm or less in diameter. Remove any moveable fluid from the sample holder.
6. Capillary sample holders with cells spanning a 2–10 mm region along its length are ready to be rapidly frozen at the beamline or plunge frozen for later imaging.

*3.1.7. Cryofixation*

There is a significant amount of literature from the electron microscopy community on cryofixation (20–23). Because SXT imaging is carried out with lower spatial resolution than electron tomography (i.e., 50 nm rather than a few nm), and with X-rays rather than electrons, not all of the commonly accepted "rules" for cryofixation developed by electron microscopists apply (see Note 5). In practice, for soft X-ray tomography success has been achieved with each of two different approaches to cryofixation: cryogenic gas streams and plunge freezing.

1. First, set up a cryostage in the soft X-ray tomography instrument to maintain the sample at near liquid nitrogen temperature

with a stream of cryogenic helium gas at atmospheric pressure (this is the set up of the XM-2 instrument at the ALS in Berkeley).

2. *First approach to freezing: Cryogenic gas stream.* Mount the sample on the sample stage so that it is outside of the cryogenically cooled region, and then quickly plunge the sample into the cryogenically cooled stream of cold helium gas. The sample is then maintained in the stream of helium gas for the duration of the experiment.
3. *Second approach to freezing: Plunge freezing.* Quickly plunge the sample into cryogenic liquid such as liquid propane, liquid ethane, or any other suitable media. In principle, this approach can be straightforward, to the extent that the capillary can be simply plunged manually into the cryogenic liquid. However, to maintain consistency, and allow the speed of plunging to be controlled, it is better if this process is carried out using instrumentation that can vary the speed with which the capillary enters the liquid. Moreover, the plunging time is specimen dependent. For some specimens the time taken to plunge into cryogen is unimportant, for others it is essential that the specimen is plunged as quickly as possible. One difficulty with this technique is maintaining the specimen at cryogenic temperature whilst it is being transferred from the cryogenic liquid onto the sample cryostage of the X-ray microscope; this requires the use of a cryotransfer device. In practice, this can be challenging, and is an additional potential cause of damage to the specimen, if the transfer device is inadvertently allowed to warm up during use. The development of new plunge methods and instruments is very much a work in progress.

In either of these approaches, the final result of the procedure is that the specimen has been loaded into a cryogenically cooled helium gas stream, and is held in place by a spring-loaded clip attached to a motion control stage. At this point, data collection can begin.

### 3.2. Soft X-Ray Microscopic Data Collection

In the following sections we will describe data collection using the soft X-ray microscope XM-2 located at the Advanced Light Source (ALS), Berkeley. However, the basic principles and concepts will apply to data collection using any soft X-ray microscope.

#### 3.2.1. Aligning the Specimen in the Microscope

The previous section described the process of mounting the specimen in a capillary holder, and cryocooling the sample, typically in situ in the XM-2 cryorotation stage. When the sample is being cooled using the gas method (most common for cells less than 5 μm in diameter), an alignment step is performed after the sample is mounted but before it is cooled. Using the plunge freezing method, the capillary is aligned off-line prior to mounting.

Alignment places the tip of the capillary such that the region to be imaged is centered on the axis of rotation. This is an important step. During tomographic data collection the specimen is rotated by 180°; during which the specimen should not leave the field of view. The specimen will stay in the field of view during this rotation only if it is centered on the axis of rotation.

1. To perform this alignment, the sample is positioned at 0 and 180°, and the position of the tube is recorded. A tip-tilt stage controlled by a joystick is used to adjust the capillary such that it splits the difference between its positions at the two angles. This procedure is repeated for the tube when positioned at 90 and 270°. Typically, two rounds of alignment are performed to confirm that the tube is exactly centered, especially in the case of large tubes (greater than 8 μm) that nearly fill the field of view.
2. The capillary is then lowered quickly into the cryogenic gas stream.
3. Once the tube is aligned and lowered into position for X-ray imaging, the tip-tilt alignment procedure is repeated a final time at finer resolution using X-ray imaging, usually using a region above or below the sample of interest to reduce the X-ray dose on the sample to be imaged.

*3.2.2. Tomographic Data Collection*

1. Look for fields of view that contain cells of interest for the given experiment. Projection images are taken along the length of the capillary. This allows the position of the optimal fields of view to be identified.
2. Set up a tomography scan. In the XM-2 control GUI, sample description is saved in a header file along with the image data; this information is also transferred into the image database. The parameters for the scan are also selected: exposure time, angular increment and total angular range. Exposure time is chosen such that the CCD receives the maximum number of counts without saturating (also known as clipping) any pixels, generally around 100 ms on this microscope. The angular increment is generally between 1 and 2°, for a total number of images between 90 and 180, spread over 180°.
3. After the scan is started, continue observing the sample. In rare cases, drift leads the specimen to leave the field of view during the scan. In this case, slightly adjust the horizontal or vertical position between images to re-center the sample in the field of view. Obviously, this adjustment should be timed so that it is not carried out during acquisition of an image.
4. Collect a set of "flatfield" images before and after tomographic image data are recorded. To do this, move the capillary out of the field of view, and collect a series of 10–20 images. By combining these images with the sample images from the tomogram,

the percent transmission can be calculated, yielding quantitative X-ray absorption data.

5. Data for a completed tomographic set is saved as 16 bit unsigned tiff stacks. There are a total of three files: one with the initial flat fields, one tomography image series, and one with the final flat fields.

### 3.3. Data Processing

#### 3.3.1. Image Correction and Normalization

1. As the first step in processing the images, subtract the detector dark counts from both the sample images and the flat field images. The dark counts can be measured by collecting an image with the X-ray shutter closed. The dark field image only needs to be measured every few months; our experience is that it does not change on this time scale.
2. After subtracting the dark counts, average each of the two sets of flat field images to produce a single initial flat field image and a single final flat field image. Average multiple flat fields to reduce the noise level.
3. Divide the first half of the images in the tilt series by the initial averaged flat field image, and divide the second half of the images in the tilt series by the final averaged flat field image. In other words, the value of each pixel in the sample image is divided by the value in the corresponding pixel of the flat field image, and the resulting value is placed in the corresponding pixel of the output image. This results in a stack of images corresponding to the percent X-ray transmission through the sample over the series of angles recorded.
4. Before going to the next step (image alignment), correct for dead pixels on the CCD. These are pixels that read out very low counts regardless of the X-ray flux on them. To carry out this correction requires a pixel-by-pixel map identifying each bad CCD pixel. For each image in the series, the bad pixels are replaced by a value interpolated from the surrounding pixels. We use a program from the IMOD package called CCDERASER to perform this step.
5. Finally, down-sample the images from 2,048 × 2,048 to 1,024 × 1,024 using a simple binning strategy. This is done to reduce the extent of unnecessary oversampling: the resolution of the experiment is approximately 50 nm (based on the zone plate outer ring radius), but the pixel size at 2,048 × 2,048 is approximately 8 nm. By reducing the image size to 1,024 × 1,024, the pixel size is increased to 16 nm, which is still sufficiently small enough to capture all relevant detail in the images.

In practice, these calculations have been implemented both as a MATLAB script and an ImageJ plugin. Either of these implementations can be used to carry out this step, yielding identical output in

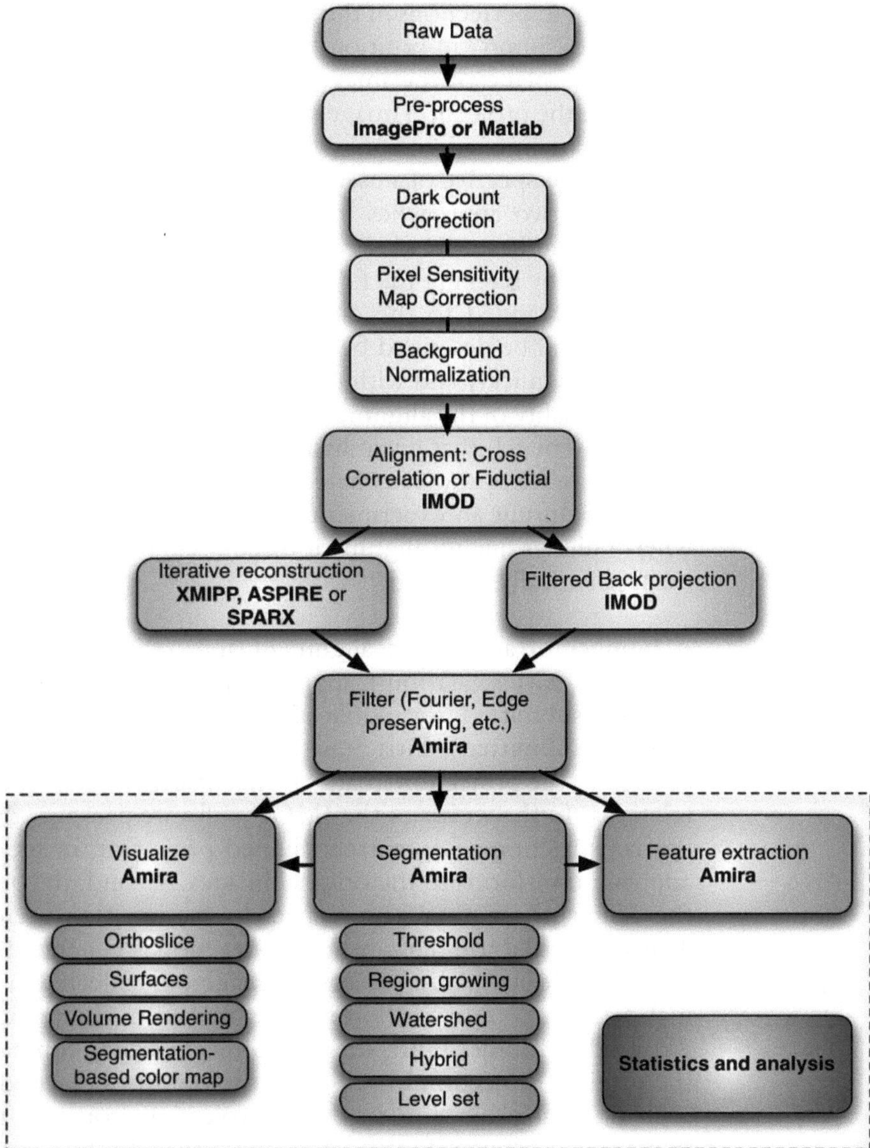


Fig. 3. A flowchart showing the steps and software packages required to process raw imaging data from XM-2, and to calculate, visualize and segment 3-dimensional reconstructions.

a similar amount of time (the bottleneck is mainly reading from and writing to disk). In both cases, versions of the script or plugin can be run in which an entire directory of images can automatically be processed. In Fig. 3 we show the software packages in common use by users of XM-2, together with their relative sequence of use and the functions performed. Naturally, this list of software is organic, and liable to change as new, improved algorithms become available, or when experimental need dictates use of other software.

*3.3.2. Image Alignment*

Because of imperfections in the specimen rotation stage, the sample appears to "jump" in position (x and y translation as well as rotation in the plane of the image) from one frame to another. The images in the projection image series must, therefore, be aligned to bring them into agreement within a single frame of reference. From the perspective of a user of the soft X-ray tomography instrument, the two approaches to this are to rely on the automated methods that have already been implemented, or to manually align the images by tracking the fiducial markers (the gold nanoparticles which were added to the outside of the capillary tubes). These methods will be described below. While automated alignment can give results quickly and with no human interaction, as of this writing, the quality of the alignment given by automated methods does not always match that given by manually tracking fiducial markers. Thus, in practice the automated approach is often used to guide decisions during an experiment and give initial results. However, when a higher quality alignment is required, manual fiducial alignment is typically required.

*3.3.3. Automated Alignment*

A number of automated alignment approaches have been published. At XM-2, a robust approach has been implemented and recently submitted for publication (Parkinson et al. J. Struct. Biology, submitted). First, each image in the projection stack is aligned to the projection images from adjacent angles by cross correlations. In the words of the IMOD software package, this allows the construction of a coarsely aligned projection image stack by applying transforms to the original images. Second, the position of the center of rotation with respect to the images is determined. Third, an initial tomographic reconstruction is generated. Fourth, at each angle at which a projection image was collected in the original data set, a re-projection is generated from the reconstructed 3D model volume. Fifth, these re-projections are compared with the original projection images, and the transform needed to align each original projection to the re-projection from the model at that angle is refined. Finally, steps three through five are repeated iteratively. On each iteration the reconstructed volume should improve as the alignment errors decrease. This process is generally termed "model-based alignment".

In practice at XM-2, a user types a single command in MATLAB, with the file name of the normalized image stack as the input parameter (alternatively, an entire directory of files can be automatically aligned). The MATLAB distributed computing toolbox is used, running on a 20-node Mac cluster, which increases the speed of the process. Typically, it only takes seconds to minutes to carry out this stage.

*3.3.4. Fiducial Alignment*

For manual alignment of fiducial markers we recommend the IMOD software package. Their Web site has extensive documentation and tutorials on this software. We note a few important details specific to soft X-ray tomography using capillary tubes, and which are not extensively covered in the IMOD tutorials.

1. In the coarse alignment phase (which is similar to the cross-correlation stage of automatic alignment, described above), check the box for "cumulative correlation" along with the box for "no cosine stretch".
2. When creating the seed model, avoid attempting to mark gold fiducial markers when they are at the edges of the tube—it is too difficult to accurately mark them in these regions. We have found that in soft X-ray microscopy the gold fiducial markers are of much lower relative contrast with respect to the surrounding sample than in electron microscopy. This is due in large part to the thickness of the samples in SXT compared to those imaged with EM. This is especially true at the extreme left and right edges of the tube, where the X-rays pass nearly tangent to the tube edge and thus transmit through more glass or quartz material.
3. In a similar vein to step 2, during fiducial model generation, avoid tracking the seed model or filling model gaps. Automatic tracking generally fails in these regions, and leads to worse overall alignment.
4. The more fiducial markers, the better the alignment. We pick a minimum of six markers, if available, which is much less than recommended by IMOD, but which we have found, nevertheless, gives a good alignment.

### 3.4. Tomographic Reconstruction

In the reconstruction step, a three-dimensional volume is produced based on the stack of aligned projections. There are a number of algorithms used to carry out this step, and this is very much an active area of research. In terms of the differences apparent to the eye, the methods mainly differ in the amount and type of noise and/or artifacts present in the final reconstruction, and in the smoothness or sharpness of object boundaries in the image. The goal of all of these methods is to give the minimum amount of noise, while still preserving all relevant image detail, including sharp edges (sometimes referred to as resolution). The challenge is that, in general, methods that reduce noise in the image also reduce resolution. The compromise that is reached between noise and resolution generally depends on what subsequent processing will be applied to the reconstructed volume. If digital slices through the volume are going to be directly visualized, different settings have been found to be optimal as compared to the case when volumes are going to be automatically segmented. The two main classes of reconstruction algorithms can be classified as Fourier methods and real-space methods, and will be briefly described below, followed by a description of their practical implementation and the comparisons between them.

*3.4.1. Fourier Methods*

The most common method in this class is known as filtered backprojection. It relies on the central slice theorem, and is implemented by Fourier transforming each image in the original stack, filtering it, performing an inverse Fourier transform, and "backprojecting" this filtered image across the volume to be reconstructed, interpolating appropriate values for all the voxels. By combining the backprojections from all of the various orientations, a final volume is reconstructed. Two strengths of this algorithm are that it is computationally fast, and it gives a unique result for a given type and level of filtering.

*3.4.2. Real-Space Methods*

In real-space methods, individual images are iteratively compared directly with the reconstructed volume, and the volume voxels are updated to bring them into better agreement with the values expected. There are a large number of strategies for how to perform this update and carry out the iterations. There are also a number of constraints that can be added during the iterations, including a requirement that all voxels must have values greater than or equal to zero (corresponding to no voxels having a physically impossible "negative transmission"). In addition, there are a number of ways to add smoothing, or regularization, to the result–for example, when updating the voxels, a penalty term can be added which encourages adjacent voxels to have similar, rather than different values. This penalty term can be made nonlinear, in an attempt to both encourage sharp edges (high resolution) while reducing noise.

A particular strength of iterative methods is that they have been shown in the literature to give the optimal tradeoff between noise and resolution. This is especially true for the relatively low number of projections used in soft X-ray tomography. In addition, iterative methods have the advantage of being able to include additional constraints, which can improve the accuracy of results. One drawback of iterative methods is that they are computationally intensive, and thus can be very time consuming in the absence of significant computational resources.

*3.4.3. Reconstruction in Practice*

A number of iterative and Fourier reconstruction methods are available to users of XM-2 through MATLAB scripts which take advantage of the 20-node cluster and the MATLAB distributed computing toolbox. The reconstructions are run by typing a single command, with inputs to indicate the selected reconstruction method, and the type and level of filtering (or regularization). Reconstructions by Fourier methods are complete in 1 min or less; reconstructions by iterative methods generally take 10–20 min to complete. Thus, for data that will be analyzed in depth in subsequent steps, users sometimes perform reconstructions with a series of methods and filtering levels to determine which best meets their needs. A challenge in comparing Fourier and real-space methods is that each method has a different approach to achieving the optimal tradeoff between noise and resolution. In addition, the original images can be filtered prior to carrying out the reconstruction, which also affects the final result. Thus, rather than to try to give a detailed comparison, we will make a few general observations:

1. Iterative methods give a clear advantage in terms of reducing "streaking" artifacts due to the presence of gold fiducial markers. Using filtered backprojection, we often find streaks coming off of these markers in one or more directions; these streaks are significantly reduced using iterative methods.
2. Iterative algorithms which use nonlinear regularization to implement edge-preserving smoothing during the reconstruction process can give volumes which are more readily segmented using automated methods, whereas volumes from filtered backprojection must be filtered by other methods before performing automated segmentation. Distinguishing which of these methods provides more accurate results for real data is difficult, and must be assessed on a case to base basis.

### 3.5. Visualization

A number of commercial, academic, and open source software packages are available for visualizing tomographic reconstructions (see Subheading 2). Amira is our preferred tool; we note that Avizo uses much of the same code base as Amira, and has similar capabilities. In this section, we will describe our workflow in Amira, but similar results can be achieved with other tools.

1. First load reconstructions into Amira. Correctly set the voxel size to the appropriate units; for example, in many cases we use volumes where the voxel size is 32.7 nm on each side. The volume is then visualized in the form of one-voxel-thick digital slices through the volume (known as "orthoslices" or "oblique slices" in Amira; Fig. 4a, b). These allow for viewing of exterior and interior portions of a volume from any arbitrary axis.
2. Adjust the contrast of the orthoslices in order to discern subtle subcellular features, for example, to help delineate the boundaries between organelles and the cytosol, or between structures inside a particular cellular space. An examination of the orthoslices is a relatively quick way to determine the cellular characteristics of the reconstructed cells, for example, to determine the condition of the cells or their stage in the cell cycle. Inspection of the orthoslices also allows for a quick comparison between control cells and variable cells. This facilitates efficient screening of the reconstructions, allowing good reconstructions with useful information to be chosen for segmentation. Since manual segmentation can be very time consuming, being able to select only the optimal reconstructions increases the efficiency of data analysis.

**a**

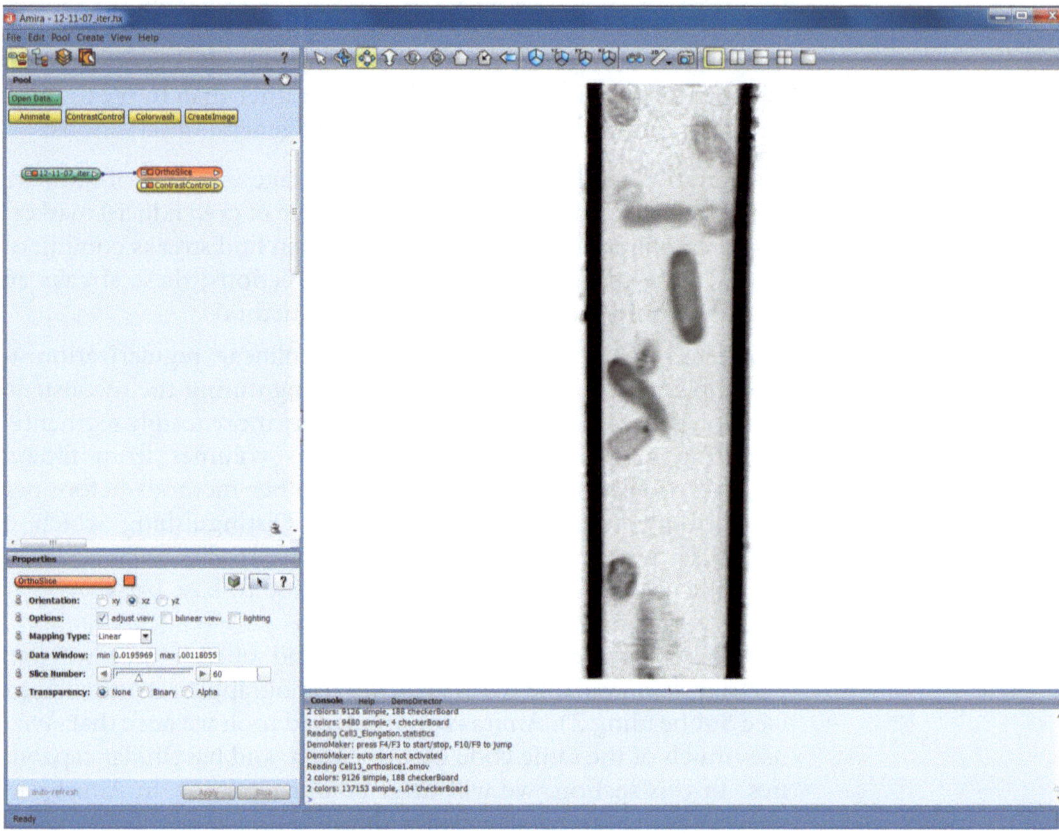


Fig. 4. Screen captures showing commonly used tools in the visualization software package Amira. (**a**) The viewer panel on the right hand side is being used to view a single orthoslice through the reconstructed field of view (in this case, this is a reconstruction of a capillary tube filled with a suspension of bacterial cells). The vertical thick black lines are from the capillary tube. Darker regions represent areas with a relatively high LAC compared to lighter regions. (**b**) When carrying out the process of segmentation it is usually beneficial to view the reconstruction from orthogonal angles. *Top half* and *bottom-left* quadrant of the viewing panel shows three slices through the reconstruction. *Bottom right* quadrant shows 3D representation of segmented cells. (**c**) View of the "magic wand" tool being used to select regions inside the cells with similar LAC values. (**d**) Using the "paintbrush" tool to edit the region selected with the "magic wand" shown in (**c, e**) The "lasso" tool is an efficient means of isolating regions of interest in the reconstruction. For example, you could use it to isolate one cell from other cells reconstructed in the same field of view. (**f**) One example of visualizing the segmented data set. Purple represents the transparent outer surface of the cells. Orange represents the internal structures of interest.

b

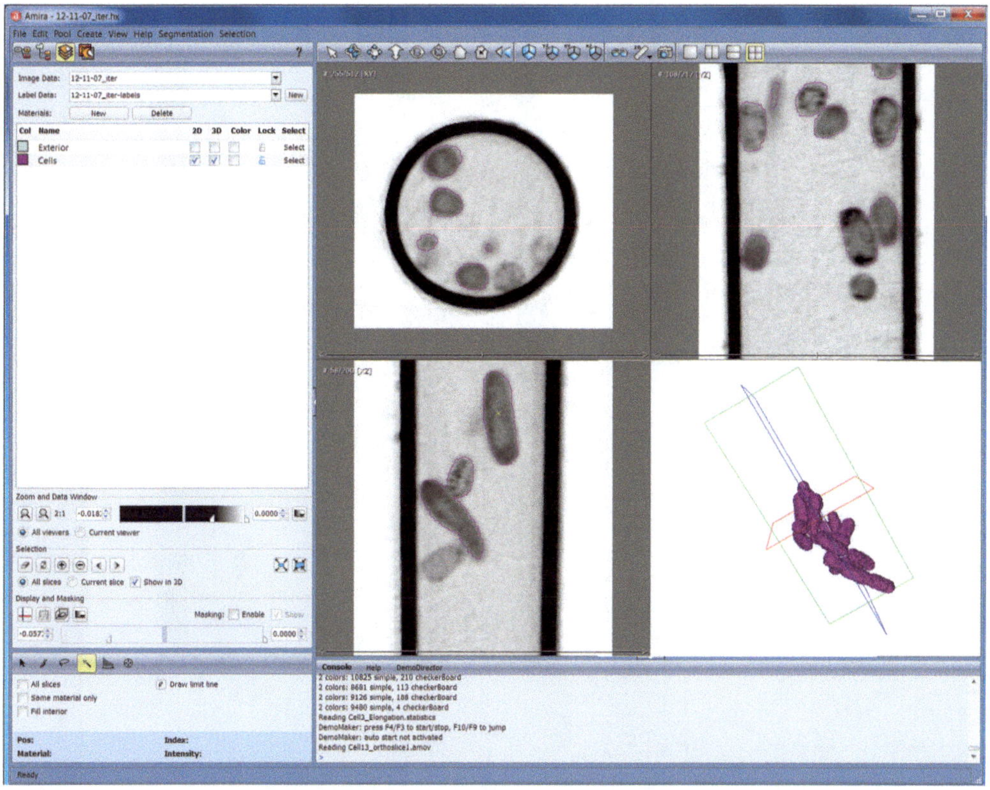

c

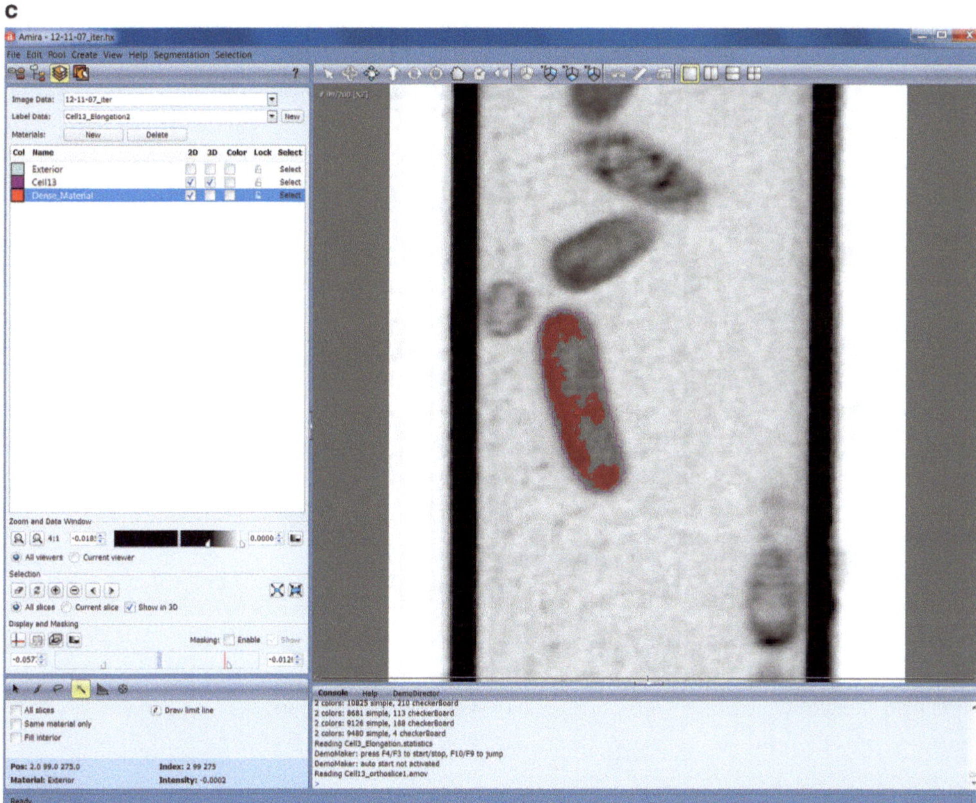

Fig. 4. (continued)

d

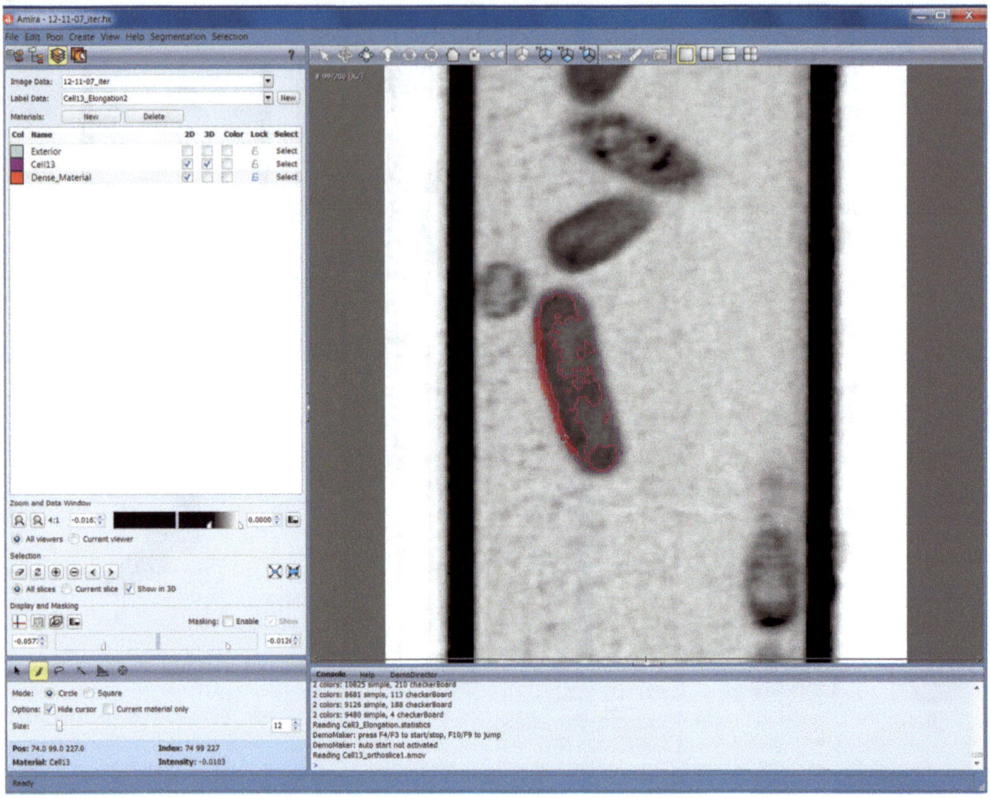


e

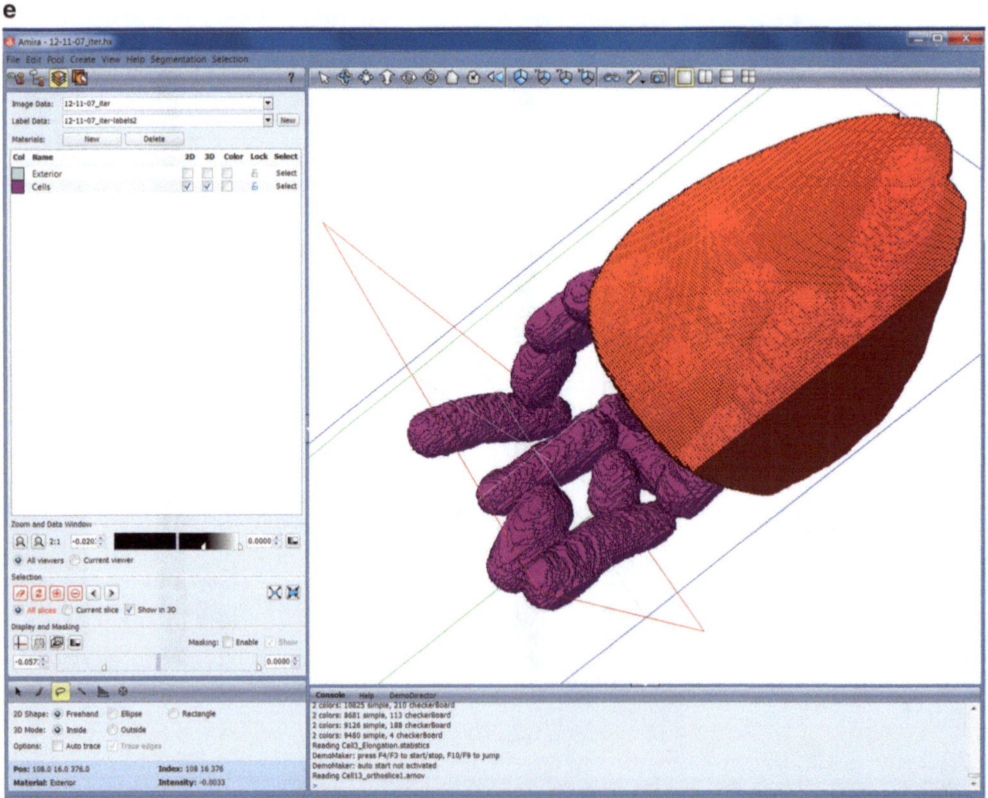


Fig. 4. (continued)

f

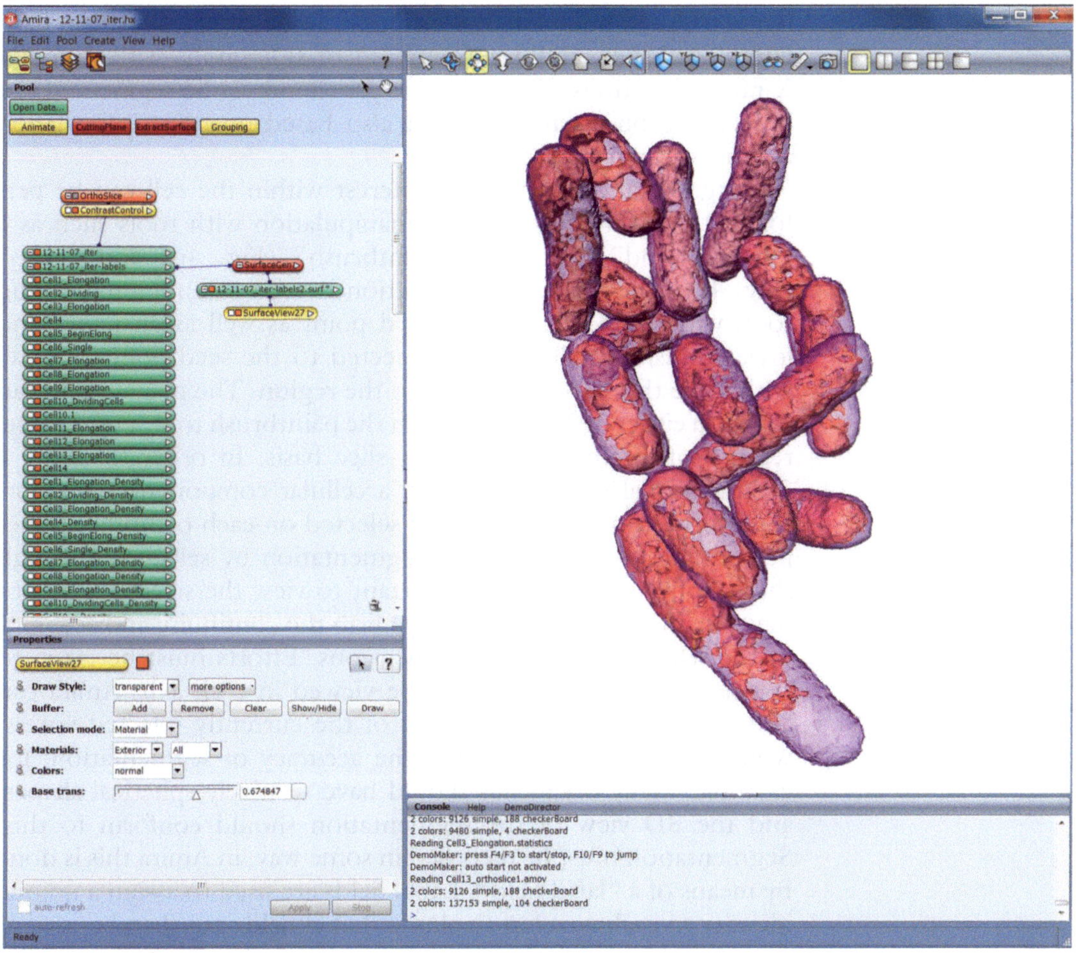


Fig. 4. (continued)

### *3.6. Segmentation and Analysis*

The purpose of segmentation is to delineate the boundaries of structures of interest. Segmentation serves as a way to isolate a specific cellular component from the data set, allowing 3-dimensional visualization of that component either alone or with other components. This segmented region can serve as the basis for a representation of a whole cell, organelle, or some other specific structure that has been selected for analysis.

Two factors serve as the basis for segmentation of cellular structures. First, different structures differentially attenuate X-rays, and thus boundaries between structures can often be found on the basis of changes in voxel values (corresponding to different X-ray attenuation). Sometimes, cells or organelles to be segmented have uniform X-ray attenuation, and thus regions of interest are separated from each other based on attenuation alone. In other cases, a second factor must be used: knowledge of the expected structure of the items to be segmented. For example, some organelles often have complex shapes, and are composed of subregions with different X-ray attenuations. Thus, these organelles must be segmented not only based on attenuation, but also based on prior information about expected shapes.

Segmentation of areas of interest within the cell can be performed manually using image manipulation with tools such as a "magic wand" (Fig. 4c), "paintbrush" (Fig. 4d) and "lasso" (Fig. 4e). The magic wand function is used as a region-growing tool, where the user picks a seed point as well as an upper and lower threshold; all voxels connected to the seed point that fall within the threshold are added to the region. The magic wand can function either in 2D or 3D. With the paintbrush tool, a user paints regions of interest on a slice-by-slice basis. In order to create a 3-dimensional representation of a cellular component, each portion of the component must be selected on each orthoslice. After performing initial paintbrush segmentation by selecting through slices along one axis, it is important to view the selected components in an orthogonal axis to confirm the continuity and accuracy of the selected region in 3 dimensions. Efforts must be taken to reconcile the regions that can be viewed in each axis. Amira can also show a 3-dimensional view of the currently selected voxels, which can aid in determining the accuracy of segmentation; for example, many organelles should have relatively spherical shapes, and the 3D view of the segmentation should conform to this. Segmentation must be recorded in some way; in Amira this is done by means of a "label field". Label fields are used to assign a unique identity to each voxel in a volume. Label fields can then be used as a mask for the original volume or can be used as the basis for a 3D surface representation (Fig. 4f).

Although segmentation of complex biological samples often needs to be done manually, it is possible to automate the process in some cases. In fact, the magic wand tool could be viewed as a semi-automatic segmentation method. If the difference in X-ray attenuation between a given structure and its neighbors is relatively large, the magic wand can be fast and accurate. In other cases, or in cases of samples with high levels of noise, selecting appropriate thresholds is more difficult; often a smaller threshold range must be used to prevent bleed-through of the selected region to the surrounding structures. This adjustment of the threshold in turn changes the overall selected volume. Thus, the ability to arbitrarily adjust the thresholds is very useful but can lead to highly subjective results if segmentation is being used as the basis for volume analysis.

A number of more automatic segmentation methods are possible, however, we have found that automated image processing "pipelines" must have a number of parameters very specific to a given problem. Unless a number of cells or organelles are being analyzed, finding and testing appropriate algorithms and then setting up the "automatic" pipeline can take longer than manually performing the segmentation.

With semi- or fully automatic segmentation, it is often necessary to go through orthoslices in each axis and make small corrections to the given selection. The paintbrush, with that ability to adjust the brush size, has proven to be useful for subtracting or adding voxels to a selection. While corrections can be time consuming, we have found that the use of a pen tablet by Wacom can make this process more efficient. Pens are useful precision tools that tend to be easier to manipulate than a mouse. Additionally, a tablet is able to provide more screen space for segmentation and the angle of display can be adjusted based on user preferences. The work environment can be further personalized using adjustable ergonomic products from companies such as Ergotron (http://www.ergotron.com/). We have used these products to mount and mobilize tablets so that different users can configure their workspace optimally. We have found that ergonomics are an important factor in the segmentation process. Improving workstation ergonomics greatly improved the productivity and throughput of this frequently laborious task.

When selecting structures that are close to one another or that overlap, it can be useful to segment the structures in different label fields. This is because any given voxel in an orthoslice can only belong to one material in a given label field. Selecting a voxel for a certain material that has already been selected will remove the voxel from the original material. This could create inaccuracies in the representative 3-dimensional volume. For example, one label field could contain the overall cell, while another could contain information about the organelle boundaries, and a third could contain information about sub-organelle structure.

In order to visualize the segmentation results for a certain material in 3-dimensions, a surface must first be generated. There are a number of methods of surface generation. In Amira, an important option is the amount of smoothing that is applied in the generation of the surface. The constrained smoothing option works well for most specimens, especially for components with detailed structures. Once a surface has been generated, it can be viewed in the viewer window. Any number of surfaces representing different materials can be displayed simultaneously in the viewer window. The method used for displaying the surfaces can have a big impact on the final product. In Amira, under the "more options" section of the "draw style" parameter on a Surface View module, the default is to use "triangle normals." We have found that the surfaces are much more pleasing when "vertex normal" is selected; this prevents the surface from showing the corners of the individual voxels in the segmentation. The surface area function will generate a spreadsheet that displays the volume and surface area of a certain material in voxel units. The material statistics function can also be used to directly calculate the volumes from the label field. If the voxel size was appropriately set when a data set was loaded, the volume and surface area values will be in the appropriate units (in our case, we use cubic or squared microns).

We note that because data is measured in terms of percent transmission, if appropriate reconstruction algorithms are used, the voxel values in the final volume correspond to the average value of the linear absorption coefficient (LAC) for the material in that voxel. In some cases, it may be necessary to scale the values in the volume by some conversion factor so that they correspond to appropriate units—in X-ray microscopy; absorption per inverse micrometer is a common measure.

## 4. Notes

1. A wide range of computing hardware can be used to run tomographic reconstruction and analysis software packages. However, as with most computationally intensive operations, access to more/faster CPUs will result in a reduction in the time it takes to process and analyze data. At XM-2, data are currently processed and analyzed using an Apple cluster with 20 nodes, each with two 2-core processors. In the near future, it appears these types of calculations will be best carried out using GPU clusters (Graphical Processing Units) as these typically produce a significant decrease in processing time (a factor of 20 or more is generally expected for this type of calculation). However, at a push tomographic reconstruction using the filtered backprojection algorithm can even be carried out on a modest laptop. Therefore, computational hardware is mostly determined by availability of funding, and required throughput.
2. Much of the pre- and post-processing can be done using commonly available commercial software packages such as MATLAB (Mathworks) or with the freeware package ImageJ (National Institutes of Health, USA; http://rsbweb.nih.gov/ij). Tomographic reconstructions can also be carried out in MATLAB or in ImageJ using readily available plugins. There are also a number of dedicated tomography software packages, and in our hands we have found these to offer some distinct advantages, such as utilities for the correction for bad CCD pixels (CCDERASER in IMOD). Many of the required computational operations can be combined into one automated "pipeline" using either a MATLAB script or an ImageJ plugin; this saves time and makes it relatively easy for a neophyte to process their own SXT data. For the calculation of iterative reconstructions we use a number of different packages, including XMIPP

(http://xmipp.cnb.csic.es/), ASPIRE (http://www.eecs.umich.edu/~fessler/aspire/), and SPARX (http://macro-em.org/sparxwiki/SparxWiki). We recommend flexibility in terms of software packages. Some packages deal with particular data sets better than others (based on how the algorithm handles noise, aberrations, and other inherent features in the data). For best results, it is usually worthwhile to carryout data processing and analysis in parallel using different software packages, and compare the final products to determine which performed best with that particular specimen/data set.

3. Amira™ (Visage Imaging, Germany) and Avizo (VSG, France) are two examples with similar functionality. Additional software packages include ImageJ, Chimera (University of California in San Francisco, USA), IMOD (University of Colorado, USA), VisIt (Department of Energy: Advanced Simulation and Computation Initiative, https://wci.llnl.gov/codes/visit/), Paraview (Kitware, Inc., http://www.paraview.org/), and VG Studio Max (Volume Graphics). The choice of package (between these, and many others available either commercially or freely) depends on funding and preference. For the most part, any of the modern software packages are sufficiently advanced and can meet most needs; choice comes down to one of individual preference, perceived ease of use, and other similar considerations.

4. The actual strength of trypsin used is cell line dependent as is the method of stopping action of trypsin. Some cell lines require lower strength of trypsin, for example 0.10%, and may require non-serum stops such as soybean trypsin inhibitor (Sigma Aldrich). To achieve cells in suspension, balance exposure to trypsin with the morphology of cells. Longer trypsin exposure detaches more cells but also negatively effects cell viability and morphology. Also, do not use forceful pipetting, knocking, or scraping to detach cells. Fast liquid movement, knocking and scraping can shear cells, producing debris that may be difficult to separate from desired samples.

5. SXT is a relatively straightforward imaging technique, and comparatively free of issues once the cells are mounted in a suitable holder and cryofixed. The primary concern is always the authenticity of the specimen—is the reconstructed cell a true and accurate representation of the in vivo state. Therefore, we encourage the use of correlated imaging methods, both as a source of additional information and as a "reality check." Towards this end we have developed high-numerical aperture cryogenic light microscopy, and installed this instrumentation into XM-2 (9, 24). As a result, fluorescent tagged-molecules or structures in the cells can be imaged using fluorescence microscopy prior to the cells being imaged using SXT. By fluorescently tagging subcellular structures with well-established

architecture—such as the mitochondria in yeast—and comparing room temperature fluorescence imaging data with the patterns seen in cryogenic light and X-ray microscopy it is possible to validate the integrity of the cell after cryofixation, and after SXT data collection. We feel this is enormously important and this correlated approach should be a routine part of imaging cells by any high-resolution modality. We have observed a number of dramatic structural changes that we determined were simply due to poor freezing, or other specimen handling protocols. In the absence of correlated imaging data these changes could easily have been attributed as being biologically significant.

## Acknowledgments

XM-2 and associated research is funded by the Department of Energy Office of Biological and Environmental Research Grant DE-AC02-05CH11231, the NIH National Center for Research Resources (5P41 RR019664-08) and the National Institute of General Medical Sciences (8P41 GM103445-08) from the National Institutes of Health.

### References

1. Larabell CA, Nugent KA (2010) Imaging cellular architecture with X-rays. Curr Opin Struct Biol 20:623–631
2. Leis A, Rockel B, Andrees L, Baumeister W (2009) Visualizing cells at the nanoscale. Trends Biochem Sci 34:60–70
3. Fagarasanu A, Fagarasanu M, Rachubinski RA (2007) Maintaining peroxisome populations: a story of division and inheritance. Annu Rev Cell Dev Biol 23:321–344
4. Warren G, Wickner W (1996) Organelle inheritance. Cell 84:395–400
5. Uchida M, Sun Y, McDermott G et al (2011) Quantitative analysis of yeast internal architecture using soft X-ray tomography. Yeast 28:227–236
6. Baumeister W (2002) Electron tomography: towards visualizing the molecular organization of the cytoplasm. Curr Opin Struct Biol 12: 679–684
7. Natterer F (1986) The mathematics of computerized tomography. Wiley, New York
8. Larabell CA, McDermott G, Le Gros MA (2005) X-ray tomography of whole cells. Curr Opin Struct Biol 15:593–600
9. McDermott G, Le Gros MA, Knoechel CG, Uchida M, Larabell CA (2009) Soft X-ray tomography and cryogenic light microscopy: the cool combination in cellular imaging. Trends Cell Biol 19:587–595
10. Uchida M, McDermott G, Wetzler M et al (2009) Soft X-ray tomography of phenotypic switching and the cellular response to antifungal peptoids in Candida albicans. Proc Natl Acad Sci USA 106:19375–19380
11. Attwood DT (1999) Soft X-rays and extreme ultraviolet radioation: principles and applications. Cambridge University Press, New York
12. Weiss D (2000) Computed tomography based on cryo X-ray microscopic images of unsectioned biological specimens. Georg-August University of Göttingen, Göttingen
13. Le Gros MA, McDermott G, Larabell CA (2005) X-ray tomography of whole cells. Curr Opin Struct Biol 15:593–600
14. Parkinson DY, McDermott G, Etkin LD, Le Gros MA, Larabell CA (2008) Quantitative 3-D imaging of eukaryotic cells using soft X-ray tomography. J Struc Biol 162:380–386

15. Larabell C, McDermott G, Uchida M, Knoechel C, Le Gros MA (2009) Imaging whole cells at better than 50 nm isotropic resolution with X-rays. Abst, Papers Am Chem Soc, 238
16. Larabell CA, Le Gros MA (2004) X-ray tomography generates 3-D reconstructions of the yeast, saccharomyces cerevisiae, at 60-nm resolution. Mol Biol Cell 15:957–962
17. Bertilson M, Von Hofsten O, Lindblom M, Wilhein T, Hertz H, Vogt U (2008) Compact high-resolution differential interference contrast soft X-ray microscopy. App Phys Lett 92:064104
18. Bertilson MC, Takman PAC, Holmberg A, Vogt U, Hertz HM (2007) Laboratory arrangement for soft X-ray zone plate efficiency measurements. Rev Sci Instrum 78:026103–026101
19. Takman PAC, Stollberg H, Johansson GA, Holmberg A, Lindblom M, Hertz HM (2007) High-resolution compact X-ray microscopy. J Microsc 226:175–181
20. Bell PB, SafiejkoMroczka B (1997) Preparing whole mounts of biological specimens for imaging macromolecular structures by light and electron microscopy. Int J Imaging Syst Tech 8:225–239
21. Quintana C (1994) Cryofixation, Cryosubstitution, Cryoembedding for Ultrastructural, Immunocytochemical and Microanalytical Studies. Micron 25:63–99
22. Ryan KP (1992) Cryofixation of tissues for electron-microscopy—a review of plunge cooling methods. Scanning Microsc 6:715–743
23. Studer D, Humbel BM, Chiquet M (2008) Electron microscopy of high pressure frozen samples: bridging the gap between cellular ultrastructure and atomic resolution. Histochem Cell Biol 130:877–889
24. Le Gros MA, McDermott G, Uchida M, Knoechel CG, Larabell CA (2009) High-aperture cryogenic light microscopy. J Microsc 235:1–8

# Chapter 26

# Secondary Ion Mass Spectrometry Imaging of Biological Membranes at High Spatial Resolution

**Haley A. Klitzing, Peter K. Weber, and Mary L. Kraft**

## Abstract

Characterization of the distributions of specific proteins and lipids within cellular membranes is currently a major challenge. Advances in secondary ion mass spectrometry (SIMS) now enable the distributions of isotopically labeled lipids within cellular or model membranes to be imaged with chemical specificity and high (≥50 nm) lateral resolution. Here, methods to image the distributions of sphingolipids within the membranes of intact cells with a Cameca NanoSIMS are described. For NanoSIMS detection, the incorporation of distinct stable isotopes into the lipid species of interest is essential. Metabolic labeling, cell preservation, imaging conditions, and data analysis are critical factors. The methods and principles described here can be extended to studying other membrane lipids or cholesterol.

**Key words:** Chemical imaging, Secondary ion mass spectrometry, NanoSIMS, Lipid organization, Metabolic labeling

## 1. Introduction

The distributions of cholesterol and specific lipid species within the plasma membrane are hypothesized to be linked to cellular function. However, characterizing the lateral variations in lipid abundance within biological membranes is especially challenging because few techniques provide the location-specific compositional information that is required for conclusive identification. Imaging secondary ion mass spectrometry (SIMS) offers a direct approach to imaging the lipid distribution within biological membranes in a chemically specific manner. In this method, a primary ion beam is used to produce neutral and ionized molecules and molecular fragments from the sample surface. The ionized species, which are called secondary ions, are analyzed in a mass spectrometer, providing chemical information about the molecules located within the primary beam. By scanning a focused primary ion beam across the

Alioscka A. Sousa and Michael J. Kruhlak (eds.), *Nanoimaging: Methods and Protocols*, Methods in Molecular Biology, vol. 950, DOI 10.1007/978-1-62703-137-0_26, © Springer Science+Business Media, LLC 2013

sample and collecting the secondary ions, the intensities of the component-specific secondary ions detected at each position can be used to construct a map of the sample's surface composition. Several different approaches have been used to image lipid membranes with SIMS (1–7). This chapter will focus on the use of high-resolution SIMS performed with a Cameca NanoSIMS 50 or 50 L. In our experience, the NanoSIMS provides the optimum combination of spatial resolution, mass resolution (mass specificity), and ion transmission (sensitivity) for these experiments. Its shortcomings are that a limited number of species can be detected simultaneously, and isotopic labels are necessary to distinguish most cellular molecules.

The Cameca NanoSIMS is a magnetic sector SIMS instrument that enables high lateral resolution imaging of sample composition with high sensitivity and specificity (6, 8, 9). Unlike the more common molecular imaging mass spectrometry approaches, such as time-of-flight (TOF) SIMS or matrix-assisted laser desorption/ionization (MALDI), a full mass spectrum is not acquired during NanoSIMS analysis. Instead, individual species are preselected for detection by individual ion counting detectors. Static magnetic field multicollection is usually performed, though magnetic peak switching is possible. Another difference from TOF-SIMS and MALDI is that only monatomic or small molecular ions are typically detected. This instrument uses reactive primary ions (typically $Cs^+$ for lipid membrane analysis) to fragment the surface molecules and enhance the yield of the preselected secondary ions that are sputtered from the top few monolayers of the sample surface. Depending on the NanoSIMS model, secondary ions with up to five or seven different mass-to-charge ($m/z$) ratios can be simultaneously detected by a magnetic sector mass spectrometer. Isobars of the same nominal mass (i.e., $^{12}C^{15}N^-$ at 26.9996 amu and $^{13}C^{14}N^-$ at 27.0059 amu) can be separated. The best lateral resolution achievable with a NanoSIMS is 30 nm (8). However, the low primary ion beam current that accompanies this small beam diameter results in relatively low secondary ion counts that may limit the working lateral resolution achievable on biological samples. In practice, we routinely image lipid membranes with a lateral resolution of approximately 100 nm (4, 7).

Proper preparation of biological membranes is critical for NanoSIMS analysis. Distinct stable isotopes or nonnative elements must be substituted into each molecule of interest so that the monatomic and small molecular secondary ions generated during NanoSIMS analysis can be linked to a specific parent molecule. In addition, biological samples must also be preserved so that the membrane retains its organization when subjected to the ultrahigh vacuum conditions within the NanoSIMS analysis chamber. Although NanoSIMS is not performed on living cells, it has advantages that are complementary to many techniques that can be performed on living cells, including fluorescence microscopy.

1. Chemically specific information is used to *directly* visualize the lipid organization.
2. Isotope-labeled lipids have the same chemical structure as the analogous native lipid molecule, so the isotope labels do not alter the lipid distribution.
3. Analysis can be restricted to the outermost 1 to 2 monolayers (≤5 nm) at the sample surface (10).
4. A lateral resolution of 30 nm can be achieved with optimization; approximately 100 nm lateral resolution is routinely acquired on biological membranes (4, 7).

This chapter describes the basic principles of assessing the distribution of a specific lipid species in the plasma membranes of intact cells with NanoSIMS. As an example, the sample preparation, NanoSIMS imaging, and image analysis required to image the distribution of sphingolipids and all cellular lipids (i.e., sphingolipids and glycerolipids) within the plasma membrane of intact mouse fibroblast cells is described. These protocols and methodologies can be extended to imaging other lipid species or cholesterol within the plasma membrane of various types of cells.

## 2. Materials

### 2.1. Cell Culture

1. Adherent mammalian cells, such as NIH3T3 fibroblast cells (ATCC CRL-183™) (see Note 1).
2. Water-jacketed $CO_2$ incubator for mammalian cell culture
3. Media A: high glucose Dulbecco's modified Eagle's medium (DMEM) supplemented with 10% Hyclone calf serum, $10^4$ U/mL penicillin G, and 10 mg/mL streptomycin.
4. Media B: high glucose DMEM supplemented with 10% Hyclone lipid-reduced fetal bovine serum and 1% Hyclone calf serum.
5. $^{15}$N-sphingoid bases: stock solution made from a 1:1 molar mixture of $^{15}$N-sphingosine and $^{15}$N-sphinganine (see Note 2).
6. $^{13}$C-Fatty Acids: stock solution made from algal lipid mixture (uniformly carbon-13 labeled, UL-$^{13}$C, 98%+) (Cambridge Isotope Laboratories, Inc.), stearic acid ($^{13}C_{18}$, 99 atom% $^{13}$C) (Isotec), fatty acid free bovine serum albumin (BSA), dimethyl sulfoxide (DMSO), and Dulbecco's phosphate-buffered saline (PBS) without calcium and magnesium.
7. 1.2 μL of ethanolamine, minimum 98% dissolved in 10 mL Millipore (18 MΩ) water and sterile filtered for a working concentration of 2 mM.

8. 0.22 μm polyvinylidene fluoride (PVDF) Millex syringe driven filter unit.
9. 0.2 μm polytetrafluoroethylene (PTFE) membrane, single use, syringe filter.
10. Vortex mixer.
11. Sonicator.
12. 5 mm × 5 mm silicon wafer chips. This size is convenient for the "Harvard" style NanoSIMS sample holder.
13. Poly-L-lysine hydrobomide, m.w. 30,000–70,000, 0.1%.

### 2.2. Evaluation of Label Incorporation

1. Centrifuge (similar to Eppendorf 5810R) (use when centrifuge is specified, with the exception of Step 9 of Subheading 3.2.1).
2. Millipore (18 MΩ) water (use when water is specified).
3. Solvents: methanol, chloroform, dichloromethane, ethanol and hexane.
4. Chemicals: trimethylsilyldiazomethane, concentrated hydrochloric acid, and potassium hydroxide.
5. Microfuge 18 Centrifuge used only in Step 9 of Subheading 3.2.1.
6. Liquid Chromatography-Mass Spectrometer (LC-MS), such as an Agilent 1100 LC/MSD Trap XCT Plus equipped with a Phenomenex Gemini 3u C6-Phenyl 110A column (2 × 100 mm).
7. Incubator capable of maintaining 70°C.
8. Gas Chromatography-Mass Spectrometer (GC-MS), such as an Agilent 6890 gas chromatograph with a ZB-WAX (30 m × 0.25 mm I.D., 0.25 μm film thickness) capillary column (Phenomenex, USA), an Agilent 5973 mass selective detector and HP 7683B (Agilent Inc, USA) autosampler.

### 2.3. Cell Preservation

1. Dulbecco's phosphate-buffered saline (PBS) without calcium and magnesium.
2. Hendry's phosphate buffer (HPB): $NaH_2PO_4{\cdot}H_2O$, $Na_2HPO_4$, pH 7.35.
3. 8% Glutaraldehyde diluted in HPB to 4%.
4. 4% Osmium tetroxide diluted in water to 1%.
5. Millipore (18 MΩ) water (use when water is specified).

### 2.4. Cell Imaging

1. Upright bright field microscope.
2. Scanning electron microscope (SEM).
3. Sputter coater.
4. Cameca NanoSIMS 50 or 50 L.

#### *2.5. Image Analysis Software*

1. Suitable software for visualizing the data. For example, the LIMAGE (L.R. Nittler, Carnegie Institution of Washington) custom software package that runs on the PV-Wave platform (Rogue Wave Software) or the OpenMIMS Data Analysis plugin for ImageJ.

## 3. Methods

### *3.1. Metabolic Labeling of Cells*

To visualize specific molecules in the outer leaflet of the plasma membrane, a distinct stable isotope must be incorporated into the target molecules in the cell. This can be achieved with metabolic labeling. The general approach is to culture the cells in the presence of the isotope-labeled precursors, which are metabolized and integrated into the cells' infrastructure. Methods have been established to selectively incorporate distinct stable isotopes or radiolabels into specific lipid species, including phosphatidylethanolamine (11), cholesterol (12–14), sphingolipids (15–17), and gangliosides (18). These methods must be optimized so that a high fraction of the lipid species of interest within the cell contains the distinct isotope, as the isotope enrichment image will reveal the location and density of only the labeled lipids.

The selection of the distinct isotope label will influence NanoSIMS detection. Species with high ionization probability (i.e., $CN^-$, $F^-$, $O^-$, and $S^-$) can be detected the most efficiently. For the detection of species with lower ionization probabilities (i.e., $CH^-$ and $C^-$), a high fraction of the abundant isotope should be replaced with the rare isotope (i.e., the use of $^{13}C_{16}$-palmitic acid is more desirable than $^{13}C_2$-palmitic acid). Here, we describe how to incorporate $^{15}N$ into the majority of the cellular sphingolipids, and $^{13}C$ into the majority of cellular lipids.

#### *3.1.1. Preparation of Cell Substrates*

1. Lightly scratch an asymmetric pattern onto a desired number of 5 mm × 5 mm silicon chips. This pattern will facilitate sample positioning during NanoSIMS analysis.
2. Wash the chips with soap and water. Rinse well with water to remove any traces of soap.
3. Sterilize the chips by autoclaving.
4. After sterilization, cover the surface of the chips with sterile poly-L-lysine solution. The poly-L-lysine solution can be sterilized by passing it through a 0.22 μm syringe driven filter unit, and depositing the effluent directly onto the sterile chips.
5. Allow the poly-L-lysine to adsorb onto the chips overnight at 4°C (see Note 3).

*3.1.2. Preparation of $^{15}$N-Sphingoid Bases (See Note 2)*

1. Dissolve the $^{15}$N-sphingoid bases into DMSO creating a 6 mM stock solution. This bulk stock solution should be stored at −20°C.
2. Sterilize the $^{15}$N-sphingoid bases by filtering through a pre-DMSO-rinsed, syringe driven, sterile 0.2 μm PTFE filter unit. In order to speed up thawing time before each use, sterile filtering can be done into a sterile 1.7 mL microfuge tube to create ~1 mL aliquots. Store the sterile aliquots of $^{15}$N-sphingoid bases at −20°C, and warm to room temperature before every use.

*3.1.3. Preparation of $^{13}$C-Fatty Acids*

1. To make a 1 mL stock solution of the fatty acid mixture, begin by weighing out 13.3 mg of algal lipid mixture into a microfuge tube. Use the outside of a hayman style microspatula scoop to obtain small portions of the lipid mixture that can then be deposited into the microfuge tube by wiping. Add 3.3 mg of stearic acid to a separate microfuge tube. In a third microfuge tube, add 125 mg of BSA.
2. To the microfuge tube with BSA, add 1 mL of PBS and vortex to dissolve. If large crystals persist, centrifuge for a matter of seconds to assist in the dissolving. Once fully dissolved, sterile filter the BSA into a sterile microfuge tube using a syringe driven, 0.22 μm PVDF filter unit that had been rinsed with PBS. Air-purge the filter unit to ensure a quantitative transfer. Set aside for later use.
3. Add 250 μL of DMSO to the microfuge tube containing the algal lipid mixture, and vortex to dissolve. Quantitatively transfer the dissolved algal lipid mixture into the microtube containing the stearic acid. Vortex to dissolve. If stearic acid crystals are visible, sonicate the solution in a bath sonicator for 10 s, or until crystals are dissolved.
4. Sterile filter the algal lipid/stearic acid mixture using a pre-DMSO-rinsed, syringe driven, 0.2 μm PTFE filter unit. Air-purge the filter to ensure quantitative transfer. To this sterile 250 μL solution, add 750 μL of the sterile BSA solution prepared in step 2. Vortex the BSA/algal lipid/fatty acid mixture, and then sonicate the solution for 15 min, or until the precipitate dissolves.
5. The resulting 1 mL stock solution has a concentration of 16.6 mg/mL, or approximately 0.0556 M. Use this solution to achieve the desired concentration in the cell culture media. Store at 4°C. This stock solution must be warmed to room temperature and thoroughly mixed by vortexing prior to subsequent use.

*3.1.4. Cell Culture (See Note 1)*

1. Grow cells in Media A for 1 passage (3 days) at 37°C in 5% $CO_2$. On a daily basis, add the stock solution of $^{15}$N-sphingoid bases via pipette in order to maintain a 3 μM concentration in the

culture. It is assumed that cells have consumed all supplemental reagents from the previous day. The cells should reach approximately 80% confluence on day 3, at which point they are ready to be passaged.

2. Plate cells onto new culture dishes and grow in Media B at 37°C in 5% $CO_2$. On a daily basis, add the $^{15}$N-sphingoid bases, $^{13}$C-fatty acids, and ethanolamine via pipette in order to maintain a 3 μM, 215 μM, and 2 μM concentration, respectively. Assume that cells have consumed all of the supplemental reagents from the previous day.
3. On the sixth day, passage cells and reset with fresh Media B and supplements in dishes that also contain sterilized and poly-L-lysine-coated silicon chips.
4. Collect the cells on the 7th day (see Subheading 3.1.5).

#### *3.1.5. Collection of Samples*

1. Collect silicon chips with attached cells and chemically fix (see Subheading 3.3).
2. Harvest all cells remaining on the dish as follows. Detach the cells and collect in a centrifuge tube. Pellet the cells by centrifuging for 5 min at 201×*g*. Remove all but approximately 0.5 mL of the supernatant from the cell pellet to avoid allowing the cell pellet to dry. Rinse the pellet twice with 10 mL PBS, pelleting the cells by centrifugation and removing excess PBS between each rinse. Finally, spin down the cells, remove excess PBS, and store at −20°C until label incorporation is evaluated (see Subheading 3.2).

### 3.2. Evaluation of Label Incorporation

In order for analysis with the NanoSIMS to be successful, a sufficient amount of isotope label must be incorporated into the plasma membrane to ensure the lipid-specific secondary ions can be detected by NanoSIMS with satisfactory signal to noise. In addition, because only the isotope-labeled lipids will be visible in the NanoSIMS image, a high fraction of the lipid species of interest within the cell should contain the isotope label so that the isotope enrichment image is representative of that lipid's distribution and density within the plasma membrane. The cells that remained attached to the dish after removing the silicon chips are used to assess the isotope incorporation into the lipid species of interest (see Subheading 3.1.5). LC-MS is used to determine the fraction of the cellular sphingolipids that contain nitrogen-15. The fraction of the fatty acids within the cellular lipids that are uniformly labeled with carbon-13 is measured with GC-MS.

#### *3.2.1. Lipid Extraction*

The instructions for lipid extraction are modified from the method of Blight and Dyer (19). Extractions are typically performed on approximately $2 \times 10^6$ cells/pellet.

1. To each cell pellet, add 2 mL methanol and 1 mL chloroform. If there is no PBS remaining on the cell pellet, add 0.8 mL of water.
2. Vortex and sonicate to break apart the pellet and mix thoroughly.
3. Let sit at room temperature for 1 h.
4. Centrifuge at 1,811×*g* for 5 min.
5. Decant the liquid supernatant into a new centrifuge tube, leaving behind the cell pellet.
6. To the supernatant, add 1 mL of chloroform and 1 mL of water. Vortex.
7. Centrifuge at 1,811×*g* for 5 min.
8. Transfer the lower, chloroform layer into a new vial and dry down under a stream of nitrogen gas.
9. Dissolve crude lipid extracts in 100 μL methanol, centrifuge at 17, 562×*g* for 5 min. An injection of 4 μL of this supernatant will be used for HPLC-MS analysis of $^{15}$N-incorporation.

*3.2.2. Evaluation of Nitrogen-15 Incorporation into Cellular Sphingolipids*

1. Analyze the nitrogen-15 incorporation into the cellular sphingomyelin with HPLC-MS. For the instrument described in the Subheading 2, perform the separation using methanol at a flow rate of 300 μL/min. Monitor the following mass peaks (retention time is 1.5 min for the stated instrument): *m/z* 703.6 (unlabeled N-palmitoyl sphingomyelin), 704.6 ($^{15}$N-palmitoyl sphingomyelin), 719.6 ($^{14}$N-palmitoyl $^{13}C_{16}$-sphingomyelin), 720.6 ($^{15}$N-palmitoyl $^{13}C_{16}$-sphingomyelin), and 735.7 ($^{14}$N-palmitoyl $^{13}C_{32}$-sphingomyelin) (see Note 4).
2. Remove the contribution of N-palmitoyl sphingomyelin isotopologues other than $^{15}$N-palmitoyl sphingomyelin from peak *m/z* 704.6 by subtracting 43.3% of the intensity of the N-palmitoyl sphingomyelin peak (*m/z* 703.6) from that of peak *m/z* 704.6. Remove the contribution of sphingomyelin isotopologues other than $^{15}$N-palmitoyl $^{13}C_{16}$ sphingomyelin to peak *m/z* 720.6 by subtracting 25.8% of the intensity of the $^{14}$N-palmitoyl $^{13}C_{16}$ sphingomyelin peak at *m/z* 719.6 from that of peak *m/z* 720.6. Calculate the fraction of N-palmitoyl sphingomyelin that contained a nitrogen-15 isotope as the ratio of the sum of $^{15}$N-palmitoyl sphingomyelin and $^{15}$N-palmitoyl $^{13}C_{16}$-sphingomyelin peak intensities to the sum of the peak intensities from all N-palmitoyl sphingomyelin isotopologues. Calculate the fraction of N-palmitoyl sphingomyelin that biosynthesized from uniformly labeled $^{13}$C-palmitic acid as the ratio of the sum of $^{14}$N-palmitoyl $^{13}C_{16}$-sphingomyelin, $^{15}$N-palmitoyl $^{13}C_{16}$-sphingomyelin, and $^{14}$N-palmitoyl $^{13}C_{32}$-sphingomyelin, over the sum of the peaks produced by all N-palmitoyl sphingomyelin isotopologues.

*3.2.3. Evaluation of Carbon-13 Incorporation into All Cellular Lipids*

1. To the crude lipid extracts after LC-MS analysis, add 2.5 mL of 95% ethanol.
2. Add 50 μL of 33% KOH (w/v), and incubate the sample at 70°C for 45 min.
3. Cool to room temperature, transfer the solution to a centrifuge tube, and add 3 drops of concentrated HCl.
4. Add 5 mL of hexane and vortex.
5. Centrifuge at 1,811×*g* for 5 min.
6. Transfer the upper, hexane layer to a new vial and dry down under a stream of nitrogen gas.
7. Resuspend sample in 100 μL of MeOH:$CH_2Cl_2$ (1:1).
8. Add 20 μL of a 2 M solution of trimethylsilyldizomethane in hexane.
9. Incubate for 1 h at room temperature.
10. Perform GC-MS on the resulting solution of fatty acid methyl esters (FAMEs). For the instrument described above, set the inlet and MSD interface temperatures to 250°C, and the ion source temperature to 230°C, and use a constant flow rate of 1.3 mL/min for the helium carrier gas. In positive electron impact mode (EI), collect the signal intensities corresponding to the FAMEs of $^{13}C_{16}$-palmitic acid ($m/z$ 286), $^{12}C_{16}$-palmitic acid ($m/z$ 270), $^{13}C_{16}$-palmitoleic acid ($m/z$ 284), $^{12}C_{16}$- palmitoleic acid ($m/z$ 268), $^{13}C_{18}$-stearic acid ($m/z$ 316), $^{12}C_{18}$-stearic acid ($m/z$ 298), $^{13}C_{18}$-oleic acid ($m/z$ 314), $^{12}C_{18}$-oleic acid ($m/z$ 296), $^{13}C_{18}$-linoleic acid ($m/z$ 312), and $^{12}C_{18}$-linoleic acid ($m/z$ 294). These intensities can then be correlated to determine a representative incorporation value.

### 3.3. Cell Preservation with Chemical Fixation

Imaging with the NanoSIMS is performed under ultrahigh vacuum conditions, so the cells must be dehydrated without perturbing the lateral distributions of lipids and proteins within the cell membrane. This is accomplished by chemically preserving the cells. The method described here is derived from protocols used to chemically fix cells in order to prepare them for analysis with scanning electron microscopy. This chemical fixation procedure produces a membrane whose morphology is retained during the inevitable drying process needed for introduction into the vacuum of the SIMS analysis chamber (20).

1. Remove chips with adherent cells from cell culture and rinse three times with HPB for 3 min each. Be careful not to disturb the attached cells.
2. Flood the chips with 4% glutaraldehyde. Allow to fix for 30 min (see Notes 5 and 6).
3. Remove the glutaraldehyde and rinse twice with HPB for 5 min each, followed by rinsing once with water for 5 min.

4. Fix and stain with freshly diluted, freshly filtered 1% osmium tetroxide for 15 min.
5. Remove the osmium tetroxide and rinse three times for 5 min each with water.
6. Remove excess water, allow the sample to air dry, and store at room temperature.

### 3.4. Pre-characterization of Cells

To ensure efficient use of time on the NanoSIMS instrument, assessment of cell preservation is performed using bright field microscopy and low voltage SEM prior to NanoSIMS analysis. An optical map that shows the locations of cells in relation to one another, the scratches on the chip, and the edges of the chip is first created with bright field microscopy. Then, low voltage SEM is used to identify the cells that are best-suited for NanoSIMS analysis. This map will facilitate locating desirable cells during NanoSIMS analysis. Cells that are ideal for NanoSIMS analysis should be well-preserved (i.e., free of cracks or debris), relatively flat, and located away from the edges of the chip. The specific cells that are imaged with SEM should be indicated on the optical bright field map of the chip, which will be used to relocate them in the NanoSIMS.

1. Map the entire surface of the chip with bright field imaging. Mapping is performed at 10× or 20× magnification. Starting at one corner of the chip, acquire a series of overlapping images of the entire chip. Stitch these images together to create the optical map of the sample.
2. Metal coat samples to ~5 nm thickness (see Note 7).
3. SEM imaging is performed using low voltage to avoid sample charging and to prevent damage (see Note 8). Acquire images of individual cells at a low magnification (500–1,100×). Indicate which cells are imaged with SEM on the optical map of the chip.

### 3.5. SIMS Imaging

The specifics of this section are based on a NanoSIMS 50. Other SIMS instruments can be used for this work, though with some loss in spatial resolution, sensitivity, specificity and/or analysis speed.

#### 3.5.1. NanoSIMS Settings

The analysis conditions are set according to the species to be detected and the spatial resolution needed. Here, the parameters are set based on detecting $^{12}C^{1}H^{-}$, $^{13}C^{1}H^{-}$, $^{14}N^{12}C^{-}$, and $^{15}N^{12}C^{-}$ with 50–100 nm lateral spatial resolution. Carbon isotopes can also be detected using $C^{-}$, $C_2^{-}$, and $^{14}NC^{-}$. Other species can be detected on additional detectors. The target mass resolving power ($MRP \equiv M/\Delta M$) is 7,000 to resolve $^{13}C_2^{-}$ from $^{14}N^{12}C^{-}$ ($\Delta M = 0.0037$ amu) and $^{11}B^{16}O^{-}$ from $^{15}N^{12}C^{-}$ ($\Delta M = 0.0041$ amu). Because of the way the NanoSIMS software calculates MRP, ~11,000 MRP must be obtained based on the NanoSIMS high mass resolution scan to achieve the functional mass resolving power of 7,000. Lower mass resolving power can be used if the $^{13}C$ enrichment is low or natural abundance and $BO^-$ counts are negligible.

1. The cesium microbeam ion source is placed at 8,000 V and is mechanically aligned in the primary column based on maximizing the current at the primary Faraday cup (FCp). The filament current on the microbeam ion source reservoir is adjusted to achieve ~30 nA $Cs^+$. The sample is placed at −8,000 V.
2. The L1 lens in the primary column is placed at approximately 6,500 V for high resolution (<100 nm), or left at zero for moderate resolution (~150 nm).
3. Select a 150 μm or smaller diameter aperture from the D1 aperture strip (primary beam limiting aperture) for the highest resolution (50 nm), 200 μm for high resolution (~75 μm), or 300 μm for moderate resolution.
4. The resulting beam current at the sample, as measured at the object Faraday cup (FCo), should be ~0.25 pA $Cs^+$ for the highest resolution or ~2 pA for moderate resolution. Adjust reservoir current or perform more advanced alignment procedures to achieve target beam current.
5. The secondary ion beam must be aligned in the column in the total ion count detector with the secondary ion beam centering deflectors. This alignment is typically only performed with changes in instrument hardware.
6. Next the desired masses are aligned in the individual detectors in the multicollection chamber. Both vertical and horizontal coincidence must be confirmed.
7. To enable 11,000 MRP in NanoSIMS units (a functional 7,000 MRP), the 10 μm × 100 μm entrance slit is aligned on the secondary ion beam by maximizing the current at multicollector detectors; the 150 × 150 $\mu m^2$ aperture slit is then aligned on the secondary ion beam in the same way.
8. The secondary ion beam is next tuned using standard Cameca procedures to achieve ~11,000 MRP.
9. Primary beam focusing is performed in parallel with secondary ion beam focusing. Primary beam diameter on the sample, which determines image resolution, should be checked at this point to ensure the imaging target is achieved (see Subheading 3.6.2).
10. Align SIMS analysis location with the CCD optical imaging reticule using a reference sample. This alignment is necessary to allow centering on target cells without SIMS imaging.

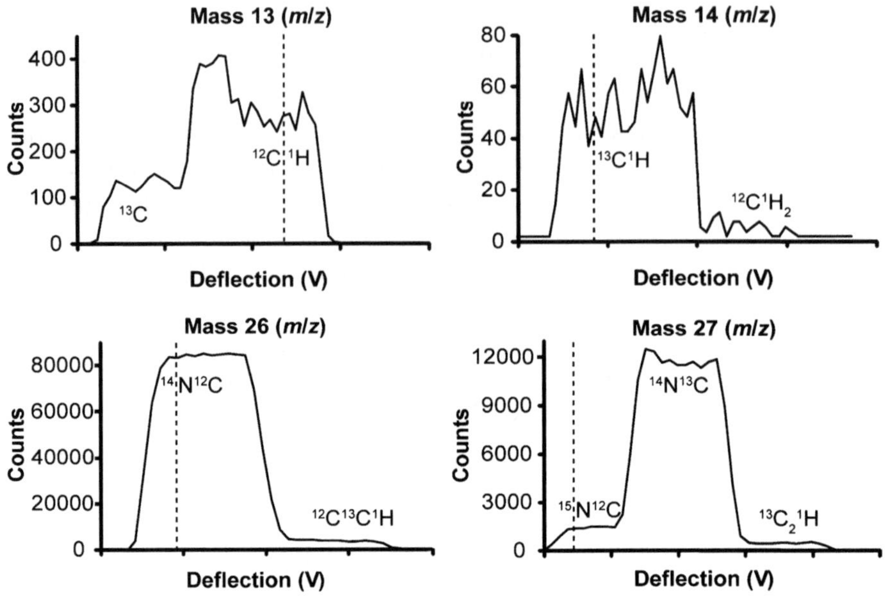


Fig. 1. Mass spectra acquired with ~11,000 MRP and 75 μm-width slits in front of the detectors. The analyzed cell was isotopically enriched in $^{13}C$ and $^{15}N$. The ions of interest and some of the interfering isobars at *m/z* 13, *m/z* 14, *m/z* 26, and *m/z* 27 are indicated. The individual masses are not resolved in these scans, but by setting the deflection on each detector as indicated with the dashed lines, the contribution of the interfering isobars is less than 0.1% of the species of interest. Several species are too low in abundance to be visible in the mass spectra (see Table 1 for the full list of interferences at *m/z* 26 and 27).

*3.5.2. Sample Loading and Analysis Locations*

1. Sample chips should be loaded into a front-justified sample holder. The sample well in the holder can be 7.2 mm diameter, 5×5 mm squares, or a ~6 mm diameter aperture in a larger hole. The samples should be held in place from behind by non-magnetic springs.
2. The sample holder is introduced into the analysis vacuum chamber after pumping in the airlock.
3. Prospective analysis locations are located in the CCD image based on the light microscopy and SEM imaging, and the coordinates are saved in the point table. Also save the location of nearby cells that can be used for tuning. All sample chips on the holder can be mapped before moving to SIMS analysis position to save time.

*3.5.3. Cell Analysis*

The primary challenge of the sample analysis is to position the sample and align the secondary beam so that the tuning accomplished above is achieved with the target cells. Control samples should be analyzed in the same session to insure analysis quality.

1. Move to SIMS analysis position and choose starting area.
2. Go to the preselected tuning cell and adjust sample height and secondary ion beam steering. Check MRP and peak alignment, ensuring that interfering isobars are not collected (Fig. 1 and

**Table 1**
**Mass Interferences**

| Calculated clusters | m/e (u) | Δ (mu) | Abundance | MRP |
|---|---|---|---|---|
| At mass 26 | | | | |
| $^{12}C{-}^{14}N$ | 26.0025 | 0.0 | 0.98538 | – |
| $^{13}C{-}^{13}C$ | 26.0062 | 3.6 | 0.00012 | 7,155 |
| $^{10}B{-}^{16}O$ | 26.0073 | 4.8 | 0.19853 | 5,443 |
| $^{1}H{-}^{12}C{-}^{13}C$ | 26.0106 | 8.1 | 0.01088 | 3,208 |
| $^{1}H{-}^{1}H{-}^{12}C{-}^{12}C$ | 26.0151 | 12.6 | 0.97783 | 2,068 |
| At mass 27 | | | | |
| $^{12}C{-}^{15}N$ | 26.9996 | 0.0 | 0.00362 | – |
| $^{11}B{-}^{16}O$ | 27.0037 | 4.1 | 0.79909 | 6,567 |
| $^{13}C{-}^{14}N$ | 27.0059 | 6.3 | 0.01096 | 4,272 |
| $^{1}H{-}^{12}C{-}^{14}N$ | 27.0104 | 10.8 | 0.98523 | 2,502 |
| $^{1}H{-}^{13}C{-}^{13}C$ | 27.0140 | 14.4 | 0.00012 | 1,872 |
| $^{1}H{-}^{1}H{-}^{12}C{-}^{13}C$ | 27.0185 | 18.9 | 0.01088 | 1,429 |
| $^{1}H{-}^{1}H{-}^{13}C{-}^{12}C$ | 27.0185 | 18.9 | 0.01088 | 1,429 |

Note that abundances are calculated relative to isotopologues and cannot be directly compared to other molecules. For example, the yield of $CN^-$ is ~10× that of $C_2^-$, double hydrides tend to have low yields, and boron is not normally sufficiently abundant in cells to be significant

Table 1). Significant changes in MRP or peak alignment potentially indicate excessive sample topography. Adjustments in tuning can be made to improve MRP, but caution is necessary to assure analysis quality.

3. Turn off the primary beam and move to the starting location on the target cell. Five to ten micrometers leeway should be allowed for stage drift and misalignment of the CCD and SIMS imaging.
4. Set the analysis conditions so that the primary beam cross section from pixel to pixel overlaps by at least 50%. Scan rate is set to enable multiple scans before eroding through the plasma membrane (e.g., 512 × 512 pixels, 15 × 15 μm², 0.5–1 ms/pixels, 3–5 scans; see below for calculation of sputter depth). Collect target species (e.g., $^{12}C^{1}H^-$, $^{13}C^{1}H^-$, $^{14}N^{12}C^-$ and $^{15}N^{12}C^-$) and secondary electrons.

### 3.6. Data Analysis

#### 3.6.1. Determination of Analysis Depth

Determination of the sputtering depth is straightforward, involving use of established formulas and procedures. To ensure that the vast majority of secondary ions are produced by the cell membrane and few secondary ions are collected from the underlying cytoplasm, the sputtering depth should be less than the sum of the thicknesses of the membrane (7.5 nm (21)) and the metal coating (~5 nm). If the sputtering depth exceeds the membrane thickness, it can be reduced by discarding some of the image planes. Because the sputtering depth increases with each subsequent image plane, discard the last image plane(s) to be acquired until the sputtering depth is less than approximately 5 nm.

1. *Calculate the sputtering depth.* Sputtering depth ($D_{sp}$) is calculated using the sputter rate ($R_{sp}$), primary ion beam current, at the sample (FCo) in pA, raster area ($A_{ras}$) in $\mu m^2$, and the sputtering time ($t_{sp}$) in seconds. The sputter rate determined on other biological samples, which ranges from 0.9 nm·$\mu m^2$/pA·s (unpublished results) to 2.5 nm·$\mu m^2$/pA·s (10) may be employed. The primary ion beam current at FCp and raster area were recorded during NanoSIMS analysis. The primary ion beam current at the sample (FCo) must be measured separately. The sputtering time ($t_{sp}$) is calculated using the dwell time ($t_{dw}$) for each pixel (pxl), the total pixels ($pxl_{tot}$) per image plane, and the total number of image planes ($N_P$) according to Eq. 1.

$$t_{sp} = \left(\frac{t_{dw}}{pxl}\right)\left(\frac{pxl_{tot}}{plane}\right)(N_P) \tag{1}$$

For example, for a 512 by 512 pixel image consisting of four image planes that were each acquired with a dwell of 1 ms/pixel, the total sputtering time is 1048.576 s. Calculate the sputtering depth in nanometers using Eq. 2.

$$D_{sp} = \left(\frac{R_{sp} \times FCo \times t_{sp}}{A_{ras}}\right) \tag{2}$$

*3.6.2. Determination of Lateral Spatial Resolution*

Lateral resolution is also determined using standard procedures. The primary challenge is to find a sufficiently small or sharp edged feature so that its width is negligible relative to the primary beam diameter. To determine the spot size, a single scan of a standard sample is used. To determine the practical resolution, multiple scans of a cell sample can be used, but with the caveat that the features may not be sufficiently small or sharp to accurately characterize the beam diameter.

1. Collect a secondary electron image of a standard sample (256×256 pixels, 5×5 $\mu m^2$, 0.5–1 ms/pixels) or use an image that you have already collected of a cell (see above). For multiscan images, align the NanoSIMS image planes to correct for spatial drift in the beam or the sample. The presence of image drift would compromise the lateral resolution of the resulting image. Images of high count rate species, such as $^{12}C^{14}N^-$ or secondary electrons, typically provide the best alignment for cells.
2. Calculate the effective diameter of the primary ion beam, to determine the lateral resolution of the NanoSIMS analysis. Make a line scan across a region in a secondary ion or secondary electron image where the sample composition changes abruptly. Determine the distance over which the secondary ion or electron signal intensity changes from 84% to 16% of the maximum intensity. This distance is the effective diameter of the analysis beam (7).

#### 3.6.3. Creation of $^{15}N$- and $^{13}C$-Enrichment Images

To accurately visualize the distribution of $^{15}N$-sphingolipids and $^{13}C$-lipids in the plasma membrane, the secondary ion signal must be normalized to remove ion intensity variations due to concentration-independent factors that affect ion yields and sputter rates (22). These factors include changes in ionization probabilities related to sample composition—which is referred to as "matrix effects"—and sample topography (8, 23, 24). Because these concentration-independent factors influence the intensities of isotopologues, which are chemical species that differ in isotopic composition (i.e., $^{15}N^{12}C^-$, $^{14}N^{13}C^-$ and $^{14}N^{12}C^-$), to relatively the same extent, expressing the lipid-specific ion counts as a ratio to the corresponding major ion accurately represents the concentration distribution of the target molecules (22). To minimize random noise in the images, the ratio images are typically "smoothed" over a 3 by 3 pixel window, which means the value at a given pixel is the average ratio of the 3 by 3 pixel region that is centered around that pixel.

1. Align the image planes.
2. For each image plane, ratio the counts of each lipid-specific ion (e.g., $^{15}N^{12}C^-$or $^{13}C^1H^-$) detected at each pixel to the counts of the corresponding major element ion ($^{14}N^{12}C^-$ or $^{12}C^1H^-$) detected at the same pixel.
3. Smooth the isotope ratio measured at each pixel using a 3 by 3 pixel window.
4. For every pixel, calculate the mean ($\mu$) isotope ratio ($^{15}N^{12}C^-/^{14}N^{12}C^-$ or $^{13}C^1H^-/^{12}C^1H^-$) of the individual image planes ($\mu = \Sigma x_i/N$, where N is the number of total image planes).
5. Divide the isotope ratio ($^{15}N^{12}C^-/^{14}N^{12}C^{14}N^-$ or $^{13}C^1H^-/^{12}C^1H^-$) by their natural abundance (0.00367 or 0.011237, respectively) to produce an enrichment factor that is a quantitative relative measure of the amount of $^{15}N$-sphingolipids or $^{13}C$-lipids in the plasma membrane compared to an unlabeled (natural abundance) cell.
6. Use a color scale to encode the value of the isotope enrichment measured at each pixel in the output image with a color (Fig. 2).

#### 3.6.4. Visual Interpretation of $^{13}C$- and $^{15}N$-Enrichment Images

The $^{13}C$-enrichment and $^{15}N$-enrichment images (Fig. 2) reveal the distributions of $^{13}C$-labeled lipids and $^{15}N$-sphingolipids in the plasma membrane. The $^{13}C$-lipid distribution is used to assess the quality of cell preservation. As shown in the $^{13}C$-enrichment image in Fig. 2, the continuity of the $^{13}C$-enrichment on the cell surface confirms that the cell membrane is intact, and the $^{13}C$-lipids are not clustered within the membrane. This continuous distribution of $^{13}C$-lipids on the cell surface confirms that the preservation process did not create holes in the membrane, or induce artifactual lipid

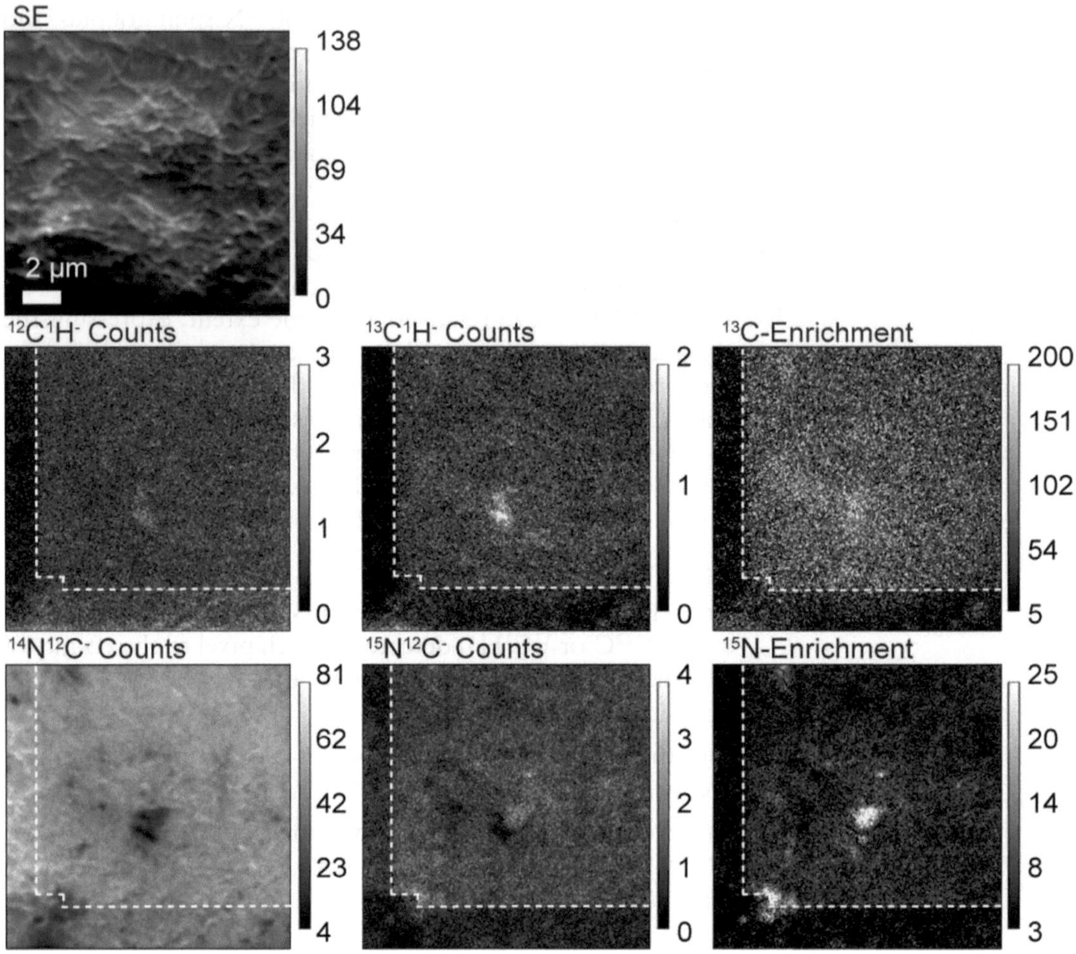


Fig. 2. NanoSIMS images acquired from a fibroblast cell. The secondary ions were acquired in parallel with the $^{12}C^1H^-$, $^{13}C^1H^-$, $^{14}N^{12}C^-$ and $^{15}N^{12}C^-$ secondary ions. To image the entire cell at this lateral resolution, multiple overlapping images are required. Low ion counts at edges of images (below and to the left of the *dashed lines*) are characteristic of previously analyzed locations. The $^{13}C$-enrichment (($^{13}C^1H^-/^{12}C^1H^-$)/0.011237) and $^{15}N$-enrichment (($^{15}N^{12}C^-/^{14}N^{12}C^-$)/0.00367) images reveal the distributions of $^{13}C$-labeled lipids and $^{15}N$-sphingolipids in the plasma membrane. The secondary ion and enrichment images were smoothed over 3 by 3 pixels.

clustering. Therefore, the lateral variations in the $^{15}N$-enrichment on the cell surface (Fig. 2), which indicate that the $^{15}N$-sphingolipids are heterogeneously distributed in the cell membrane, cannot be attributed to preparation-induced artifacts. Instead, the local elevations in the $^{15}N$-sphingolipid enrichment that are dispersed within a continuous matrix of lower $^{15}N$-sphingolipid abundance signify the presence of $^{15}N$-sphingolipid domains within the plasma membrane. By using this approach to image large regions on the cell surface, the long-range organization of the sphingolipid microdomains in the plasma membrane may be visualized.

## 4. Notes

1. Cell lines used in these studies included NIH3T3 (ATCC CRL-1658™) and a NIH3T3 transfected mouse fibroblast cell line that stably expresses influenza hemagglutinin (Clone 15 cell line). Other cell lines can follow similar procedures for metabolic labeling, but adjustments in the labeling conditions (concentrations, labeling period, media, etc.) may be required to obtain high levels of isotope incorporation.
2. The $^{15}$N-sphingoid bases, $^{15}$N-sphingosine and $^{15}$N-sphinganine, are synthesized from $^{15}$N-serine following reported methods (25–29). Deuterated sphingosine (Avanti Polar Lipids) can be used in its place, and $^{12}C^{2}H^{-}$ and $^{12}C^{1}H^{-}$ secondary ions can be collected with the NanoSIMS for the detection of the deuterated sphingolipids. Note that we have not evaluated the intensity of the secondary ions that are specific to the deuterated sphingolipids.
3. These are suggested conditions for coating the substrate with poly-L-lysine to promote cell adhesion. Satisfactory results may also be obtained with other temperatures and incubation times.
4. For the cell lines used, no significant signal was detected for other sphingolipid species. However, different cell lines may metabolize the lipid precursors into different sphingolipid classes, so the presence of ion peaks corresponding to other sphingolipid species should also be evaluated. If present, include these peaks when determining the incorporation values.
5. An easy way of fixing and rinsing without wasting large amounts of material is to place the chips in a parafilm-lined dish and to use pipettes for transferring the liquids. The parafilm will encourage the liquids to pool onto the chip, and the pipettes can be easily used for placing a drop of liquid onto the chips or suctioning off and disposing used material.
6. Different chemical preservation protocols can be used as long as they involve a heavy fixation that can withstand ultrahigh vacuum conditions.
7. Sputter coating should be optimized for the individual sample and imaging conditions. Employ a thicker coating if sample charging occurs. Gold, gold-palladium, iridium and other metals may be used. Carbon coating should not be used.
8. SEM imaging at higher than 5 kV damages the sample. This damage is visible as cracks on the cells. Making small rasters on the sample, such as for higher magnification, can cause the decomposition of molecules at the surface of the cell.

## Acknowledgments

We thank K. Lou for synthesizing the $^{15}N$-sphingoid bases that were used for metabolic labeling, A. Ulanov for his valuable comments on the evaluation of isotope incorporation, C. Ramon for technical assistance and L. Nittler for software development. MLK holds a Career Award at the Scientific Interface from the Burroughs Wellcome Fund. Work at Lawrence Livermore National Laboratory was supported by Laboratory Directed Research and Development funding and performed under the auspices of the U.S. DOE under contract DE-AC52-07NA27344. This material was supported in part by the National Science Foundation under CHE—1058809.

### References

1. Ostrowski SG, Van Bell CT, Winograd N, Ewing AG (2004) Mass spectrometric imaging of highly curved membranes during tetrahymena mating. Science 305:71–73
2. Piehowski PD, Davey AM, Kurczy ME et al (2009) Time-of-Flight Secondary Ion Mass Spectrometry Imaging of Subcellular Lipid Heterogeneity: Poisson Counting and Spatial Resolution. Anal Chem 81:5593–5602
3. Kurczy ME, Piehowski PD, Van Bell CT et al (2010) Mass spectroscopy imaging of mating Tetrahymena show that changes in cell morphology relate to lipid domain formation. Proc Natl Acad Sci USA 107:2751–2756
4. Anderton CR, Weber PK, Lou K, Hutcheon ID, Kraft ML (2011) Correlated AFM and NanoSIMS imaging to probe cholesterol-induced changes in phase behavior and non-ideal mixing in ternary lipid membranes. Biochim Biophys Acta 1808:307–315
5. Vaezian B, Anderton CR, Kraft ML (2010) Discriminating and imaging different phosphatidylcholine species within phase-separated model membranes by principal component analysis of TOF-SIMS images. Anal Chem 82:10006–10014
6. Boxer SG, Kraft ML, Weber PK (2009) Advances in imaging secondary Ion mass spectrometry for biological samples. Annu Rev Biophys 38:53–74
7. Kraft ML, Weber PK, Longo ML, Hutcheon ID, Boxer SG (2006) Phase separation of lipid membranes analyzed with high-resolution secondary Ion mass spectrometry. Science 313:1948–1951
8. Williams P (2006) Biological imaging using secondary ions. J Biol 5:18
9. Slodzian G, Daigne B, Girard F, Boust F, Hillion F (1992) Scanning secondary ion analytical microscopy with parallel detection. Biol Cell 74:43–50
10. Ghosal S, Fallon SJ, Leighton TJ et al (2008) Imaging and 3D elemental characterization of intact bacterial spores by high-resolution secondary Ion mass spectrometry. Anal Chem 80:5986–5992
11. Zhu L, Johnson C, Bakovic M (2008) Stimulation of the human CTP: phosphatidylethanolamine cytidylyktransferase gene by early growth response protein 1. J Lipid Res 49:2197–2211
12. Keller P, Simons K (1998) Cholesterol is required for surface transport of influenza virus hemagglutinin. J Cell Biol 140:1357–1367
13. Campbell S, Gaus K, Bittman R et al (2004) The raft-promoting property of virion-associated cholesterol, but not the presence of virion-associated Brij 98 rafts, is a determinant of human immunodeficiency virus type 1 infectivity. J Virol 78:10556–10565
14. Scheiffele P, Rietveld TW, Simons K (1999) Influenza viruses select ordered lipid domains during budding from the plasma membrane. J Biol Chem 274:2038–2044
15. Chigorno V, Riva C, Valsecchi M, Nicolini M, Brocca P, Sonnino S (1997) Metabolic processing of gangliosides by human fibroblasts in culture. Formation and recycling of separate pools of ganglioside. Eur J Biochem 250:661–669
16. Dolo V, D'Ascenzo S, Sorice M et al (2000) New approaches to the study of sphingolipid enriched membrane domains: the use of electron microscopic autoradiography to reveal metabolically tritium labeled sphingolipids in cell cultures. Glycoconjugate J 17:261–268

17. van Echten G, Birk R, Brenner-Weiss G, Schmidt RR, Sandhoff K (1990) Modulation of sphingolipid biosynthesis in primary cultured neurons by long chain bases. J Biol Chem 265:9333–9339
18. Schwarzmann G, Hoffmann-Bleihauer P, Schubert J, Sandhoff K, Marsh D (1983) Incorporation of ganglioside analogues into fibroblast cell membranes. A spin-label study. Biochemistry 22:5041–5048
19. Blight EG, Dyer WJ (1959) A rapid method of total lipid extraction and purification. Can J Biochem Physiol 37:911–917
20. Cheng J, Fujita A, Ohsaki Y et al (2009) Quantitative electron microscopy shows uniform incorporation of triglycerides into existing lipid droplets. Histochem Cell Biol 132:281–291
21. Edidin M (2003) Lipids on the frontier: a century of cell-membrane bilayers. Nat Rev Mol Cell Biol 4:414–418
22. Peteranderl R, Lechene C (2004) Measurement of carbon and nitrogen stable isotope ratios in cultured cells. J Am Soc Mass Spectrom 15: 478–485
23. Kumar S, Aaron J, Sokolov K (2008) Directional conjugation of antibodies to nanoparticles for synthesis of multiplexed optical contrast agents with both delivery and targeting moieties. Nat Protocols 3:314–320
24. Rangarajan S, Tyler BJ (2006) Topography in secondary ion mass spectroscopy images. J Vac Sci Technol A 24:1730–1736
25. Garner P, Park JM, Malecki E (1988) A stereodivergent synthesis of D-erythro-sphingosine and D-threo-sphingosine from L-serine. J Org Chem 53:4395–4398
26. Garner P, Park JM (1987) The Synthesis and Configurational Stability of Differentially Protected β-Hydroxy-α-amino Aldehydes. J Org Chem 52:2361–2364
27. McKillop A, Taylor RJK, Watson RJ, Lewis N (1994) An improved procedure for the preparation of the Garner aldehyde and its use for the synthesis of N-protected 1-halo-2-(R)-amino-3-butenes. Synthesis 1:31–33
28. Roush WR, Hunt JA (1995) Asymmetric Allylboration of 2-N,3-O-Isopropylidene-N-Boc-L-serinal: Diastereoselective Synthesis of the Calicheamicin gamma 1I Amino Sugar. J Org Chem 60:798–806
29. Peters C, Billich A, Ghobrial M et al (2007) Synthesis of borondipyrromethene (BODIPY)-labeled sphingosine derivatives by cross-metathesis reaction. J Org Chem 72: 1842–1845

# INDEX



Alioscka A. Sousa and Michael J. Kruhlak (eds.), *Nanoimaging: Methods and Protocols*, Methods in Molecular Biology, vol. 950, DOI 10.1007/978-1-62703-137-0, © Springer Science+Business Media, LLC 2013

MIX
Papier aus verantwortungsvollen Quellen
Paper from responsible sources
FSC® C105338

If you have any concerns about our products,
you can contact us on
**ProductSafety@springernature.com**

In case Publisher is established outside the EU,
the EU authorized representative is:
**Springer Nature Customer Service Center GmbH**
**Europaplatz 3, 69115 Heidelberg, Germany**

Printed by Libri Plureos GmbH
in Hamburg, Germany